The Savvy Business Traveler's Guide to Customs and Practices in Other Countries

The Dos & Don'ts to Impress Your Hosts and Make the Sale

By Dan Blacharski

THE SAVVY BUSINESS TRAVELER'S GUIDE TO CUSTOMS AND
PRACTICES IN OTHER COUNTRIES: THE DOS & DON'TS TO
IMPRESS YOUR HOSTS AND MAKE THE SALE

ISBN-13: 978-1-60138-013-5   ISBN-10: 1-60138-013-5

Library of Congress Cataloging-in-Publication Data

Blacharski, Dan, 1959-
   The savvy business traveler's guide to customs and practices in other countries : the dos and
don'ts to impress your hosts and make the sale / by Dan Blacharski.
       p. cm.
   Includes bibliographical references and index.
   ISBN-13: 978-1-60138-013-5
   ISBN-10: 1-60138-013-5
   1. Business etiquette. 2. Business travel. 3. Intercultural communication. 4. Corporate
culture. I. Title.
   HF5389.B58 2008
   395.5'2--dc22
                            2007049124

INTERIOR LAYOUT DESIGN: Vickie Taylor • vtaylor@atlantic-pub.com

Printed in the United States

Printed on Recycled Paper

We recently lost our beloved pet "Bear," who was not only our best and dearest friend but also the "Vice President of Sunshine" here at Atlantic Publishing. He did not receive a salary but worked tirelessly 24 hours a day to please his parents. Bear was a rescue dog that turned around and showered myself, my wife Sherri, his grandparents Jean, Bob, and Nancy and every person and animal he met (maybe not rabbits) with friendship and love. He made a lot of people smile every day.

We wanted you to know that a portion of the profits of this book will be donated to The Humane Society of the United States.

–Douglas & Sherri Brown

THE HUMANE SOCIETY
OF THE UNITED STATES ©

The human-animal bond is as old as human history. We cherish our animal companions for their unconditional affection and acceptance. We feel a thrill when we glimpse wild creatures in their natural habitat or in our own backyard.

Unfortunately, the human-animal bond has at times been weakened. Humans have exploited some animal species to the point of extinction.

The Humane Society of the United States makes a difference in the lives of animals here at home and worldwide. The HSUS is dedicated to creating a world where our relationship with animals is guided by compassion. We seek a truly humane society in which animals are respected for their intrinsic value, and where the human-animal bond is strong.

Want to help animals? We have plenty of suggestions. Adopt a pet from a local shelter, join The Humane Society and be a part of our work to help companion animals and wildlife. You will be funding our educational, legislative, investigative, and outreach projects in the U.S. and across the globe.

Or perhaps you'd like to make a memorial donation in honor of a pet, friend, or relative? You can through our Kindred Spirits program. And if you'd like to contribute in a more structured way, our Planned Giving Office has suggestions about estate planning, annuities, and even gifts of stock that avoid capital gains taxes.

Maybe you have land that you would like to preserve as a lasting habitat for wildlife. Our Wildlife Land Trust can help you. Perhaps the land you want to share is a backyard — that's enough. Our Urban Wildlife Sanctuary Program will show you how to create a habitat for your wild neighbors.

So you see, it's easy to help animals. And The HSUS is here to help.

The Humane Society of the United States
2100 L Street NW
Washington, DC 20037
202-452-1100
www.hsus.org

Table of Contents

Section 3: The World In-Depth: Asia & the South Pacific......... 57

Section 4: The World In-Depth: Central & Eastern Europe ____ 111

Section 5: The World In-Depth: Western Europe —————————— 145

Section 6: The World In-Depth: The Americas ———————— 213

Section 7: The World In-Depth: Africa & the Middle East _____ 245

Section 1

The Savvy Business Traveler

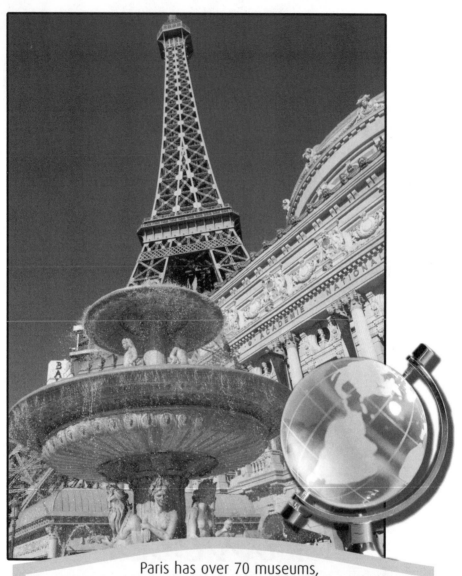

Paris has over 70 museums, monuments, and cultural tourist stops. You can see some of the world's best art in Paris.

1
Introduction: Business Is Never Just Business

Sitting in a restaurant with my wife and several of her colleagues in Bangkok, I unintentionally made a rude suggestion to the waitress.

We were with several of my wife's coworkers at a publishing house, and I wanted to make an impression. Unfortunately, I believe I did. The Thai language is a tonal one, and saying the same word with a slightly different inflection can yield an entirely different meaning. I thought I was merely asking the waitress to bring me coffee with cream, but it came out quite differently. Fortunately, she was a good sport and knew what I was trying to say, and she, my wife, and her coworkers had a good laugh at my expense.

My grandfather came to the United States from Poland and never learned to speak English well. Although we lived in a predominantly Polish neighborhood, it was difficult for him to communicate. After traveling to several foreign countries, I know now what it must have been like for him, struggling just to speak a foreign language in a broken manner, making mistakes, and being thought of as the "quaint foreigner with the funny accent." You may laugh at the Chinese lady at your favorite takeout place who wants to know if you want to eat some "flied lice," but you are going to be providing the comic relief when you go overseas, so be prepared.

Doing business in a foreign country is so much more than merely doing business. As Americans, we are more accustomed to making a business presentation and striking a deal as quickly as possible. In most other countries, it just is not done that way. Business, pleasure, family, and, indeed, life in general are intertwined so that, before someone does business with you, they want to see you in a social setting, and they want you to accept not only their business, but their culture as well.

The old phrase, "It is not personal; it is just business" does not apply in many foreign cultures where business is not considered separate from the rest of life. Business is personal. It involves much of your waking day and affects your life style and image. In many cultures, the entire family is part of the business in one way or another, whether it is actually working in a family-owned shop or acting as host for a dinner party. In Asian cultures, if you lose face in a business deal, you have lost face to the entire world. There is no clean separation. At the end of the day, this intertwining of business and life keeps our business practices more humane.

America Is Only a Place

Americans tend to have an unfortunate point of view about our native country that makes us unpopular in the rest of the world. That point of view is that "America is the best." We believe in "good old American know-how," "American ingenuity," and "the land of the free and the home of the brave," as if citizens of all other countries are somehow inferior. We speak of the "American Way" and the "American Dream." We think that America somehow has an edge on the rest of the world, whose customs, religions, and political infrastructures are not acceptable. We think we have the best political system, the best social services, the best workers, and the best universities. We think soccer is not "real football." We think that our food is good but other countries' food is weird.

That is pure ego.

Countries are nothing more than artificial constructs created for the purpose of social organization. The United States is one such artificial construct. We have some good schools, and people tend to do their jobs to the best of their abilities. Then again, the same can be said about dozens of other countries. India produces some of the brightest engineers in the world. Switzerland is more democratic than the United States by design (and they make better chocolate). While we may enjoy our steaks, there is nothing deadly about eating a plate of fried beetles.

The most important element of being a successful business traveler is adopting an open attitude, and not bragging about oneself or the United States. You may have an excellent product at a fair price, but your hosts will not assume that your proposal is the best just because you are American. They do not share your same point of view that America is the best in the world at everything. They just want a good deal, and they want to make it with someone they like. In short, acquire some humility. While you are in a foreign country, do not look for what is the same. Seek out what is different and savor it.

Will They Reject Me Because I Am American?

There is a perception that Americans are not welcome abroad, but this is for the most part incorrect. I have visited many countries throughout Europe and Asia and have always been treated regally. It is true that foreign countries see some American foreign policy and politics as aggressive, but for the most part, people you encounter will distinguish between you as a person and the country you come from.

I have felt safe everywhere in the world, probably safer than many places I have been in the United States. I have always had freedom of movement, even in Communist countries, and I have found that locals are willing to go out of their way to accommodate me. People, especially in countries that do not get many outside visitors, are very eager to talk to you and learn about where you are from.

Having the Best Deal Is Not Enough

Being "all business" is not going to work in most foreign countries. You may have a superior business plan, a well-engineered proposal, a crew of the finest and most talented individuals ever assembled, and you may also have the best deal financially. If you were in an American setting, that might be enough to win you the contract; but it is not going to be enough elsewhere. Foreign bids are not always won based solely on who offers the best terms. Business may be about more than the bottom line.

In most other cultures, it is just as important to make a deal with someone you can get along well with, as it is to make a deal with someone offering the best bottom line. If you are in Japan and your business associate invites you out to a karaoke bar and you do not sing, you refuse to taste the sake, you insist on a Budweiser, and loudly proclaim "I only eat my fish breaded and fried," you are not going to get the deal, no matter how much better your proposal is than the next guy's. The guy from Germany who hates sushi but eats it anyway is going to get the deal.

2
Food: Fried Fish Guts, Lard Sandwiches, & Duck's Blood Soup

I knew an American businessman who traveled throughout Asia. Upon arriving, the first thing he did was open his suitcase and take out canned tuna and other canned items that traveled well. They were his entire diet for the duration of his trip. "I cannot eat the food here. It is nasty," he complained. When he did manage to venture out for restaurant food, he went no further than the local McDonald's. He never got far in business because, in many countries, more business gets done in informal situations where food is present.

Food is a big part of our social lives in America, but in foreign countries it is often a big part of business as well. People take pride in either serving you their favorites in their homes or taking you to their favorite restaurants. When I took my son Shanti with me to spend the summer in Poland a few years back, I was struck by the importance of accepting unfamiliar food. My son and I took the train to several different cities and, no matter where we were, one relative or another sent us off with a bag of kielbasa, some bread, lard sandwiches, cheese, and other goodies. As a special treat, we might even get a container of czarnina (duck's blood soup).

My dad loved to tell stories of how his mother made "lard and onion sandwiches on black bread." I could not conjure up any delightful visual and figured he was just telling tall tales. Then I got to Warsaw and found that they really do eat that. What was even more surprising was that I enjoyed it. Smalets, or seasoned lard, is often used instead of butter. It may be a heart attack waiting to happen, but as my cousin Macek (whose medical knowledge is a bit suspect) always said, "The vodka burns away the cholesterol."

Part of the delight of traveling to foreign countries is experiencing new cultures, and that means eating the food, no matter how strange. As a businessperson, you are not just in it for the experience; you are there to make some money. To do that, you have to accept your host's culture and country and be a gracious guest. In many parts of the world, the peculiarly American phrase, "It is just business," does not make any sense. Business, life, and enjoying good food are all intertwined. Enjoying a plank steak in Sweden, a full English breakfast in London, or a snack of fried fish skins in Vietnam is so much more than just trying out new foods. It is accepting and respecting your host's culture. In many locations, before any business gets done, there is socializing, which often centers on food (and sometimes alcohol). If you reject your host's choice of cuisine and insist on a Big Mac, your negotiations just may hit an impasse.

Thai food is strange to the uninitiated. Before I learned how to speak Thai, my wife (who is a native of Thailand) took me to a footpath restaurant. They are small, local food stalls, usually right on the sidewalk, where you can get good, authentic, and inexpensive food. She ordered for us, and I did not really know what anything was that came to our table a few minutes later. I was enjoying eating a plate of small, fried crunchy things, and asked her what they were. "Oh, those are fried fish organs," she said casually. Now to be truthful, if she had told me that ahead of time, I may not have eaten any, but they were quite good.

Your host takes pride in serving you native dishes or bringing you to his favorite restaurant; to refuse is an insult to his country. Suppose, for example, you were entertaining someone from France, and you took him or her to your favorite Texas steakhouse for a Porterhouse and a cold beer. Suppose then, that the Frenchman picked up the American side dish called "French fries," and informed you, "These are not French!"; and then further told you that he did not want the steak, could not possibly drink that American beer and could you please ask the waiter to bring some Bordeaux? You would think him presumptuous and arrogant. Americans are often thought of as being presumptuous and arrogant because of stubbornly clinging to the "everything American is best" misconception and refusing to try something different.

When you are in Scotland, you had better sample the haggis if you do not want to be seen as an arrogant outsider. When in China, try the birds' nest soup (it is delicious, by the way, and yes, it is made out of birds' nests). In Poland, make every effort to at least take a few spoonfuls of the czarnina; and when you are in Scandinavia, do not turn down a snack of rotten shark meat on crackers. There is simply no polite way to avoid it and save face.

While visiting some of my in-laws in the South of Thailand (where the food is especially fiery), I was able to make a good impression at dinner. My wife's aunt asked her if her farang (foreigner) husband was able to eat spicy food, and I barged in and answered for her, "Gin aharn pet dai khrap" ("Yes, I can eat spicy food"). The food was beyond spicy; it was pure fire. Nonetheless, I happily ate it. And even though I spent most of that evening in the bathroom, my in-laws loved me.

Na Zdrovie! Alcohol and Business Do Mix

Americans tend to reserve alcohol for after work, while in many other countries it is an integral part of the business day and an important part of negotiations. The concept of abstinence from alcohol is foreign to people who enjoy it as a part of life. To them, "I do not drink wine" may sound the same as, "I do not drink water." If you turn down a glass of wine with dinner because you think it is immoral, you are saying, "I am better than you," to your host. This may appear insulting. If you belong to a religious persuasion that does not permit you to patronize a restaurant that serves alcohol, you should avoid doing business in most foreign countries.

Having a glass of wine with a colleague in France or a glass of vodka with a business associate in Russia is part of everyday life. It is not vile, and to refuse may be awkward and difficult. It is part of the negotiating process, part of gaining your associate's trust, and part of the celebration after the deal is made.

In many countries, age restrictions against alcohol are less important. This is something my son, who was then 16 years old, found quite delightful when we were traveling around Europe. While on board a ship in the Baltic, we took the opportunity to have dinner in a very elegant restaurant with some friends and relatives. Following the customary formalities with the sommelier, wine was poured for everyone, and there was no question of whether or not the younger members of the group should be included — they were. Similarly, during the Krakow Jazz Festival, my son and I were sitting at an outdoor café, I ordered a fine Polish beer, and the waitress asked if he wanted one as well.

In Poland, there is no such thing as half of a bottle of vodka. Your host will pour for everyone, and you are expected to drink the entire serving in one shot. When your glass is empty, it will be refilled again as a matter of course. This will continue until the bottle is empty, and then, more often than not, another bottle will follow.

The French take pride in their wine, as well they should. To refuse it at a meal is unforgivable.

If you refuse to have any alcohol, the best approach is to accept a glass, take a sip or two when toast time comes around, and then drink from your water glass with your meal.

3
A Little Common Sense

Courtesy does not always express itself the same way, but a little common sense goes a long way.

Language

In major cities in Western Europe, English is spoken almost universally. In Stockholm, I prepared myself with a few key phrases, figuring that, although businesspeople may speak English, taxi drivers, waitresses, and salespeople probably do not. To my delight, almost everyone who worked with the public spoke enough English to provide me with good service. I was quite relieved, since I had quickly forgotten the few words of Swedish I had learned. In Central and Eastern Europe, and in major Asian cities, English is spoken less frequently.

Regardless of the country you visit, it pays to come prepared with at least a few key phrases, or if you have time, take a course. Be sure to know how to say, "I do not speak it well." You may be surprised when you utter one of your few memorized phrases, and your host smiles and then starts talking to you a mile a minute and you do not understand a single word. Even understanding the bare minimum of "please," "thank you," and the standard greeting will impress your host, just for having made the effort.

After you have traveled around a bit, it is easy to get confused, and language similarities can cause some amusing situations. When my cousin moved from Warsaw to Stockholm, everyone thought he was polite, because he often said "tak" in conversations. "Tak" in Swedish means, "thank you." In Polish, "tak" simply means, "yes." In an even more bizarre coincidence of language, upon visiting Hong Kong my wife discovered, to her horror, that the standard greeting in Cantonese was "jiu sahn." When a Hongkongese person says "jiu sahn," they are simply wishing you a good day, but in Thai, it means, "I am shaking my penis." And if you have ever wondered why Polish people are often shown the bathroom when visiting Thailand, it is because "jak sie masz" (pronounced "yak shay mahsh") is a simple Polish greeting that means, "How is it going," but in Thai, "yak shi maak" (pronounced almost the same) means, "I have to pee a lot."

Bribes

In some countries, the bribery of officials is an unfortunate fact of life. While you do not want to offer one if it is unexpected (doing so could get you in trouble), it does pay to have a keen understanding of when it is appropriate.

In some countries, giving an extra payment ("tea money," in Asian countries) to a government official is ingrained in the system; it is not considered corruption. In Cambodia, for example, government officials are low paid, but the positions are nonetheless highly sought after, because of the bribes involved. They see it as the same as a "tip" that a waitperson receives in a restaurant.

In crossing the border into Cambodia, I was faced with the necessity of bribery. The Cambodian immigration office in Poipet was an open-air center under an awning, with several hard, plastic chairs, and some officials behind metal desks. The temperature was 100 degrees, and the wait for getting stamped into the country was more than four hours. I was hot, tired, and hungry.

Nonetheless, I was strolling beyond the gates of the Kingdom within 15 minutes, because of the efforts of a young entrepreneur who had an "in" with the immigration official. Passing the young man a tip, including an extra bonus to slip inside my passport (which was discreetly removed and pocketed by the immigration official) got my passport moved to the top of the stack and stamped immediately. Bribery often goes beyond gaining extra conveniences and a quicker passage through immigration. It is often necessary to obtain the government permits you require and the cooperation of those who make your business work for you.

Bribery is usually reserved for government officials and bureaucrats. Your business acquaintances make their decisions based on other criteria, not the least of which is your understanding and acceptance of their culture, your level of friendliness, and your adherence to the standard cultural rituals. Before you start handing out bribes to government officials, get a clear understanding of what is expected, because you may get yourself in trouble. In China, for example, you may give a government official a token gift, but bribery accomplishes only embarrassment.

Gifts

Corporate bribery is not a part of the package, but gift giving is frequently acceptable. In most countries, bringing some small gifts from your own country is welcome and seen as a gesture of friendliness. Find something that is not overly expensive or ostentatious but reflects your own country. I bring some Notre Dame caps because I am from Indiana. If you are from Switzerland, bring some of your country's excellent chocolates, and if you are from Napa Valley, bring some California wines (unless you are going to a Muslim country). I have had good luck bringing several boxes of jellybeans to give away. Bringing gifts that are expensive can be awkward, because your host may feel obligated to return the gesture. If your host gives you a gift first, you should give a gift in return, but never one of greater value than the one you received. In addition to giving small gifts to your host and your target company's top executives, you may also wish to bring a small gift for the secretary. In many countries, including the United States, the secretary functions as the gatekeeper — and if you displease the secretary, he or she could make things very difficult for you.

Religion

"We base our business on Christian principles." Those words have merit when bidding on contracts among some people in America, but they sink your business deal in Bangkok or Bahrain. (The majority of the world is not Christian and has no desire to be.) To say those words is to imply that a business based on Buddhist principles, Islamic principles, Judaic principles, Bahai principles, Confucian principles, or the principles of any other faith, is inferior. Proclaiming that your business is based on Christian principles when trying to do business in an Arab state, in China, Thailand, or any other of the many non-Christian countries is an insult to your host.

Most religions claim to be the only true path, and suggest that a business based on their particular set of holy books is inherently superior. I say "most" here, because some religions, such as Buddhism, have no notions of superiority.

Religions are more similar than we want to believe. What is a business that is "based on Christian principles?" It is one that strives to be fair in all dealings, to treat employees well, and perhaps to give something back to the community. We could just as well say we are a business based on Islamic principles, and have the same result; in fact, an Islamic business has more laws relating to how a person can conduct business. Shari'ah law, for example, prohibits hoarding a commodity, making it scarce in the market so as to command a higher price. To say we have a business based on "moral principles" or "ethical principles" is more accurate and more culturally acceptable. Following the "Golden Rule" is similarly not unique to Christianity:

- "All things therefore that you want people to do to you, do thus to them." ~ Christianity

- "Hurt not others in ways that you yourself would find hurtful." ~ Buddhism

- "That which you want for yourself, seek for mankind." ~ Islam

- "Do not impose on others what you do not desire others to impose upon you." ~ Confucianism

The majority religion in the Far East is Buddhism. In China, Buddhism is mixed with Confucian beliefs. Christianity and Islam are exclusive religions. That is to say, Christians and Muslims both believe they have the only true path and the only true prophet. Buddhists are not exclusive, and are quite open. You may ask a Buddhist about his or her beliefs without them taking offense. It can even be a good conversation starter. Buddhists are not evangelistic and are happy to discuss religion with you without pressuring you to attend religious events. You may be invited to go to a Buddhist temple. This is not an invitation to convert to Buddhism; it is merely a desire to share a philosophy with you, and allow you to see one of their beautiful temples.

When in other countries, make every attempt to respect local religions. In a Muslim country, remember to make accommodations for your associates to say their daily prayers, even if it interrupts a business discussion (and do not ask for booze; you will not get any). Muslims must stop what they are doing to pray five times a day. If you happen to walk into an office and see a Muslim in prayer, just walk back out quietly without saying anything. Dress appropriately; women should cover their arms and legs at all times. It is not necessary for a Western woman to cover her face with a veil.

In a Buddhist country, be respectful of Buddha images. Do not lean against them or sit on them for picture-taking opportunities and, if you are a woman, do not touch or sit next to a Buddhist monk. If you are at a temple, remove your shoes before entering. If you want to have a picture of one of the magnificent Buddha images housed in a temple, ask the presiding monk if it is okay. Usually they are accommodating; but remember to be respectful in your photography. One exception, the "Emerald Buddha" in Thailand, is accorded an exceptionally high level of respect and photographs are never allowed. Similarly, some Christian cathedrals in parts of Europe date back to the middle ages, are strikingly beautiful, and contain rare works of art; but you are not always allowed to take photographs. Ask first.

Toilets of the World

You may go to your destination well prepared by studying the language, knowing the appropriate gifts to give, and understanding how to greet someone; but one thing that travelers often overlook is the different customs related to bathroom behavior. Nonetheless, it is important, and lack of understanding can lead to embarrassment. The standard American bathroom, which is equipped with a toilet, sink, towels, and bathtub, is not a universal standard. In many countries, bathrooms are divided into two separate rooms: one with a toilet and another with everything else. In France, for example, if you ask for the "salle de bain," which literally means "bathroom," you will be shown a room that has a bathtub in it, but no toilet. You must ask specifically for "la toilette."

Some toilets may seem primitive to you as a Westerner, but you must adjust your expectations. While traveling on a train between Gdansk and Krakow with my son, I noticed that, when I pushed the flush handle on the toilet, all that happened is that a flap on the bottom of the toilet opened up and I could see the train tracks underneath it. I let my son know that it is not a good idea to walk along the train tracks in Poland!

In many countries, it is customary to pay to use a public toilet; there may be an attendant that collects money and, in some cases, dispenses toilet paper individually. Do not expect there to be toilet paper in the stalls — you may be disappointed. I got into the habit of carrying a small packet of tissue in my pocket, as well as making sure I have change with me.

Lastly, be aware that toilets do not always look like toilets. It took me a while to figure out the Southeast Asian "squat" toilet, which is low to the ground and has a ceramic platform with ridged edges. Do not sit down on it and put your feet on the floor as you do with a Western toilet; instead, put your feet on the ceramic platform and squat down over it. It is a little awkward, but you will get used to it. There is usually no automatic flushing mechanism; instead, there is a large vessel of water and a small scoop. You scoop out some water and pour it into the toilet when you are finished. There may be no toilet paper, but there will probably be a small hose with a sprayer for cleaning yourself off.

In Japan, toilets are high-tech, sometimes with multiple buttons for flushing, depending on the degree of water and rinsing that is required, as well as an extra button for a bidet function. In many South American countries, water pressure is weak and toilets clog easily; therefore, do not flush toilet tissue or anything else down the toilet. Next to the toilet you will find a basket, which is where to discard used tissue.

What Is Polite?

Many cultures take politeness and courtesy much more seriously than Americans do. In many Asian cultures, there are elaborate rituals that are done out of respect. In Europe, Continental manners are the rule. In most of the world outside America, men are much more gracious toward women. In the U.S., for example, it is less common for a man to hold a door for a woman or to allow her to go first, and many women are offended if this is done for them. In many parts of Asia, a man walks ahead of a woman and enters a room first — it is impolite for the man to do otherwise. Similarly, if a man in America greets a woman by kissing her hand, he may well get slapped, or at least thought of as somewhat of an oddball. But in much of Europe, this is still practiced and appreciated.

Personal style is also held to a higher standard in many other countries; while it may be perfectly acceptable in the U.S. to go out in public or greet guests to your home in torn jeans and a sweatshirt, you should avoid this almost everywhere else in the world. Dress modestly. In many countries, shorts are unacceptable, even in extremely hot weather. In Muslim countries, if you are a woman, be prepared to be hot, because you have to cover up your arms and legs. In much of Asia, people also dress modestly. In tropical Thailand, for example, despite temperatures that regularly exceed 100 degrees Fahrenheit, most people still wear long pants and long-sleeved shirts. If you are visiting Bangkok, leave the short-shorts and breezy sheer tops at home, and get ready to sweat.

Simple things that are never taken into account in the U.S. may cause consternation abroad. Gum chewing, for example, should be left on the plane. Many people see it as low class and vulgar to chew gum while speaking with someone. Smoking, on the other hand, is much more prevalent in many other countries, although there is a definite etiquette attached to it. Generally in Europe, for example, if you wish to smoke, you should offer a cigarette to other members of your immediate group, and light it for them.

In general, when traveling abroad, you should have a heightened awareness of manners and etiquette. If you are in doubt, just observe your host and others around you.

Style of Dress

Your attire makes a big difference in your ultimate success, probably much more so than it does in the United States. Your manner of dress reflects who you are and what your company represents.

When in doubt, err on the conservative side. Men should wear dark business suits with polished shoes and a fresh haircut. Avoid the "business casual" look of the sport coat with no necktie, unless you see your counterparts dressed the same. If you have facial hair, make sure it is very neatly trimmed. When you are attending a social occasion or just walking down the street, pay more attention to your appearance than you otherwise do. In many countries, dress stylishly for social occasions; often a suit and tie is appropriate when going to a colleague's house for dinner or out to a restaurant. If you are in another country on business, avoid the jeans and tee-shirt look. In some countries, such as France and Italy, fashion is extremely important at all times. The standard black suit, red tie, and no accoutrements is too bland; instead, opt for a trendier suit with a colorful tie and cuff links.

For women, it is a little trickier. The same advice of avoiding casual dress applies. In many cases, even stylish pantsuits are unacceptable, especially in very conservative countries or areas where there is a large Muslim population. In Muslim countries, a Western businesswoman should cover her arms and legs. In countries where there is more attention to high fashion (France, Italy), you may accessorize more.

Some Universal Tips

Some pieces of advice for business travelers are universal. Following are a few bits of advice you can use anywhere:

- Be sure to have your business cards printed with English on one side and the language of your country of destination on the other side.

- It helps if you can develop a "cast-iron stomach," so you can respect your host's culture by sampling the local cuisine.

- For standard of dress, err on the conservative side. Do not try to "go local" by wearing sarongs, dashikis, or Nehru jackets to business meetings. There are a few exceptions, however, and local dress may be required for some ceremonial events. For example, you will wear a kimono if your host takes you to a geisha house in Japan. Western businesswomen may feel free to wear the traditional sari while in India. In Muslim countries, women should cover their legs and arms, although it is seldom necessary for non-Muslim women to cover their faces. Avoid the temptation to wear shorts and flip-flops even though you may be in a tropical country and it is 115 degrees.

- Avoid "talking with your hands" or making gestures in any way. Common gestures made in the United States may not be understood in other countries, or may be considered rude.

4
A Changing World

In the '50s, America ruled the world. India was not yet engaged in the business of IT outsourcing, and China was still closed off to the rest of the world, remaining a place of deep mystery. Japan was only just beginning to flex its postwar muscle economically, but still had a long way to go. The Kingdom of Siam was just something out of a play. The Iron Curtain kept the Soviet states off of the world stage economically, and the rest of Europe was still just finding its way.

The days of American dominance are rapidly dwindling, and it is vitally important for Western companies and their CEOs to undertake a fundamental change in how and where they do business. While Americans have certain ways and beliefs that govern business and life, they are not universal. A global businessperson must come to the humble acknowledgment that the American way is just one way, and not necessarily superior or inferior to other ways.

What Goldman Sachs referred to as the BRIC countries (Brazil, Russia, India, and China) are all becoming a major force in the world economy and are likely to be larger than the G6 (France, Germany, Italy, Japan, United Kingdom, and United States) in less than 40 years. Today's dominant countries may not be tomorrow's dominant countries. In fact, it is a pretty good bet that the short list of countries that control most of the world's resources will look different in the near future. Japan is a stunning example. After World War II, Japan literally reinvented itself.

In the late '40s and '50s, the United States went through a growth spurt and a period of new prosperity. Now other countries are having their growth spurts. India and China's Gross Domestic Products (GDPs) are growing at rates far beyond that of the United States. Although it is not practical to believe that a supercharged GDP growth rate will continue indefinitely, it may continue long enough for India and China to become major forces in the world stage — to the point that they will be equal in wealth and power to the United States. Then the "American Way" of doing business will no longer be relevant. The "Global Way" is the way of the future.

This is what has been called the "New Normal." The old model of large American corporations ruling the world is over; today's economy is entrepreneurial and nimble. It is less dependent on large corporate bureaucracies and more dependent on a global interconnect of businesses and individuals. Like it or not, if a corporation wants to survive, it has to concentrate on a core mission and outsource noncore functions, often to offshore locations, to remain competitive. Despite what protectionists and "America first" believers may say, the result is win-win: citizens in emerging countries are able to raise their standards of living, domestic companies are able to sustain themselves and remain profitable, and American workers are shifted to higher-value tasks.

This shift is happening quickly. At the World Economic Forum 2007 held in Switzerland, Nandan M. Nilekani, Co-Chairman of the Board of Directors of Infosys Technologies Limited, an Indian IT outsourcing firm, noted, "there is a significant shift happening toward emerging economies as they start to integrate more rapidly into the global trading system. In 2005, they made up more than 50 percent of the increase in world GDP and are creating a new class of consumers, manufacturers, entrepreneurs, and policymakers. New opportunities have opened up in these economies and they are benefiting from competing at the global level."

When those emerging economies open up and jobs come in from the West, an interesting thing also happens: those third-world countries develop a consumer economy, and — as we are already seeing in China — they become not just exporters; but also importers of American goods.

For many years, the focus has been on the world dominance of the United States, Japan, and Western Europe, but this is changing. The global economy is a fact of life, and other emerging marketplaces are figuring into the mix. Rachid M. Rachid, the Egyptian Minister of Foreign Trade and Industry, noted at the World Economic Forum that "The most significant global trend today is the fast-growing shift in the world economy towards stronger trade and investment between emerging markets. The emergence of China and India, the fast-growing role of Asia in the world economy, the influx of oil-fueled liquidity into the Arab world, from the Gulf: all these are examples of the growing power of emerging markets."

Today, we are faced with a global shift of power, and it is up to the United States and the businesspeople who stand behind American industry to adapt if they want to continue to be major players in the world stage.

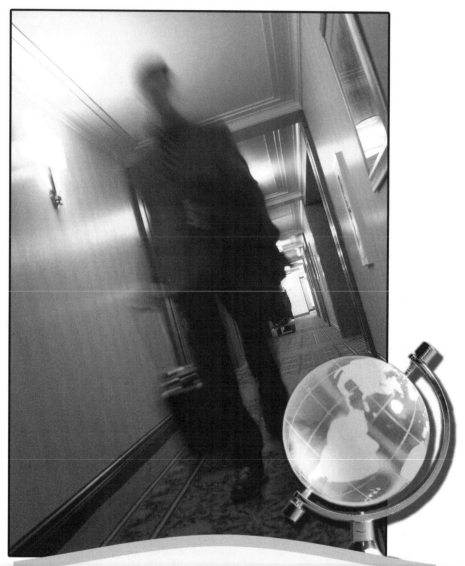

Asia is the largest continent, and
China has the fastest growing economy in the world.

Section 2

The New World Leaders: China, Japan, & India

The dynamics of the world are rapidly changing, and the United States can no longer expect to dominate the world in every area. Today, the United States still ranks on top of the world in terms of Gross Domestic Product (GDP), with the European Union (taken as a whole) following closely on its heels. Following the EU, the next in rank are China, Japan, and India, respectively.

The dynamics of China, Japan, and India are powerful. Taken together, the GDP of these three countries far outweighs that of the United States, and each of them is in high-growth mode. If you are looking for the fastest growth and the best opportunities for business, look no further than China, Japan, and India. If you are not doing business in one of these three countries now, the day is probably coming. The Asia-Pacific region, according to Merrill Lynch's World Wealth Report, has surpassed Europe to become the second most popular area for international investment, just after North America.

Goldman Sachs' well-known 2003 research paper, the BRICs report, predicted that by 2039 China will be the world's largest economy. India will surpass Japan, becoming the world's third largest economy after China and the United States. Recent revisions to the 2003 BRICs report are even more impressive, with predictions that India will move from the number three to the number two spot, pushing the United States down another notch. This is not a negative reflection of the United States, but indicates different stages in each country's economies. The United States is a mature economy, and so has slower growth. India, for example, is where the United States was in the '50s, when our economy was burgeoning, jobs were plentiful, and Americans thought growth was never ending.

5
China

In China, as in most Asian countries, business transcends bottom-line economics. Business decisions are based on business metrics, but there is much more involved. If you are going to China, or any other Asian country, get out of your bottom-line mode, and be prepared to figure in some intangibles. For example, the concept of "face," or personal pride, is highly important and must be considered in all social and business interactions. Never do anything that could cause someone else to lose face; direct criticism is not acceptable.

If you are opening an office in China, calling your local supply store and ordering some cubicles and desks will not be adequate. A feng shui consultant is necessary to ensure that your office is congenial to your visitors.

In our home, which looks Asian even though it is located in the American Midwest, we have several strategically placed water fountains and mirrors; and Lord Buddha and the King of Thailand both look down on our parlor from the highest positions on the wall. Placing any picture higher than the King, with the exception of Lord Buddha, is disrespectful.

Some Western businesspeople make the mistake of regarding other religious observances as superstition, thinking that educated businesspeople do not bother with such matters. These are cultural traditions rooted in deep spiritual beliefs that go back thousands of years. They are not simply traditions of common peasants. Even the most educated and worldly Chinese businessperson avoids signing a contract on the fourth day of the month, because the number four, or "si", sounds like the Chinese word for "death."

When a Chinese businessperson tells you that a contract will be signed on a given day, the person is not doing so out of mere scheduling concerns. Do not take out your Blackberry and say, "No, that does not work for me. How about a week from Tuesday?" The date they are setting is important. Before a contract is signed or a deal is made, a Chinese businessperson may consult a fortune teller or a monk to determine the most auspicious day for signing. As Westerners, we may not understand these traditions, but they are part of the Chinese business and social culture.

Economics and Opportunities

Although we may think of Japan as being the most modern and industrialized of the Asian countries, China is the driving force throughout Asia's economy today. The economies of all Asian countries, including Japan, are dependent on China. As a major economic force in the world, China is a country that must be addressed, and any large company that seeks to join the global economy will have to do business in China at some point.

Economic reform in China started in the '70s, with President Nixon's famed visit. In the 30 years since, China has made rapid progress in opening itself economically to the rest of the world. Communism notwithstanding, there is a significant private sector and market economy in China, and welcoming foreign trade and investment has made China the powerhouse it is today. Annual GDP per head has risen from Rmb379 in 1978 to Rmb10,502 in 2004; but there are large disparities in income and lifestyle, with much of the Chinese wealth concentrated in the coastal areas and major cities.

The most obvious opportunity in China may be importing goods into the West, but despite the many "Made in China" stickers you see at your local department store, this is not the only opportunity in today's China. At the World Economic Forum 2007, Group Executive Vice President Min Zhu of the Bank of China made a comment about China's economy that indicates China's tremendous growth potential. "There is a perception that China's growth is export-driven but last year's exports only accounted for 15 percent of growth, whereas domestic consumption contributed around 20 percent and investments were responsible for over 40 percent." As China's economy continues to grow and personal wealth becomes a greater reality, domestic consumption will give a greater boost to the economy and provide opportunities to Western companies. Instead of importing goods from China to sell to Western consumers, Western companies will be sending goods to China to sell to Chinese consumers. In fact, Beijing houses the biggest shopping mall in the world, the Golden Resources Mall, sometimes called the "Great Mall of China." It is known for upscale Western shops, brand names, and refined elegance.

Greetings

China actually has several dialects, so make sure you study the right one before going. Standard Chinese is Mandarin. It is the most commonly spoken language, and is referred to as "the common language." In Hong Kong and parts of southern China, however, Cantonese is spoken. There are also several other minor dialects that are spoken in different parts of the countryside, but learning a few words of Mandarin and Cantonese will get you far. Written Chinese is the same in both Mandarin and Cantonese, although the verbal language is different. Within China, written Chinese is now "reformed," written left to right instead of in vertical columns, with some of the more complex symbols simplified.

A slight bow — not as deep as the Japanese bow — or even a nod, usually without smiling, is common, although many Chinese businesspeople now use the Western style handshake when greeting someone, especially someone from the West. Touching, kissing, and backslapping are never done.

It is important to be on time to all meetings. Upon arriving, you may also wish to bring small gifts. Expect the gift to be refused at first out of politeness, and then offer it again. When you are offering it, hold it out with both hands. Pay attention to the basic rules of feng shui when choosing a gift. Something that comes in a pair or in even numbers is good luck (unlike most of the rest of the world, where odd numbers are the custom). Black, blue, or white colors are associated with funerals, so avoid gifts with those colors. Red and pink are better colors, indicating happiness; yellow or gold colors indicate good fortune. Sharp items (which the airlines might not let you take on the plane anyway) are not a good idea, because they indicate "cutting off" a relationship. Expensive gifts are inappropriate, especially if given to a government official.

Food

Chinese use chopsticks, and visitors are expected to as well. Although you can ask for and receive a fork and spoon at a restaurant, it does not make a good impression. During dinner, wait for your host to pick up his or her chopsticks before you begin. You may notice that some Chinese rub their chopsticks together before eating; this is to make sure there are no loose splinters. Do not do this if you are a guest at an upscale restaurant or in a home, because it implies that your chopsticks are substandard, and the gesture is insulting.

In China, as in many Asian countries, food in restaurants is served communally. Do not attempt to fill up your plate right away; instead, take small quantities several times, continually dipping in to the communal dishes. Be careful not to use your personal chopsticks to dip into the communal serving dishes. Your host may serve you, or there may be a separate pair of chopsticks in each communal dish just for serving yourself.

Nearly every city in the United States has a Chinese restaurant, and so you may think that you know Chinese food. Chances are, you do not. Food in American Chinese restaurants is usually made to suit American tastes. China has many unusual dishes. For your first course, you may be served shark's fin soup, which is a delicacy, and it is rude to turn it down. I have eaten grilled jellyfish, and I must admit I will try not to sample it again unless I am obliged to; it is tough and tastes like shoe leather. Another rare delicacy that you should sample if offered, but avoid if you can, is the "thousand-year-old egg." The one offered to me had a stench so bad, I had to remove the plate from my table. Despite my aversion to these two dishes, however, I have found Chinese cuisine to be very delightful and varied. China is a very large country, and the cuisine differs from region to region.

You must also pay attention to where you are sitting at the table. A great deal of attention is placed on status, and the seating arrangement is made accordingly. Do not take a seat at random. The host sits facing the door. If there is a second host of lesser importance, that person sits on the other side of the table, facing the first host. A guest of honor sits beside the main host. Other guests sit, according to rank, alongside the guest of honor, toward the second host.

Do not expect nonsmoking areas in restaurants; however, if you wish to smoke, ask permission from the host. It is unlikely that anyone will object, and if someone should ask you if you mind if they smoke, they are asking out of courtesy. Also, do not leave tips in a restaurant in China.

Although it pays to be a good conversationalist, do not be surprised if there are periods of silence during a meal. This is not considered impolite or awkward, as it is in the West; so do not feel that you have to fill any void with idle chatter. Do not discuss business during dinner.

There are likely to be toasts made. Unlike in Poland or Russia, where you are expected to down your entire glass of vodka upon each toast, you only need to raise your glass and take a small sip.

Religion

As a Communist country, China is officially atheist, although spiritual beliefs are pervasive. The majority of Chinese are Buddhists and follow Taoist and Confucian philosophies. Christians and Muslims make up a small minority in China. Confucianism in particular has a major influence on Chinese philosophy and business. Confucianism does not have a god, and one can be a Confucianist as well as a Buddhist, Christian, or anything else at the same time. Confucius is not considered a god, but a wise philosopher whose teachings have great influence over Chinese life, including how business is conducted. Confucian philosophy affects all types of relationships and focuses on preserving harmony by setting rules for interaction and respect for all parties. It is the Confucian philosophy that has resulted in the intense respect you see in China for the elderly and for authority figures. It is the Confucian outlook that promulgates rigid chains of authority and hierarchy in Chinese business.

Etiquette

Although China is not a touch-oriented culture, you may find a person standing near you without touching, simply because major Chinese cities tend to be extremely crowded, and they have become accustomed to having little personal space. Nonetheless, physical touching usually is not done outside the basic handshake, especially with business acquaintances and people who are either older than you or more senior in position. You may see close friends of the same sex walking hand in hand as a gesture of friendship.

Chinese do not engage in conversation that is critical to the Communist government in any way; therefore, it is best to avoid any potentially controversial talk. Do not ask about or mention government repression, shootings at Tiananmen Square, censorship of the press or Internet, or Taiwan.

Titles are important in China, and an associate may be introduced to you by his or her proper title, such as, "Division Chief Lo." Use an individual's proper title.

Bargaining and Deal-Making

As is common in many Asian countries, Chinese people do not like to say "no" outright. They indicate disagreement in different ways. When presented with an unacceptable proposal, a Chinese businessperson may silently shake his or her head from side to side.

It may take longer to make a deal in China than in the West; decisions are not made lightly. Patience is critical.

A Chinese businessperson is aggressive, seeing business as a zero-sum game. They may make unrealistic offers or attempt to change the deal at the last minute without alerting you. They consider such subterfuge to be an aspect of bargaining, and it should not be taken as an attempt to gain an unfair advantage. A last-minute addendum that changes the deal to the Chinese businessperson's advantage is a bargaining ploy.

In business as well as in social occasions, you must pay attention to China's hierarchical order. Confucian ideals have a strong influence on how business is done. Older businessmen are shown greater respect. Do not send a younger, junior member of your team to negotiate.

Other Common Customs

- Use both hands when presenting your business card or a gift. (True also in Japan.)

- Do not point with one finger, but with your entire hand without any fingers outstretched.

- It is common for Chinese to spit or blow their noses on the street; this is not impolite.

- Defer to your elders or to those who hold a higher rank. Allow them to go first into a room, sit at the head of the table, and do not do anything that might cause them to lose face.

Traveler's Tip

Although this is an issue you do not discuss while in the country, the Chinese government is impassive when it comes to freedom of the press, and more to the point, the Internet. The Internet is used in business, and its use is growing in China, but it is by no means the same Internet we are used to in the West. It is heavily censored by the government; many Web sites are not available, and Internet cafés are monitored.

6
Japan

Japan has a highly educated population at all levels. Unlike some lesser-developed Asian countries, the literacy rate is 99 percent. Japan has undergone significant changes since World War II, before which it was the most powerful military force in Asia. Its industrialization dates back to the Treaty of Kanagawa with the United States in 1854, after which the country's ports were opened. Since the War, however, Japan's focus on military strength has given way to a focus on industrial strength. Perhaps because only 1 percent of Japan's GDP is spent on the military, there are more resources for investing in the country's prosperity.

Avoid criticizing the Japanese way of doing business, but try to understand it. Although the highly hierarchical and rigid structure in Japanese business may seem foreign and inefficient to Westerners, Japan has become an economic leader since the end of the War. Their ways have worked well for them.

One important factor in Japanese business is the keiretsu. In Japanese business, a keiretsu is a closed circle of businesses that do business with one another. Although each business is autonomous, it is also a member of this group, which may consist of manufacturers, suppliers, and distributors, all of whom work together without competition from outside the country. In the West, a company may do business with many different suppliers and is constantly in search of the best deal, but in Japan, the relationship itself is more important. A Japanese business may be using a supplier from its own keiretsu, even though a better deal is available elsewhere. In America, we call this the "Old Boy Network," and even in Japan, some compare the system with the feudal lords and shoguns of the past.

Although the keiretsu is for the most part a closed system, do not let this discourage you from doing business in Japan. The underlying philosophy of the keiretsu is important to understand, because it governs virtually all business dealings, whether within the keiretsu or not. The governing factor in business is the relationship, which requires a basis of trust and friendship. As in most Asian countries, Japanese businesspeople want to know you as a person before they do business with you. After you have gone, your Japanese business associates will remember you — not just as an American businessperson, but also as "that American businessperson who sang out-of-tune karaoke after enjoying sake." Singing karaoke out of tune is far better than not singing it at all. All attempts to observe Japanese culture are appreciated, even if you fail.

Economics and Opportunities

Since the War, economic growth has been outstanding in Japan, averaging 10 percent in the '60s, 5 percent in the '70s, and 4 percent in the '80s. Growth has slowed since the '90s, partly due to the fact that much of the industrial build-out has already occurred, but also due in part to the Asian economic crisis of the '90s and the slowdown in European and American economies.

Japan is well known for being one of the world's largest producers of automobiles and electronic equipment.

Greetings

Japanese businesspeople are much more formal than Americans. Although they are well acquainted with the Western handshake and often use it when greeting people from America, the traditional bow is customary. When a Japanese person shakes hands, it is lighter than the firm, Western counterpart. A strong grasp is too aggressive. The "look 'em straight in the eye" custom also seems challenging.

The bow is an elegant greeting. It does not involve touching. Japan, like most Asian cultures, avoids most touching in public, except for in the subway, where the rules of personal space are abandoned in favor of practical necessity. The Japanese bow is subtle and has many meanings, depending on who bows first, how deeply you bow, and for how long. The bow indicates rank and status: the person with the lowest rank or the youngest person bows first and deepest. The lower and longer you bow, the greater degree of respect you show.

The exchange of business cards is also part of the initial greeting. Present it with both hands outstretched while bowing. When you receive a card in return, take a moment to look at it; do not just stick it in your pocket. If possible, lay the card directly in front of you on the table as you sit down. Under no circumstances should you ever put it in your back pants pocket! To do so is a great sign of disrespect.

Food

Dining at a Japanese restaurant is an experience that should not be missed. It is elegant and refined, and the presentation of the food is an art form. Although you can request a fork and spoon, you make a better impression by trying to use chopsticks.

At some restaurants, you sit on the floor around your table; women must either sit with legs tucked underneath or to the side. Avoid wearing short skirts.

If you are invited to a tea ceremony, follow the lead of your host. These ceremonies are beautiful, highly stylized, and one of the most graceful experiences you will ever have. Remember, do not talk business during the tea ceremony.

Business dinners are common, so expect to be taken out to dinner after working hours. The host pays for the dinner. Do not try to haggle over the check as you might in America. If you offer to take them out to dinner, you pay. Tipping at restaurants is seldom done.

Religion

Most Japanese observe Buddhism, Shintoism, or Confucianism. The three are not mutually exclusive, and a Japanese may observe the beliefs and traditions of all three. Shintoism in particular is important to understand, as it is indigenous to Japan. Shinto belief holds that all beings and objects have a life force, and there is a strong focus on achieving harmony and balance in all things. The Japanese focus on structure and hierarchy is also derived from Confucian beliefs, as Confucius stressed the importance of respect for a person's role and the need for structure and order in all relationships.

Etiquette

The Japanese are a polite and respectful society. Details are important. For example, showing an open mouth is impolite. Chewing gum when talking to someone is considered low class. Most Japanese women cover their mouths when they laugh. Pointing is also rude. However, it is acceptable and expected for people to make slurping sounds when they eat noodles.

Using first names for new business acquaintances is an American custom neither practiced in Japan nor in most of the world. The respectful way is to address a person by their last name, followed by "san." For example, I am "Blacharski-san" when in Japan.

Addressing someone as "you" is inappropriate. Do not ask, "What do you think about this offer?" Instead, say, "What does your company think about this offer?"

Bargaining and Deal-Making

In an American business meeting, silence is deadly, and if the conversation lulls for even a few seconds, a good salesperson takes the opportunity to inject a comment about the company or the deal to avoid an awkward moment. In Japan however, silence during a business meeting is not negative.

As with many Asian societies, the Japanese seldom say "no." Rather, they say something other than "yes." If they say, "I will consider this," they could mean "no."

The Japanese work long hours and are punctual. There is a strong teamwork ethic and a sense of company loyalty, to the point of singing a company song every morning. If you wish to say something nice, compliment the group or the company, not an individual worker.

The senior member of your business group does the negotiating, and the younger ones defer to that person, only providing information when asked. The Japanese team will be insulted to receive a presentation from or engage in negotiations with a junior member of your company.

It is unusual to find Japanese women at the upper echelons of business. Western women may participate in negotiations, however, and are accorded the respect of their position within the company.

Common Customs

- Public displays of emotion are unusual. Displays of anger are discouraged.

- If you tend to talk with your hands, try to show restraint. Japanese see it as distracting, and might misinterpret your gestures.

- Avoid the masculine, Western style of leg crossing, resting the ankle of one leg on the knee of the other leg. (Showing the bottom of your shoes is rude in most of the world.) Instead, cross your legs either at both knees or both ankles. Although in Western society crossing legs at the knees may be considered effeminate, that is not the case in Japan.

- Finger pointing, even to point out something or attract someone's attention, should be avoided.

- Remove your shoes when you enter a home or a restaurant, and of course, before entering a Buddhist temple.

- Avoid touching others, backslapping, and touching shoulders. Males and females do not show affection in public.

- Special note to women: avoid high heels so as not to be taller than your Japanese business associates.

- If you are speaking or making a presentation, your Japanese counterpart may say "hai" during your speech. While "hai" literally means "yes," in this context, it does not necessary indicate agreement, but merely that they acknowledge your point.

- Do not refill your own sake cup. Your host will attend to it.

Traveler's Tip

When a male Japanese associate offers to take you to a geisha house, your reaction may vary from absolute horror to an eager expectation of sex. This misunderstood custom needs some explanation. Geishas are not prostitutes, and a geisha house is not a house of ill-repute. You are not going there for sex. A geisha is a professional and highly skilled hostess, performer, and artist; she is an exceptional dancer, conversationalist, musician, and an expert in performing the tea ceremony.

Although it is a good idea when traveling to keep your wallet in your side pocket or jacket pocket to avoid pickpockets, you will not see many thieves in Japan. One of my more worldly friends who lived in Japan for a year told me that you could leave your wallet on the turnstile in the subway, come back the next day, and find it waiting there with all the money and credit cards still intact.

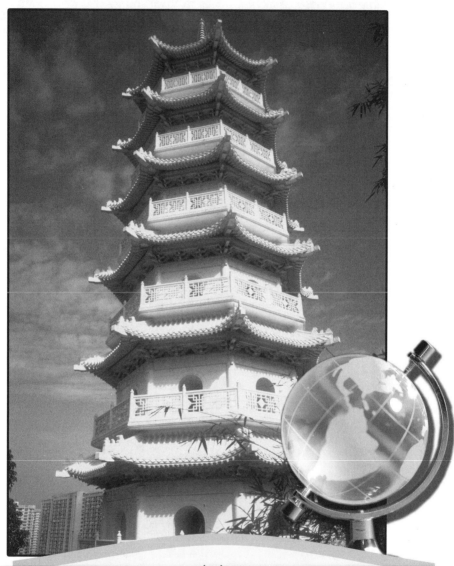

In Japan,
it is considered rude to blow your nose in public.

7
India

If there is a country the international businessperson should focus on, it is India. India's rise to prominence is a success story like none other. It is the second largest country by population, and currently the fourth largest economic power. Its economic prowess is likely to surpass Japan's in the near future. India's economy is growing at about 8 percent per year, the fastest growth rate in the world. The country's middle class is increasing rapidly, which translates to a tremendous demand for comfortable housing, consumer goods, telephones, and automobiles. India is creating millionaires at a breakneck pace, with their millionaire population increasing by nearly 20 percent in 2005.

China is still considered the fastest growing economy in the world, but between the two, India has more long-term potential, because it has a younger, better educated population. There are two distinct Indias: one of upward mobility, prosperity, and modern conveniences; and one of desperate rural poverty. Much foreign investment goes to six urban states, with wealth concentrated in only a few areas; but India's technology and telecommunications infrastructure is advancing, and technology centers will become dispersed.

Economics and Opportunities

The caste system has been formally prohibited by the government, but it nonetheless persists in every aspect of Indian society. The caste system reflects the Hindu belief in reincarnation and in different stages of human existence. The highest caste is the Brahmin, followed by the kings and warriors class, the merchant class, and farmers and laborers. At the bottom is the "untouchables," who are excluded from mainstream society.

India is known for being self-contradictory. Even though the caste system has not successfully been outlawed, the Indian workforce is highly educated and speaks British English. Since the '90s, there has been a great deal of economic liberalization, although it is still difficult to do business there because the bureaucracy is top heavy and tariffs are still high. Reforms continue, and foreign investment is welcomed, but there are complex restrictions on foreign direct investments, even though India joined the World Trade Organization in 1995.

A huge IT infrastructure has developed in India, and it is a major center of outsourced IT services. Much IT development in the world is done in India, and many Western companies have call centers there. Do not think that outsourcing is the only industry in India, however. A focus on education, economic reforms, and a rapidly developing infrastructure have led to a large consumer class. There are significant opportunities in the areas of infrastructure build-out and housing, as India's rapidly growing middle class seeks out modern homes with modern conveniences. Retail is another area to look to in the future. At present, India's entire retail segment is in the form of small, locally owned shops; but American and European chains are knocking at India's door, eager to change the character and nature of retail. Telecommunications is also growing rapidly.

Greetings

Although an Indian may offer the Western handshake, the traditional greeting is the namaste, which is similar to the wai in Thailand. You put your hands together in a prayer position, raise your joined hands to the level of your chest, and bend your head downward, and you say, "namaste." A subsequent handshake may be offered to a Western male, but it is a light handshake, not the Western grip.

India is not a touch-oriented culture. Avoid touching whenever possible, especially between sexes. Wait to be introduced to someone before speaking, instead of introducing yourself. Indians usually greet using the first name. You may be greeted with a garland of flowers around your neck, but do not keep it on; acknowledge the gift with a namaste, then remove it and set it next to you.

Food

Most restaurants in India are small and locally owned, and standard American food is rare. Indian food is a real treat, with wonderful sweets and desserts. My favorite is gulab jamun. If you visit an Indian at home, you will probably be served a sweet with a cold drink or tea. Lunch is the biggest meal of the day. At the end of a meal, you may be served tea, and it is drunk in the British style, with milk and sugar. You will also eat with spoons, forks, and knives, although in some less formal situations people may eat with their fingers, depending on the region. At the end of the meal, you may be served caraway seeds. Take a few of them with a spoon, put them in your hands and chew them. It is similar to an after dinner mint and is quite refreshing.

Much Indian food is spicy, and the degree of heat varies depending on the region. There is no beef in restaurants here, since the cow is a sacred animal. Many Hindus are vegetarian.

There is not much in the way of hard liquor, but beer is common. Regardless of what you are drinking, do not refill your own glass; your host or the waitstaff does that for you. You may make a toast after the host makes one, or at the end of the meal, but there is not a constant barrage of toasts and drinking as there is in Russia or Poland.

Spouses do not accompany a businessperson to a business meal, although the spouse may be invited along to an associate's home. After the meal, you do not offer a verbal "thank you" for the meal, but merely offer a "namaste" to the host. Note that your host will probably have household servants to prepare and serve the food, so do not offer to help clean up or wash dishes after the meal is over.

Tipping in restaurants is expected.

Religion

The dominant religion is Hinduism, which is adhered to by about 80 percent of the Indian population. There is also a significant number of Jains, Sikhs, and Muslims. There are many gods in Hinduism, and the religion affects all parts of Indian life, including business. Before you reject Hinduism as a primitive set of beliefs with bizarre rituals, consider how Christianity looks to a Hindu. It just may appear to be a primitive set of beliefs with bizarre rituals. Do not make the mistake of thinking that India is going to adopt Christianity. The Hindu faith works well in Indian society and is ingrained. Hinduism has an advanced, highly civilized belief structure that has helped to maintain order, unity, and peace among its believers.

Traveler's Tip

If you have an opportunity, go to see a performance of the Ramayana. This beautiful, ancient epic is about an exploration of Hindu ideals and tells an ancient tale of the gods. Even if it is performed in the Hindi language, it is beautiful and worth seeing.

Etiquette

Punctuality is observed, and meetings start on time. The weather is extremely hot and wet in this tropical climate, but avoid the temptation to dress casual. It is acceptable, however, to wear lightweight clothing, so long as it is modest; men should feel free to go without the traditional jacket. Avoid the conservative, black and white combination; these are funeral colors. Wear dress pants, avoid shorts, and wear dress shoes, but not leather ones. Most women wear the traditional sari. Western women may dress in modest Western clothing, or they may wear the sari if they wish.

A small gift may be offered to your host or associate, but do not make it expensive.

Bargaining and Deal-Making

Indians tend to be well versed in Western customs and often speak multiple languages. Although an Indian businessman may wear a Western suit and tie, know how to shake hands, speak flawless English, and even eat the occasional cheeseburger, do not make the mistake of thinking that you are freed from your own burden of knowing their culture. Indian business has gained a prominent position in the global economy, and Indians recognize the need to understand other cultures, including the American culture; but they are by no means abandoning their own. Get to know the Indian ways just as well as they know American ways.

In keeping with the traditions of Indian society, business tends to be fairly rigid, with a strict hierarchy. Titles are important, and subordinates defer to their seniors. It is a top-down business structure, so be aware that you must conduct negotiations with the highest level possible, and you should send your own highest-ranking person to conduct the negotiations.

Common Customs

- Although women are increasingly seen in professional roles, they are still seldom seen at the highest levels of business. Nonetheless, female businesspeople from the West are accepted as equals once their authority is established.

- The common naming structure for Indians is to use the first initial of one's father's first name before one's own given name; for example, R. Subramanian's father's first name started with the letter R. When addressing him, you precede "Subramanian" with "Sri" or "Mister."

- If you are at a meeting or a dinner, wait for the host to seat you, as the seating arrangement reflects the status of each participant.

- Indians love to talk. Unlike in Japan, where silence may be an integral part of a business meeting, Indians love conversation and may become animated.

- If an Indian shakes his or her head from side to side, it does not mean disagreement, as in the West; it simply indicates that he is listening to what you are saying.

- Remove your shoes when visiting a home or Hindu temple.

- Do not point with one extended finger; point with your whole hand or gesture with your head.

- Out of respect for the religion, try to avoid wearing leather while in India.

Section 3

The World In-Depth: Asia & the South Pacific

I had already spent a considerable amount of time in Europe and thought I was a seasoned traveler, but I was unprepared for the alien feeling I felt when I first landed in an Asian country. Most Western European countries have languages that have the same Latin root as English, so even if you do not understand French, Spanish, or Italian, the structures and some of the words are similar. It is easy to pick up a few words along the way to make yourself understood. The alphabet is based on the Latin alphabet, and you can read the signs.

When I was on the plane and saw the Bangkok airport just below me, I noticed the big sign and felt like I was a lost child. Presumably it was telling me it was the Bangkok International Airport in Thai characters. In Paris, you land at "l'aeroport," and that is pretty easy to figure out, even if you do not speak French. But to me, the big sign on the side of the Bangkok airport was just a bunch of scribbles. For the first time in my adult life, I knew what it was like to be illiterate. The feeling got worse when I asked my wife where the bathrooms were, and she pointed me in the general direction. I walked over and there were two doors with only Thai writing, and not the usual stick figures. I did not have a clue which one to enter until I saw someone else go in. Even today, I feel more like I am four years old instead of 47, as my wife attempts to teach me to read Thai, and I point to a character and say "that letter makes the 'aah' sound."

In some of the more isolated areas of Asia, you may find yourself being stared at; do not take this as rudeness. In such areas, people seldom if ever encounter a Caucasian or Negro, and you are a novelty — a curiosity that they want to get a good look at. Of course, in the big cities, Asians are used to Westerners, and in Bangkok, Tokyo, and Beijing, I am just another American. But in Burma, where few Westerners venture, it was another story, and the natives were just as curious about me as I was about them. While it struck me as unusual that even the men there wore colorful sarongs (a type of long skirt), I am sure that they stared in wonder at my pants, which they thought must be quite uncomfortable.

When you first get to Asia, everything is different. There is nothing that relates to Western culture. The food is strange, and they eat with either chopsticks or with their hands. The customs are peculiar, and all the religions are different. The architecture is like nothing else. People on the street sell fried insects. Whole ducks, with heads and beaks intact, hang from shop windows. Grand, majestic statues of Buddha make it feel even more otherworldly. An elephant with its master lumbers down the sidewalk, ignoring the passing cars. It is a land of mystery, and it gives the feeling of leaving behind all familiar things.

Women and Equality

There is an appearance of subservience in many more traditional Asian countries. Many a feminist from the West has taken great umbrage at the sight of an Asian woman walking behind her husband, but one must see this from an Eastern point of view. In some Asian countries, women walk behind their husbands and the husbands enter a room first (instead of adhering to the polite European custom of holding the door open and allowing the woman to enter first), but it is not done out of a sense of male dominance or superiority. This custom began out of chivalry in deep jungles in countries like Thailand and Vietnam. The man went first because of the threat of wild animals and poisonous snakes.

Nevertheless, there are different outlooks toward women and equality in Asian countries, where equality of the sexes is either nonexistent or a new concept. Female businesspeople are treated with the same respect as their male counterparts. In Japan, higher-echelon business is a male domain; however, rank is more important than sex.

In Thailand, you may be surprised at hiring practices. The following help wanted ad is representative: "Attractive, well-dressed girl under 25 needed for front office." The logic behind such hiring practices is that a place of business should be fun and attractive to visitors.

Common Asian Customs

There are some customs that are pervasive in almost all Asian countries. For example, you should remove your shoes when you enter a home. Also, never show the bottoms of your feet — that means not crossing your leg Western style.

Coffee is a staple of the Western business meeting, and it is often served in disposable foam cups, something you avoid in an Asian country. Equip your office in Asia with a complete set of porcelain china with a full tea set.

Most Asian cultures are not touch oriented, and public displays of affection, even between husband and wife, are seldom seen. My wife had to ask permission from her father before I was allowed to hold her hand in public; and even then, it was only permissible in more cosmopolitan Bangkok. When we went to smaller towns, it was forbidden.

Chopsticks are not universal in Asia. Outside of China, Japan, and Korea, forks and spoons are used, holding the spoon in the right hand and using it to eat with; the fork is used to push food onto the spoon. Knives are seldom seen on the table, because they are considered aggressive. Having a knife at your place setting is like having a gun by your plate in America.

Food is an important part of business, as a way for both parties to develop a rapport and establish trust. It is not common to discuss business over meals. Asian social life means food, and a common greeting in many Asian countries is, "Have you eaten yet?" It is a rhetorical question, like "How are you?" in America. The polite way to answer is that you have eaten. In many Asian countries, you do not ask, "How are you?" when greeting someone. In China, you ask "How are you?" of someone who has been ill. A more acceptable alternative is to greet someone by saying, "You are well." In Mandarin Chinese for example, you greet someone by saying "Ni hau," which means, "You are well." Saying "Ni hau maa" on the other hand, means, "Are you well?" You ask that to someone who has just gotten over a cold. The same is true in Laos, where you say, "sabai dee."

At restaurants, the American fashion of grabbing or arguing over the bill is not done. Usually, the highest-ranking or eldest person pays, or in some cases, the one who has made the invitation. Splitting the bill, or "going Dutch," (which in Thailand is referred to as "going American,") is seldom done, especially for a business meal. Interestingly, this is not really a custom among the Dutch.

Gift giving is common in business situations; gifts are opened later, in private. The gift should not be extravagant.

The Confucian Work Ethic

Confucianism pervades throughout much of Asia, determining how business is done. More attention is paid to rank, status, and respect for authority. It is assumed that a person in a high position should be respected at all times and deferred to in all matters. There is very little dissent, at least obvious dissent, among the rank and file. However, Confucianism imposes a level of respect for each person's role and contribution, no matter how large or small. If the janitor is diligent about sweeping the floor and emptying the trash cans, then that janitor has "face" and is accorded respect, even by his superiors.

Confucianism also places a high value on ritual and formality. Learn the rituals for greeting people in your host country and you will gain face there. Another principal tenet of Confucianism is trustworthiness. Your Asian counterparts may want to get to know you on a personal level before they get down to business.

In accordance with Confucian principles, when you are doing business in

Asia, be sure to send a senior-ranking person to negotiate. Sending lower-ranking employees is an insult to your hosts. Also, be aware that in most cases, negotiations take place with your Asian counterpart's highest-ranking individuals as well. There may be lower-ranking individuals present at the negotiating table, but they are more likely to be there for support, or as observers.

Traveler's Insight: Traveling in the Far East

Mary McEvoy, Director of Public Relations at SonicWALL, a leading provider of network, Web, e-mail security, and backup and recovery solutions, has traveled throughout the Far East and most of Europe, and has had a chance to observe the subtle differences among Asian countries. She has spent time in Singapore, Malaysia, Australia, Japan, Thailand, England, France, Germany, China, and Korea.

Mary has noticed that in many Asian countries, while members of the opposite sex often do not touch in public, it is not the same between two friends of the same sex, whose touching is seen as a gesture of friendship. "During the closing party for a conference in Bangkok with our Indian channel partners, all the men danced together," she recalls. "It was just the cultural norm for India, with no other overtones, but some of our U.S. executives were a little disconcerted to find themselves doing close disco dancing with another man."

Local foods are another big difference, and Mary advises you to "leave your diet plans at home," noting that a tolerance for hot food is essential just about anywhere in Asia. "It does not help business if you restrict yourself.

to a lettuce leaf while your French distributor is tucking into cassoulet. Let your local contact advise you on the menu, show you how to eat whatever it is, and even serve you. If you absolutely cannot eat something, be plain and clear but polite about the fact that you greatly regret that you are unable to digest whatever it is: chicken feet, snake, or snails." Mary, who currently works out of SonicWALL in California, a state where smoking in office buildings is prohibited, also adds that "smoking is still common" in other countries. "You need to get used to it in France, England, and throughout Asia, apart from Singapore."

And what about alcohol? Every country has a different custom, it seems. "In Seoul, I was surprised by the importance of drinking as a business tool and the acceptability of throwing up in the street after a long night on soju." It may be worthwhile to learn a few words of the

Traveler's Insight: Traveling in the Far East

language, at least enough to say, "No more beer, please." Mary has found that "even one or two words of the local language are incredibly helpful in a social setting. In business, if your usage is limited, it is best to confine yourself to one or two phrases of greeting and thanks, and make use of an interpreter; however, it never hurts to understand the translation of words closely related to your business, such as 'network,' 'manufacturing,' 'silk,' and 'discount.' Doing so allows you to look intelligent during long bursts in a foreign language that you do not speak."

Bargaining and deal-making are also quite different, especially between Asia and Europe. "It helps to be clear about what terms you need for the deal to be a success," Mary adds. "Never lose your temper unless you really do not care about the deal. Beware of the word 'yes' in many cultures. In Asia, it often means, 'Yes, I have heard you,' rather than, 'Yes, we have a deal.' In England, the culture of self-deprecation and understatement can be misleading. "I have done this a few times" could mean that the speaker is widely acknowledged as a worldwide authority. "Quite nice" can signify ecstatic approval." Gift giving is also a big part of the ritual that takes place before getting down to business. "In Asia," she says, "we tend to exchange small personal gifts on visits. For major customers, we may put together golf trips or special commemorative trophies."

What goes on before the negotiations is just as important as the negotiations themselves. Mary advises, "It is always worth trying to follow the local custom; it signifies you are serious about a permanent relationship in that country, unless you are a senior VIP, in which case your reputation carries all before you. It is important in Muslim countries: in Malaysia, Muslims go through a purification ritual before going to the mosque on Fridays and do not shake hands with women after completing their ablutions. It is also the custom for men and women to wear traditional, modest dress when visiting government agencies.

"You can get away with a moderate version of some customs (a slight bow in Korea rather than a deep waist bend) but you may not be able to dodge three kisses on alternate cheeks in Switzerland, even if you do not greet people that way yourself. There are some charming customs that are easy to follow, such as the Malay greeting of touching hands and then bringing the hand to your heart, and offering a business card with both hands throughout Asia." The most common mistake she has seen throughout all of her many travels is "making assumptions based on knowledge of a different country or personal beliefs."

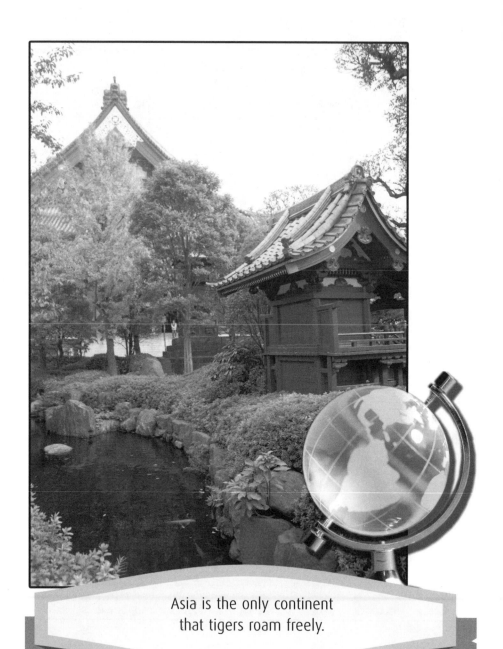

Asia is the only continent
that tigers roam freely.

8
Thailand

Mention the city of Bangkok to the uninitiated, and you are likely to get the "wink and a nod," or an elbow to the side, as if to say, "Oh, I know what you are up to." Entire books have been written and devoted to Bangkok's sex industry, and they have given rise to the unfortunate perception that Bangkok is overrun with houses of prostitution.

Bangkok has a red light district, known as "Patpong," where most of the illicit sexual activity goes on, just like many other major cities around the world. Technically, some aspects of prostitution are illegal (soliciting and pimping), and while enforcement is liberal, the penalties are severe. It is not illegal to pay for sex or to pay a "bar fine" (a fee the bar collects if you want to take an employee away), but if you bring a prostitute to a hotel outside of the Patpong district, your guest is likely to be refused entry.

Westerners often congregate in such parts of Thailand and bring their Western misconceptions with them; much of the debauchery that goes on is initiated by the Westerners themselves. In fact, this is only a small part of Bangkok, which for the most part is very traditional, highly spiritual, and full of rare beauty and fascinating history.

Economics and Opportunities

The Asian economic crisis of the early '90s hit Thailand hard, but recovery has been successful. Most recently, a military "bloodless coup" that ousted Prime Minister Thaksin Shinawatra threatened the stability of the economy, including the baht, but the economic impact has been minimal. This may be because of the relief many middle-class Thais felt at Thaksin's ousting. Khun Thaksin was considered to be the head of a corrupt government, and the coup was inevitable.

As an emerging economy, Thailand presents several business opportunities for Western cooperation. Retail chains are making significant inroads with large European retail chains such as Tesco, a U.K.-based international grocery and general merchandising retail chain, and Carrefour (literally "crossroads"), a similar global chain founded in France. Carrefour is the second largest retail group in the world, in terms of revenue and sales figures, after Wal-Mart.

Greetings

The proper greeting is the wai, similar to the Indian namaste. To wai, press your hands together with your thumbs facing yourself. Lower your head, and then raise your joined hands up to the level of your mouth. You may say "sawatdee khrap" if you are a man, or "sawatdee kah" if you are a woman. There are numerous rules about who wais first; generally, the younger person or lower-ranking person is first, and then the elder or person of superior rank returns the wai. Servants, grocery store clerks, and maids may wai you, but do not wai them back; a simple nod of the head is sufficient.

Although Thai people have been using last names for a more than 100 years, they seldom use them outside of written correspondence. Instead, they use the first name preceded by the honorific "Khun" for both men and women. (I am "Khun Dan.") If a Thai speaks English, he or she may address you as "Mister," followed by your first name, as in, "Mister Dan."

Guests in a person's home are offered tea or some other drink. It is polite to accept. Remove your shoes before going into a person's home or into a Buddhist temple. It is not necessary to remove your shoes at a restaurant.

Food

Thai food is fiery, influenced by both Chinese and Indian cuisine. At a restaurant, you may request that your dish be mild, but it may still be hot. "No chili peppers" to a Thai chef means "fewer chili peppers." Once you have become accustomed to it, the food is fabulous and includes a rich variety of tastes and flavors. There is nothing else in the world like it.

When you are invited to a business dinner, meals are usually served communally, as opposed to each person's getting an order. Several dishes of different types of foods are ordered, and each person can then sample each one. This is especially good for adventurous Westerners, because you can try everything. Do not pile on a large helping of any one food at a time; instead, take small quantities throughout the dinner. You may dip into the dishes on your own, but you may also find your host serving you as well. Similarly, do not refill your own drink; that is the task of your host or the waitress. Thais do not use chopsticks, except at Chinese restaurants. Use a fork and spoon, with the spoon in your right hand and functioning as your primary food deliverer to your mouth. The fork is used to move food onto the spoon. Drinks may include a sweetened iced tea or iced coffee, which is refreshing in the hot, tropical climate.

Tipping is seldom done, although rounding up or leaving the change is appreciated.

Religion

The majority of Thais are Buddhist and tolerant of other religions; there is a small Christian minority. There is also a large Muslim population in the southern border provinces of Yala, Narathiwat, and Pattani, where there is a long-standing separatist movement and considerable violence. (Stay away from these provinces if possible.)

There are many markets where you can purchase Buddha images, but if you are not Buddhist, avoid buying one as a souvenir. They are meant as objects of reverence.

Etiquette

If you have seen "The King and I," you may have a certain perception of what Thailand is like. Forget everything you learned from this play, and do not ever mention it to anyone there. The story is forbidden because Thais feel that King Mongkut, one of their most beloved monarchs, is portrayed as a barbarian.

The Royal family is highly revered in Thailand. The King is accorded the highest level of respect, and is deferred to in all matters. In England, making fun of the Windsors is a national pastime; in Thailand, it is different. For example, although the government is a constitutional monarchy with elected officials and a prime minister, the King may still make decisions that affect the government. You may hear the Royal anthem played over loudspeakers in the morning and evening, especially in rural areas. When this occurs, stop what you are doing and stand quietly. The Royal anthem is also played in theaters before movies; everyone stands while it plays. Never speak against the King or royal family in any way. Do not make speculations about their personal lives, and do not do anything at all that shows disrespect. Thai people act this way not out of fear, but out of respect and sincere admiration for the King, who has done much for the benefit of his people during his long reign.

Bargaining and Deal-Making

Little attention is paid to punctuality, and because Bangkok's traffic is considered to be just about the worst in the world, you should never expect anyone to show up to a meeting on time. Business decisions may take a long time. As in much of Asia, business and social lives are intertwined, and entertainment is a big part of business. Avoid attempting business ventures during the festival of Songkran (Buddhist New Year) in April; very little business is done during that time.

During the exchange of business cards, you may see your acquaintance's name on the card and attempt to say it; however, do not try this unless you are familiar with the language. The name may be long. However, most Thais also have short, monosyllabic nicknames. They will let you know how they wish to be addressed.

Many businesspeople in Thailand are Thai-Chinese and observe a mix of both Thai and Chinese customs. Most Thai are soft-spoken. Loud, aggressive talking is considered rude.

Traveler's Tip

The uniquely Thai expression "mai bpen rai" literally means, "it is okay," or "it does not matter." But more than just an expression, it reflects a gentle, relaxed way of life that implies acceptance of the unchangeable. "Road rage" is virtually unknown in Thailand, despite the horrific traffic jams in Bangkok. If Thais are caught in traffic and are late for an important meeting, instead of getting angry, they shrug their shoulders, say "mai bpen rai," and roll down the window to get some food from a nearby vendor.

Common Customs

- The style of dress in this traditional country, despite the exceedingly hot and humid weather, is modest but casual. Although it may be 110°F, do not wear shorts. Most Thais do not wear short-sleeve shirts. A jacket and tie are required in formal situations; otherwise, if you do wear a jacket, feel free to take it off and drape it around your chair.

- Remove your shoes when going into a Thai home or a Buddhist temple.

- Do not point with a single outstretched finger. Gesture with your whole hand if you must point.

- Small gifts are appreciated.

- Avoid stepping on doorsills.

- Displays of anger are rare.

- Do not take the last bit of food from the common plate, but accept it if you are offered it; it is considered good luck.

- The Western, Gregorian calendar is not used in official documents; instead, the Buddhist calendar is followed. The year 2008 is equivalent to the year 2551 B.E. (Buddhist Era).

9
Vietnam

In the '40s, Japan was America's sworn enemy, but today it is a strategic trading partner. The same transformation is now happening with Vietnam.

In the '60s and '70s, America was in a bitter war in Vietnam, determined to stem the tide of Communism. Vietnam is still Communist, but there are no Vietcong marching down Main Street. Instead of shooting at us, the Vietnamese are now trading with us.

Many Americans are reluctant to travel there, fearing that the people might resent them because of the war. This fear is unfounded. While the Vietnamese may be resentful, they welcome Americans, American businesses, and American money.

Economics and Opportunities

The Tax Relief and Health Care Act of 2006 promoted free trade around the world, and authorized permanent trade relations with Vietnam. One month after its signing, Vietnam joined the World Trade Organization. The law and the reforms in Vietnam bode well not only for Western companies doing business in Vietnam, but in the entire ASEAN (Association of Southeast Asian Nations) region.

Labor costs are quite low in Vietnam, but the infrastructure is limited. Despite a large, unyielding bureaucracy, the Vietnamese government is anxious to have outside foreign investment, and offers many opportunities for business.

Greetings

Although greeting your associate in Vietnamese is appropriate, you may find some Vietnamese, especially in the South, speaking French; and you may be greeted in restaurants and public places as "Monsieur" or "Madame." Younger Vietnamese may also speak English, but the language is not pervasive.

Like many Asian languages, Vietnamese is tonal, and many of the words sound alike to the untrained ear. You greet a man who is older than you by saying "Jao on," or a man who is your own age by saying "Jao an." The two sound almost exactly the same, but saying the wrong one could cause your acquaintance to take offense.

As in Thailand, Vietnamese have not been using surnames for that long; instead, use an honorific followed by the first name when addressing someone. Mister is "u," and Miss or Mrs. is "daw." If they speak English, they may address you as "Mister" followed by your first name, as in "Mister Dan." When the full name is written out, the given name comes last.

Vietnamese do not usually touch upon meeting; instead, they make a small bow with their hands joined together at about waist level.

Food

The Western habit of loading up your plate at the beginning of a meal is rude in Vietnam (as in much of Southeast Asia). Instead, take small amounts of each dish multiple times throughout the meal. Your host may serve you.

Vietnamese food is delightful and unique, and has many influences. As in most of Southeast Asia, some of it may be unusual and unseemly, especially in rural areas. However, there is a great variety of international cuisines in Vietnam, especially in the larger cities, including French, Chinese, Thai, and even American.

Religion

Most Vietnamese are of the Mahayana Buddhism belief system, which follows the philosophies of Taoism and Confucianism. Vietnam enjoys freedom of religion, and there are several religious beliefs that are practiced, including some that are indigenous to Vietnam (such as Hoa Hao, and Cao Dai). Some people in more remote areas practice animist beliefs; that is, they believe in individual spirits that inhabit natural objects and phenomena. There is a minority of Roman Catholics, primarily in the South. There is also a small Islamic minority, mostly along the Central coast.

Etiquette

The Confucian philosophy has a major impact on daily life and business in Vietnam. This philosophy, which prizes order and respect for position and authority, is behind the Asian tradition of having respect for one's elders and superiors in all situations. All are deferential to the eldest person in a group.

Bargaining and Deal-Making

Do not be deceived by the fact that Vietnam is a Communist nation; the Vietnamese tend to be astute businesspeople and skillful traders. Business conversations are subtle and similar to interactions you may have in China. Do not put a Vietnamese businessperson in the position of having to say "yes" or "no." Instead, ask an open-ended question. As in Japan, if a Vietnamese says "yes" to you as you are making a proposal, it only means,

"I hear what you are saying."

Women maintain positions of authority in Vietnam, and female businesspeople from the West encounter no resentment in Vietnam.

You may encounter some confusion when it comes to licenses, taxes, and duties, and there may be more than one bureaucracy that you have to deal with; more than likely, it may be redundant, but you must pay their fees nonetheless. You may encounter corruption and the occasional need to provide "tea money" (bribes).

When you are doing business in Vietnam, it is best to include a senior member of your company in the negotiation process. Vietnamese tend to be aggressive bargainers; do not take this as hostility but just part of the game.

Common Customs

- Ho Chi Minh is considered to be the "father of Vietnamese independence" and is highly respected throughout the country. Despite what you may have been told during the American occupation, you must show respect to his memory.

- Show up on time for business meetings. It is acceptable to be up to 30 minutes late for social events.

- Before business is transacted, social interaction is essential. The Vietnamese must have an opportunity to get to know you before doing business with you.

- Be prepared for karaoke, which has become a common form of entertainment throughout Asia.

- Dress is modest, despite the overwhelming heat and jungle humidity.

10
Indonesia

When people think of Indonesia, they think of the island paradise of Bali and having tropical drinks on the beach while watching exotic topless women stroll by. In fact, most of the business end of your trip will take place in urban areas.

With nearly 250 million people, Indonesia is one of the most populous nations in the world, and although it is comprised of many small islands, the total area is quite large. One factor that has hindered Indonesia economically is the many different cultures throughout the islands. Although the sheer size and population of Indonesia cannot be discounted, its economic future and position on the world stage could be significant.

Economics and Opportunities

Formerly the Dutch East Indies, Indonesia is no stranger to trade, but it has not fared as well as some of the other Asian countries in recovering from the financial crisis of the '90s. The country is also still recovering from the devastating effects of the 2004 tsunami. There is substantial inflation and high interest in Indonesia, and the government is striving to reform the investment climate.

Greetings

Although the Indonesian language is dominant, English and Dutch are spoken occasionally. A handshake with the right hand, with a slight nod of the head or bow, is the appropriate greeting. The handshake is not a firm, Western handshake, but a light grip. In a few areas outside of the large urban cities, however, you may see people greeting you with the Thai-style wai. Outside of the handshake, touching is seldom acceptable, and public displays of affection are not done. When leaving, do not give an American wave of the hand; instead, make a short bow to each individual.

As in other parts of Southeast Asia, last names are a relatively recent innovation, but names are structured as in the West, with the given name first and family name last. Address people with an honorific, followed by the given name. Although Mister, Miss, or Mrs. is acceptable, you may also address people with the Indonesian honorific, which is "bapak" for Mister, or "ibu" for Miss or Mrs. If you are addressing a Muslim who has completed the Hajj, you address them as "Hajji" for a man or "Hajja" for a woman.

Food

When at a restaurant or eating at a home, leave a little food on your plate when you are done; if you finish everything on your plate, it is an indication that you want more. Tipping at a restaurant is not expected, but if you leave change or 5 percent of the bill, it is appreciated. You may be able to find Western food, but try to avoid bringing your Indonesian associates to Western restaurants.

Indonesian cuisine has many Asian influences. While some of it is spicy, there is a wide range of flavors, and you are sure to find something that suits your palate. Alcohol is available; however, before ordering, follow the lead of your associates. If they are observant Muslims, they may refrain from alcohol, and thus you should as well. Although business lunches are common, avoid inviting your Muslim associates out for lunch during Ramadan, the time of fasting. If possible, do not plan your business trip during the month of Ramadan.

If there is alcohol, it is customary for the waitress or your host to fill your glass for you. As for utensils, chopsticks are used at Chinese restaurants; otherwise, forks and spoons are used. Eating with fingers is also common.

Religion

The majority of Indonesia is Muslim (about 88 percent), with small Christian, Hindu, and Buddhist minorities. There is a Chinese minority in Indonesia, although many businesspeople are Chinese and Buddhist. The majority religion, Islam, has taken a form that is unique to Indonesia and incorporates some of the other traditions, including Buddhism and animism, found in other parts of the country. Indonesian Islam tends to be less rigid than Islam in the Middle East. The hostile division between Sunni and Shia branches of Islam is not seen in Indonesia, where there is more of a spirit of unity and tolerance.

Etiquette

It is appropriate to bring flowers to someone's house for dinner. Small gifts for business associates are also appreciated, although they should not be too expensive or extravagant. When choosing gifts, remember this is mostly a Muslim country, so gifts of alcohol are inappropriate. If a married man is giving a gift to a female associate, it is given with the understanding that it is from him and his wife together.

When sitting, do not cross your legs with your ankle sitting on your other knee; this shows the bottom of your foot, a discourtesy.

Indonesia is made up of thousands of islands and has 300 distinct ethnic groups, so customs may vary, especially outside of the major cities.

Bargaining and Deal-Making

It is important to be punctual, but traffic is a problem in some of the larger cities, and tardiness is accepted in such cases. Similar to other Asian cultures, Indonesians are reluctant to disagree in public or say "no" outright, so do not put your associate in a position to have to answer a question with a direct yes or no answer.

If you have lunch or dinner with your Indonesian business associates, do not expect business to be discussed during the meal. Discussion is light, since it is considered impolite to talk too much while eating.

Women often have high-level positions, especially in government, and female businesspeople should have no difficulty in being accepted on equal footing.

Common Customs

- It is unusual for women to go out alone.

- Remove your shoes before walking into a mosque or other holy place, or into a carpeted room. You can tell whether you are supposed to remove your shoes by observing others.

- As is the case in many Muslim countries, use of the left hand should be avoided whenever possible.

- At a meeting or a dinner, do not seat yourself. Seating plans are worked out in advance, and people are seated according to rank and importance.

- The style of dress is usually Western, but not informal. Avoid jeans, sweats, and other casual wear. A lightweight suit may be appropriate. Going without a tie is acceptable, and it is common to remove your jacket once you are seated.

- Women should dress modestly, with legs and arms covered. It is not necessary for a non-Muslim woman to cover her face; the practice is optional even among Muslim women.

Indonesia consists of over 300 different ethnic groups, and is the world's most populated Muslim nation.

11
Malaysia

Malaysia is one of the most diverse countries in Southeast Asia. Only about half the population is native Malay, with about a quarter Chinese, a significant minority Indian, and a mix of other cultures as well. Malaysia is relatively prosperous, with a growing middle class and modern, gleaming cities.

Economics and Opportunities

Malaysia enjoys a diverse economy, and the country has become a major center for manufacturing, tourism, and raw materials exports. Much of Malaysia's impressive growth has been driven by exports and the electronics industry. Although hit by the global recession in 2001 as well as the Asian economic crisis of the early '90s, Malaysia has done well with its recovery, with a 5.5 percent GDP growth in 2006. Its growing middle class is transforming Malaysia into a consumer center, bringing opportunities for Western companies interested in both imports and exports.

Greetings

Though many different cultures and traditions are present, Malaysia is generally not a touch-oriented culture. Most businesspeople are familiar with Western customs, so the handshake is sometimes appropriate. Malays avoid shaking hands with people of the opposite sex, however. Contact between sexes in public should be avoided at all times. The traditional Malay greeting is a slight bow, followed by the salaam, which is to briefly touch the hands of the other person and then touch your hands to your heart. Use both hands when greeting someone who is older than you. Among the significant Indian minority, the traditional greeting is the namaste.

Gift giving is a little difficult to figure out, because there are so many different cultural groups. Consider small gifts that are not too extravagant; avoid alcohol so as not to offend Muslims, and leather so as not to offend Hindus. There are strict antibribery laws, so make sure that your gift is appropriate. Also, be careful if you are a man and want to give a gift to a woman; if you do, be sure to indicate that it is from both you and your wife, or from your company.

Food

Use a fork and spoon, holding the spoon in the right hand, and using the fork to move food into the spoon. Chopsticks are only used at Chinese restaurants. Malaysian food is similar to Indonesian and reflects a blend of several Asian cuisines, especially Chinese and Indian. Among the Chinese population, using the left hand for eating is acceptable; however, Malays eat only with the right hand. There are many delightful tropical fruits in Malaysia, and you may be served some of them from time to time. The durian is a delicacy throughout Southeast Asia. Called the "king of fruits," it is large, spiky, and has an overwhelming, unpleasant odor. The first time I tried it, I was sitting in a shopping mall with my wife, and she had just purchased some at the grocery store. I was reluctant, because the smell was so overpowering, but I gave it a try and immediately attracted the attention of everyone around me. I could not help but make a face and shout out loud, as though I had taken a big drink of sour milk. It is an acquired taste, so prepare yourself before biting into it.

Religion

Several religions are practiced in Malaysia. Although the majority is Muslim and Malaysia is officially an Islamic state, there are also significant minorities of Buddhists, Hindus, Christians, and Sikhs. In Eastern Malaysia, Shamanism is practiced. Prime Minister Abdullah Ahmad Badawi has promoted "Islam Hadhari," or "civilizational Islam," that emphasizes education, social harmony, and economic progress. While still strict, Islam as observed in Malaysia is not as divisive as in some areas of the Middle East.

Etiquette

Avoid pointing with your finger or showing the bottom of your feet, as both are considered impolite. Avoid eating or shaking hands with your left hand (although this is acceptable among Chinese).

There are strict censorship laws in Malaysia.

Bargaining and Deal-Making

Although the official language is Malay, Thai is also spoken, particularly around the border areas, as well as English and Chinese.

If you are negotiating business with Malaysian Indians, be aware that if they shake their head from left to right, it does not mean "no" as it does in the West; instead, it indicates agreement. Malaysians are friendly. Although you should try to be on time to a business meeting, adherence to schedule is often not observed. Many Chinese do business in Malaysia, so be aware of traditional Chinese customs as well.

As is the case in much of Asia, business is done only after a relationship has been cultivated. Do not jump right into business negotiations on the first day. The pace of business is much slower than in the West.

Common Customs

- The style of dress is modest; women should keep arms and legs covered.

- Take off your shoes before entering a home or a mosque.

- In some homes, people may sit on the floor. Men may sit with their legs crossed, but women should sit with their legs tucked underneath. A long skirt is advisable.

- Avoid littering at all costs. Even dropping a cigarette butt on the sidewalk can bring a severe penalty.

- Astrology and other beliefs may influence business decisions, such as determining the most auspicious day for signing a contract.

12
Philippines

The Philippines has a reputation for corruption, but English is widely spoken, making it easier to do business there.

Economics and Opportunities

The Philippines was not affected as much by the Asian financial crisis of the '90s as other Southeast Asian nations, and a GDP growth of about 5 percent since 2002 shows good resilience and a strong service sector. However, there is still significant poverty, unequal distribution of income, and higher inflation than in neighboring countries.

Greetings

There are two official languages, English and Filipino (Tagalog), with eight major dialects of the Filipino language. Although many of the surnames in the Philippines sound Spanish, few people speak Spanish. Handshaking is the common greeting, although it is softer than in America. A man should not offer his hand to a woman first for a handshake. Wait for the woman to offer her hand first. If you are a foreign businesswoman, it is acceptable for you to initiate the handshake. Filipinos do not bow or wai. Giving gifts is common, and a small token gift is appropriate. If you are invited to a social gathering, you may give a gift of flowers when you arrive, or send a gift or a thank-you note the day after.

When Filipinos speak English, you may notice pronouns being used generically; that is, a woman may be referred to as "he." You may address people as Mister, Mrs., Miss, or Ms., followed by the family name, in all situations. Alternately, you may use the Tagalog "mang" for Mister or "aling" for Mrs., especially with older people.

Beer is a common alcoholic drink, and you should make a toast if you are the guest of honor (though wait until the host makes a toast first).

Food

Filipino food is spicy. There is a major Spanish influence in the cuisine, such as stews with pork or chicken, and heavy use of seafood. There are plenty of fast food places all around.

Forks and spoons are used in the style of other Asian countries, with chopsticks being used only in Chinese restaurants. It is best to leave a little food on your plate when you have had enough; otherwise, your host will continue giving you more.

Unlike in many other Pacific Rim countries, where food is served communally, here each person is served individually, in the Western manner.

If you are fortunate enough to be invited to a home for dinner, do not discuss business. Be aware that your host probably has servants. It is not necessary to address or compliment the servants. At dinner, you will be told where to sit. At the end of the meal, the host will give you a bag of leftovers that you should accept graciously. Sometimes after the meal, there is group singing, and it is polite to participate. Do not worry about having a poor singing voice; it is the participation itself, and not the quality, that matters.

Religion

About 80 percent of Filipinos are Roman Catholic, with Christian Protestant groups making up a minority. There is about a 5 percent Muslim population, and there is a significant Muslim separatist movement in the Southern part of the country that sometimes erupts into violence.

Etiquette

There are similarities between the Philippines, Malaysia, and Indonesia, and there are some ethnic commonalities. Be aware that some Chinese and their descendants are there. There are also many Filipino-Spanish (mestizo), and there is a European cultural influence. Although smoking is common, women rarely smoke in public.

Bargaining and Deal-Making

Socialization is a big part of business, and Filipinos may invite you to an evening of group singing or some social event. Punctuality is important in both social occasions and business. There is a big emphasis on relationships and connections, so that even a tenuous connection — friend of a friend of a friend — is helpful. An emphasis on connections and family is important in Philippine offices, where employment is often secured through such networks. Negotiations are slow and formal.

Although there are not likely to be Filipina women in higher echelon positions, Western female executives are accepted as equals. Filipino businessmen are familiar with Western business practices and culture.

Common Customs

- Avoid pointing with your finger outstretched. Instead, just gesture with your head toward the object you wish to indicate.

- Remove your shoes before you enter a home.

- Seating arrangements at dinners or meetings are according to rank and status.

- Dress conservatively, but because of the heat, it may be acceptable to just wear a dress shirt without a jacket.

- Do not pour your own drink at dinner; your host or the waitress will do it for you.

- Smoking is pervasive, but you should not smoke at the table until the meal is over.

13
South Korea

South Korea is an early Asian success story. Although not coming close to the power of India or China, the country presents many opportunities for foreign investment and trade. Growth has been impressive since the '60s, and in 2004, South Korea joined the "trillion dollar club" of world economies.

Economics and Opportunities

South Korea has come a long way, from a poor, agricultural economy in the '50s to a modern and strong industrial base with a good export business. Although open to outside business, South Korea is somewhat protectionist and favors domestic businesses over foreign competitors. The infrastructure is fairly well built-out, with roads and telecommunications. South Korea is comparatively wealthy and has developed a consumer economy that may present import as well as export opportunities for Western companies. The economy is generally stable, with strong growth, moderate inflation, low unemployment, and an educated workforce.

Greetings

The traditional Korean greeting is a slight bow, but Korean businesspeople usually greet a Westerner with a light, Asian-style handshake. Often, the handshake follows the slight bow. Korean women do not shake hands, but Western women may do so. When shaking hands, the senior-ranking or eldest person offers his or her hand first. When bowing, however, it is customary for the junior person to bow first. When departing, bow again. When departing, do not wave good-bye, because this gesture is taken to mean you are calling the person to come over to you. Instead, the bow is more appropriate when departing.

Gift giving is appropriate. Take a gift when visiting a home, and business gifts are also frequently exchanged. The gift is not opened in the presence of the giver.

Food

Korean food can be spicy and makes heavy use of local vegetables and seafood. Kimchi, a pickled dish, is served with most meals. Korea is a chopstick country; practice using them ahead of time. During a meal, there may be a common glass of wine or alcoholic beverage that is passed around the table; you take a sip from it and pass it along. Do not pour your own drink at the table; it will be done for you. It is customary to lift your glass a little when someone is pouring a drink for you. After dinner, there may be group singing, so go along with it if possible. The quality of your singing matters less than the mere fact that you are participating. Although overindulgence in alcohol is seen virtually anywhere in the world, an aggressive or rude drunk is not appreciated here.

Restaurant bills are seldom split; one person pays the bill, usually the eldest or highest ranking. Tipping is not customary.

Religion

There are about an equal number of Christians and Buddhists — 26 percent each — with a large percentage of the population not professing any religion. South Koreans follow Confucian beliefs, which form the backbone of their business community and daily life. Doing so means respecting elders and those of higher rank, as well as a general aura of respect employees have for their employers.

Etiquette

Koreans are typical Asians, quiet and nonconfrontational. Laughing out loud is rude, and women cover their mouths if they laugh. Do not put a Korean in the position of having to say "no." Pay careful attention that you do not cause your host to lose face in any way. Direct eye contact is considered aggressive and should be avoided, especially with an elder or superior. Avoid showing the bottom of your feet; do not cross your legs with your ankle on the other knee, and do not put your feet on a table. Respect for rank and age is pervasive, and it is appropriate to stand when an elderly person enters the room.

First names are not used except by invitation; use the family name with the honorific. After you have established a relationship, it is acceptable to use the family name without the honorific. Naming conventions are the opposite of those in the West. Place the family name first, a generational name, and the given name last. The honorific is added to the end of the family name, in the same way "san" is added to the family name in Japan. In Korea, you add "ssi," so a person with the family name "Lee" is formally addressed as "Lee-ssi." Women do not use their husbands' family names. Titles are also often used. If Lee-ssi is an engineer, you may address him as Engineer Lee, or Engineer Lee-ssi.

When giving or receiving objects of any kind, hold the object with both hands. In less formal situations, you may use one hand to hold the object; however, your other hand should still be involved. For example, if you are giving or receiving something with your right hand, place your left hand somewhere on your right arm (anywhere between the wrist and shoulder, though higher on the arm shows more respect). If one of your hands remains by your side, or in a pocket, it is a sign of disrespect.

Korean guests tend to be prompt, often arriving early.

Bargaining and Deal-Making

Business is formal and focuses on the group as opposed to the individual. The entire business group must reach consensus, taking much longer than in the West. There are few women in decision-making roles, although Western businesswomen have no problem being accepted.

Contracts tend to be associated with the signatories. If an executive of your company signs a contract with a Korean firm and then leaves your company's employ, Koreans may void the contract, requiring the departed executive's replacement to sign another one. As in much of Asia, saying "no" directly is usually not done; instead, a Korean businessperson may just say that something is "difficult."

The Asian custom of business and social interaction is common. You can expect to be invited to a restaurant or bar. Spouses are seldom included.

Common Customs

- Schedule business meetings ahead of time and send everyone involved a memo that describes the issues to be discussed.

- The most senior member of your group should make the presentation, as much attention is paid to rank. Having a junior member give a presentation is an insult to the host.

- Standard formal business dress is appropriate.

- As in much of Asia, a personal connection is important, and there is likely to be social interaction with your Korean partners outside of work.

- Men enter doors before women.

- Remove your shoes when entering a home or a restaurant.

You are in luck if you are a coffee fan.
Visit Myungdong, South Korea where
you will find a 5-story Starbucks.

14
Taiwan

The division from Mainland China is a sensitive one, best not discussed. There is a historic dispute between Taiwan and the People's Republic of China. The Taiwanese Kuomintang, or Nationalist Party, claims to be the legitimate government of the Chinese people, but the Communist regime in Beijing sees Taiwan as just another province of China. The situation is at an impasse, but Taiwan is an economic power and one of the wealthiest nations in Asia. Relations with other countries, except China's mainland, are generally good. Taiwan is neither independent, nor completely subjugated by China.

Taiwan is a crowded, noisy, and exciting place to visit. The streets are jam-packed, and the country moves at a faster pace than in much of Asia.

Economics and Opportunities

Despite the gray area that exists between it and China, Taiwan functions as a capitalist nation. It does not have official relations with many countries, although they do maintain offices that are unofficial diplomatic stations. For example, the Taipei Economic and Cultural Representative Office has offices in Washington, D.C., and other world cities.

Greetings

The bow in Taiwan is less pronounced and may be manifested in just a nod of the head. If meeting a person in authority, a slight bow is appropriate. While you are bowing, do not make direct eye contact. Handshakes are customary, but the grasp is much lighter. When greeting an elderly person, a more respectful and traditional bow is customary. With your left hand, pull in your fingers as if to make a fist. Cover the fist with your right hand outstretched, and raise it to your heart, and then bow. Outside of the light handshake, Taiwan is not touch oriented, and physical contact, even between husband and wife, should be avoided in public. Some older Taiwanese speak Japanese in addition to their native tongue, which is a dialect of the Fuji Province. However, Mandarin is the official language. English is also widely spoken. Many Taiwanese businesspeople have Anglicized first names and give their names Western style, first name first and family name last, at least when dealing with Americans.

Food

Food is similar to that found in Mainland China, although meals tend to have multiple courses. In Taiwan, chopsticks are used. Taiwanese are used to Westerners though, so if you cannot use them, feel free to ask for a spoon and fork. Business dinners are often elaborate and may go on all night, complete with singing and plenty of alcohol. Toasting occurs throughout the meal.

Religion

There is no official religion, but about 90 percent of the people are Buddhists, with observation of Confucianism and Taoism mixed in. About 4 percent are Christian.

Etiquette

Taiwanese customs are similar to those of nearby countries, including Mainland China. Gift giving is common but not as formal as in Japan. Present a gift that was made in the United States, not one made in Taiwan.

Shoes are usually removed before you enter a home; if you visit a home, bring a small gift, such as flowers or candy. Present the gift with both hands, and send a thank-you note for the evening the next day. Small business gifts are also appropriate, such as an item from your home country. Expect the Taiwanese businessperson to decline the gift out of courtesy up to three times before accepting it. When arriving at a social occasion or business meeting, punctuality is expected.

Confucian beliefs form the basis of much social interaction. Taiwanese are modest and conservative, and you should dress appropriately.

Bargaining and Deal-Making

Entertaining outside of business is common, but can take place in a restaurant. Taiwanese businesspeople rarely entertain in their homes. Feng shui and astrology may figure into business decisions, and an astrologer may be called in to determine the best day for signing a contract or carrying out an important task. Many Taiwanese people vacation around Chinese New Year (a date that varies every year according to the Gregorian calendar), so little business gets done around that time.

The Taiwanese have a strong Confucian work ethic that focuses on success. Because of their Confucianist beliefs, much attention is paid to rank and status, meaning your chief negotiator should be a senior, high-ranking member of your team. Despite their capitalist nature, your Taiwanese associates have traditional beliefs and want to take time to get to know you before doing business to establish a relationship and mutual trust.

Entertainment and socialization are important to business. Expect some evening entertainment that may last quite late; therefore, you may not want to schedule early morning business meetings. Business dress is conservative, with suits and ties, or conservative skirts or suits for women.

Common Customs

- Remove your shoes when entering a home or temple.

- Pay attention to your feet. Do not show the foot bottom or cross your legs Western style.

- Do not point using an index finger, but with the whole hand instead.

- Taiwanese are reluctant to discuss illness of any sort.

- Be punctual. Traffic is heavy; allow plenty of time.

- Taiwanese do not say "no" directly.

- It is best to avoid aggressive salesmanship and criticism of competitors.

- The concept of "face" is important. Strive to avoid embarrassing any other individual for any reason.

15
Bangladesh

Originally part of British India, Bangladesh became independent in 1947 and was first known as East Bengal.

Economics and Opportunities

Bangladesh is a poor, inefficient country; if you are looking for opportunity, try India first. More than half the country's GDP is generated through the service sector, and two thirds of Bengalis work in agriculture. Floods and cyclones have hindered growth as much as the inefficiency of state-owned enterprises and an inadequate infrastructure. Economic reforms are slow in coming. Bureaucrats and other interest groups have blocked many necessary reforms. Nonetheless, growth has been steady at around 5 percent for the past few years.

Greetings

The handshake is an acceptable greeting, although it should be lighter than the Western handshake. If you are being introduced to a woman, wait for her to extend her hand first. The traditional greeting is the salaam, a touch of the right hand to your heart. It is common to shake hands first and then offer the salaam. Men do not shake hands with women; a simple nod of the head is sufficient.

There is a Hindu minority in Bangladesh, so if you are greeting a Hindu, offer the namaste greeting, which is pressing the hands together at chest level and bowing slightly. The main languages are Bengali, and to a lesser degree, British English. The honorific "sahib" goes after the name, instead of before; for example, "Blacharski sahib." "Hajji" precedes "sahib" if the individual is a Muslim who has made the Hajj.

Food

Food is similar to that found in India. Goat and lamb are common meats. The lucky guest is offered the goat's eye at a meal, considered an honor. As in India, when you visit someone at home, you are served a cold drink with a sweet of some sort. Tea is a common drink. There is not likely to be alcohol, because it is predominantly a Muslim country. Spoons and forks are used Asian style, with the fork held in the left hand to push food onto the spoon. In a traditional restaurant people may eat with their fingers. Among Muslim diners, use of the left hand in eating should be avoided. Among Hindus, avoid moving food between one another's plates.

Although spoons and forks are used in most restaurants, it is often acceptable to eat with one's fingers. It is common for men and women to eat in separate rooms. Tip about 10 percent in restaurants.

Religion

The majority of Bengalis — about 85 percent — are Muslim. There are both Sunni and Shiite Muslims, and although Shari'a law dominates, Bangladesh is more liberal than its neighbors. Most of the remaining 15 percent of the population is Hindu.

Etiquette

Giving a Western style "thumbs up" gesture as a positive indicator is a bad idea; in Bangladesh it is a rude gesture. Punctuality is appreciated.

Bargaining and Deal-Making

It is important to be on time to meetings in Bangladesh; punctuality is a sign of respect. Expect business entertainment, either at a hotel restaurant or a club. Spouses are usually invited to these social functions. Bangladeshi women do not have positions in business, and Western women should establish their authority before making the trip. Business dress is Western style, but because of the hot tropical climate, businessmen often do not wear a suit and may opt for a lightweight jacket. Western businesswomen may wear lightweight skirts, but must be careful that they are not above the knees; arms should be covered below the elbow. Although some Bengali women cover their faces, Western women are not expected to do so.

Business is usually a top-down affair in Bangladesh, with high-ranking people making the most important decisions. Hierarchy is important, and executives work in a separate area, usually on a higher floor in the office.

Common Customs

- Use only the right hand for eating.

- Remove your shoes before entering a mosque or Hindu temple. You may notice people also wash their feet before entering a mosque; follow their lead.

- Wait to be seated at a meeting or a dinner gathering.

- When you leave, acknowledge each person individually.

- Do not pour your own drink; wait for your host to fill your glass for you.

16
Hong Kong

There is an interesting blend of East and West in Hong Kong, with many British customs as well as the more dominant Asian elements. Hong Kong is busy and fast paced.

Economics and Opportunities

Hong Kong is one of the great commercial and financial centers of the world that has grown rapidly over the past century. Beijing promotes Shanghai as its commercial center. Nonetheless, Hong Kong enjoys a strong economy due to the "one country, two systems" philosophy of the Chinese government, which holds that China's Communist system is not to be imposed on Hong Kong. China agreed to provide Hong Kong with a great deal of autonomy for 50 years following its return to China from Great Britain in 1997. Hong Kong's market economy is dependent on international trade, and it has a strong service industry, with most manufacturing moved to the mainland.

Greetings

Because of the heavy Western influence in Hong Kong, the standard Western handshake is the most common greeting and signal of departure. The main dialect is Cantonese, although English is spoken in many cases.

Many businesspeople also speak the Mandarin dialect. Greet people as you do in China, although many people here have adopted Anglicized names.

Food

Chopsticks are common, but spoons and forks are also used (forks are used to push food onto the spoon). Toasting is common and appreciated, and do not forget to make a toast if shark's fin soup is served. Slurping your noodles and other food is not considered impolite, and smacking your lips at the table is an indication that the food is good. Many meals involve fresh seafood of several different varieties.

Religion

The religious makeup of Hong Kong is similar to that of China, with about 90 percent following traditional Asian religions, including various forms of Buddhism. About 10 percent are Christian.

Etiquette

Gift giving is appropriate. If you are invited to someone's house, take a small gift of flowers or candy and present it with both hands. Be aware that feng shui is common here, so pay attention to the nature of gifts. Sharp objects, such as decorative pocketknives, may sound like a good idea but indicate the severing of a relationship. Similarly, auspicious dates and numbers are important in business decisions; for example, many hotels do not have a fourth floor, because the number four in Chinese sounds like the word for death.

Like most other Asian countries, Hong Kong is not a touch-oriented culture, and displays of affection are not made in public. You may sometimes see two individuals of the same sex touching or holding hands as a gesture of friendship.

Bargaining and Deal-Making

Business deals are more efficient here than in many other parts of Asia, and business decisions may be made faster here than in the rest of China and other Asian countries. Formality is key in Hong Kong, and confrontation is not an effective strategy. Be on time to meetings, although being 10 or 15 minutes late is acceptable. Hong Kong is relatively modern, and there are more women in business here than in the rest of China. Gift giving is a good idea, and the gifts to be presented are higher-end than those given in Mainland China. Western luxury goods are appropriate.

During a business meeting, tea may be served. If you are hosting, have tea served in porcelain cups with a proper tea service, rather than in paper or foam cups.

Common Customs

- Remove your shoes when entering a home or a temple.

- Pay attention to your feet. Do not show the foot bottom or cross your legs Western style.

- Do not point using an index finger, but with the whole hand instead.

- Arrive on time to business meetings. Traffic is heavy; allow plenty of time.

- Hongkongese do not say "no" directly.

- The concept of "face" is important; strive to avoid embarrassing anyone for any reason.

17
Australia

Economics and Opportunities

Australia has moved from being largely agrarian to being an internationally recognized market force. During the '90s, it was named one of the fastest-growing economies by the OECD (Organisation for Economic Co-operation and Development), and it still has an advanced, Western capitalist market economy. Continued economic strength is derived not only from strong exporting of raw materials, but also from a renewed interest in tourism due to natural wonders such as the Great Barrier Reef.

Greetings

Australia is one of the few countries outside of America where the firm, bone-crushing handshake is still used, and direct eye contact is considered positive. Greet someone by name and shake hands. After you have become better acquainted and perhaps spent some time together at a local pub, your Australian business partner may refer to you as "mate" or by your first name. Women, if they are friends, may kiss each other as a friendly greeting. When meeting someone in business, make direct eye contact. It is considered important.

Mr., Mrs., or Miss are commonly used, although in written form in

Australia, there is no period. They are written as "Mr Blacharski" or "Mrs Blacharski." If someone has a title, such as "doctor," you may use it to address that person; however, you should not use your title when referring to yourself.

Gift giving is seldom done between business associates, although holiday cards or thank-you cards are appropriate. You may give a gift if invited to a private dinner party, and something like flowers or a bottle of wine is appropriate.

Traveler's Tip

The phrase "g'day" is well known throughout the world as a common Aussie greeting, mostly thanks to the efforts of Hollywood movies and television. When in Australia, you are sure to hear it said, but it is not nearly as pervasive as Hollywood suggests: do not overuse it.

Food

Australians eat in the Continental style, with a fork in their left hand and a knife in their right. Breakfast can be heavy, as it is in England. Dinner is usually late, about 8 o'clock, and usually starts with alcohol. You may also be invited to tea, which is a light meal in late afternoon. A favorite style of gathering is the barbie, or barbecue, an informal affair usually held outdoors, with plenty of seafood. If you are invited to a business dinner, spouses are usually included.

Religion

Australia is highly secular, and there is no official religion; but Christianity is predominant, both Roman Catholicism and Anglicanism.

Etiquette

Australia is known for being friendly and informal, with little attention paid to class or rank. The culture is egalitarian; relying on your position or rank is rude. You may hear the expressions, "cutting the tall poppy" and "tall poppy syndrome," which refer to the tallest poppies in the field having their heads cut off, meaning, "do not try to be better than everyone else." Citing your qualifications, degrees, and titles will not get you far. Gift giving between business associates is seldom expected, although a small, business-related gift can be presented. The "thumbs up" gesture is considered rude, unlike in America where it indicates something positive. It is acceptable to use the term "Aussie" to refer to someone from Australia.

When seated at a meeting or at a restaurant, the guest of honor sits to the right of the host.

If a third party does not introduce you, feel free to introduce yourself.

Bargaining and Deal-Making

Despite an apparent air of informality, Aussies tend to be prompt and direct in business. Most members of your associate's hierarchy are approachable, and you might talk and negotiate with people from all levels. Business dress is more casual than in other countries, although they do have a high standard of cleanliness and order, so you should be neat and tidy, and err on the conservative side (no high fashion). Standard business dress is good to start, and then follow the lead of your associates, who may be less formal. In social occasions, less attention is paid to formality.

Australian businesspeople are direct and to the point. Present the facts, keep your presentations short and avoid too much salesmanship. Reputation and brand are less important than quality and facts.

Common Customs

- If you are in a pub, do not turn your glass upside down when you are done. It is a challenge to fight.

- It is common to socialize at a bar or restaurant after working hours.

- If you ride in a cab by yourself, you should sit in the front next to the driver.

- Despite the prevailing air of informality, you must show up to meetings on time. However, do not be surprised if no one else does.

Section 4

The World In-Depth: Central & Eastern Europe

The former Soviet Union and its satellite countries are distinct from Western Europe in many ways. The culture, outlook, and sentiment are uniquely Eastern European. London, Paris, and Rome are Western Europe. Cultures are different in Minsk, Prague, and Krakow. Remember, for many years, many of these countries were isolated from the West, and have only within the past two decades been opened to the rest of the world.

Russians and Eastern Europeans have a unique relationship with suffering, which you may understand if you have ever read Tolstoy or Dostoevsky. People in the coldest parts of Europe have had a hard history that has permeated their psyche. Their subtle humor is tinged with universal suffering. In Poland, my relatives tell jokes left over from the Communist days: A Communist party chairman was strolling in the countryside one day and met a farmer. "Comrade Farmer!" he said. "Are you true to the Party?" "Yes," said the farmer. "Would you do anything for the good of the Party?" "Yes," said the farmer. "Would you give your cow for the Party?" "Yes, of course," said the farmer. "That is good," said the party chairman. "Would you give your horse for the good of the Party?" "Yes, of course," said the farmer. "Excellent!" exclaimed the chairman. "Comrade Farmer, would you give your sheep for the good of the Party?" The farmer got a worried look on his face. "No, Comrade Chairman!" he said. "Why not?" said the chairman. The farmer looked him in the eye and said, "I have a sheep."

Today the free market is starting to take hold. It is a slow, difficult process, but 20 years from now, Eastern Europe will look much different. Now is an excellent time for moving into this market. In many countries here, the infrastructure is not yet built out to a satisfactory level, and there are great opportunities for those investing in telecom build-out and housing construction.

The business mind of the former Soviet Union and former Communist countries is tasting capitalism for the first time. For some who have been accustomed to depending on the government for everything, capitalism is bitter, and in many of these countries, a minority of people want to return to Communism. "Better the devil we know than the devil we do not know," is their attitude. However, capitalism has improved the lives of many in these countries, and whenever there is a transition, there is opportunity.

The complexity of this region has increased since the breakup of the Soviet Union, which has given way to the emergence of new and distinct cultures.

The former Soviet states of Central Europe, including Poland and the Czech Republic, are mostly Roman Catholic, and use the Latin alphabet. The former Soviet states of Eastern Europe, including Russia and Ukraine, are Orthodox, and use the Cyrillic alphabet.

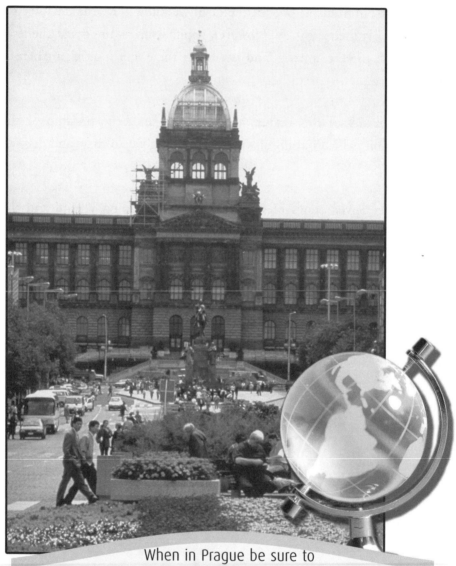

When in Prague be sure to check out the John Lennon Peace Wall. This wall was built after Lennon's death and became a place for Czech youth to write their views.

18
Russia

Russia became a dominant country under the Communist rule of Joseph Stalin's brutal regime, although its economy and society were both stagnant. General Secretary Mikhail Gorbachev's glasnost ("openness") and perestroika ("restructuring") presented a partially successful attempt to bring Communism into a more modern era, but it led to the breakup of the USSR. Under Vladimir Putin, there is some centralization of power. Russia is very diverse, with many subcultures and unique customs in different parts of this considerable country. Russians are eager to talk to Westerners, if for no other reason than to learn more about that part of the world from which they were isolated for so long. The Russian educational system under Communism was inefficient, did not prepare students for the future, and focused more on propaganda than literature or science. Those who grew up in that era are thus unprepared for a free market economy.

Economics and Opportunities

Economic reconstruction presents both difficulty and opportunity. Ever since the fall of the USSR, Russia has been struggling to build a market economy, but creating a new system takes time. This has not stopped Russia from enjoyed eight consecutive years of growth. Consumer demand is rising, and the Russian middle class — the consumer class needed to drive the economy — is making significant gains. Foreign direct investment is up. However, the country remains vulnerable to commodity prices, as much of its wealth comes from exports of oil, gas, metals, and timber. The manufacturing base is also out of date and needs modernization to be competitive.

Greetings

Although Russian is the official language, many younger people speak a little English, and some of the more educated speak French. Russians shake hands firmly, with direct eye contact, and state their names on the first meeting. Russians are more touch oriented; friends may greet one another with hugs and kisses on the cheek. Unlike some Western European countries, where people may kiss the air directly above the cheek, Russians make contact. Some older men may kiss a lady's hand. Males may greet each other with a firm handshake and hug; for business acquaintances, however, offer a handshake and then follow the other person's lead.

Russians have three names: a given name, a middle name (referred to as the patronymic), and a family name. The patronymic is an individual's father's first name with an added suffix that means either "son of" or "daughter of"; for example, Vladimir Vladimirovich Putin. Russians do not always use an honorific; instead they may state a person's entire name, including the patronymic. You may, however, use the honorific "gaspodin" for Mr. and "gaspazhah" for Miss or Mrs. (The old honorific "tovarich," which means "comrade," is no longer used.) If you are well acquainted with someone, you may refer to him or her by the first name and patronymic, as in "Vladimir Vladimirovich." Good friends may use the first name.

Food

Expect long waits at restaurants, and do not be surprised if they are closed at lunchtime. After you are seated, it may be an hour before you get a waiter. You may be tempted to wave or call out to a waiter, as it becomes a lengthy battle between you, who wants service, and the waiter, who wants to ignore you. However, it is considered rude to wave at a waiter; instead, eye contact is the preferred method of summoning. In the meantime, you will have a bottle of water and a bottle of vodka at your table. The menu may look extensive, but many restaurants have only a few dishes available. More modern restaurants with better menus are starting to appear in larger cities, but prices are high. Under Communism, tipping at restaurants was illegal (it was seen as a symbol of classism), but it is becoming standard practice now, and 10 percent is acceptable. A pack of American cigarettes also makes a nice tip for a waiter, or anybody else you wish to get better service from. When going to a restaurant with others, the host who made the invitation, or possibly the highest-ranking individual, pays the bill. Russians do not split the bill.

Vodka is the drink of choice and may accompany meals as well as social occasions and even business meetings. After a toast, the entire glass is drained and then refilled by the host; more toasts are made until the bottle is empty. A toast can be made to the host, to Russia, to business, or anything else. The most common toast is to health ("na zdrovie").

After vodka, your meal starts with appetizers or a small plate of zukuski, meat (pig fat), and pickles. Eating is Continental style, with the fork in the left hand and knife in the right. In addition to vodka, tea usually accompanies the meal. Lunch is the main meal of the day, and consists of multiple courses, starting with soup (such as borscht, or beetroot soup).

Religion

About 15 to 20 percent of the population is Russian Orthodox, 10 to 15 percent Muslim, and 2 percent other Christian. Under the old Soviet rule, Russia was officially atheist and there remain a large number of atheists in Russia.

Etiquette

Russians enjoy giving as much alcohol as possible to visitors; prepare yourself for an evening saturated in alcohol. Gift giving is appropriate. If you visit a home, you should take flowers or a bottle of liquor. Vodka is drunk on almost all occasions, straight and in shots, all in one gulp. Empty glasses are refilled until the bottle is empty.

Style of dress is not trendy, but standard business-formal. Women, if they happen to visit an Orthodox church, must wear long sleeves, a skirt that goes below the knees, and a head covering.

If you become friendly with a colleague, you may be invited to their home. Russians are very gracious hosts. Your host may go to great lengths to provide extravagant delicacies for you, including Russian caviar and the best vodka. Their homes are likely to be very small apartments. It is often customary to remove your shoes at the door.

Bargaining and Deal-Making

It is important to look the part of an executive. Negotiators in casual attire are not taken seriously. As for women in business, despite the fact that the Communists preached egalitarianism, women have had a difficult time reaching executive positions under Communism, and now again under reconstruction. Business and the social order tend to be hierarchical with respect to authority. Little gets done at the lower echelons of a corporation without first gaining multiple approvals. Above all, remember that Russians are new to the concept of capitalism and Western business practices; do not expect them to be open to or familiar with Western ways.

Be on time for business meetings, but do not be surprised if your Russian counterparts are late. It may be difficult to get any concessions from Russians who resist any form of compromise.

Traveler's Tip

Under the Communist regime, employment was guaranteed and getting fired from a job never happened. Work practices became lax, and Russian workers paid less attention to punctuality. Some such tendencies persist, and it is still common for Russians to be late for business appointments.

Common Customs

- Contrary to most countries in Asia, Russians display their emotions freely.

- Plan well ahead of time; it may take several months to get through the bureaucratic tangles of visas and transportation.

- Take care not to drop anything on the street or litter; it is offensive and illegal.

- It is common to see people of the same sex walking arm in arm as a gesture of friendship.

Traveler's Insight: From Russia With Love

Eric and Angie Lupa, of Granger, Indiana, went on a trip to Russia to adopt two young brothers from an orphanage in Murmansk. The trip was a first for Angie. When he was still young, Eric had gone to Medjugorje with his father, but this was his first trip to Europe as an adult, and both of them found Russia exotic and confusing.

English was not spoken nearly as much as in other European countries, and from the start there were problems. Angie recalls, "Ours was one of the last flights into the Moscow airport. That in itself was terrifying, because you walk in at 11:30 at night, nobody speaks English, and you cannot read the signs. We just kind of followed the crowd." Unlike other international airports, there were no signs in English and none of the airline staff spoke English. "We were not told on the plane that we were supposed to fill out an immigration card. Eric was at the

Traveler's Insight: From Russia With Love

passport check, and the lady there was yelling at him in Russian to go get this piece of paper. They were trying to get us through passport control, and we did not even know how to fill out these forms." Fortunately, the couple got through eventually, but once they retrieved their luggage, everything in the airport was closed.

They discovered the only people at the airport who could speak English were the taxi drivers. They were able to find a driver, tell him that they wanted to go to their hotel and show him the location. He told them it cost about $20. "I said, 'let's go,'" Eric recalls. "It is late. I do not care what it costs. Just get us there." As it turned out, the hotel was just at the edge of the airport, less than two miles away, but as Eric notes, "At that point, it was worth it." In fact, being grateful just to have an English-speaking driver, he tipped him an extra $5.

Tipping had been illegal under Communism. When they went to a restaurant and left a tip, the waiter rushed to correct them, thinking that they had mistakenly paid too much.

Although getting to their hotel without a guide was difficult, they were able to meet their Russian coordinator and a translator. "They helped guide us through this process," said Angie. "But when we went back to Moscow after our trip to see the boys, we were on our own again." They again found it difficult to read the signs or a map, which were in Russian, in the Cyrillic alphabet. "The first morning we went out and we made it to McDonald's two blocks away," Eric said. Relieved to see something familiar, they walked in, thinking someone there spoke English, but not so. The menu and signs were all in Russian. Fortunately, McDonald's franchises the world over have picture menus, so they had little trouble ordering.

One of the best resources, notes Eric, was the hotel concierge. The concierge was bilingual, knew everything in the city, and was able to help a great deal.

The accommodations were not quite up to standard. Angie recalls the hotel in Murmansk, a city in the Arctic Circle: "The hotel was drafty. They had two single beds just pushed together, and the wind felt like it was just coming in through an open window." They had stayed at a therapy center, a sort of spa in Murmansk where locals go for special massages, mineral baths, and other luxuries, although to Angie, "It reminded me of an old college dorm with single beds and rugs for our top blanket. The inside was smelly, the water was dirty when you turned on the faucet, and the shower had a strong mineral smell."

There was a big difference between Murmansk and cosmopolitan Moscow. Eric observed that several people said, "Moscow is its own country inside of Russia. It is like New York or Chicago." Although they were able to get around Moscow once they became familiar with

Traveler's Insight: From Russia With Love

it, through the helpful hotel concierge, Eric noticed that people make up their own rules in traffic. He said, "As crazy as they drive compared to the United States, there was no road rage. Nobody honks the horn. They just nosed their way through."

McDonald's notwithstanding, Eric and Angie did venture into a few more traditional restaurants and enjoyed Russian food, especially in Murmansk. There was soup, usually borscht (beet soup), with every meal. Bread was served, but without butter and usually with tea, coffee, or fruit juice. Although the locals usually do not drink it, Eric was relieved to find that you can get Coca-Cola. Russian vodka is ubiquitous and cannot be avoided. "On our last night, we had our first experience with Russian vodka," he said. "We were sitting with three other couples in the hotel common area, talking, and these guys came out and spoke English well, and they started talking to us. One guy said, 'Let's celebrate our friendship with some vodka!' And they grabbed some vodka and glasses from their room and brought it out." In true Russian tradition, drinks were poured until the bottle was empty, at which point their new friends went to look for another one. Angie was relieved to discover there were no more.

Eric and Angie found Moscow restaurants a mixed experience with plenty of international cuisine available. Restaurants in Moscow tend to be expensive. Though they found an Italian restaurant that offered excellent service, they discovered that restaurant service in Moscow is not one of the country's strongest points. At a different restaurant, Eric recalls, "The service was horrid. They took forever to just get to our table and take our drink order that was then slow in coming out." The food took even longer to arrive, and the whole experience took nearly three hours.

They both found Russians friendly and helpful, although at first their demeanor was deceiving. "They do not always smile back," said Angie. According to Eric, "They can seem cold at first, but they are friendly once you talk to them."

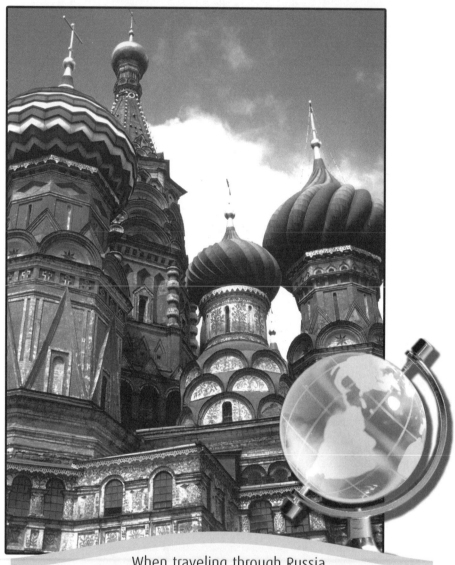

When traveling through Russia
be sure to update your watch. You will pass through
seven different time zones.

19
Poland

Of all the former Communist countries in Europe, Poland has adapted the best to the open market. That is not to say there are no problems for the traveler. Social and economic adjustment remains an ongoing process. Over the centuries, Poland has been dominated from time to time by its neighbors, especially Germany and Russia, although its culture has remained distinct.

Poles are hospitable, gracious hosts, eager to entertain foreign guests. At the home of my cousin, Stash, in the city of Elblag, the evening meal was an open-door event, with various cousins, grandmothers, and even the amateur accordionist from next door all wanting to visit the American relative. Needless to say, there was plenty of vodka at all times. As Polish-Americans, I and my son were especially welcomed, and the Poles were eager to share stories with us and hear about life in the U.S.

Economics and Opportunities

Between the end of the Second World War and the fall of Communism in 1989, Poland was ruled by Moscow. Poland's economy has been modernized since then, and stands in stark contrast to the sagging Russian economy. The consumer class is on the rise — the GDP grew 5.8 percent in 2006 — and there is a flourishing private sector. Becoming a member of the European Union in 2004 continues to help Poland adapt to a market economy, and direct foreign investment in consumer electronics and the automotive industry is increasing. As in all post-Communist states, there has been a move to shift state-owned enterprises to the private sector, which has been accomplished to some degree; however, some areas such as coal, oil refineries, and railroads are still in state hands.

Greetings

Western handshaking is appropriate, both on meeting and departing. If a man is greeting a woman, he should wait for the woman to extend her hand first. If a group of men and women are meeting, the women shake hands with each other first; then the women greet the men; and the men shake hands with each other last. Men, especially older men, may kiss a woman's hand.

There is less touching than in Russia, where the firm hug is common; however, Poles, especially women, may greet each other with a kiss on the cheek. It is common to greet each person in the room individually, and say "dzien dobrie" (pronounced "jane dough-bree," which means "good day"), or in less formal situations, "czeszcz" (pronounced "chay-sh-ch," which means "hi").

Polish names are given in the same way as in the West. There is no patronymic as in Russia, just a conventional middle name. Poles tend to be formal, so greet business acquaintances with an honorific followed by the last name ("pan" for Mister and "pani" for Mrs. or Miss). Titles are often used, which is another holdover from Communism. The head of a factory, for example, may be referred to as "direktor," as in "Direktor Blacharski." Surnames are feminized for women, usually by changing the last letter to an "a"; for example, my wife is "Pani Blacharska."

Gift giving is appropriate, especially when you are invited into a home; flowers are a good idea (but only in odd numbers). Businesspeople in the major cities, such as Warsaw or Krakow, may speak English or German. The Russian language, which is very similar to Polish (although with a different alphabet), is also widely spoken.

Food

Vodka is common during heavy meals, and toasts may be proposed before, during, and after eating. The standard toast is "na zdrowie" (which means, "to health"), although you can toast to anything you like. You may also toast to a long life, saying "sto lat," which means "one hundred years." The host pours, and it is customary to drain the entire glass. This is continued until the available bottles are empty. Polish beer is also common. It is flavorful, robust, and stronger than American beer. Although coffee is available, tea is served more frequently.

For business meals at restaurants, the one who has made the invitation pays the bill, although as a foreign guest you should offer to pay. Tipping is not mandatory, but leaving a 10 percent tip is common. If you are a woman, some of your more traditional male Polish counterparts may feel uncomfortable seeing you pay the bill. You may want to allow them to pay, or pay ahead of time.

Polish food is rich and heavy. Lunch, or "obiad," is traditionally the main meal of the day, though it is served later in the afternoon than American lunches. Similar to an American dinner, it may contain multiple courses, starting with soup. The traditional supper, or "kolacja," is a light snack late in the evening. However, many businesspeople in Poland are adopting a more American eating schedule during the workweek, having a light snack in the afternoon and then a large meal after work, around 5 or 6 p.m.

The most popular soup is beetroot soup, called barszcz (pronounced "bar-sh-ch"). There are many different styles, and every cook has a different recipe, but it is usually fresh grated beets, cooked with spices, cabbage, and other vegetables, sometimes with meat and sometimes not, and occasionally with a hard-boiled egg in it. In the country, you may also be served czarnina (duck's blood soup). Sausages ("kielbasa") are common at meals, along with smoked eel and pierogi, which are dumplings stuffed with either potato, meat, or mushrooms and cabbage. Seasoned lard is used frequently in place of butter for bread.

Religion

Poland is nearly 90 percent Roman Catholic, and Poles have a reputation for being strong in their beliefs. Unlike Russia, which became predominantly atheist as a result of Communism, Poles clung to their faith even when it was outlawed, and the Catholic Church has maintained a strong influence on Polish life.

Etiquette

Business dress should be conservative, with jacket and tie for men and conservative dresses for women. Manners and etiquette are traditional, and men rise when someone enters the room.

Offer to take off your shoes when entering a person's home. Many Poles consider it impolite to ask guests to remove their shoes; however, it is sometimes rude to enter a home without taking them off. Your host may have slippers for you to wear.

Bargaining and Deal-Making

Expect business to take longer than in the West. There is still a rigid, top-down bureaucracy, which is marred by corruption, and tends to be complex and inefficient. As in many of the former Communist states, it is less common for women to be in the highest levels of business. It is important to establish a personal relationship with your Polish associates, spending hours drinking vodka together. Business is a relationship-oriented process, and if you do not know anyone, you should have a Polish representative who is more familiar with the processes and acquainted with the right people. Rules in business and government tend to be complicated; it helps to have someone on the inside to help smooth the way for you.

Facing the Bureaucracy in Eastern Europe

Here is a little story of my own bureaucratic struggles in Poland. I, my son Shanti, and my cousin Macek wanted to take the train from Elblag to Gdansk. We walked to the train station in the morning. It was a beautiful day and I was looking forward to seeing the old city of Gdansk.

We bought our tickets at the counter and started to walk to the track for boarding. A few minutes later, who should happen to walk by but my other cousin, Pawel. We told him about our plans, and he suggested that he drive us to Gdansk in his car. It sounded like a great idea, allowing us a little more flexibility.

We returned to the ticket counter. The same lady who sold us our tickets ten minutes ago was still there, and we told her that my cousin was driving us, and we no longer needed the tickets, and wanted a refund.

The lady never cracked a smile. She sat behind the glass and recited the policy. Macek, who spoke better Polish than I, pleaded with her and explained to her, "But we just bought these tickets ten minutes ago! Surely, you remember us!" She did indeed, but it did not matter. She explained to us that there was a procedure that we had to follow; she could not just hand over a refund. She made a phone call, and one of her superior officers came over. He wore a uniform and had an officer's cap on. We had to explain everything to him again, and then he went into the back again, and re-emerged ten minutes later with a form for us to fill out. He slid the form under the window and we complied. We all had to sign it; then he read it over carefully, the lady clerk at the counter signed it, then he signed and stamped it with an official seal.

"Now do we get our refund?" we asked. No, not so simple. "Take this form over to the commissar's office," he explained, "and he will tell you what you need to do." We went to the other side of the train station, and entered a small office by the tracks. It was an official-looking place with a Polish flag on the wall. The man behind the desk also wore a uniform, but it was more impressive-looking than the other gentleman's; he wore a jacket with his uniform, and had a cap and big epaulets on his shoulders. We had to explain everything to him. He took the form, read it over, and asked us all several questions. He looked at our passports, and asked why we had the same last name, as I had an American passport and my cousin had a Polish one. The interrogation continued. I imagined that we were going to be sent to Siberia. Finally, he signed it. Then he had to call someone out from his back room, who wore even bigger epaulets, and he read it over, then signed it as well and stamped it with an official seal. A clerk took it away and made several copies and brought them back. He gave one of the copies to us, and informed us that we had to return to the clerk's window and present it for our refund. Returning to the clerk's window (by now over an hour and a half had passed), we presented the form, and she refunded us 80 percent. She could not give us back the full price, she explained, because the train had already left the station.

Common Customs

- It is important to be on time for appointments, but do not be surprised if Poles are late.

- Poles may be reserved at first, but display emotions freely once familiarity is achieved.

- Plan well ahead of time; it may take several months to get through the bureaucratic tangles of visas and transportation.

- Take care not to drop anything on the street or litter; it is offensive and illegal.

- As with most of Europe, the first floor is referred to as the ground floor, and what Americans call the second floor is referred to as the first floor.

- It is common to see people of the same sex walking arm in arm; it is just a gesture of friendship.

- Poles are very friendly and soft-spoken.

20
Ukraine

Ukraine, a former Soviet satellite, is behind Poland in economic reforms, although a protest in 2004 brought about an internationally monitored vote for a new Parliament. The country is poor and suffers from excessive state control and corruption. There has been modest growth, especially after the so-called "Orange Revolution" of 2004, when Viktor Yushchenko became president. Ukraine is seeking membership in the European Union and World Trade Organization.

Economics and Opportunities

Ukraine was important in agriculture and industry when it was a Soviet satellite, but has suffered since gaining independence in 1991. There was much resistance to privatization and market reforms, and industrial output quickly dwindled. In 2005 Ukraine eliminated many tax and customs privileges, resulting in more economic activity; however, the government plans to reinstate those privileges and maintain restrictive grain export quotas which could threaten the small growth it has enjoyed.

Greetings

It is customary to shake hands both upon greeting and upon leaving. Upon the first meeting, shake hands and state your name. A man should wait for a woman to extend her hand before shaking it. Close friends of the same sex may greet each other with kisses on the cheek.

The primary language is Ukrainian, but many people also speak Russian. The languages are similar. English is not widely spoken. It is best to present a business card printed in English and Ukrainian — not in Russian, as there are some hard feelings between the Ukrainian-speaking and Russian-speaking population. Ukrainian names are organized similar to Russian ones, with a first name, patronymic, and last name. Females feminize their last names, usually by adding the letter "a" to the end. Do not expect to use first names unless you are good friends; instead, use an honorific or title, plus the surname. Once you have become well acquainted, you may address someone with his or her first name and patronymic, as in "Vladimir Ilyich."

Food

Ukrainian food is similar to Russian, although because of Ukraine's rich agriculture, there are likely to be more fresh fruits and vegetables. There are more restaurants of better quality here than in Russia, and your Ukrainian associates may take you to a few. Ukrainians are heavy drinkers, but in addition to vodka, they also drink a honey liquor called medivka, which is tasty but quite strong. When you are seated at a table in a restaurant, you are immediately served with two bottles: one water, the other vodka. It is customary to drink the vodka until the bottle is empty; the host refills the glasses. Expect plenty of toasts.

Religion

While about 38 percent of Ukrainians report no religious affiliation, those who do tend to follow one of several Eastern Orthodox Christian faiths.

Etiquette

Refer to the country as "Ukraine," and not the older "the Ukraine." Referring to "the Ukraine" implies that the country is still a Soviet satellite.

Gift giving is a good idea for business and when you are invited into a home. Gifts for home visits may include flowers or a bottle of liquor.

Bargaining and Deal-Making

Similar to Russia, the business community is male dominated, with few women in positions of high authority. Western women doing business in Ukraine may have difficulty being accepted as an authority figure, and should establish their credentials ahead of time.

Socialization is an important part of business. Meetings may be held over drinks (alcohol) or meals. Business laws are under development and are subject to frequent change. Because of corruption, proceed with caution and with the help of a local advisor. Business tasks may take a long time, with business meetings often lasting several hours. As with most of the former Soviet states, business is hierarchical and somewhat rigid, although a bit less so than in Russia.

Rank and status are important in business negotiations, and you should be represented by a high-ranking individual from your company. Business dress tends to be conservative.

Common Customs

- The Eastern Orthodox, or Julian, calendar is used, which is different from the Western calendar.

- The Eastern Orthodox Christmas is celebrated on January 7.

- Although you should be on time for appointments, do not be surprised if Ukrainians are late.

- It is common for a Ukrainian to say "no" to almost any proposal immediately, and they are tough negotiators who do not wish to compromise.

21
Czech Republic

Prague is well known as the cultural center of Central Europe, with the famous Prague Castle, beautiful landscapes, and first-class amenities. It was part of the former Czechoslovakia, which was long dominated by the Soviet Union and fought in 1968 for democracy under the banner of "socialism with a human face," an idea that the Soviets quickly crushed. Today, the country is a member of NATO and the European Union. Although the Czechs are Slavic, they are much more influenced by Western Europe than the rest of the Central and Eastern European nations, because of their geographical position. It is sometimes said that Czechs "face west," while the neighboring Slovaks "face east."

Traveler's Tip

As Prague is a major tourist destination, be on close guard against professional pickpockets, especially on the subway, or when in the center or old town at night. Also, do not trust taxi drivers to give you a fair rate. Agree on a fare ahead of time and, if possible, call a taxi service instead of using one you find on the street.

Economics and Opportunities

The Czech Republic is one of the more prosperous post-Communist nations of Central and Eastern Europe. There was significant economic growth as a result of its close ties with and exports to the European Union before the Czech Republic joined the EU. Economic growth has created more prosperity and consumerism, increasing domestic demand as well. Privatization of former state enterprises has been a successful, ongoing process.

Much of the business activity takes place in Prague, a world-class, vibrant, cosmopolitan city. There are many Americans in Prague, and English is widely spoken.

Greetings

Handshaking is the appropriate greeting, and you should shake hands with each person in the room, both upon meeting and departing. Other than the handshake, however, touching is less frequent except among good friends, and the Russian "bear hug" is not practiced. You may see good friends greeting each other with kisses on both cheeks. Older men may kiss the hand of a woman. Gift giving is a good idea, and if you are invited to a home, take an odd number of flowers or a bottle of alcohol. Unlike people of other countries in Central and Eastern Europe, Czechs are not vodka drinkers as much as they are beer, wine, or whiskey drinkers.

The Czech language is similar to Polish, and if you speak some Polish you can understand some Czech as well. The honorifics are the same, with "pan" for Mister and "pani" for Mrs. or Miss. It is best to use the honorific plus surname and avoid first-name status unless you are well acquainted. If the person you are addressing has a title, you may use the honorific plus title, as in "pani director."

Food

The Czech Republic is well known for its excellent pilsner beer, which originated in the Czech town of Pilsen. Expect frequent toasts, with the host making the first one. In many countries, setting your silverware crisscrossed on the plate indicates that you are finished; in the Czech Republic, however, this is how you generally set down your silverware when you are not using it. To indicate that you are finished eating, you set them at the side of the plate. Tipping (10 percent) in restaurants is practiced. Coffee is usually served Turkish style, which is rich and sweet. (Beware — the last half inch of coffee is mud.) It is common to hold a sugar cube in your mouth while you are drinking tea.

Religion

About 60 percent of the population is unaffiliated with any religion, while 27 percent is Roman Catholic.

Etiquette

The host introduces everyone in a group; the women are introduced first, and then the men, starting with the oldest members of the group. Men stand up when someone enters the room. You may be assigned a place to seat, based on your rank.

Bargaining and Deal-Making

Czechs tend to be egalitarian by nature. Your title is not as important as your knowledge and experience. Dress is more formal and stylish than in the West; sloppy clothing, such as shorts and tee-shirts, are not respected. You may find people in business wearing "business casual" attire frequently, but not jeans. In business, the egalitarian nature is also evident; although you may negotiate directly with a higher-up, there may be lower-level employees involved as well, and they should be accorded an equal amount of respect.

Traveler's Tip

The Czech Republic, part of the former Czechoslovakia, has always been egalitarian in nature, going back to the days of King Wenceslas IV (the "good king" in the Christmas carol), who often went into the countryside in disguise to talk to commoners and seek their input and advice.

Another difference between the Czech Republic and other Central and Eastern European nations is the treatment of women in business. Women are seen as equals in business, and you may see women in positions of authority here. However, there is still an element of traditionalism, and women remain the traditional caretakers of children and the home.

Business meetings may include alcohol, such as vodka or slivovitz, a strong drink made from plums. Toasting is customary (the standard is "na zdrovie"), as in Poland and Russia. The host will refill the glasses.

Common Customs

- Czechs tend to be modest; a compliment is appreciated, but the appropriate response is to disagree.

- Czechs tend to be soft-spoken and do not appreciate someone who shouts or speaks aggressively.

- As with most of Europe, the first floor is referred to as the ground floor, and what Americans call the second floor is referred to as the first floor.

- Making direct eye contact is a sign of respect.

22
Romania

Romania is known for more than being the ancestral home of the historical Dracula, a boon to the country's tourism. Formerly known as the principalities of Wallachia and Moldavia (along with Transylvania), Romania has been in existence since 1859. After World War II, Romania became a Communist state under the brutal rule of Nicolae Ceausescu, who was executed in 1989. Communists fell from power in 1996, and Romania joined NATO in 2004 and the European Union in 2007.

Economics and Opportunities

The transition from Communism has been a hard one, and the country's industrial base was obsolete and inefficient when the Communists were in charge. Romania remains a poor country, but the middle class is gaining ground. More recently, economic growth has been moving at a reasonable pace, with a record 6.4 percent growth recorded in 2006.

Greetings

First names are not usually used in business settings, except between close friends. You may use the honorific or title ("domnule" for men, "doamna" for women), followed by the surname. However, honorifics are less often used than titles; sometimes the honorific is used with the title and without the family name.

Gift giving is appropriate when meeting people in business or visiting a home. If you give flowers, give them in odd numbers. Smaller gifts are appropriate early in your business dealings, with larger gifts being given for significant events, such as the signing of a contract.

The Romanian language is similar to Italian, though with a heavy Slavic influence. Hungarian or German may be spoken, and English is not widespread. The elite often study abroad and may speak French.

Wait for an introduction before approaching someone. Shake hands firmly upon greeting and departing; remember to shake hands with each individual. Romanians may shake hands every time they meet someone, even if they have already done so earlier. Men may also hug upon greeting, and women may kiss each other on the cheek. Some Romanian men kiss ladies' hands.

Food

Romanian food is rich and hearty, with lots of soups and stews. Romanian wines are highly regarded and of good quality. Tipping in restaurants is practiced; the standard is 10 percent. As in the Czech Republic, coffee is strong and sweet, and usually served Turkish style, which means that the bottom of the cup is full of coffee grounds. Beer or wine may be served with dinner, although at business meetings there may be brandy or schnapps. Toasts are offered throughout the meal. Silverware is used Continental style. Despite the country's poverty, Romanians take great pride in serving a delicious meal to a foreigner; be sure to be gracious and sample everything.

Lunch is the main meal of the day and may last for two hours or more.

It consists of multiple courses, starting with soup, and wine is usually served.

Religion

About 87 percent of Romanians are Eastern Orthodox, with Roman Catholic, Protestant, and Muslim minorities. If visiting an Orthodox church, women should wear a long dress and cover their heads.

Etiquette

Men should rise when someone enters a room. Wait to be introduced; your host will introduce you to the women in the group first and then the men. If you are introduced to a husband and wife, greet the wife first. If a man is seated, he should stand up when being introduced to a woman, but a woman does not need to rise when making an introduction. Third-party introductions are important here. In some parts of Romania, especially in the east, it is customary to remove your shoes when entering a home.

Traveler's Tip
Toilets in Romania do not have locks on them; knock first before entering.

Bargaining and Deal-Making

Punctuality is important, so be on time for meetings. You may be taken out frequently, as entertainment is a big part of business and the decision-making process. The business environment in general is hindered by the country's poverty; expect corruption. The bureaucracy is riddled with red tape. Women are not likely to occupy positions of authority, and Western businesswomen must take extra steps to be taken seriously; dress and act strictly professional at all times. "Business casual" dress is often the rule, though you should follow the lead of your Romanian associates. Invitations to business dinners and other events are common, but spouses are seldom invited. Business lunches are long and usually involve alcohol.

In doing business in Romania, keep in mind that until recently, this was a Communist state, and older people are accustomed to doing business in a certain way. For example, many Romanians do not answer questions directly, but rather tell long, complicated stories.

Common Customs

- Romanians are soft-spoken; loud, aggressive behavior is rude.

- Do not cross your legs at the knee.

- Direct eye contact is preferred.

- As with most of Europe, the first floor is referred to as the ground floor, and what Americans call the second floor is referred to as the first floor.

- Smoking is common everywhere, even at restaurants; there are no separate sections for nonsmokers.

23
Hungary

Hungary is a graceful country, and Budapest is a pleasant, cultured place to do business. Hungary became Communist during World War II. The country started liberalization in 1968, when it introduced what became known as "goulash Communism"; multiparty elections came about in 1990, and a free market economy ensued. Hungary joined NATO in 1999 and the European Union in 2004. Culturally, Hungary is quite different from other countries in the region. Hungarians are not Slavs, and the language is unlike most other European languages, though it is similar to Finnish.

Economics and Opportunities

Hungary has adapted well from Communism to a market economy, with a strong per capita income and good economic growth. The private sector accounts for 80 percent of the GDP, and foreign ownership and investment are common. Inflation has been brought under control, moving from 14 percent in 1998 to 3.7 percent in 2006. Even before the fall of Communism, there were early experiments in free market enterprise, and today Hungarian businesspeople are accustomed to capitalist practices.

Greetings

Hungary is the only European country where they reverse the order of names in greetings; say the family name first and the given name second, as in "Blacharski Dan." Do not use first names unless you are good friends; instead, use the honorific and professional title, along with the surname. It may also be appropriate to use just the honorific and title, as in "Mister Director." If using the honorific, it is placed after the surname. For Mister, the suffix "ur" is appended to the last name, as in "Blacharski-ur." A married woman adds the suffix "ne" to her last name; for example, my wife is "Blacharski-ne." It is common for women who are well acquainted to kiss each other on the cheek. However, Hungarians are not as touch oriented as other Europeans, and the handshake is more appropriate for business meetings. A woman extends her hand first, which a man should shake lightly. Gift giving is appropriate, and if you are invited to a home, a small gift such as a bottle of liquor or flowers is a good idea.

Many Hungarians are bilingual, but their second language is often German, not English.

Food

Hungarian wines are among the most delightful, and wines from Tokaj are well known the world over. When you go to a restaurant or bar, tipping is expected. Although you can get by with just rounding up the bill, tipping 10 percent is more appropriate. Eating is Continental style. Lunch is the main meal of the day and may last for two hours or more. It is usually made up of multiple courses, starting with soup; beer or wine is usually served. Many of the dishes are savory, but not overly spicy, with heavy use of paprika.

Hungarians are exceptionally hospitable, and your host will probably pay the bill at a restaurant. Your dinner may be quite lengthy; expect to stay in the restaurant well into the evening. There may be entertainment. In some cases, gypsy musicians may come in and play at your table, and it is customary to tip the leader of the group.

If you like strong coffee, you will like Hungary. The coffee at breakfast is sweet and often served Turkish style, with a layer of fine grounds at the bottom of the cup. Evening meals are usually accompanied by beer or wine. Liquor, especially schnapps, is often served during business meetings, even in the morning. As in many Central and Eastern European countries, drink your liquor neat, all in one shot, after toasting; your host will refill your glass. If you do not want more liquor, leave a little bit in your glass. The same goes with food; your host will refill your plate or insist that you take more.

Religion

Hungary is about 52 percent Roman Catholic and 16 percent Calvinist.

Etiquette

Continental manners are the rule in Hungary, and it is customary to treat women with courtesy, standing when they enter a room, and walking on the street side and entering a restaurant ahead of a woman. Hungarian may be the most formal of all the Central and Eastern European cultures, though they have a reputation for being temperamental.

Bargaining and Deal-Making

Although women in Hungary do hold important positions, higher levels of business continue to be dominated by men. Although men in Hungary treat women courteously (standard European manners), they may dominate the conversation in business, forcing an American businesswoman to make an extra effort to stake her claim as an equal. Hungarian business dress is conservative with men wearing stylish, dark suits and ties, and women wearing business suits or dresses. On social occasions, people dress stylishly, with attention to fashion (no tee-shirts).

Punctuality is important, and appointments should be made ahead of time. Be aware that holidays and vacations occur in July and August, and between about the middle December to the middle of January. Little business goes on during those times.

Any business deal includes plenty of socializing, including drinking, dining, and entertainment. Spouses are seldom included. Restaurant meals with your business counterparts are important, but little business is discussed during these events; do not be too aggressive in bringing up business matters during a meal. However, you may find that business is done more quickly and efficiently than in many of the other post-Communist states.

Common Customs

- Hungary is exceptionally clean; do not litter or throw cigarette butts on the street.

- Clinking glasses when toasting is considered impolite.

- Offer a toast to your host's success, but never to your own.

- As with most of Europe, the first floor is referred to as the ground floor, and what Americans call the second floor is referred to as the first floor.

- A man walks on the left side of a woman when accompanying her unless on a street with traffic, in which case he would walk on the street side.

Section 5

*The World In-
Depth: Western
Europe*

Western Europe offers a far different worldview than Central and Eastern Europe. American businesspeople may feel more at home in Western Europe because of similarities in language and custom, but do not be deceived — there are still many cultural differences, and many American customs may seem strange here.

Europe has eight of the ten most expensive cities in the world, and many of the major European Union cities continue to rise in the ranks.

In most Western European countries, etiquette is far more important than in the United States, and Continental manners rule. In most Western European cities, people eat Continental style, with a fork in the left hand and a knife in the right. European society is mostly polite, with much attention paid to protocol. Automatically calling someone by his or her first name is seldom done. First names are considered too informal for business and are reserved only for close friends. Depending on what country you are in, you should use the local honorific: in France, for example, do not call your associate "Mr."; he is "Monsieur." Handshakes are the standard greeting, although the handshake is lighter than the American handshake and brief, often with no pumping. In some European countries the kiss on the cheek is common among friends.

Do not stand with your hands in your pockets. More attention is paid to dressing with style in Europe than in America. The standard of dress tends to be more formal. Avoid the "jeans and tee-shirt" look, even on social occasions. In many European countries, wear business or "business casual" clothes to social events.

Gift giving is common in many European countries, both for business and social occasions, with one exception: Many Europeans take wine seriously; bringing a bottle of wine to a dinner party can be an insult, as if to say your host's selection is inadequate.

As for the treatment of women, it is more traditional in Europe. A man in Europe is expected to hold the door open for a woman and should walk on the street side. In most countries, when entering a restaurant, the man walks in ahead of the woman and then holds the door for her.

There are plenty of business opportunities for Americans in Western Europe. Although Western European cultures are diverse, and you have to revise the rules of etiquette every time you cross a border, there is a new prosperity and a new era of opportunity and cooperation. The European Union has emerged as a profound force not only in Europe, but in the world.

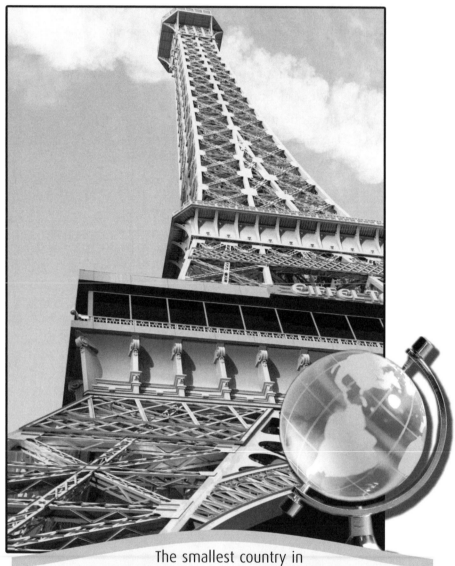

The smallest country in
Europe is the Vatican, which is considered a separate country
from Italy even though it is in the middle of Rome.

24
United Kingdom

The first thing to remember is that the United Kingdom is not just England. The United Kingdom of Great Britain is made up of England, Scotland, Wales, and Northern Ireland. Calling a Scotsman "English" is an insult. While the British Empire at one time covered about a fourth of the known world, its range has lessened after two world wars and the withdrawal of the Irish Republic. Regardless, its world standing remains considerable.

As an American in the U.K., you may mistakenly expect to adapt easily, sharing a common history and language. The fact is, British English is quite different than American English, and although you can converse in most situations, different dialects and simple uses of the language may be confusing.

Economics and Opportunities

With a trillion dollar economy, the United Kingdom is one of the greatest financial centers of the world and a major power in trade. Public ownership has been reduced significantly over the past 20 years, giving a greater role to the private sector. Agriculture is large but highly mechanized, and natural resources such as coal, gas, and oil are a major part of the economy. The United Kingdom's economy is among the strongest in Europe, with steady but mature growth patterns and healthy rates of unemployment and inflation.

Greetings

Within the United Kingdom, there are many different cultures, but in general, people are more formal than in America. Saying "hi" to someone, as is common in America, is informal, and should only be used between good friends. Instead, use "good morning," "good afternoon," and "good evening." A more formal greeting is "How do you do?" The proper response is to repeat the expression. "Hello" is said when answering the phone, not when meeting someone in person.

Wait to be introduced before speaking, and wait to be seated by your host at a gathering. Small gifts and follow-up thank-you notes are good ideas. If your host has a title of nobility, use it. Address nobility by their title and their first name (e.g., "Lady Di"). In general, honorifics are used in the U.K. more than in the U.S., although the honorific "Ms." is not commonly used. Degrees or titles should be used when addressing somebody.

The United Kingdom is a nontouching society, although there may be more touching in Scotland where social customs are less formal than England. The European kiss on the cheek is not common. Handshaking is customary. The handshake is firm, but not as strong as the American grip, and without pumping; a woman extends her hand first. Shake hands with everyone in the room, both upon greeting and departing. Do not wave goodbye as you do in America.

English is the dominant language throughout the U.K., although Wales is bi-lingual (with both English and Welsh as official languages). Gaelic is spoken by a significant minority of people in Scotland. In the countryside, you may hear Cornish in the Southwest of England, or Manx on the Isle of Man.

Traveler's Tip

If you are meeting a member of the royal family, be on your best behavior. Bowing (or curtsying, for a woman) is optional for a foreigner, but great deference should be shown. At the very least, a man should offer a slight bow with the head down, and a woman should bend one knee slightly. Do not offer to shake hands with a member of the royal family; wait for them to extend their hand.

Food

You cannot visit the United Kingdom without enjoying some time at a pub. More than just a place to drink beer, a pub is a place to enjoy yourself and socialize.

English beer is excellent, much richer than American beer, and usually with a higher alcohol content. English beer is usually served at room temperature, though lagers are sometimes served cold. It is also served in pints. Be cautious, because the standard pint in England is much larger than a glass of beer served in an American bar. You can order a half pint, although men may get laughed at; usually just women order the half pint.

At a formal dinner, the host may say grace, and after the main course someone may toast to Her Majesty's health.

At restaurants, people eat Continental style. At a restaurant, tipping is customary at 10 to 15 percent, but not more.

There are several meals during the day. Breakfast is a large meal: toast and breads, juice, bangers (sausages), fried potatoes, bacon, kippers, and scones. It is served with coffee or tea. Lunch is available between noon and 2 p.m., and this is followed by tea.

Tea is not just tea, but a light meal which may include sandwiches, meat pies, and cakes. There are two different types of tea: regular tea and high tea. There is plenty of food at both, although high tea has more, including a hot dish in addition to everything found in the standard tea.

Dinner is usually late and begins with an alcoholic drink, such as sherry. Dinners can be quite long.

Religion

The United Kingdom is predominantly Christian, with 72 percent of the population being Anglican (the Church of England), Roman Catholic, Presbyterian, or Methodist. There is a small Muslim minority of about 3 percent. The Church of England is the State church, although there is religious openness.

Etiquette

Manners and self-discipline are important. The British are known for understatement and their dislike of boastfulness. While Americans may be known to be casual, the British have a reputation for formality. Americans say, "Hi," "Howdy," "S'up?" or "Yo"; Brits say, "How do you do?" While some young men in America walk around with their pants down below their underwear, their young British counterparts are still wearing bowler hats.

Avoid loud conversations in public, standing with your hands in your pockets, and other minor infractions. Remove your hat when indoors, cover your mouth when you yawn, and do not cross your legs at the knee. Women should cross their legs only at the ankles.

Bargaining and Deal-Making

Keep your business appointments on time. Avoid displays of anger or emotion, and pay close attention to protocol. In Wales and Scotland, there is a less strict adherence to formality. You may be taken to a restaurant or pub, but the meal is strictly social; avoid talking business over dinner or on social occasions. Business gift giving is seldom practiced, but holiday cards and thank-you letters for dinners are appropriate. If you are invited to a home, it is appropriate to bring a small gift of flowers (in odd numbers).

Traveler's Tip

As an American, you may think that language is not be a problem when visiting the United Kingdom, but be assured that it may cause problems. Many terms are different. Furthermore, working-class people tend to be from other countries, and thus have two accents. You may even hear rhyming slang, which is often unintelligible to Americans. Of course, Americans have their own regional dialects: Someone speaking Ebonics in London will get blank stares. Business expressions are also different. For example, in America, to "table" something means to delay discussion of it; in the United Kingdom, to "table" something means to bring it out for discussion immediately.

Style of dress is conservative, both during business meetings and in daily life. Casual dress is still stylish and not sloppy (no jeans or tee-shirts).

It is common for colleagues to socialize at a pub for an hour or two after working hours, and you may well be invited. It is customary for each person in a group to buy a round of drinks in turn.

Common Customs

- Soccer is popular, though it is called "football"; avoid making any derogatory statements about any team or player.

- If possible, do not talk about England if you are in Wales or Scotland.

- English people are proud of the region they come from; for example, do not call someone from Cornwall "English"; call them "Cornish."

- Although the United Kingdom is in Europe, do not call them Europeans.

- As with most of Europe, the first floor is referred to as the ground floor, and what Americans call the second floor is referred to as the first floor.

- Avoid wearing a striped tie, because it may be similar to one of their school ties.

- If you smoke, offer cigarettes to others before lighting one of your own.

25
Germany

Germany has a beautiful countryside and is an impressive economic success story. A vital part of the continent's political landscape, Germany has Europe's largest economy and its second largest population. In 1949, it was divided into two Germanies: the Federal Republic of Germany and the German Democratic Republic. The latter was aligned with the Soviet Union. The country reunified in 1990, and there have been significant efforts to bring the former East Germany up to Western economic standards. The two regions remain culturally distinct and have vastly different levels of economic development.

Economics and Opportunities

Germany's economy is among the most powerful in the world, ranking fifth in terms of GDP. Although there was some stagnation between 2001 and 2005, the country is now seeing good growth levels. Modernization of the former East Germany continues to consume a large amount of resources. Unemployment is still high, due to declining investment and flat domestic consumption, as well as high interest rates. There is a large immigrant population laboring as guest workers at entry-level jobs, mostly in agriculture and the service sector. This has become a sensitive issue in Germany, as there is a perception that these immigrants burden the country's social service infrastructure and take jobs away from natives.

Greetings

Third-party introductions are best, and you should wait to be introduced to someone before introducing yourself. A common informal greeting is "Wie gehts" (pronounced "vee gates"), roughly meaning "How are you?" or "How is it going?" However, many take it literally and respond by telling you all about their problems. In Bavaria, you may be greeted with "gruss Got" (meaning "Great God"). When departing, Germans may say "ciao" (Italian), but on formal occasions they may say "auf wiedersehn."

Shake hands with everyone in the room, both on greeting and departing. The German handshake is firmer than the standard European shake, but still brief. A woman extends her hand first, and some older Germans may kiss a lady's hand lightly. When shaking a lady's hand, a man should offer a lighter handshake than with another man. Address people formally, with an honorific and title, and never by first names. Use "Herr" for men and "Frau" for women, followed by the surname. The honorific "Fraulein" is used with young girls. Use the honorific along with a person's title, as in "Herr Director." Business gift giving is seldom practiced. If you are invited to a home, bring flowers in an odd number, and follow up with a thank-you note.

English is widely spoken as a second language, although Russian is more common in the East.

Food

Germany is known for its beer, as well as its excellent wines. Their white wines are known throughout the world for their delicate flavor. Toasting is expected, and glasses are clinked. Desserts are also well known and delicious, and Germans have dozens of different ways of preparing sausages.

Waiters often ask if you enjoyed the meal, and they are sincere in their questioning; if you liked it, respond enthusiastically. If you want to summon a waiter, the best way is to make eye contact, instead of waving.

In the afternoons, you may be invited to a home for coffee and cakes, especially on weekends. When at a restaurant, tipping is not required, but is appreciated, and 10 percent is appropriate. Tipping is practiced less in the post-Communist, eastern part of the country. (Communists tend to frown on tipping as a symbol of classism.) Eating is Continental style. When you are at a dinner, do not eat until the host says, "guten appétit."

Breakfasts tend to be substantial and filling. While not as extensive as the English breakfast, it includes many items. Lunches are also large, usually the day's main meal, and served with beer or wine. Dinner is served late.

Religion

Germany is evenly divided between Protestant, Roman Catholic, and unaffiliated, with a 4 percent Muslim minority. East Germany was officially atheist, though since reunification there has been renewed interest in religion.

Etiquette

It is appropriate to rise when a woman enters the room. When entering a restaurant with a woman, it is appropriate for the man to enter first. As in the Orient, where a man sometimes walks ahead of a woman out of chivalry, a German man enters a restaurant first to determine whether it is suitable for a lady. Pay attention to small details of etiquette and manners. When walking with a woman, walk on her left side, unless you are on a street with traffic, in which case you should walk on the side nearest the street. Also, in business meetings, restaurants and even on some social occasions, wait to be directed to a seat. Your host will seat people according to their status and position. Clothing styles tend to be formal on both business and social occasions. Dress tends to be more conservative, as compared to France or Italy, for example, where stylish flamboyance is acceptable. A standard dark suit with a white shirt is appropriate here.

Bargaining and Deal-Making

Germans expect punctuality. If you are late, a meeting may start without you and a group may depart before you have arrived. Make sure your dealings and proposals are highly organized and well prepared.

Germans are less vulnerable to marketing and salesmanship; be prepared to bring plenty of facts and evidence to the table to support your initiative. Make direct eye contact and maintain constant respect for their famed sense of order and organization.

There are several important differences between the regions that made up East and West Germany. Although it is one country now, the business cultures remain distinct. The Eastern area has a workforce which is unfamiliar with Western practices, and, following their Communist predecessors, lacks a stringent work ethic.

Common Customs

- Germans recycle whenever possible. In public places, be careful when throwing out trash; take care to use recycle bins whenever appropriate.

- Joining one's hands together and raising them above one's head is a gesture of gratitude, similar to "thank you."

- If you are seated at a table or in a conference room and it is inconvenient to greet someone formally, you may hit the table with your knuckles as a greeting.

- Though it means, "OK," in America, joining the thumb and forefinger together is a rude gesture in Germany.

- As with most of Europe, the first floor is referred to as the ground floor, and what Americans call the second floor is referred to as the first floor.

- Germans tend to be neat and orderly (except when driving on the autobahn, where rules are wanting).

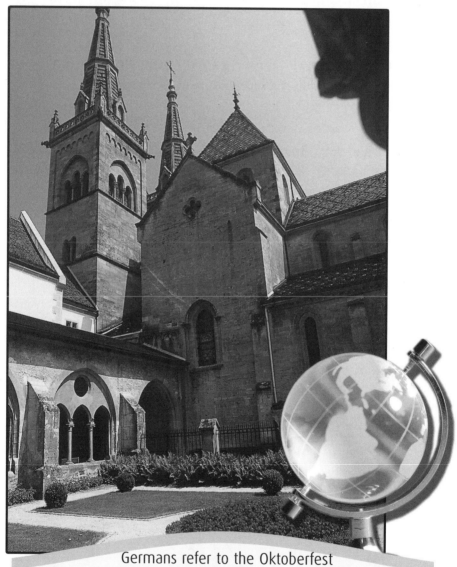

Germans refer to the Oktoberfest as "die Wiesn," meaning meadow or field. The original Oktoberfest was actually a wedding held in the fields of Munich.

26
France

Despite some negative stereotypes, the French are very friendly and open. France is a great place to do business, if for no other reason than you get to visit one of the world's greatest cultural meccas. It is one of the most modern countries in the world and stands out in Europe for its relative economic and political stability.

Economics and Opportunities

France is moving away from government ownership and intervention in business, to a more market-oriented economy and business environment. Several large companies and financial institutions have been privatized, and the government has sold its stakes in major companies such as Air France, France Telecom, and Renault. However, it maintains a strong presence in power, public transportation, and defense. Telecommunications will present an opportunity in the future, as the government gradually opens it to private competitors. Although capitalist, France focuses on social equity. The French oppose labor reforms, despite the government's attempt to make the labor market more flexible. Taxes are the highest in Europe.

Greetings

It is hard to visit France without being kissed. The French kiss on the cheek (first on the left, then on the right) upon greeting and then again on departing. This is practiced between women, between men and women, and sometimes between men if they are well acquainted.

Wait to be introduced before speaking, and third-party introductions are better than the direct approach. Handshakes are light with no pumping. The American handshake is considered low class. The higher-ranking person extends his or her hand first, and a French woman may offer her hand, as well. First names are only used between friends; use Monsieur or Madame followed by the surname. Mademoiselle, which is equivalent to "Miss," is not used often; most adult women are "Madame," even if they are single. Small gifts are appreciated, and if you go to a home, bring a box of chocolates or flowers. Unless you are knowledgeable about wine, do not bring a bottle. French are serious about their wine, and they take some effort to choose the most appropriate wine for every occasion. Direct eye contact is important, and sometimes French eye contact is so intense that it may be intimidating to people from other cultures, especially Asians, who avoid eye contact as a rule.

Traveler's Tip

The French are much like Americans, in that they expect people in their country to speak their language. Americans tend to believe the English language is the global standard for doing business, and to a degree, it is; however, French runs a close second, and is spoken by about 300 million people in over 50 countries throughout the world. Make an effort to speak the language well if you plan on doing business in France.

Food

France is well known for its food and wine, and while in Paris you must indulge in a traditional French "dejeuner," or midday meal. Bread in France does not come presliced in a plastic wrapper, as it does in America. It is much more than something to wrap around a sandwich. The baguette is a delicacy, and an important part of a meal. Lunches tend to be long; expect to be taken out to business lunches that last two to three hours. Although you may discuss business, the long lunch is also a time to relax and get to know one another. Feel free to linger in the restaurant; although service may be slow, waiters will not push you out. Tipping is standard at about 10 percent, and this is often included in the bill. If the tip is included in the bill, it is still appreciated if you leave the odd change or a few francs on the table. French eat Continental style.

Religion

Although there is no official religion, the French are about 85 percent Roman Catholic, with Muslims at 10 percent and growing. The increasing Muslim population is largely due to immigration from the former French colonies in North and West Africa.

Etiquette

The French are fairly formal, both socially and professionally. The French take etiquette seriously, especially at the dinner table. Fruit is eaten with a fork, and the French seldom eat with their fingers, except bread and rolls. Do not use toothpicks or have loud conversations. Avoid gum chewing or talking in an loud voice. Dress is never casual; even in social situations, "casual" means stylish and dressy. Paris is often seen as the center of the fashion industry; feel free to go beyond the plain black suit and white shirt and add a few flourishes to your wardrobe.

Bargaining and Deal-Making

French businesspeople tend to be formal and conservative. Be on time for appointments. However, they may often be late for an appointment.

No disrespect is intended by being late. Pay attention to formality during business meetings; perceived informality may be seen as disrespectful. Do not remove your jacket during the meeting.

Expect them to voice every possible objection to your proposal. This is not because they are opposed to it or want to pick it apart, it is just their nature of doing business. They want to examine a proposal from all sides before coming to an agreement. Also, they do business based on personal contacts and long-term relationships, so it is vital to have a viable network. While you will be doing business with high-ranking officials, be aware that French managers usually have good relationships with their employees, often treating them like family. That attitude plus their strong labor laws result in decisions that take their employees into account. Also be aware that employees have stricter job descriptions; employees resist working outside the scope of their duties. Thus, be familiar with individual responsibilities and expect a strict business hierarchy.

Common Customs

- Standard of dress is more stylish and formal.

- As with most of Europe, the first floor is referred to as the ground floor, and what Americans call the second floor is referred to as the first floor.

- Like much of Western Europe, the French take long vacations, usually in July and August.

27
Italy

One of the most romantic countries in Europe, Italy may be challenging to a foreign businessperson, but it is a fascinating and beautiful place to visit. The country has a rich cultural history. Although Italy had a parliamentary government early on, it fell in the 1920s with the rise of Fascism. In 1946, a democratic republic was put in place, and the country enjoyed some economic revival after the War. Italian culture is hard to define, as its many independent regions enjoy unique customs, cuisines and styles.

Economics and Opportunities

Italy's economy is diversified and industrial, comparable to that of France and the United Kingdom. The North is particularly well developed industrially, with a strong private sector, while the South is largely agricultural and suffers from high unemployment. Several reforms are under way, designed to improve competition and growth, although structural reforms have been slow in coming. The tax burden is high, and the union-dominated labor market is rigid.

Greetings

They always say "ciao" in the movies, but you are not Cary Grant; it is best to avoid this common greeting except with good friends. This is often accompanied with kisses on each cheek, right first and then left, but again, only between good friends. Among business acquaintances, use the more formal "buon giorno" or "buona sera" ("good morning" or "good evening") when greeting, and say "arrivederci" when departing. Handshakes are customary, both upon greeting and departing. Although the American firm grasp and pump are not done, there may be a lingering, two-handed handshake between good acquaintances.

Remain formal in addressing people and refrain from using first names. Use a person's title or honorific (signore, signora, or signorina) when addressing somebody. Titles are also important, and should be used at all times.

Although Italian is the official language, there are multiple dialects. Many businesspeople speak English.

Food

Italy is by no means a "fast food nation." Meals are enjoyed over a long period of time. Lunch and dinner are pleasant, relaxed social occasions. Lunch, the main meal of the day, may last for two or three hours. Tipping is customary; even if a gratuity is included in the bill, leave a tip on the table.

Traveler's Tip

Italy is well known for its varied coffee beverages. Espresso ("caffe") is strong and served in a small cup. The first time I had espresso, I wondered if they were serving it in a tiny cup to be cheap; but when I tasted it, I understood. There are dozens of coffee preparations, including caffe ristretto, which is even stronger than espresso; caffe macchiato; cappuccino; caffe latte; caffe Americano; and many more. There is a different time of day for each one. Cappuccino, for example, is drunk only in the morning, and never with a meal.

The cuisine itself is fascinating and varied. Italian food goes far beyond the American's impression of pizza and spaghetti. There is plenty of fresh seafood prepared in complex ways, and each region has its own style of cooking.

Religion

Home to the seat of the Roman Catholic Church, that religion is dominant. However, there is no official religion in Italy, and there are small Protestant, Jewish, and Muslim communities.

Etiquette

Italians are hospitable and enjoy treating their guests to a restaurant meal; do not refuse if you are offered. Italians eat in the Continental style, and seldom eat with their fingers. Wine is pervasive throughout the day, and is meant to be savored. Your host will pour the wine and refill your glass; however, it is not customary for a woman to pour wine. Furthermore, while it is enjoyed on a daily basis, it is rude to overindulge.

Exchanging business gifts is customary, but do not give gifts that feature your company's logo. Instead, bring a high-quality item from home, such as a special handicraft. If you go to a home, take a bottle of wine or flowers in an odd number.

Bargaining and Deal-Making

Although you should be on time for appointments, do not be surprised if your Italian counterparts are late, especially if they are higher in rank. No disrespect is intended. Italians pay much attention to doing business with good form, or with "bella figura," a philosophy of acting and dressing appropriately, with style and grace. Dress fashionably for both business and social occasions. Business suits are more flamboyant than the standard black suit with white shirt; men may also wear accessories, such as cuff links. Businesswomen also dress stylishly, with accessories and makeup. Italian businesspeople want to get to know you well before transacting any business. Introductions are important; if possible, arrange for an introduction from a representative ahead of time.

When doing business in Italy, you may find the bureaucracy ponderous and complicated. Business meetings may seem unorganized, with everyone talking at once. However, business does get done. Within the corporation, the hierarchy tends to be more horizontal than vertical, with many people involved on any given project. There is due respect for authority and status.

Common Customs

- Italians dress with style and are much more formal than Americans.

- Italians talk with their hands and have dozens of different gestures with subtle meanings. Unless you understand Italian gestures, it is better not to copy any of them.

- Do not exchange business cards at social occasions.

- As with most of Europe, the first floor is referred to as the ground floor, and what Americans call the second floor is referred to as the first floor.

- Special visitors are introduced first.

- Italy is a touch-oriented society, and you may see male friends walking arm-in-arm.

28
Spain

Traveling in Spain is easy, and the people are relaxed. Although Spain was neutral in both world wars, its civil war in the 1930s devastated the country. It became democratic following the death of dictator Francisco Franco in 1975, and it quickly adopted a modern economy. Spain has several distinct regions and cultures. The Catalonia region for example, has its own distinct language (Catalan), and the people there are independent. The Castile region is centered on Madrid.

Economics and Opportunities

Spain enjoys a strong economy and a vibrant free market. The economy grew about 5 percent each year between 1986 and 1990, then suffered from the European recession, but growth resumed in 1994. Spain has a mixed capitalist economy, although former President Aznar introduced privatization, deregulation, and tax reforms. The socialist President Zapatero has had limited progress in carrying out the structural reforms necessary to support Spain's continued economic growth.

Greetings

Although Spanish is the primary language, Spain has millions of tourists each year, and the citizens are multilingual. You are likely to hear English in the larger cities.

People in Spain have a given name (usually two names that are said together), followed by two surnames, with the father's first surname first, followed by the mother's first surname. Stick with standard Continental manners, and refrain from first names. Use "señor" for men and "señora" for women. "Señorita," which is the equivalent of "Miss," is infrequently used, except for young girls.

Socializing is more likely to occur in restaurants, but if you are invited to a home, it is considered a great honor. It is customary to politely decline the first invitation, and then accept if a second is offered. If you do visit a home, bring a small gift; include a small item for each child in the home as well. Greetings tend to be affectionate. Men shake hands, but women kiss on the cheek. Men who are well acquainted may hug upon greeting. The Spanish handshake is lighter than the American version, but may linger and be accompanied by a light embrace, or abrazo.

Be aware that the Spanish spoken in Spain is a different dialect than those found in Mexico and Latin America, which are most often taught in American schools. Nonetheless, the Spanish people appreciate any effort to speak their language, regardless of dialect or proficiency.

Food

Tipping at a restaurant is not common, although it is acceptable to round up the bill. Lunches tend to be long and relaxed, as this is often the main meal of the day. The evening meal is usually quite late, and restaurants may not open until after 9 p.m. Spanish eat Continental style and use utensils. Do not even eat a sandwich with your hands, and never drink straight from a bottle. Spanish food is not the same as food from Mexico — there is greater use of seafood in Spain, and it is less spicy.

Religion

Spain is one of the most Catholic countries in Europe, with 94 percent of the population Roman Catholic; 3 percent is Muslim.

Etiquette

European manners are observed. For example, men wait for women to be seated before they take their own seat. Never start eating until everyone in the room has taken a seat. Men cross their legs at the knees, and never with the ankle on the opposite knee. Women do not cross their legs at all. Rise when a lady enters the room and hold doors open.

Refer to people from Spain as "Spaniards," not "Spanish."

Bargaining and Deal-Making

Business dress is more formal than in America, and some older people dress in suits or dresses just to go out to the store or for a stroll. Jeans or sweat pants are seldom worn, even for social occasions. While men should wear standard business attire, women should be more stylish. During meetings, Spanish businesspeople do not engage in lengthy small talk. They are not afraid to state their opinions. Be punctual, but do not be surprised if your Spanish counterparts are late; be prepared to wait.

The Spanish have respect for authority, but are also independent. Businesses tend to be bureaucratic. Although there are social events, there is less focus on socializing with business partners before business. Lunch meetings are less common, and many Spanish prefer to go home for a long lunch and siesta. There are many women in business, but they still face difficulties in achieving equality at the highest levels of business.

Common Customs

- In many cities, everything closes for a few hours in the afternoon for the siesta; only restaurants remain open.

- The host sits at the head of the table; the guest of honor is to the right of the host.

- As with most of Europe, the first floor is referred to as the ground floor, and what Americans call the second floor is referred to as the first floor.

- There may be several different wines served throughout dinner, each one selected to accompany each course.

- Direct eye contact is important, especially in business.

29
Ireland

The Irish landscape is legendary, and the pubs are not to be missed. Be sure not to confuse Ireland with Northern Ireland: in 1921 a guerrilla revolt resulted in the southern part of Ireland gaining independence from the United Kingdom. The six northern counties remain part of the United Kingdom. Despite a history of violence, there is effort toward a peaceful reunification.

Economics and Opportunities

Ireland has shown steady growth levels averaging about 6 percent a year. In part due to the influence of the European Union, Ireland's economy is growing. The country has become an attractive place for manufacturers from all over Europe to establish plants, and the country has an increasingly large high-tech sector. The economy depends on trade, and agriculture has become less important than industry and service. Exports, including machinery and equipment, computers, chemicals, and pharmaceuticals, are the main engines for growth, while consumer spending is on the rise.

Greetings

English is generally used throughout Ireland, although Gaelic (the official language) is the primary language in some areas along the western coast.

Most Irish are bilingual, and Gaelic is taught in all schools.

Exchanging business gifts is seldom done, although if you are invited to a home, it is appropriate to bring flowers or chocolates. Handshaking is customary and somewhat firm, though physical contact is uncommon.

Standard English honorifics are used ("Mr.," "Mrs.," and "Miss"), though "Ms." is not widely used. In abbreviated form, the honorifics do not have periods after them. If a person has a title, use it during conversation. When referring to yourself, do not include your title. Business gift giving is not commonly practiced, but holiday cards and thank-you cards are appropriate. Bring a small gift for social occasions such as dinner parties.

A light handshake is common. As in England, the standard greeting is, "How do you do?" Ladies extend their hands first. There is more touching than in England, and once you become familiar with someone, there may be backslapping or prolonged handshakes accompanied by a pat on the back or a grasp of the elbow. Eye contact is appropriate.

Food

Tipping is customary, and 10 percent is standard. Breakfast is a heavy meal, similar to that of England, and accompanied by tea (with cream and sugar) or coffee. Dinner is not served as late as some other European countries, starting at around 6 o'clock. A dinner with business associates may start with liquor. Beer or wine accompanies the meal. Irish also observe the custom of afternoon tea, with high tea offered around 4 o'clock and consisting of a light meal, usually with a hot dish. Standard tea includes sandwiches and cakes. Irish food is similar in many ways to English food, with some regional differences.

Eating is Continental style.

Religion

Ireland is about 88 percent Roman Catholic, with minorities in other Christian religions, such as Church of Ireland (Anglican).

Etiquette

Continental manners are observed. Do not cross your legs with one ankle over the knee. Rise when a lady enters the room and allow women to be seated first. The Irish are friendly, forgiving, and substantially less formal than the English; they are not likely to take offense if you commit a faux pas. When in a pub, it is customary to order rounds in turn; women do not buy rounds. Turning down a drink is a relatively serious offense. If your tolerance to alcohol is not well developed, you can save face by ordering a shandy (half beer and half lemon-lime soda). Be prepared to sing if you are invited to a dinner party. Group entertainment is customary, with everyone participating.

Bargaining and Deal-Making

Make your business plans in advance, but expect your Irish counterparts to be late. Interpersonal relationships are vitally important; take time to get to know your Irish associates on a personal basis. Expect social events where business is not discussed. There are few women in executive positions, but those women who have achieved positions of authority are treated as equals. The standard of dress is less formal than in England, but more formal than in the United States. It is seldom necessary to wear a jacket and tie to dinner, for example, although you should refrain from jeans and tee-shirts.

Your Irish business associates may invite you to a pub after work. Beer is available in pints or half pints, but usually only women order half pints. Be aware that the alcohol content of Irish beers is higher than that of American beers. The standard Gaelic toast is "slainte."

Common Customs

- In smaller towns, you may find restaurants and inns closing from October until Easter.

- If you use someone's phone, it is customary to offer to pay for the call.

- Irish love to joke, sometimes at the expense of others, so do not take it personally if you are the subject of a joke or prank. Join in the fun. It is expected in a pub.

- Women are seated first in business and on social occasions.

- To be referred to as "plain" is not an insult; it means that you are one of them, the "salt of the earth."

30
Netherlands

Dutch society tends to be quite liberal, although you may find people to be reserved. Netherlands has a long history, going back to its declaration of independence from Spain in 1581. The Kingdom of the Netherlands, or "Netherlands," includes the Netherlands, the Netherlands Antilles and Aruba. "The Netherlands" refers only to the European country, and not the entire Kingdom. It is incorrect to refer to the Netherlands as "Holland," which is only a specific region of the country.

People in Netherlands are referred to as "Dutch", which should be distinguished from "Deutsch," the German word for "German."

Economics and Opportunities

Today Netherlands is a highly industrialized exporter of agricultural products. It was a founding member of NATO and the EEC (now the EU). The economy of Netherlands is prosperous, open, and highly dependent on foreign trade and investment. Unemployment and inflation are under control, and agriculture is highly mechanized. It is the most aggressive European nation, in terms of attracting foreign direct investment. Netherlands has a large social services infrastructure, which provides a safety net for all citizens, and taxes are among the highest in Europe.

Greetings

The European cheek kiss is common in Netherlands, sometimes accompanied by a handshake. Men usually do not kiss each other. Upon greeting and departing, it is customary to shake hands with everyone in the room firmly, without pumping, and with eye contact. Although you may use someone's title in written correspondence, in person there is less attention to such formality. The language spoken is Dutch, although many people are bilingual and English is common.

Avoid using first names until a close relationship is established. The Dutch are familiar with other cultures and may offer a first name to Americans to make them feel more comfortable. Use honorifics when appropriate, using "mijnheer" for men and "mevrouw" for women.

Gift giving is appropriate in business and on social occasions, but do not offer anything ostentatious or expensive.

Food

In Netherlands you eat Continental style, using utensils for just about everything, rather than the hands. There is often a light snack at midmorning, as well as a 4 p.m. tea break. Expect to be invited to lunch or dinner, usually at a restaurant. Tipping at a restaurant is expected at around 10 percent. Breakfasts tend to be large (although not as large as the English breakfast). Lunch is usually about an hour, and comes with beer or wine. Dinner is between 6 and 7 p.m. Dutch food includes plenty of dairy products and cheeses, with meats and sausages, and fresh or smoked fish. Although hard liquor is seldom served before dinner, you may be served an aperitif, which should be finished before the meal starts. The evening meal may be served with wine.

Traveler's Tip

The American phrase "going Dutch" means splitting a bill at a restaurant. However, there is nothing Dutch about it. In some countries, people refer to the practice as going "American style." In Netherlands, the host or senior-ranking person customarily pays the bill.

Religion

Netherlands is a highly secular society, with 41 percent of the population proclaiming no religious affiliation. Otherwise, the population is 31 percent Roman Catholic, 13 percent Dutch Reformed, 7 percent Calvinist, and 6 percent Muslim.

Etiquette

Continental manners are observed in Netherlands. Allow women to be seated first; if a man is accompanying a woman in public, he should walk on the side next to the street. Men should stand when a woman enters the room. Netherlands is known as a tolerant society with a great focus on equality. While soft drugs and prostitution are legal, if you partake in either of these, keep it discreet. They may be legal, but not openly condoned.

Bargaining and Deal-Making

The Dutch are friendly and easy to do business with. They are well acquainted with trade and American business practices, and most speak English. When doing business, make eye contact. They may be blunt, but this is not meant to be offensive. Unlike many European countries, meetings in Netherlands are direct, and do not begin with socializing. Expect strong organization, with highly planned meetings and initiatives. Be on time, and expect your Dutch counterparts to be on time as well. In business, the group is the focus, so that success is rarely seen as the work of a single person.

Although hierarchy is respected at all times, Netherlands is egalitarian by nature, and every person is expected to participate. Men and women are treated equally in business.

Business dress is conservative, with a standard business suit and tie for men and business suit or dress for women. It is common for a man to take off his jacket at the office.

Common Customs

- The Dutch tend to be a little reserved and "making a fuss" is impolite.

- The Dutch tend to avoid ostentation.

31
Austria

Austria is an exciting vacation destination, and has tremendous opportunities for those in the tourist industry. Once a large empire, it was annexed by Nazi Germany in 1938 and then occupied by the Allies after the War; its independence was recognized in 1955.

Austria is very different from Germany, with a unique history and different customs. Austria entered the European Union in 1995 and today is democratic and prosperous.

Economics and Opportunities

Austria has a well-developed market economy and a relatively high standard of living. The country has close ties to other EU members, especially Germany, and its economy has strong industrial and service sectors. Foreign direct investment in Austria has gained momentum, and there has been significant economic reform, streamlined government, and a focus on increasing the competitiveness of the country's business environment, making it an attractive target for foreign investment.

Greetings

It is customary to greet everyone in the room, even strangers. If you are invited to a home, bring a small gift, such as flowers in an odd number, or chocolates. Handshaking is customary, and the handshake should be firm, done both upon greeting and departing. Women extend their hands first, and a man should shake the woman's hand before greeting other men. Men are deferential toward women, and some Austrian men kiss the hand of a lady. Use honorifics plus surnames; never first names.

Gift giving is commonly practiced in business and in social occasions.

The official language is German, although English is spoken in most corporations and in cities. The Austrian dialect of German is somewhat different from that of Germany, although Austrians understand German.

Food

Tipping is customary, and a 10 percent tip may be included in the bill. Austrian food is like German food. There may be a light meal in late afternoon, around four o'clock, similar to the British tea. Tea is served in a glass and drunk in the style of the Middle East, with a sugar cube held in the mouth. Eating is done European style, with the fork in the left hand.

Religion

Austria is 74 percent Roman Catholic, with small Protestant and Muslim minorities.

Etiquette

Although there are some similarities between Austrian and German culture, do not confuse the two; Austrians are not Germans, and do not want to be considered Germanic.

The old Austrian aristocracy helped to define "old world" European etiquette. Continental manners are the rule, and Austria tends to be formal and traditional. For example, men should stand up when a lady enters the room. Avoid chewing gum in public, and avoid keeping your hands in your pockets.

Bargaining and Deal-Making

Austrians are punctual; make every effort to be on time for meetings and social events. "Fashionably late" does not exist in Austria. Hierarchy is respected and important in business. It is also important to have good connections before doing business. Business dinners are less common, although there may be one held upon completion of a project or the signing of a deal. Business lunches are more common. Austrians do not respond well to high-pressure sales; business is done at a more relaxed pace. Confrontation is never a good idea. Although consensus is sought and input from all members of the group is appreciated, the highest-ranking person makes the final decisions.

While there is no official nobility, Austria has an active class system with most business executives coming from the upper class.

Traveler's Tip

If your business counterpart makes fists with both his hands and pounds the table, do not be alarmed; it is only a friendly "good luck" gesture.

Common Customs

- It is polite for a man to enter a restaurant ahead of a woman.

- Business meetings tend to be led by the highest-ranking person.

- Making eye contact is customary.

In Austria, the hills really are alive with the sound of music. Mozart was born in Salzburg, which is also where *The Sound of Music* was filmed.

32
Belgium

Belgium is the cultural centerpiece of the region, with beautiful castles and landscapes throughout. The country has a short history, gaining independence from Netherlands only in 1830, and having been occupied by Germany during both World Wars. Since World War II, however, Belgium has gained considerable prosperity and has become modern. It is a member of NATO and the European Union. With Brussels the center of the EU, the country is significant for all Europeans.

Economics and Opportunities

Belgium's economy is modern and private enterprise flourishes. The country has a key geographic location and a strong transportation network, which makes this small country an important linchpin for international organizations doing business in Europe. Most of the country's industry is located in the North. Belgium has few natural resources of its own, and therefore imports a great deal of raw materials. It is also a large exporter of finished goods, mostly to other countries in the European Union.

Greetings

Belgians speak either Flemish (a dialect of Dutch) in the North, or French in the South, with many speaking both languages. The northern Flemish may kiss cheeks three times, alternating each side; embracing, even between two men, is a common greeting. Handshaking should be done both upon greeting and departing, at both social and business meetings, although the shake is light and not prolonged. Women extend a hand first. If you wish to bring a gift, avoid chrysanthemums, as they signify death. Although gifts are appropriate at social events, gift giving is not common between business colleagues.

Avoid first names unless invited; address a person with an honorific and their family name. The appropriate honorific depends on whether you are in French-speaking or Flemish-speaking Belgium; in French-speaking areas, use "Monsieur" and "Madame"; in Flemish areas, use "Mijnheer" for men and "Mevrouw" for women.

While third-party introductions rule in many parts of Europe, you may introduce yourself in Belgium.

Food

Belgium is well known for its chocolate, so do not forget to visit a chocolatier while you are there. Belgian dishes are heavy, with meat (horse is a very common dish). Eating is Continental style. Breakfasts are usually quick and small, with lunch being the main meal of the day, taking at least an hour, and often two.

Tipping is standard at 10 percent.

Religion

About 75 percent of the population is Roman Catholic, and 25 percent Protestant, with Catholics dominating the South and Protestants the North.

Etiquette

Belgians are very serious about privacy, which one should be cautious to respect. Old-world European manners rule. Men should allow women to enter ahead of them and hold doors open. Dress is formal, with a standard dark suit and white shirt. Dress is conservative in most of the country, although in Brussels businesspeople are more stylish.

Bargaining and Deal-Making

There is some tension between the North (Flanders) and the South (Wallonia), and both regions have a degree of autonomy. As such, do not expect much cooperation between parties on opposite sides, and be aware that the customs of the Flemish and Walloons are distinct. Legally, the entire country is bilingual, so signage and documents must be both in French and Flemish. Punctuality is important, so ensure you are on time to all meetings and social events.

Belgians have a strong sense of family and belonging, and it may be difficult to make a personal connection while you are there. When meeting with your Belgian counterparts, make eye contact and maintain a sense of formality. Avoid emotional displays and keep your presentations straightforward.

Common Customs

- Belgium is a constitutional monarchy, and the king is highly respected. Avoid offering opinions about the monarchy, including negative opinions about the royal family.

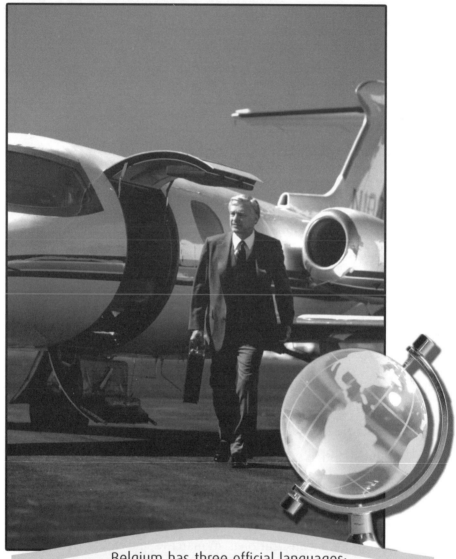

Belgium has three official languages:
Flemish (Dutch) in the north, French in the south, and
German in the east, spoken by a small minority.

33
Switzerland

A peaceful country, Switzerland has a strong reputation for its neutrality, which has helped establish its position as a world center for trade and commerce. The country's neutrality is honored by all major European powers, and the country has close ties with all of its neighbors. Switzerland joined the UN in 2002. The culture of Switzerland is unique and blends German, French, and Italian elements, although the German influence is particularly evident.

Economics and Opportunities

Switzerland enjoys a low rate of unemployment, a skilled labor force, and a stable, internationally competitive economy. Banking secrecy has long been a tradition in Switzerland, which makes it an international banking and financial center, although bank secrecy laws have been tempered to some degree in conformity with EU regulations.

Greetings

The Schwyzertütsch dialect of German is spoken by a majority of the population, although French is also widely used. Giving small gifts is appropriate, although red roses should be avoided, since they indicate a romantic interest. Handshaking is done with everyone present, both upon arriving and departing. The handshake is short and firm. When good friends meet, there may also be a short embrace or cheek kissing.

Avoid using first names unless you are invited to do so, in which case you should use the entire first name. First names are often two names that are hyphenated, so use both of them. Use honorifics when addressing people; in most of the German-speaking areas of the country, use "Herr" for men and "Frau" for women. "Fraulein" is only used to address young girls. In the French-speaking part of the country, address people as "Monsieur" and "Madame."

Food

It is a polite compliment to take additional helpings of food. Smoking at the table is not done. Fondue is a popular dish; a quaint custom is that, if you drop your bread into the fondue, you must pay for the wine. At a restaurant, it is common for strangers to be seated at your table if you have unused seats. Tipping at restaurants is standard at 10 percent.

The main meal of the day is lunch, which usually lasts for two hours and has multiple courses.

Religion

Switzerland has a mix of religions, with Roman Catholics making up 42 percent of the population, and Protestants 35 percent. There is a small Muslim population as well, making up about 4 percent.

Etiquette

Style of dress is formal and conservative; even on informal occasions, be neat and clean.

Bargaining and Deal-Making

Punctuality is extremely important in Switzerland; it is vital to be on time for all business and social events. July and August are vacation times, so these months are best avoided if you want to transact any business.

Common Customs

- Swiss tend to be humble; overt displays of wealth are unacceptable.

- The Swiss are very tidy; avoid littering at all costs.

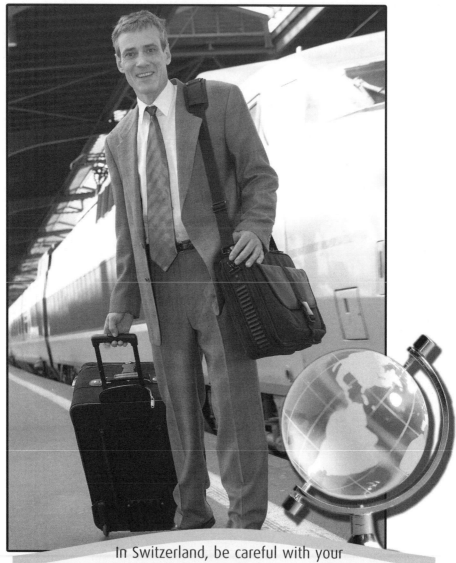

In Switzerland, be careful with your math. Traditional American commas are written as periods, and periods are written as commas. Also, when counting on your fingers, one is your thumb.

34
Greece

The country of Greece grew during the first half of the 20th century, as it added several islands and territories. During the Second World War, Greece was invaded by Italy and then Germany, and also suffered a civil war between Communists and supporters of the King. After defeating the Communists, Greece joined NATO in 1952 and the EC (now the EU) in 1981. There still exists a Greek royalty, although the monarchy has no influence or authority.

Economics and Opportunities

Despite a history that includes a Communist insurgency and a military dictatorship, Greece now enjoys a strong capitalist economy. Tourism is a major industry in Greece, making up 15 percent of the country's GDP. About 20 percent of the workforce is immigrant (mostly in unskilled and agricultural jobs), and the country is a beneficiary of European Union aid programs. Nonetheless, the economy is growing, due in part to increases in the availability of credit and consumer spending. Although public debt, inflation, and unemployment are above average for countries in the Eurozone, these statistics are falling. Government spending is high, the public sector is large, and the current government is seeking to reform labor and pension systems. However, there is much opposition from the country's powerful labor unions.

Greetings

Greek is the official language, although in the major cities you may hear French and English. Use the honorifics ("Keereeoss" for men and "Keereeah" for women) with the last name. Although most residents are ethnic Greeks, there are a growing number of migrants from other nearby countries, including Albania, Poland, Russia, Ukraine, Bulgaria, and Romania.

Wait to be introduced before speaking. Third-party introductions are preferred. Use titles when addressing someone, especially if they are of higher rank or are older. Handshaking is common, as is embracing and cheek kissing. The handshake is firm and with eye contact.

If you are invited to a home, take a small gift, such as flowers, candy, or a cake.

Food

The largest meal of the day is served midday, between noon and 2 p.m. Greek food is a delight. Appetizers are commonly served at restaurants, and you may eat them with your fingers; otherwise, it is appropriate to use utensils. Cigarette smoking during meals is common.

Breakfast is usually a quick, Continental breakfast served with Greek coffee, which is served Turkish style and is very rich, thick, and sweetened. Remember, there are thick grounds at the bottom of the cup.

Meals may be served with beer or wine. The Greek specialty wine, retsina, is made by adding pine resin for additional flavor. It is unusual and may take some getting used to. After an evening meal, you may be served ouzo, which is a strong, anise-flavored alcoholic liqueur. It is traditional to drink a shot of ouzo in one gulp, although you are not obligated to do so. Be aware that drinking ouzo in Greece is like drinking vodka in Russia or

Poland; as soon as your glass is empty it is automatically refilled.

Tipping is standard at 10 percent.

Religion

Ninety-eight percent of residents are Greek Orthodox, and there is a small Turkish-speaking Muslim community in northeastern Greece. Greek Orthodoxy is the official state religion, and is prevalent throughout Greek life.

Etiquette

The American-style wave of the hand is an insult in Greece. Never wave to get somebody's attention, or to greet anybody. Dress is formal and stylish.

Bargaining and Deal-Making

Greeks are punctual, so be on time to business and social events. Women here take on very different roles, and while they are present in the workplace, the higher echelons of business are male dominated.

When responding in the affirmative, do not say "okay." The similar-sounding Greek word, "oh-kee," means "no."

Common Customs

- Nodding the head means "no"; turning the head from side to side means "yes."

- A person who is older than you should be shown respect at all times.

- Avoid excessive admiration or compliments on specific objects, as the host may feel obligated to give it to you as a gift.

- Avoid waving your hand with your palm outward, because it may be mistaken for a rude gesture. Similarly, do not ever use the "okay" sign of joining the index finger and thumb; this may be construed as a sexual gesture.

- The Eastern Orthodox calendar is observed.

35
Portugal

Portugal has a long seafaring tradition, and was a major world power during the 15th and 16th centuries. Its status declined in the centuries after, however. For most of the 20th century, following the 1910 revolution, it was governed by repressive governments. A left-wing military coup installed some reforms in 1974. Portugal was a founding member of NATO and joined the European Union in 1986.

Economics and Opportunities

Portugal has in recent decades become quite diversified in its economy, and has focused on service-based industries, especially since joining the European Union. More recently, many state-run firms have been privatized; the economy, including the financial and telecom sectors, has been liberalized. Economic growth was above average for a few years during the 1990s, but fell between 2001 and 2006; the GDP per capita today is two-thirds of the EU25 average. Portugal has a substandard education system, which has hindered further growth and productivity. The government faces a tough battle in becoming economically competitive while still meeting EU mandates.

Greetings

Men may greet each other with a firm and enthusiastic handshake, which may be followed by an embrace and even backslapping. Women may greet each other with a kiss on both cheeks. Gift giving is less common than in other countries; it is more common to express appreciation by inviting hosts to a restaurant. The Portuguese language is not Spanish. Many Portuguese speak English or French as a second language.

Use honorifics when addressing someone ("Senhor" for men, "Senhora" for women). It is best to wait for an introduction, rather than introducing yourself.

Food

Tipping is practiced (10 percent), and it is often included in the check. Eating is Continental style, and very few items are eaten with the hands. Breakfast is Continental, a quick affair with coffee and rolls. Lunch is the main meal of the day and may last for two or more hours. There is much use of fresh seafood. There may be different wines served with each course at dinner, and port should be expected. Be aware that port wine is fortified, and therefore stronger than ordinary wine. It is very flavorful and sweet, and usually served after a meal.

Religion

Over 90 percent of the population is Roman Catholic, with a Protestant minority.

Etiquette

It is common to arrive at social events a half hour late. For business meetings, however, you should be on time, although others may be late. Dress is formal on both business and social occasions. Men wear stylish suits with accessories to work; women also dress stylishly and accessorize tastefully.

Bargaining and Deal-Making

Everything tends to close down between noon and 3 p.m., so avoid making appointments during this time. Business takes longer here than in many other countries, and Portuguese are not eager to adapt to new business practices. You may have to work hard to gain the trust and acceptance of your Portuguese counterparts. Foreign businesswomen have to work even harder to gain acceptance; although there are women in business, the highest levels are male dominated. The Portuguese workplace is strictly governed by hierarchy.

Common Customs

- Portuguese enjoy taking life slowly, and are rarely in a hurry.

- If you wish to call someone to you, do not use fingers to gesture. Rather, wave to them with your whole hand, palm down.

- To compliment your host on dinner, there is a unique gesture. First, kiss the side of your index finger, and then squeeze your earlobe with your thumb and index finger.

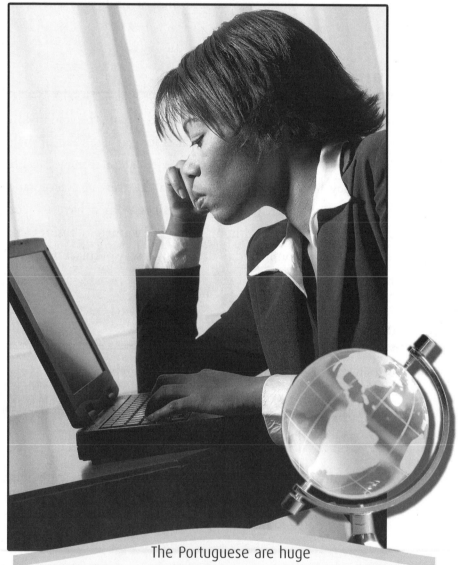

The Portuguese are huge football fans, but football in Portugal is not the same as football in America. Football in Portugal is really what Americans refer to as soccer.

36
Sweden

Compared to other EU countries, Sweden is affordable. Swedes take pride in keeping their cities and towns spotless. Stockholm, the capital city, is large and progressive, with many small, residential communities. A far cry from the Viking days of old, Sweden has not been at war in about 200 years, having been neutral during both world wars.

Economics and Opportunities

Sweden's economy is a mix of capitalism and welfare socialism. Due in part to its neutrality, Sweden has created a good standard of living for all of its residents. The labor force is highly skilled, and thanks to generous sick-leave benefits, Swedish workers call in sick more than in any other European country. Sweden enjoys a brisk foreign trade, and private firms account for 90 percent of industrial output. Sweden is strong in technology, and a major center for startup technology firms.

Greetings

English is a widely spoken second language. Gift giving is common, and flowers are appropriate when visiting a Swedish home. The handshake is the standard greeting, with a firmer grip than in some European countries, with one or two pumps. Touching outside of the handshake is not customary. Shake hands both on arriving and departing, and with every person present. Use honorifics when addressing someone, saying "Herr" for men and "Fru" for women. "Fru" applies to both married and single women, and is the equivalent of the English "Ms." First names may be used after becoming acquainted. It is best to wait to be introduced, and it is best to wait to be seated at a gathering.

Some Swedes may refer to you in the third person, as in, "Would Mr. Blacharski like a beer?"

Food

Toasting is common and formal. Do not toast to the host or someone senior to you, unless they have toasted to you first. Do not take a drink until your host has said "skoal." The proper way to toast is to move your glass upward to eye level, look the other person in the eyes, and say "skoal"; then take a drink, wave your glass toward the other person, and set it down. The person to the host's left should make a toast during dessert. The person to the host's right should make a short speech.

Eating is Continental style, and eating with your fingers should be avoided, even with sandwiches or other traditional finger foods. In restaurants, tipping is standard at 10 percent.

Lunch is extensive and includes beer or wine. The food is hearty, with a heavy use of seafood, including fresh, smoked, and pickled fish. A delightful culinary tradition is the smorgasbord. Avoid overfilling your plate, but feel free to make several trips back to the smorgasbord, and try to sample everything.

Religion

Sweden is about 87 percent Lutheran, with minorities of Roman Catholic, Orthodox, Baptist, Muslim, Jewish, and Buddhist.

Etiquette

Be on time to social events. When you are sitting, you may cross your legs with one knee atop the other one, but resting one ankle on top of the other knee is rude. Some Swedes prefer that you remove your shoes when you enter their home. Gift giving is not usually practiced in business, but a small gift on social occasions or when visiting someone's home is appropriate. Swedes, and most Scandinavians, are not outwardly emotional, so avoid aggression and speaking in a loud voice.

Bargaining and Deal-Making

Punctuality is very important in Sweden. Eye contact is a good idea when meeting with a business associate. If your associate crosses his or her arms, do not be alarmed. While many countries see this as a gesture of boredom or defensiveness, in Sweden it shows that the person is listening to you. Swedish business culture is not too concerned with hierarchy, and decisions are often made with the support and input of subordinates. Swedes tend to be egalitarian, and women in business are treated as equals.

Business dress is on the conservative side. Even in social situations, dress neatly and stylishly, avoiding the casual, American style of jeans, shorts, or tee-shirts.

Common Customs

- Swedes take cleanliness seriously and are proud of their beautiful environs; trash or litter is seldom found on the streets, even in the larger cities.

- Avoid planning for business in July, as it is vacation time; businesses are also often closed for the Christmas season.

37
Norway

Norway, with its beautiful and rugged landscape, gained its independence from Sweden in 1905, and maintains much of the same mentality as Sweden. Its cities are unclogged and clean, and the country is built on an economic blend of capitalism and socialism. Norway was neutral during both world wars, although it was occupied by Germany between 1940 and 1945. Norway joined NATO in 1949 and abandoned neutrality. Norway has not joined the European Union, rejecting their conditions for membership.

Discovery of oil and gas in adjacent waters in the late 1960s boosted Norway's economic fortunes. The current focus is on containing extensive welfare spending and planning for the time when petroleum reserves are depleted.

Economics and Opportunities

The system of "welfare capitalism" combines free market principles with government intervention, and the government still controls the lucrative petroleum sector and other natural resources. Some efforts at privatization are underway. Norway is a major oil exporter, with oil and gas making up a third of all exports.

As with Sweden, the Norwegian social welfare system is highly developed and relies on high taxes. However, Norway differs from Sweden in some ways, such as the greater sense of personal responsibility and individualism found among Norwegians.

Greetings

A short but firm handshake is customary. Other types of touching, such as cheek kissing and backslapping, are not common. Stand up when you are introduced to someone. Do not use a person's first name until a close relationship has been established. When visiting a Norwegian home, bring a small gift for the host.

The Norwegian language is similar to Swedish, with the same honorifics used: "Herr" for men and "Fru" for women. English is commonly spoken, especially in business and in the larger cities.

Food

Norwegian food is similar to that of other Scandinavian countries. Toasting is usually done at a meal or when sharing drinks. Toasting is done as it is in Sweden. Tipping is standard at 10 percent.

Dining customs and meals are similar to those found in Sweden. Norwegians also enjoy the smorgasbord, which should be approached as it is in Sweden. Regional delicacies may include whale or reindeer meat.

Religion

Most Norwegians are members of the Church of Norway (Lutheran), with minorities in other Christian denominations. There is also a small Muslim minority.

Etiquette

Rules for etiquette are similar to those found in Sweden. However, dress is less formal, and men can wear a sport coat with a tie. Women can wear dress slacks.

Bargaining and Deal-Making

As in Sweden, punctuality is vital in business. Avoid going to Norway on business during popular vacation times: July and August, and during the entire Easter season (the ten days after Palm Sunday). Like most of Scandinavia, Norway is egalitarian in nature, and women are treated as business equals. Business is somewhat less hierarchical, and input from all parties is valued.

The business lunch is common, but alcohol is usually not available until after 3:00 pm.

Common Customs

- It is common to be served aquavit, which is customarily drunk all at once.

- Norwegians do not usually work late; expect them to quit promptly at 4:00 pm.

- Norwegians tend to be soft-spoken; avoid loud speech and aggressive behavior.

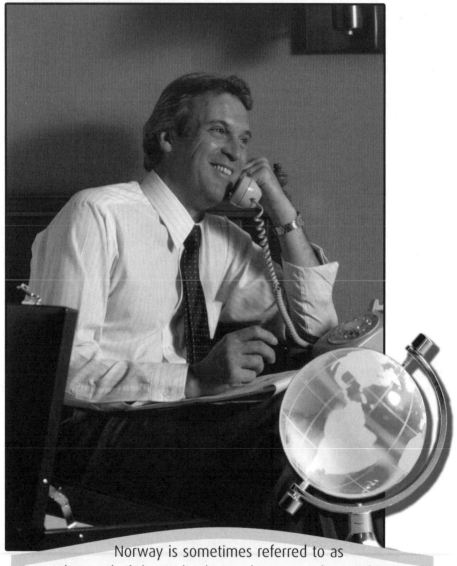

Norway is sometimes referred to as
The Land of the Midnight Sun because it lies so far to
the north that the sun does not set from mid-May to
the end of June.

38
Denmark

Like many other Scandinavian countries, Denmark has evolved from a Viking warrior nation into a peaceful and prosperous capitalist-socialist community. Denmark joined NATO in 1949 and the EEC (now the European Union) in 1973; however, Denmark has opted out of a few elements of the EU, including the European Economic and Monetary Union, European defense cooperation, and other justice-related and home affairs issues.

Economics and Opportunities

The Danish economy has expanded in recent years due to a strong consumer economy, a high level of confidence and a good level of exports. The market economy includes much small-scale industry and a high-tech agricultural sector. There is significant dependence on foreign trade as well, although Denmark has a healthy balance of payments surplus. The standard of living for Danes is among the highest in the world. Welfare measures are extensive, although there is a strong sense of personal responsibility. There are attempts to privatize state assets and streamline the bureaucracy. Unemployment is low, although long-term concern has been expressed over the declining worker/retiree ratio.

Greetings

Handshakes are common for both men and women. The handshake is firm and short, without lingering. Stand up before shaking hands, and when in mixed company, shake hands with the women first. Although Danish is the primary language (similar to Swedish), English is widely spoken. The same honorifics are used in Denmark as in Sweden: "Herr" for men and "Fru" for women. Start by addressing people by the appropriate honorific followed by their last name; you may be invited to use a person's first name after a short time.

Food

Toasting is common and formal. Do not toast to the host or a senior-ranking person, unless they have toasted to you first. Do not drink until your host has said "skoal." Toasting etiquette is the same as in Sweden.

Although you may have enjoyed the smorgasbord in Sweden and other parts of Scandinavia, this is not a custom in Denmark.

Formal dinners are more common in Denmark than in the U.S., and you may be invited to dinner at your host's home.

Tipping in restaurants is usually included in the total price, so no extra tip is necessary.

Religion

The vast majority of Danes (95 percent) are Evangelical Lutherans. Protestants and Roman Catholics make up 3 percent of the population, and Muslims the final 2 percent.

Etiquette

Old-world manners are important. Men hold the door and allow a woman to go first. However, Danes tend to be less formal than other Scandinavians. Dress is neat and clean, but less formal than in Sweden. As in Norway, it is acceptable for a man to wear a sport coat and tie, and women can wear dress slacks. As in other Scandinavian countries, avoid talking too loudly or exhibiting aggressive behavior.

Bargaining and Deal-Making

Avoid doing business during the July and August vacation season. Punctuality is of vital importance in business. As in Norway and Sweden, women are treated as equals in business. Danish businesses are less hierarchical, and input is accepted from all parties.

Common Customs

- Saint Hans' Eve on June 23 is a special holiday with large feasts.

- The "okay" sign, formed by making a circle with your forefinger and thumb, and extending your remaining fingers, is a rude gesture in Denmark.

- Danes stop working promptly at the end of the workday.

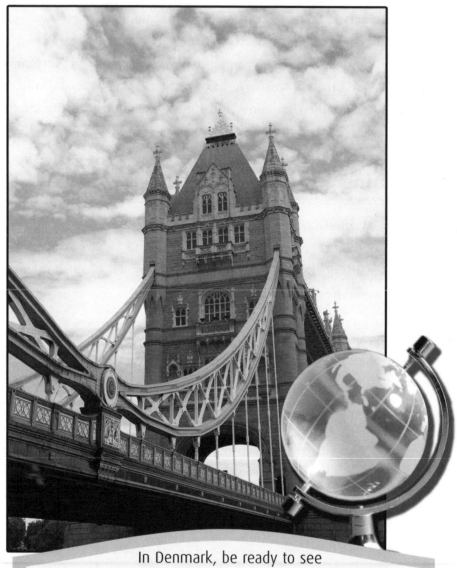

In Denmark, be ready to see many flags. The Danes raise flags outside their windows to announce their birthdays.

Section 6

The World In-Depth: The Americas

In most Latin American countries, punctuality is not a virtue, and businesspeople tend to be a little more relaxed than businesspeople from the United States. Meetings seldom start on time, so it is acceptable to arrive late. Business is also a social practice, so do not begin business negotiations too soon. Start a business meeting with friendly small talk. Your associates want to know you as a person and feel comfortable with you before talking business.

Another common factor among all Latin American countries is the greeting. Handshakes are common, but you may also see men greeting one another with an abrazo, a vigorous hug, which may involve backslapping. This is not appropriate when meeting someone for the first time, but can be expected once a close relationship has been established.

Latin Americans tend to have many names. It is customary to have a hyphenated last name that comes from the father's and mother's names, with the father's name coming first. You do not need to address a person with both names; use the honorific plus the first of the two hyphenated names. The exception is Brazil, where the order is reversed, and the mother's name comes first in the hyphenation.

A significant amount of business transpires between North and South America.

Traveler's Tip

People in the United States often refer to themselves simply as "Americans," but when doing business in South American countries, one must realize that America is not just the United States, but the whole of two continents.

39
Mexico

Mexico has a long history, and became independent early in the 19th century after Spanish rule. Economic and social upheaval continues today, and there are few advancement opportunities for much of the population. The southern states are particularly impoverished. The 2000 elections put an opposition candidate in party for the first time since the revolution.

Economics and Opportunities

Following the devaluation of the peso in 1994, Mexico went into a major recession. Currently, the nation has been making an economic recovery, although there are continuing economic concerns. Much of the population is unemployed or underemployed, and wages are low. Although there is some modern industry, much of it is still out of date. Most of the economy is run by the private sector, and there is free market competition in many public areas, such as seaports, railroads, and telecommunications. The 1994 NAFTA agreement has tripled trade with the U.S. and Canada, and Mexico has free trade agreements with several other countries as well.

Greetings

Handshakes are the customary greeting; it is firm, with only one quick pump. Women may greet one another with a kiss on the cheek, and two males that have known each other for a long time may greet each other with a vigorous embrace (the abrazo). Gift giving is not common, but appreciated. Be cautious when giving flowers, however, since the colors of flowers have specific meanings. English is widely spoken, but knowing a little Spanish is appreciated and shows respect for your host. The Spanish in Mexico differs from the language spoken in Spain. Use the honorific "Señor" for men and "Señora" for women, plus the family name. "Señorita" is used for unmarried women. The honorific "Don" is a title of respect, and used before a person's first name.

Waiting for a third-party introduction is best. When introductions are made, the women are introduced first, and then the highest-ranking or eldest men in the group.

Food

Some people call a waitperson by making a kissing sound with their lips, or by making a sort of hissing sound. Although common, these are impolite; instead, raise your arm and gesture for the waitperson to come to your table.

The standard alcohol is tequila. Shots are often taken with salt and a wedge of lime. First, put a small amount of salt on the back of your hand, between the area between your forefinger and thumb; then quickly lick the salt, take a drink of the tequila, and suck on the lime.

Tipping in restaurants is expected and standard at about 10 percent. Breakfast is usually quick — Continental style with rolls and strong coffee. Lunch is the main meal of the day and may last for several hours. Lunches consist of multiple courses, starting with soup, then a meat dish, rice and beans, and a dessert. Tortillas accompany most meals, and you can use them to scoop up food on your plate. Mexican food in general is varied, spicy, and more diverse than the usual fare served in American Mexican restaurants. Dining habits are more European, with utensils used Continental style.

Drinks may include mescal, which is similar to tequila and made from a cactus. Traditionally, a mescal bottle has a worm in it, and the worm is eaten by whoever is served the last drink out of the bottle.

Religion

Mexico is 89 percent Roman Catholic, with a 6 percent Protestant minority.

Etiquette

Mexicans are gracious and easy-going, both in social and business environments. If you are invited to a social occasion, you should arrive an hour late. Business meetings also seldom start on time, but you should be prompt. Dress for business is conservative and stylish, with accessories to highlight the outfit. Fashions are more like Continental Europe (France and Italy, for example) than the United States.

Traveler's Tip

A word about toilets: Plumbing in Mexico is very limited. Water pressure is weak, and the toilets accommodate very little refuse before clogging and overflowing. There is a small can next to the toilet; this is where your used tissue should be discarded.

Business gift-giving is not expected, though a small gift from your home country is pleasantly regarded. Gifts should be given for social events, however.

Bargaining and Deal-Making

Traffic in the major cities, especially Mexico City, is congested. Attempt to be on time to meetings, though be aware that traffic can make this impossible. Although Mexicans are easy-going and friendly, there is a formal business manner which should be followed. There are women in business throughout Mexico, but less so at the higher levels of management. It is important to cultivate a personal relationship with your Mexican counterparts; they want to get to know you and gain your trust before getting down to the details.

Common Customs

- The siesta takes place at midday and lasts two or three hours.

- When shopping at small shops, it is common to bargain over the price.

40
Argentina

The many immigrants from Europe have given Argentina a European style, which helps make traveling in Argentina easy for Europeans and North Americans. Independent from Spain since 1816, Argentina has suffered dramatic political conflicts, alternating between civilian and military authoritarian rules. Democracy has been in place since 1983, although challenges continue.

Economics and Opportunities

Although once a wealthy a country, Argentina has been steeped in economic disaster for the past century. The economic crisis of 2001 led to much protest throughout the country, and recovery continues to this day. However, Argentina paid off its IMF (International Monetary Fund) obligations in 2006, marking a new era in the country's economy. Argentina is rich in natural resources, and the population is well educated. The large agricultural sector is oriented toward export, and there is a rich industrial base as well.

Greetings

It is common for men to hug (the abrazo), especially if they are well acquainted or have not seen each other for a long time. Women may kiss each other on the cheek. Upon meeting for the first time, however, a handshake is best. The handshake may be warm and lingering, sometimes accompanied by a touch on the elbow. Small gifts may be appropriate.

The primary language is Spanish (spoken with a different accent than in Mexico), with many residents also speaking Italian. English is spoken to a lesser extent; it is more common among educated businesspeople. Use the honorific "Señor" for men and "Señora" for women, plus the family name. "Señorita" is used for unmarried women. "Don" is a title of respect, and used before a person's first name.

Waiting for a third-party introduction is best. When introductions are made, women are introduced first, and then the highest-ranking or eldest men in the group.

Food

Argentine red wine is a source of national pride, and is served commonly with meals. There are toasts during and after the meal. For a simple toast, or as response to a toast, you may raise your glass and say "salud." Cuisine is meat-intensive, and many forms of beef are served often.

Tipping at restaurants is standard at 10 percent. The breakfast is Continental, with rolls and strong coffee; lunch is the main meal of the day. After lunch, there may be a tea hour, with tea or coffee, and sweets. Dinner is late in the evening, and light.

Religion

Although the country is 92 percent Roman Catholic, less than 20 percent are practicing. There are small Protestant and Jewish minorities as well.

Etiquette

Men cross their legs with one knee over the other, not with an ankle over the knee. Women do not cross their legs. Etiquette is similar to that in Mexico. Arrive an hour late to social occasions, but be prompt to business meetings. Dress for business is conservative and stylish.

Bargaining and Deal-Making

Eye contact is important during negotiations. Argentina tends to be hierarchical. Meetings may be scheduled late, even into the evening. There is a large bureaucracy, and business deals may require attention to details that may seem unnecessary to an outsider.

Common Customs

- Toilets may not accommodate discarded tissue; use the basket next to the toilet.

- Smoking is prevalent. If you smoke, offer a cigarette to others before lighting your own.

Argentina is actually bi-continental, with portions of the Republic stretching into Antarctica. Argentina is also the eighth largest country in the world.

41
Colombia

There is a long-standing conflict between the government, insurgent groups, and paramilitary groups, most of which are funded by the drug trade. Although insurgent attacks against civilians continue, the violence has been decreasing since 2002. The official government is reasserting control, but there is still violence throughout the country.

Geographically, it may take travelers one or two days to adjust to the very high altitude.

Economics and Opportunities

The past few years have seen some growth in Colombia's still-fragile economy. Unemployment remains high and is a focal point of the current government. Economic rescue measures include austerity programs, reduction of public debt, and export-oriented growth strategies. These new measures, along with a free trade agreement with the U.S., are helping renew confidence in the economy.

The drug trade is ominous, and has its own hierarchy, militia, and set of rules, operating independently from the official government. There are occasional episodes of civil unrest; check with the State Department for travel warnings before traveling to Colombia.

Greetings

Use the title or honorific, along with last name, when addressing someone. Handshakes are customary upon greeting for the first time. Shake hands with everyone in the group, both upon greeting and departing. Between men, the handshake may also be accompanied by the abrazo. Women do not shake hands with other women; instead, they grab each other's forearms lightly. When you go to someone's home, send a small gift before you arrive. As in Mexico, use "Señor" for men and "Señora" for women, plus the family name. "Señorita" is used for unmarried women. "Don" is also used, as in other Spanish-speaking countries.

Colombian Spanish is much closer to that spoken in Spain than in other South American countries. Wealthier families often send children to school in Europe or the United States; and the educated class is likely to speak some English.

Food

Expect an abundance of rich, strong coffee before and during business. However, the coffee may not be as strong as it is in some other Latin American countries. Eating is Continental style, with the fork in the left hand. If wine is served at dinner, there will be toasts; raise your glass, say "salud," and take a sip. Breakfast is small, Continental style, with rolls and coffee. Lunch is the main meal of the day, and may last for two hours or more, including multiple courses, dessert, and black coffee. Unlike in Mexico, the tortilla is not a standard addition to Colombian meals, and the food is not as spicy. The liquor of choice is called aguardiente, which is made from sugarcane. It is very strong, and drunk straight, in shots.

Tipping in restaurants is standard at 10 percent, and this is often included in the bill. You may see people summoning the waitstaff by clapping their hands, or by making a hissing sound; but this is considered rude. The best way to call on a waitperson is to make eye contact. The host or the senior person pays.

Religion

Colombia is 90 percent Roman Catholic.

Etiquette

Colombia is one of the more formal cultures in South America. When meeting someone for the first time, wait for a third-party introduction. Family is very important, and personal questions about family are common in business and social situations. If you are invited to a social occasion, arrive an hour late. Business meetings also seldom start on time, but you should be prompt. Style of dress is important, and you will be assessed on your appearance. Dress conservatively, with style, and feel free to accessorize tastefully. Your clothing should be of high quality, and your grooming impeccable at all times. Clothing styles are more European in design.

Gift giving is common, but gifts are not opened when they are presented.

Bargaining and Deal-Making

Negotiations tend to go slowly, but are relaxed. Punctuality is not as important. There are women in business, but usually not at the highest levels of decision-making. Foreign businesswomen must take extra care to present themselves professionally, so as to gain respect. Business is hierarchical, with strict protocols for doing business and sharing information.

Traveler's Tip

Be cautious at all times and be aware that violent crime is prevalent, especially outside the main tourist areas.

Common Customs

- Avoid making the Western "okay" signal by joining the first finger to the thumb in a circle. Here, when that sign is made close to the face, it is a means of calling somebody a homosexual.

- Toilets may not accommodate discarded tissue; use the basket next to the toilet.

- Smoking is prevalent. If you smoke, offer a cigarette to others around you before you light up.

42
Peru

Peru is an ancient country, once the center of the Inca Empire. The country was declared independent from Spain in 1821. After several years of military rule, democratic leadership regained power in 1980. An insurgency continues, however, and there is some violence. Remember that Peru is geographically at very high altitude, and it may take you a day or two to become acclimated.

Economics and Opportunities

Much economic recovery occurred after the 1990 election of President Fujimori, who was ousted in 2000. Peru's economy is varied and dependent on many resources which are, in turn, dependent on the fluctuating world market. There has been positive economic performance since 2002, although unemployment and poverty are still high.

Greetings

Peruvians tend to be formal yet conversational. Shake hands upon meeting and departing. Men may greet their familiars with the traditional abrazo. Familiar women greet each other with a kiss on the cheek. A small gift is appropriate, and flowers are welcome. Spanish is the main language, with some Indian languages also being spoken. Use "Señor" for men and "Señora" for women, plus the family name. "Señorita" is used for unmarried women. "Don" is also used.

Peruvians are very gracious hosts, eager to welcome you into their country.

Food

Dinner starts late, usually around 9 p.m. It is appropriate to arrive up to 30 minutes late. Tipping is standard at 10 percent. In Peru, people take the equivalent of the British tea, a light meal between lunch and dinner. If you are invited to dinner at someone's home, expect large quantities of food.

A drink called "mate de coca" is very common. It is a tea made of coca leaves, which is legal in Peru. Be aware that processed coca products (cocaine) are still illegal, and the penalties for using or possessing them are severe.

Religion

Peru is 81 percent Roman Catholic.

Etiquette

You may sometimes observe a man walking with a woman, with her hand under her bent elbow. When sitting, you may cross your legs at the knees, but not ankle-on-knee. Smoking is common; remember to offer a cigarette to others near you.

If you go to rural areas, be careful about photography. In some Indian villages, photography, especially of children, is not appreciated; there is a lingering superstition that photography robs people of their souls. Dress conservatively, with style, and accessorize tastefully. Clothing should be high quality, and grooming impeccable. Clothing styles are more European.

Bargaining and Deal-Making

Strict punctuality is not followed, and it is common for people to show up for meetings or social events a half hour late. If you are invited to a bullfight, however, you must be prompt. Direct eye contact is important. Society is hierarchical. There are women in business, but seldom at the highest levels. Gift giving is appropriate in business and on social occasions.

Common Customs

- If you want to photograph the indigenous people, ask permission first, and offer a tip for the favor.

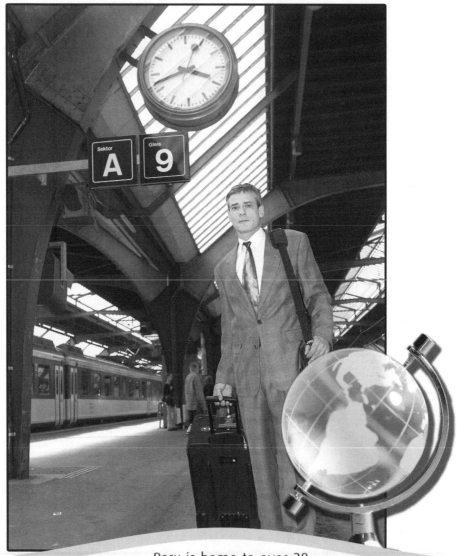

Peru is home to over 28
different climates, making it one of the top five
biggest biodiversities in the world.

43
Chile

Chile was under Inca rule before the Spanish took over in the 16th century. Independence was achieved in 1818. The Marxist government of Salvador Allende was overthrown in 1973 by a military coup, after which Augusto Pinochet was elected president. Today the country is stable and democratic.

Like Peru and Colombia, Chile is at high altitude.

Economics and Opportunities

Chile is an attractive destination for foreign direct investment. Since the 1980s, Chile has enjoyed steady growth and reduced poverty. The country's market economy is based largely on foreign trade. The country has a strong reputation for sound financial institutions and economic policies. Inflation and unemployment are relatively low, as compared to neighboring nations. Chile signed a free trade agreement with the U.S. in 2004, and similar agreements with China and India in 2006.

Greetings

A handshake is standard upon greeting, and you may also receive a kiss on your right cheek. The handshake for men is firm but brief, with a single pump; it is softer with women. Males that are well acquainted may greet each other with the abrazo, and women that are well acquainted kiss each other on the cheek. There is less attention to titles in Chile; for example, a Ph.D. is not referred to as "Doctor." "Señor" is used for men, and "Señora" for women. "Señorita" is used for unmarried women. Use "Don" as you do in other Spanish-speaking countries.

Food

Avoid eating with your fingers. Pour wine with your right hand. Chilean wines are well reputed throughout South America, and there may be wine with dinner. Toasting is customary; raise your glass and say "salud" before taking a sip. Chilean food uses much seafood. Lunch is the main meal of the day, and dinner is usually served late in the evening. Breakfast is usually a large meal as well, often with meat. Between lunch and dinner, Chileans have the equivalent of the British tea, with breads and sweets served with tea or coffee.

In restaurants, unlike in much of the rest of South America, tables are private, and strangers are not seated together.

Religion

Chile is 89 percent Roman Catholic and 11 percent Protestant. There is a very small Jewish minority.

Etiquette

It is customary for a man to stand when a woman enters the room. If the woman offers her hand, the man should shake it gently. Women do not need to stand when another woman enters. In Chile, manners and etiquette are similar to Great Britain; the are reserved and formal. It is inappropriate to be overly aggressive or loud. Style of dress is conservative, without much attention to fashion. A plain black suit without accessories is appropriate.

Bargaining and Deal-Making

It is best not to dive right into business negotiations. Start with small talk and familiar conversation. Strong eye contact is normal. Businesspeople tend to be rather conservative and direct. There is a reputation for honesty both in business and government, so although token gifts may be appropriate, bribes are not. There is also a strong focus on hierarchy, and the bureaucracy may take effort and patience. There are women in business, and the authority of foreign businesswomen is accepted easily, perhaps more so than in other South American countries.

Common Customs

- Unlike much of South America, it is not customary to bargain over prices in shops.

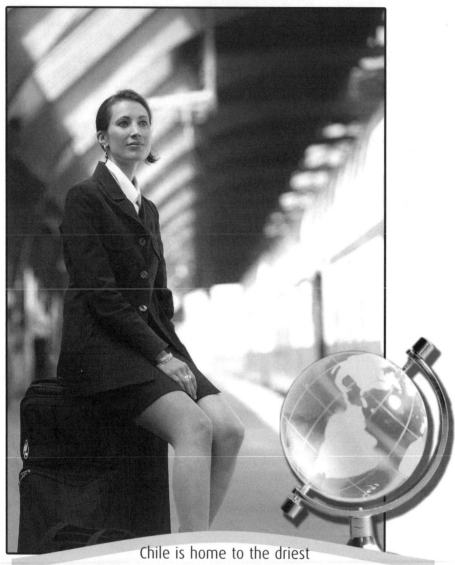

Chile is home to the driest place on earth — the Atacama Desert. Here you will find virtually no vegetation. There are parts of this desert where it has never rained.

44
Brazil

Formerly ruled by Portugal, Brazil gained independence in 1822. The largest country in South America, their military ceded power to civilians in 1985. The country has significant natural resources, and has become a regional economic leader. Brazil is culturally unique; it is not Hispanic, but is rather derived from a combination of Portuguese and African roots.

Economics and Opportunities

Brazil's economy is growing in the world marketplace, though poverty and an uneven distribution of income remain problematic. Between 2001 and 2003, wages shrank even as the economy grew; however, since 2004, continued growth has resulted in increased real wages.

Greetings

Portuguese is the primary language in Brazil. Many wealthier people have studied at universities in the United States, so English is widely spoken. Spanish, although not the primary language, is usually understood.

Shake hands with everyone in the room both upon greeting and departing. The handshake is warm and prolonged; between males, there may also be an embrace. Women may greet each other by making a kissing gesture toward the other woman's cheek without touching. If you go to someone's home, send flowers the next day with a thank-you note. Remember, never send purple flowers; they are bad luck.

In Brazil, last names combine the father's name and the mother's name; however, unlike most Latin American countries, the mother's name comes first. It is not necessary to address someone by both names; using the first of the two last names is acceptable. Address people with an honorific ("Senhor" for men and "Senhora" for women) and the last name. After you have become well acquainted, you may use the honorific followed by the first name (for example, "Senhor Dan"). Titles are very important, and a person who holds a university degree of any kind may be referred to as "Doctor."

Food

Brazilian food is very flavorful and somewhat spicy. Brazilians are proud of their coffee, which is quite strong and robust, and often it is included in business and social meetings. During dinner, do not pick up food with your fingers. Use a knife to cut your food, and never a fork. Smoking is not done during the meal, but is permitted afterwards. Ask first, and offer a cigarette to other parties near you. Wine is common with dinner, as is toasting. When a toast is offered, say "salude" and raise your glass.

Tipping is standard in restaurants at 10 percent. Breakfast may be a Continental style with coffee and rolls, and lunch is the biggest meal of the day. It may last several hours and include multiple courses.

Religion

Brazil is 74 percent Roman Catholic and 15 percent Protestant, with small minorities of Bantu/voodoo and other religions.

Etiquette

Avoid crossing your legs with an ankle over the knee. Men may cross knee over knee, while women should only cross their ankles. Manners tend to be formal. Avoid gum chewing, keep your hands out of your pockets, and make eye contact. Arrive at social events an hour late.

Bargaining and Deal-Making

Promptness is not always a virtue in Brazil, and there is a casual attitude toward punctuality. It is standard practice to arrive ten or twenty minutes late. Business dress is formal and stylish, with accessories. Also, when doing business, start out slow, with small talk. Business meetings are somewhat social. Rules for eye contact are subtle. Sustained eye contact between peers is expected, but a person should not make prolonged eye contact with an elder or superior. Business transactions may take a long time to complete, and you have to establish trust with your counterparts.

Traveler's Tip

Although you should not try to do any business during Carnival, it is an event that should not be missed.

Common Customs

- If you are in someone's home, it is appropriate to pay attention to his or her children.

- Avoid giving the Western-style "okay" sign by making a circle with your thumb and forefinger; in Brazil, this is a rude gesture.

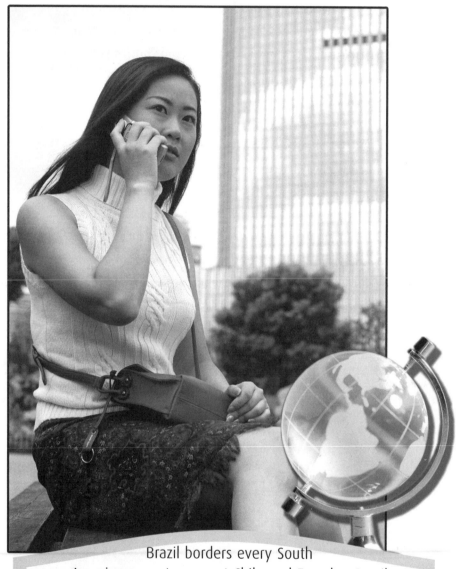

Brazil borders every South American country except Chile and Ecuador. Brazil is also the only country to have played in every World Cup Soccer Tournament.

45
Venezuela

Venezuela was under military rule for the first half of the 20th century, although democracy was established in 1959. President Hugo Chavez implemented democratic socialism in 1999, and while the intent was to relieve social problems, it has undermined stability and limited Venezuela's participation in the global economy. There is drug-related violence along the Colombian border.

Economics and Opportunities

Venezuela is very dependent on the petroleum industry, with oil revenues making up about 90 percent of its exports. There has been a boom in the consumer economy due to recent economic growth and raises in the minimum wage, along with better access to domestic credit. However, inflation remains high.

Greetings

The standard Latin American abrazo, following a warm handshake, is common among men who know each other well. Women may greet each other with a kiss on the cheek. Handshaking should be done with everyone in the room, both upon meeting and departing. The giving of small gifts is appropriate. Flowers are acceptable to give to a female business associate, or to bring to someone's home; especially orchids, the national flower. Spanish is the primary language, but English is widely spoken in business.

Food

During the meal, wait for everyone to be served before you start eating. Tipping in restaurants is standard at 10 percent. Breakfast is a quick, Continental-style affair. Lunch is the main meal of the day, and dinner is served late in the evening. Coffee is served on most occasions.

Religion

Venezuela is 96 percent Roman Catholic, with a small Protestant minority.

Etiquette

Business dress is somewhat conservative and very stylish, with attention to quality.

Bargaining and Deal-Making

Unlike many other Latin American countries, expect to start business right away. Promptness is more important here than in other countries in the region. Maintain eye contact during a business meeting. Although the telephone may be used when necessary, business is done in person.

Common Customs

- Use the basket next to the toilet for all refuse.

- Smoking is prevalent. If you smoke, offer a cigarette to others before lighting up.

46
Canada

Canada is steeped in French and British traditions. The country became self-governing in 1867, but has retained ties to the British crown, and the Queen is considered the chief of state. There are two distinct cultures in Canada: In the Western and Atlantic provinces, English is spoken and etiquette is similar to that of the United States; in French-speaking Quebec, however, etiquette is more European.

Economics and Opportunities

Canada has developed along with the United States. Its major economic challenges lie in meeting demands for improved healthcare and education. Canada has strong energy resources and is relatively affluent. Its market economy is similar to that of the United States; since World War II, it has seen significant growth in manufacturing and industry. An increase in trade resulted from the 1989 U.S.-Canada Free Trade Agreement and the 1994 North American Free Trade Agreement (NAFTA). The country enjoys an educated and skilled labor force with a modern infrastructure.

Greetings

English is the primary language of about 60 percent of Canadians, with French being the mother tongue of 23 percent of the population. Cantonese is widely spoken, due to large numbers of immigrants from Hong Kong.

Shake hands upon meeting and departing, as you do in the United States. There is not much casual touching, and usually no hugging or cheek kissing. In French-speaking Quebec, the greetings may be more personal, with a light embrace or kiss on the cheek. Address people with an honorific followed by the last name, unless you have been invited to use first names. Use "Mr." and "Mrs.," except in Quebec, where you say "Monsieur" and "Madame."

Food

Some Canadians eat Continental style, although many eat American style as well, with little attention paid to which hand is used to manipulate the fork.

Religion

The majority is divided between Roman Catholic and Protestant, with small Jewish and Eastern Orthodox minorities.

Etiquette

Canada is more like Great Britain than the United States when it comes to etiquette; when in doubt, follow British customs. If you are invited to someone's home, bring a small gift of flowers, but avoid white lilies — as in the United States, these are associated with death. Business gifts are acceptable, but should be simple and tasteful, and not overly expensive. It is polite for a man to stand when a woman enters the room. More attention to etiquette is paid in French-speaking Canada, with Continental manners being the standard.

While seated, all forms of leg-crossing are practiced, though the ankle-on-knee style is reserved for men.

Bargaining and Deal-Making

Punctuality is important, so be on time to all business meetings and social events. Direct eye contact is important during business meetings. Business practices and negotiations are similar to those in the United States. Business is conducted in a straightforward manner. In Quebec, all business and documentation is in French. Dress is similar to that of the United States, with standard business suits on the conservative side.

Traveler's Tip

Since the beginning of 2007, U.S. citizens must have a valid passport to travel to Canada.

Common Customs

- Canadians tend to be a bit reserved; avoid abundant hand gestures.

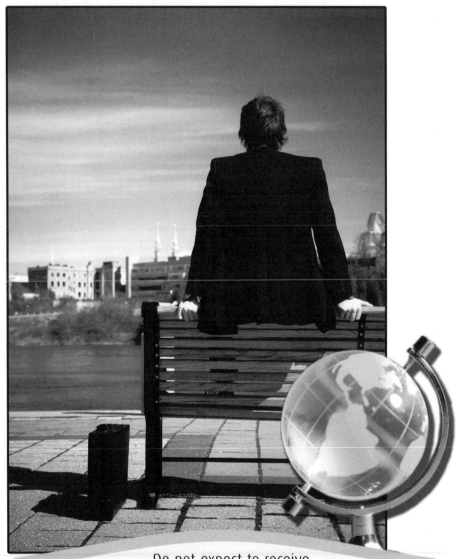

Do not expect to receive mail on Saturdays in Canada. The Canadians have not received mail on Saturdays for over 35 years.

Section 7

The World In-Depth: Africa & the Middle East

While I was visiting the home of an Israeli acquaintance, I noticed several beautiful calendars featuring exquisite reproductions of art from his homeland. I admired them, commenting on their beauty. He immediately took one from the wall and presented it to me as a gift. I was surprised, but later learned that this behavior is common among Middle Eastern people. Be careful what you admire, and come prepared with a gift of your own. Your Middle Eastern associate may have prepared a gift for you; be prepared to reciprocate. Gift giving is a high art in the Middle East, and there are many protocols. If you are invited to a Muslim home, bring high-quality gifts, but do not bring a gift for your colleague's wife.

There is a great division between men and women in most of the Middle East, and most business associates are men. If you are invited to a home in an Arab state, for example, your colleague's wife (or wives; they may have more than one) will not be part of the conversation, and will instead busy herself (or themselves) in the kitchen. If you are introduced to one or more wives, a polite nod is sufficient — do not shake hands or touch them in any way.

Traveler's Insight: Caught Off Guard in Africa

www.cameroonenterprise.com

Sammy Kale is the publisher of Cameroon Enterprise, which features information about entrepreneurial opportunities related to the Central African country of Cameroon.

Sammy often travels to Cameroon for business and is therefore well acquainted with the country. However, even an expert can be caught off guard. One of the most embarrassing moments during his travels to Cameroon, he recalls, was when he was called on to serve as chairperson of a particular event. "I was caught off guard when I was called to unveil the cake, food and drinks, not knowing I had to put money in a basket each time I went up. I later found out I was named chairperson of this event because the host expected someone from America to give huge amounts of money. It was somehow embarrassing when people noticed I did not give what they expected and a few people tried to challenge me by giving more than they could afford in a bid to push me to give more."

Traveler's Insight: Caught Off Guard in Africa

Sammy's experience also provides another valuable lesson: People in other countries often expect Americans to be wealthy.

As it is throughout Africa, food in Cameroon is a delightful experience. "Cameroon is the place to be if one is looking for good foods in Africa," Sammy says, and there is tremendous variety. "All 250 ethnic groups in Cameroon have a signature traditional dish and I enjoy the variety of local food in Cameroon. Dining with a business associate in Cameroon is always fun, it usually takes place at a nice restaurant or hotel cafeteria and the food is mouth-watering. You can never be dissatisfied with any local food in Cameroon."

If you are looking for the Golden Arches, however, you will be disappointed. "There is no such thing as fast food in Cameroon, that's what's great about it. Every meal is like home cooked and in most cases healthy." Food there is a great experience, and plays a big part in business negotiations. "In Cameroon, one can easily be carried away by the food and entertainment while dining with an associate to discuss business. Dining with an associate in Cameroon is a social and entertaining event, and business is not at the forefront during dinner or lunch."

When he started his magazine, he saw the importance of food in business first-hand. "I had a friend in Cameroon put together a team of writers and journalists to discuss the project and brainstrom some ideas. I was shocked when I got pictures of the event. They had at least five different meals prepared, with drinks, and it looked like a birthday party." This differs quite a bit from business lunches in America, Sammy observed. "In the United States having lunch or dinner with associates presents an opportunity to get out of the board room, create an environment where fresh and new ideas can flow and the meal is often light — either salad, a slice of pizza, or just coffee is enough."

When bargaining, food plays a big part in the negotiations, but it does not stop there. "Bargaining or deal making in Cameroon often requires you to cater to the other party, taking them out to dinner or lunch and in most cases giving cash as an incentive. People are very open to work with someone with an idea they believe has some financial benefit to them. The key is to pitch the financial rewards to the other party when trying to bargain or make a deal in Cameroon. In the United States, giving financial incentives or other gifts in a deal-making process comes after the deal has been made, while in Cameroon, an incentive is often required before the deal is made."

Kale also sees a big difference between the U.S. and Cameroon in terms of business gift-giving: "In the United States, a gift card to a restaurant is often good enough. In Cameroon, giving someone cash as a gift is the way to go."

Traveler's Insight: Caught Off Guard in Africa

Sammy was born in Cameroon, moved to the United States at age 15, and continues to travel between the countries regularly. He has thus observed many differences between Africa and the West. One thing that strikes him is the relative degree of informality in America. Whereas Americans often refer to him by his first name right away, his African business associates refer him to as "Mr. Kale."

Language is also important. He says, "I think it is very important to be able to speak and understand the local languages where you operate a business In Cameroon, French is the dominant language because the English-speaking part of Cameroon that I come from makes up just 20 percent of the population. I am trying to improve on my French as I venture into Cameroon. I highly recommend anyone venturing in Cameroon to understand some French or find a business partner who speaks and understands French."

African culture is unique and complex, with distinct traditions among most of the African countries. For example, although the standard handshake works in Cameroonian cities, "when you travel to tribal areas in Cameroon and meet traditional rulers, you are required to follow their traditional greetings." Sammy warns that the biggest mistake anyone can make before doing business in Africa is "not doing enough research before traveling there. Also, not having a general idea of the cultural differences might cause some headache, but Cameroonians are very welcoming people and a visitor will always find someone willing to help them."

47
Iran

Formerly known as Persia, Iran became an Islamic republic in 1979, after the monarchy was overthrown. The theocratic government places political authority in religious hands. Iran is considered a sponsor of terrorism, and is subject to U.S. sanctions and export restrictions.

Economics and Opportunities

Iran is overly reliant on the oil industry, and the country's statist economy is bloated and inefficient. The private sector is generally limited to small-scale operations and agriculture, with the state controlling the bulk of major industry. Although high oil prices have given Iran significant foreign exchange reserves, economic hardship continues, and inflation and unemployment are high.

Greetings

Greet someone with a light handshake followed by a slight bow. Among good friends, there may be cheek kissing. Avoid first names unless you are well acquainted. Address a person with an honorific or title, followed by the last name.

Food

The Iranian people are exceptionally gracious and hospitable. It is customary for an Iranian, when receiving a visitor, to offer tea and sweets. Iranian tea is strong and drunk while holding a sugar cube between the teeth. If you are invited to dinner, expect a lot of food and arrive hungry. Do not expect to see your associate's wife or wives present at dinner.

Religion

Iran is 98 percent Muslim, with the majority belonging to the Shi'a sect. There are very small Zoroastrian, Jewish, Christian, and Baha'i minorities.

Etiquette

While it is good form to be on time to all business meetings, it is less important to be prompt on social occasions. Bring a small gift if you are invited to an Iranian's home. It is common for people of the same sex to stand close together. However, affection between the sexes is never shown in public.

Bargaining and Deal-Making

Punctuality in business is important. Dealings are mostly with men.

Common Customs

- Muslims must stop to pray five times a day. It is not necessary for you to join them, but do not interrupt.

- Remove your shoes before entering someone's home.

48
South Africa

Post-apartheid South Africa is still defining itself. Despite economic progress, South Africa is subject to violence and suffers from a massive AIDS epidemic. There is a 21.5 percent HIV/AIDS adult prevalence rate, and over five million South Africans live with HIV or AIDS. South African culture mixes African tribal cultures with Dutch and English elements. The harsh era of colonialism was recent enough to be remembered by many people, so the topic should be approached with caution. The region of South Africa is made up of several countries: Madagascar, Malawi, Mozambique, Zambia, and Zimbabwe are in the north; Botswana, Namibia, Lesotho, the country of South Africa, and Swaziland are in the south.

Economics and Opportunities

South Africa is the greatest industrial power on the African continent, and one of the greatest success stories as well. Economically, South Africa is an emerging market with many natural resources, and a well-developed infrastructure. It has one of the ten largest stock exchanges in the world, and efficient distribution networks to major urban centers. However, unemployment remains very high. The country's economic policy is geared toward controlling inflation and liberalizing trade to increase employment and the average household income. Education is limited, although it is valued highly by South Africans. Early schooling is done in local dialects, but primary classes are taught in English. Afrikaans is taught as a second language in school.

Greetings

English is the official language used in government and in business. Most South Africans are bilingual, speaking both English and Afrikaans. Many also speak Swahili as a common trade language. Zulu is also widely spoken.

While South Africa is generally a nontouching society, the handshake is standard. Women extend their hands first; Muslim women do not shake a man's hand. Among the white population, the customs are similar to those of England and Netherlands. Greet people with an honorific or title. After you get to know someone, the honorific with the first name is acceptable (as in "Mister Dan"). A person may greet you by asking you about your health, the health of your relatives, and the state of your livestock. Asking how you are is not a rhetorical question; the greeting may be long, but it is sincere. Take the time to respond properly, and then ask him or her the same in kind.

Food

Continental manners are appropriate in most of South Africa; avoid eating with your fingers. In rural areas, however, South Africans may eat in a traditional way, with a common bowl and using their fingers. Tipping is customary. There are many different ethnic cuisines in South Africa, although Western food is also widely available. Alcohol is available, but if your host is Muslim, you should avoid it. Meals include root vegetables, such as yams and cassava, and meat or fish if it is available. Dinner is served late and is the main meal of the day. The English tea is practiced in the afternoon. If you are invited to dine at a South African home, you may be served tea English style.

Religion

South Africa has a large mix of predominately Christian religions, including Zion Christian, Pentecostal/Charismatic, Catholic, Methodist, Dutch Reformed, and Anglican. There is a small Muslim minority. Many also follow local animist traditions, which are sometimes combined with Christianity.

Etiquette

Even in today's post-apartheid South Africa, there are still strict conventions regarding race. Follow the lead of your host. Although it is best to avoid discussing politics when doing business in a foreign country, it may be hard to avoid this in South Africa.

Do not show the bottom of your shoe, and avoid crossing your legs with an ankle on the knee. Your manners should be more Continental than American. Gift giving is a good idea on social occasions; in business, you may present a gift to a colleague or the primary decision-maker.

Your style of dress should be conservative Western. South African businesspeople may combine Western attire with traditional, colorful African garments. If you are invited to someone's home, remove your shoes before entering.

Bargaining and Deal-Making

Punctuality is important in South Africa. With the brutal regime of apartheid still fresh in people's memories, there is much to overcome, and some distrust remains between races in South Africa. Nonetheless, the country is struggling to embrace a multicultural government. Most businesspeople have a good understanding of Western practices and customs. In South African business, hierarchy is important, and business is a top-down process.

While White South Africans are more eager to get down to business, Black South Africans are more likely to engage in personal discussion beforehand.

Common Customs

- South Africans tend to be soft-spoken. Avoid speaking loudly or aggressively.

- Decisions are not made quickly.

- It may be common for women, even women in business, to be referred to as "girls."

49
Pakistan

In 1947, British India was separated into Muslim Pakistan and Hindu India, although the division was not clean, and the two countries have fought wars over the disputed Kashmir territory. The dispute continues today, although tensions have decreased. Culturally, Pakistanis are very different from Indians, and have more in common with Bangladeshis. Pakistan is nominally a democracy, though subject to the laws of Islam.

Despite free and mandatory elementary schooling, higher education is not universal, and many Pakistanis go abroad for higher education. They are more likely to speak British English. The official language is Urdu, although there are also some regional dialects spoken.

Economics and Opportunities

Pakistan's population is largely impoverished and underdeveloped, and the literacy rate is a mere 48.7 percent. Foreign investment is minimal, and the ongoing conflict with India continues to drain Pakistan's resources. Some recovery has resulted from foreign assistance and access to global markets. Reforms include privatization of the banking sector. Inflation remains high, although the central bank follows a tight monetary policy.

Greetings

Light handshakes are appropriate among men, both upon meeting and departing, although close friends may embrace. A man should not shake hands with, or touch, a woman, even if they are well acquainted. The traditional Salam greeting is begun when one party says, "As-Salamu `Alaykum" (pronounced "ah-salam alaykum," it means "peace be upon you"); the other party responds, "wa `Alaykum As-Salam" (pronounced "wa-laykum ah-salam," this means "and on you be peace"). When shared among men, the Salam is accompanied by a light, right-handed handshake, which may be followed by touching your hand to your heart.

First names should not be used. Traditionally, the honorific "Sahib" is placed after the surname, as in "Blacharski Sahib," for a man; or "Begum" for a woman (as in the case of my wife, "Blacharski Begum"). If a male has made the Hajj, the honorific "Hajji" is added. For a woman who has made the Hajj, the appropriate honorific is "Hajja."

Avoid the American hand wave; instead, when you are departing, shake hands with each person in the group. If you are greeting a Hindu, you may instead greet them with a namaste; however, be careful not to use this greeting with a Muslim.

Food

Dinner invitations are common between business associates, but wives are not included. Food is often eaten in the traditional style, using the right hand only. Western utensils are commonly used, though in some traditional restaurants, people eat with their hands. Avoid discussing business during the dinner. Tipping (10 percent) at a restaurant is expected. Food is similar to Indian and Bangladeshi food, with meats including lamb, goat, and chicken. Pork is not served, as it is against Muslim custom. If you are invited to dinner, as guest of honor you may be offered the most favored part of the animal, which regrettably may be a sheep's eyeball. If you can, accept it with gusto and you will be a hero. The main meal is at midday and is often elaborate and lengthy. Food is moderately spicy, not as spicy as in India or Thailand.

Alcoholic drinks are not usually served, but tea is drunk frequently. Do not pour your own drink; instead, allow your host to refill your cup for you. Women may eat in a separate room.

After the meal, expect mints or caraway seeds to chew for fresh breath. Betel may also be served after dinner.

Religion

Pakistan is 97 percent Muslim, with 77 percent of Muslims belonging to the Sunni faith and 20 percent Shi'a. There is a small Christian and Hindu minority. Islamic Shari'a law holds tremendous influence in the country.

Etiquette

At a meeting, whether business or social, do not seat yourself. Instead, wait for the host or hostess to seat you, as seating arrangements are often arranged according to rank or status. Touching between sexes is never done, not even between married couples. Similarly, do not touch anyone on the head, and do not show the bottoms of your feet or the soles or your shoes. Dress is conservative, and women should dress modestly, with skirts or dresses below the knees, and arms covered below the elbow. You may notice Pakistani women covering their heads and faces, but it is not necessary for Western women to do so.

Bargaining and Deal-Making

It is important to be on time for meetings in Pakistan, although you may find your Pakistani counterparts often arriving late. Businesses are closed on Fridays. Pakistanis tend to speak quietly but rapidly; silence, eye contact, and intense staring are sometimes used as part of a conversation. Do not fill every silence with conversation. Pakistani women have an unequal role in society, and are unlikely to found in business roles.

Common Customs

- In response to some questions, a common response is "insh allah," which means "God willing."

- If you are handing something to another person, never use your left hand.

- Muslims consider dogs to be unclean animals, so avoid telling cute stories of your pet Boston Terrier back home.

- If you are the guest of honor, it is customary to make a small toast after the host makes one.

- Women must avoid touching food that will be served to a Muslim male.

- Do not take photographs of Pakistani women without permission.

50
Saudi Arabia

Saudi Arabia, the birthplace of Islam and home to the holy shrines of Mecca and Medina, is an ancient location, but was founded in its modern state in 1932 when Abd Al-Aziz bin Abd al-Rahman Al Saud unified the Arabian peninsula. After Iraq invaded Kuwait in 1990, Saudi Arabia allowed entrance to 400,000 refugees, and allowed Western troops into the country to prepare for the liberation of Kuwait. U.S. troops left Saudi Arabia in 2003. King Abdullah has undertaken a cautious reform program, and there has been some increased political participation. Elections were held in 2005 for half the members of the municipal councils, with the King appointing the other half.

Economics and Opportunities

Saudi Arabia, an Islamic monarchy, is a major producer of oil and natural gas, holding about 25 percent of all the world's proven oil reserves. Unlike most other monarchies in the world, the King has absolute authority, and there is no representative government. The only law that is higher than that of the King is the law of Islam. Under King Abdullah, the country has pursued economic reform and diversification; the country joined the WTO in 2005. Foreign investment is encouraged, and the country is a strong ally of the United States. There are heavy government controls over economic activity. The petroleum sector makes up 75 percent of budget revenues, 45 percent of GDP, and 90 percent of exports. The heavy reliance on oil has been somewhat relieved by reforms, promotion of the private sector, and foreign investment in other sectors such as power generation and telecom. Due to its oil wealth, Saudi Arabia has been able to invest heavily in job training, education, and development of the country's infrastructure. The population is well educated and the country enjoys a high level of literacy, as schooling (including higher education to qualified students) is free. Still, many Saudi families send their children abroad for higher education, often to the West. Thus, many younger Saudis have a strong knowledge of Western culture and business. Arabic is the national language, but many Saudis are highly educated and speak English as well.

Greetings

Greetings are formalized and very detailed. Upon meeting someone, the Salam is appropriate. While you are shaking hands, say "kaif halak." Following the handshake, your Saudi counterpart may put his left hand on your right shoulder, and may kiss you on both cheeks. He may then touch or hold your hand for a prolonged period of time. Keep in mind that this takes place between men. Do not shake hands with, kiss, or touch a Saudi woman, though they are rarely encountered in business.

Two men holding hands is a show of friendship and nothing more, so do not be surprised if a Saudi businessman takes your hand while you are walking. If in the unlikely event that your Saudi business associate is accompanied by his wife or wives, they may stay in the background and might not be introduced. If they are, however, keep in mind that the above greeting is for men; when introduced to a wife, a simple acknowledgment or nod of the head is sufficient.

Use titles and honorifics at all times. "Sheikh" is used often; it is not a title of nobility, but rather an indication of importance and honor. Women usually do not take their husbands' names, but should be referred to as "Mrs." followed by her first name. In any case, you are not likely to be introduced to women within the context of business. If a man has made the Hajj, he should be addressed with the honorific "Hajji" before his name; a woman should be addressed with "Hajja."

Traveler's Tip

In Saudi Arabia, Arabic speakers tend to be flowery in their speech and use many religious phrases and references. Be cautious in your speech; a casual "Oh my God" might be seen as disrespectful. Do not use religious references casually or disrespectfully. Religious phrases within the context of everyday speech are taken seriously, with a sincere "insh allah" ("if God wills it") added often to statements.

Food

A rare delicacy is a sheep's eyeball; if you are offered one, politely eat it. Dinner is lengthy and large, with a lot of food in several courses. The dinner invitation may not include your wife; if it does, the wives eat separately, usually in another room. If you are invited to a Saudi home, expect a sumptuous meal with multiple courses. There may be an enormous amount of food, so pace yourself, take small helpings and try to sample everything. When invited to a home for something other than a meal, cool drinks and sweets are served. You may see the carcass of a sheep in front of the house; it has been slain in your honor, and its presence is a compliment. During Ramadan, meals are after sunset. Avoid eating in front of your hosts during their fast.

Coffee is strong, black, and served with additional spices. Your cup will be refilled for you; do not pour your own drink. Alcoholic drinks are never served and are not available. If you are the guest of honor, it is appropriate to make a toast after the host does. Toast to the host's health and to your mutual prosperity, and make your language as flowery as possible.

Spoons, forks and knives are standard, and the right hand alone is for eating. In some circumstances, there may be no utensils, and the food is eaten with fingers. At restaurants where eating with hands is customary, there is a washbasin in plain view to use before and after eating. Do not be embarrassed at having messy fingers. Tipping at 10 percent is customary in restaurants, and usually the host or highest-ranking individual pays the bill. Saudis do not split the bill.

A woman should never touch food that is to be served to a man who is not a relative. Since men and women eat separately, this may not be an issue, but it is something to be aware of if you are a female visiting Saudi Arabia.

Religion

Saudi Arabia is exclusively a Muslim state, with Islamic Shari'a law holding influence over much of society and business.

Etiquette

Gift giving is important in Saudi Arabia. Be careful in choosing your gifts, however, and be sensitive to Muslim law and customs. Liquor, anything made from pigs or pigskin, and pictures or sculptures of women are forbidden. If you openly admire something, your Saudi acquaintance may offer it to you as a gift. If they do, it is impolite to refuse it. A man gives a gift only to another man; a woman only to another woman. If you are a female businessperson presenting a gift to a Saudi business associate (who is male), offer it as a gift from your husband or from a male associate in your company.

Bargaining and Deal-Making

Education is a priority in Saudi culture, and girls are educated as well as boys, although they attend separate schools. While some Saudi women have careers as teachers, or in the field of medicine, they are rarely found in the business world.

Traveler's Tip

When traveling to Saudi Arabia, keep Muslim holy days in mind. Do not expect to do any business during Ramadan.

The Saudi business organization adheres to a strict hierarchy, and executives are placed on different floors in an office building according to rank. Do not be offended if, in the course of a business meeting, your Saudi host allows people to come in and out of his office in need of his attention. Also be aware that the business day must be adjusted for prayers (five times a day), and business meetings are interrupted for this purpose.

Serve refreshments if you are hosting a business meeting, but never serve drinks in foam or paper cups. Use a formal porcelain or silver tea set.

Common Customs

- Saudi Arabia does not follow the Western, Gregorian calendar.

- In response to some question, a common response is "insh allah," which means "God willing."

- Direct eye contact is common. However, eye contact with a supervisor or superior is inappropriate.

- It is common for two men to walk arm in arm down the street.

- Stores are closed during the Muslim Sabbath, which begins on Thursday evening and continues through Friday.

- Smoking is pervasive throughout Saudi Arabia, although you should refrain from smoking during Ramadan.

- Do not refer to Muslims as "Mohammadans."

- Never photograph or videotape anybody without his or her permission.

51
Egypt

Egypt is a great ancient civilization that has been a unified kingdom since 3200 B.C. Egypt became important to the industrialized world with the completion of the Suez Canal in 1869. Britain gained control of Egypt in 1882, though it became fully independent after World War II. The population is growing rapidly, and there is limited arable land. Heavy dependence on the fertile Nile River Valley, which has become the center of economic activity, has caused the country's resources to be overtaxed. The country has undertaken some economic reforms, and has invested in its communications and other physical infrastructures.

Economics and Opportunities

The Egyptian government, a parliamentary democracy, has been highly centralized, although there are some reforms underway. Most recently, in 2005, personal and corporate taxes were reduced, and several enterprises were privatized. Afterward, the stock market surged and GDP grew by 5 percent a year. However, the standard of living has not improved, and necessary subsidies continue to drain the economy. The population is less educated than in other modern societies, with a literacy rate of only 57.7 percent. Foreign direct investment is low.

Greetings

Egyptians are warm, friendly, and accepting of most foreigners. Arabic is the official language, although the upper class speaks English and French as well. Greetings are sometimes elaborate, as in Saudi Arabia. Handshakes are more expressive, although the grip is not as firm. It may involve both hands, shaking with the right hand and touching the elbow with the left. The language is an Egyptian dialect of Arabic. Younger people speak at least some English, as it is taught in most schools, and some older Egyptians speak French.

Use honorifics and titles when addressing someone. The title "Sheikh" is not an indicator of nobility but rather a high level of importance; it is used less often in Egypt as it is in other Arab countries.

Food

Dinners are extravagant and usually start late in the evening. When you are invited to a dinner, take a small gift. Present it either with your right hand or with both hands, and never with your left hand alone. Alcoholic drinks are available, but your Egyptian acquaintance may not appreciate them; Egypt is largely a nondrinking, Muslim country. As such, follow the lead of your counterparts. If you are at a restaurant together, do not order an alcoholic drink unless your associate orders one as well. As in most Muslim countries, use the right hand only when eating.

Egyptian food is savory and varied, but not overly spicy. Lunch is the main meal of the day and may consist of several courses, starting with several appetizers. Tea is usually served, and coffee is often available, strong and sweet.

The guest of honor is served first, but should not start eating until everybody has been served. As in Pakistan, the honored guest in an Egyptian's home may be served the favored part of the animal, such as a sheep's eyeball or goat's head. If you cannot stomach it, acknowledge your host's generosity before refusing. However, eat it if you are able, to please your hosts.

Depending on where you are and what is being served, you may or may not have forks and spoons; follow the lead of your host in figuring out how to eat. Many Egyptian foods are eaten with the fingers. If your hosts are observant Muslims, men and women may dine in separate rooms.

Religion

The population is 90 percent Muslim, primarily of the Sunni sect. Of the rest, 9 percent are Coptic Christian, and 1 percent belong to other Christian sects.

Etiquette

Egypt is among the more liberal of the Islamic states. Veils for women are not mandatory, though they are still in practice. While husbands and wives walk side-by-side and even hand-in-hand, it is still common for a wife to walk slightly behind her husband.

It is more common for people of the same sex to stand closer together than they do in America and European countries, while men and women stand farther apart. Avoid crossing your legs, because showing the bottom of your feet or sole of your shoe is insulting. It is best to wait for a third-party introduction, as opposed to introducing yourself. Also, at a restaurant or business meeting, wait for the host to seat you, because seats are assigned based on rank or status.

Smoking is quite common in Egypt, even in public places, although you should offer a cigarette to others in your party.

Bargaining and Deal-Making

Egyptians like to get to know a person before doing business with them. Your Egyptian colleagues will be well educated. Schooling, including university-level education, is free. About half of university students are women, and it is common to see women in positions of authority. You may see your Egyptian colleagues dressing in traditional clothing, but it is not appropriate for a Westerner to dress in those garments. Conservative, Western business clothing is appropriate. Women should not wear pantsuits, and should cover their arms and legs at all times. Men may wish to wear loose-fitting pants, as sitting cross-legged on the floor is common.

Unlike other Muslim countries, work is permitted on Fridays; however, assume that the Muslim Sabbath is observed, and do not schedule meetings on Fridays unless your Egyptian counterparts suggest it first.

Common Customs

- It is customary to leave a little food on your plate.

- In response to many questions, expect to hear "insh allah," which means "God willing."

- Eye contact is common and often intense. However, avoid making direct eye contact with a supervisor or superior.

- It is common for two men to walk arm in arm down the street.

- Smoking is pervasive throughout Egypt, although you should refrain from it during Ramadan.

- Do not refer to Muslims as "Mohammadans."

- Do not photograph or videotape anybody without his or her permission.

52
Nigeria

Nigeria, Africa's most populous country, is a fascinating, sometimes tense mixture of different cultures, with over 250 different peoples and many different languages and religions. The most influential ethnic groups are Hausa and Fulani (29 percent), Yoruba (21 percent), Igbo (18 percent), Ijaw (10 percent), Kanuri (4 percent), Ibibio (3.5 percent), and Tiv (2.5 percent).

The British controlled Nigeria until 1960, when the country became independent. Military rule followed for 16 years, until the establishment of a new constitution and civilian government in 1999. Ethnic and religious tensions persist, hindering economic growth and stability. The country's physical infrastructure is still inadequate.

Economics and Opportunities

Nigeria's economy is primarily petroleum based, although much of the revenues have disappeared due to corruption and mismanagement. The economy has not diversified, with oil providing 95 percent of foreign exchange earnings and 65 percent of budgetary revenues. Agriculture is primarily subsistence based, and food must be imported. Nigeria received a billion dollar credit from the International Monetary Fund in 2000, contingent upon economic reform; but it pulled out of the program in

2002, after it did not meet spending and exchange rate targets. More recently, Nigeria has shown some initiative in modernizing its banking system and curbing inflation by preventing excessive wage demands, and by resolving disputes concerning oil revenues. Deregulation of fuel prices began in 2003, and the country's refineries were privatized. Education is limited and the literacy rate is 68 percent.

Greetings

English is the official language, and is spoken by most businesspersons and government officials, although in a strong dialect that may be difficult to understand at first. Other languages are spoken as well, including Hausa, Yoruba, Igbo, and Fulani. Because there are so many different cultures, greetings vary a great deal; it is best to follow the lead of your acquaintance. However, shaking hands is common, though Muslim women do not shake hands with men. Use honorifics and titles whenever greeting someone, as status is very important here. People often have nicknames they use in place of their given names. If you address somebody by his or her nickname, precede the nickname with the appropriate honorific.

Food

Tipping is expected in all situations, and not just restaurants. Nigerian food is varied, as each ethnic group has its own specialties. Meals may include root vegetables, such as yams and cassava. Tea and coffee are usually available. Alcohol, while available, should be avoided if your business associates are Muslim.

The traditional way of eating is to use your right hand to eat from a communal bowl, and be sure to allow the elders of the group to take from the bowl first. Spouses may be invited to dinners, although if your hosts are Muslim, the women eat in a separate room. Your host will not directly invite your spouse, but will instead say, "My spouse invites your spouse to dinner with us."

Religion

About half the population is Muslim, with another 40 percent Christian, and the remaining 10 percent following various indigenous beliefs. Animism is also widely practiced by both Muslims and Christians.

Etiquette

Wait for a third-party introduction, but if none is forthcoming, it is acceptable to introduce yourself. Greetings are elaborate and filled with friendly discussion, including inquiries into your health, the health of your relatives and your livestock. This is not small talk; take time to respond to their questions, and then ask similar questions in kind.

When at a restaurant or business meeting, wait to be seated, because seating is assigned according to rank and status. Usually, women are seated separately.

Bargaining and Deal-Making

Travel is difficult in most of the country, so it may be difficult to be on time. In general, Nigerians are less preoccupied with timeliness. Nigerian corruption is known the world over, and one must proceed with extreme caution when doing business in Nigeria. In addition to the danger of falling victim to scams, there are other dangers as well; tensions and resentment between ethnic groups may lead to unrest. As education is limited, and instructional materials and teachers are hard to come by, it is difficult to find skilled, literate workers.

The protocols of hierarchy are important when making deals. Women have more freedom in Nigeria than in some other countries with large Muslim populations, and women are not required to wear a veil.

Conservative, Western dress is appropriate, and most Nigerians dress in Western clothing as well; however, some may include more traditional Nigerian designs. Nigerian clothing is very colorful and beautiful, but Westerners should not wear it. Women (especially in Muslim areas) should not wear pantsuits or tight-fitting clothing.

It is common for Nigerians to bargain over prices, so expect strong negotiations.

Common Customs

- Do not photograph or videotape anybody without his or her permission.

- Public displays of affection should be avoided.

- Avoid showing the sole of your shoe or foot; do not cross your legs ankle-over-knee.

- Avoid using your left hand when giving somebody something.

- Remove your shoes before entering somebody's home.

- Smoking should not be done until after the meal, and not where the food is being served. Women do not smoke.

53
Algeria

Algeria is quite large, but sparsely populated. Ruled for more than a hundred years by France, the country gained independence in 1962. The National Liberation Front (FLN) remains the largest political party, although the Islamic Salvation Front (FIS) came close to political dominance at the end of 1991. The FIS's success caused the Algerian army to intervene, postponing further elections to prevent Islamic extremists from taking over politics. An insurgency followed, with fighting throughout much of the 1990s. The FIS's army disbanded in 2000. Some small bands of militants still persist today.

Economics and Opportunities

Persistent economic problems in Algeria include high unemployment, a housing shortage, unreliable electricity and water, and an inefficient and corrupt government. Although the petroleum economy generates large amounts of cash, it is not used to mitigate any of Algeria's social problems. Education is inadequate, with a 70 percent literacy rate. Some recent policy reforms, supported by the International Monetary Fund (IMF), have improved the country's economy. Also, Algeria runs a trade surplus and is building up foreign exchange reserves. The country's external debt was reduced to less than 10 percent of the GDP after it repaid its Paris Club and London Club debts in 2006.

Greetings

Arabic is the official language, with some educated citizens also speaking French. Berber dialects are also present. Shake hands when you meet someone and also when you depart. The handshake is prolonged, without marked pumping, but with a slightly lingering grip. People may also greet one another by kissing on both cheeks; this is common between men as well as women. Greet people by an honorific or title, followed by last name. Gift giving is appropriate, but usually reserved for the second meeting.

Food

There is a considerable French influence on Algerian cuisine.

Religion

Algeria is a Muslim state, with citizens observing the Sunni sect of Islam.

Etiquette

Algeria tends to be one of the more conservative Muslim countries. Many Algerians dress in traditional clothing which indicates where in the country they are from.

Bargaining and Deal-Making

Punctuality is not common here, so expect your associates to be late. Begin meetings with small talk.

Common Customs

- When young people greet an elder, they may refer to them as "aunt" or "uncle" even if they are not related.

- Some women wear gold jewelry as a sign of wealth and status.

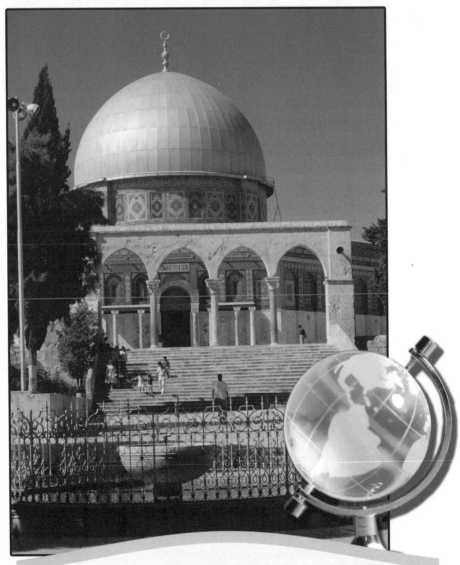

Over 90 percent of
Algeria is considered to be part of the Sahara desert.

54
Turkey

Turkey is an ancient land with both European and Middle Eastern elements, and in many ways functions as a connection between East and West. Modern Turkey was founded in 1923, initially under the one-party rule and authoritarianism of Mustafa Kemal, who ushered in many social, legal, and political reforms. Multiparty politics began peacefully in 1950. Since then, there have been three military coups, each of which ended with the return of power to civilian governments. There is a separatist insurgency, led by the People's Congress of Kurdistan, which has caused some violence. Turkey joined the UN in 1945, NATO in 1952, and the European Community (now the EU) in 1964.

Economics and Opportunities

Turkey's economy is dynamic and complex. Although industry and commerce are modern, the agricultural sector is quite traditional. The private sector is flourishing and continues to grow, although the state continues to rule several basic industries, including banking, transportation, and communications. The textile industry is large, making up a third of industrial employment; however, there is stiff international competition in this area. The automotive and electronics industries are gaining importance in the country.

Greetings

Most of the population (80 percent) speaks Turkish, while 20 percent speak Kurdish. Turkish people are very hospitable. If you visit someone's home, they might prefer that you remove your shoes before you enter. Although some people in the larger cities no longer follow that custom, take a moment to observe before you enter, and follow the lead of your host. Educated Turks often go to school in the U.S. or Europe, and are multilingual.

Some people have Arabic-sounding names and others, Turkish names. Only recently have people been using a first name and surname in the Western style. Married women take their husbands' last names.

Use titles and honorifics. You may be addressed by your entire name (as in "Mr. Dan Blacharski") in conversation. A third-party introduction is best. Avoid the American-style handwave when leaving; instead, take your leave of each person individually. Western-style handshakes are appropriate, and can be firm, like in America. Sometimes an associate may grasp your hand with both of theirs while shaking. Greetings between men and women should be less physical, and if a woman is a Muslim, she may not touch or shake hands.

Food

Invitations to dinner at a restaurant are more common than in a home, but if you are invited to a home, bring a small gift. Tipping (10 percent) at restaurants is expected. Turkish food is excellent and exotic, containing elements of both Eastern and Western cuisine. Beer and wine is available at many restaurants, but observe your hosts before ordering; if they are observant Muslims, they do not drink alcohol, and it is disrespectful to do so in front of them. Visitors to a Turkish person's home are served tea and coffee, even if it is not mealtime. Tea is not drunk with milk. Coffee is strong, with fine grounds on the bottom, so be careful when drinking it. Turkish coffee is thick and very rich, usually already sweetened. It is polite to at least take a few sips, even if you find it too strong. Often drinks are not be served until the meal is over, as many Turks believe that drinking while eating is unhealthy. After the host makes a toast, you should make one to health and prosperity.

Eating is done Western style, with forks, spoons, and knives, and with the fork held in the left hand. Never eat with your fingers. Your glass will be refilled by your host. You will be offered seconds; if you do not want more, leave a bit of food on your plate. At dinner, do not eat or drink until the eldest person has started eating.

Religion

Turkey is a secular republic, and although it is mostly a Muslim nation, residents are more modern and tend not to embrace the fundamentalist version of Islam found in some Middle Eastern countries. Although respect for Islamic rules should be observed, life is not quite as strict as in, say, Iran. Alcohol is allowed (although Muslims do not partake of it), and women do have some opportunities in business.

Etiquette

The people in Turkey are very hospitable and generous. Avoid body language or talking with your hands. Turks tend to use a lot of nonverbal gesturing, and you may make a rude gesture without meaning to. For example, a closed-fist gesture of any kind, or crossing your arms while talking, are rude. Smoking is common, but courtesy rules; avoid smoking on the street or in a crowd, and ask permission before lighting up. When you do, offer a cigarette to those in your group. Public displays of affection are avoided, and in smaller towns, women might not talk to a man unless they have been formally introduced.

Bringing a gift to someone's home is appropriate as a thank you for a meal, and it should be something small, such as pastries or sweets, or flowers (but not chrysanthemums). Flowers should be given in odd numbers.

Bargaining and Deal-Making

It is important to be on time to business and social events. When meeting with associates, direct eye contact is expected, but do not make eye contact with a person of the opposite sex. Standard Western-style business dress is appropriate, with conservative suits and dresses. Summers are hot, so Turkish businessmen may remove their jackets and ties. Feel free to do so as well, if your counterpart does so first. Avoid casual dress or jeans, and never wear shorts, despite the summer heat. Women should dress conservatively out of respect for Muslim tradition, so that their legs are covered below the knee, and arms covered below the elbow.

Gift giving is common in social situations, but less so in business. A single token for the highest-ranking individual is adequate; you do not need to bring gifts for everyone.

Avoid doing business during Ramadan (called "Ramazan" in Turkey). Fridays are also bad days for business, as most practicing Muslims do not work on the Muslim Sabbath.

Common Customs

- Contact between men and women in public is generally frowned upon.

- Shaking the head up and down is a negative sign. Shaking left to right means "I do not understand."

- Smoking is pervasive, and it is common to smoke between courses at dinner.

- Pointing your shoe or feet toward someone is considered an insult.

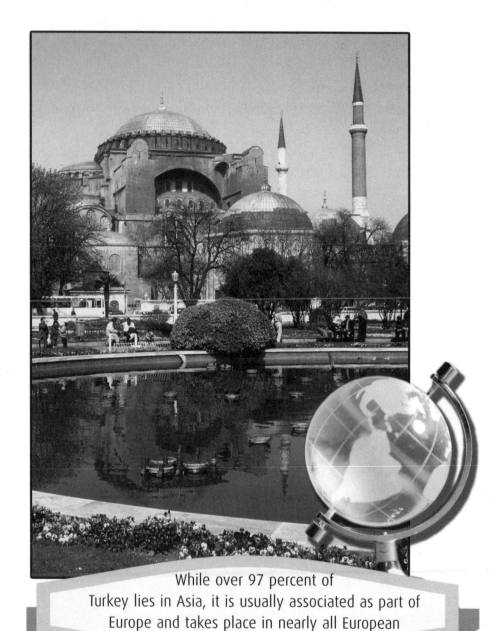

While over 97 percent of Turkey lies in Asia, it is usually associated as part of Europe and takes place in nearly all European contests and events.

Conclusion

Congratulations, you are a world traveler! Whether you are traveling for business or pleasure, you will always enjoy your trip more (and have greater success in your business dealings) if you take time to learn a bit about the country's culture, language, and people.

Whenever I get on a plane, the first thing I do is to empty my mind of all the preconceived notions and habits that I, as an American, have accumulated over the years. When I get to my destination, I am no longer just an American, I am a citizen of the world, and new cultures, new beliefs, and new ways of doing things are opened up to me. I stay receptive. There will be American food almost everywhere, but then again, why travel half way around the world just to eat hamburgers and pizza? The American Way is, after all, just one way out of many.

There is an old expression that goes, "When in Rome, do as the Romans." This is more than just a trite saying, it is wisdom that every traveler must keep to heart. I believe that if everyone took time to travel to different countries, learn about different people, and get some first-hand experience of different cultures, the world would be a far more peaceful place to live.

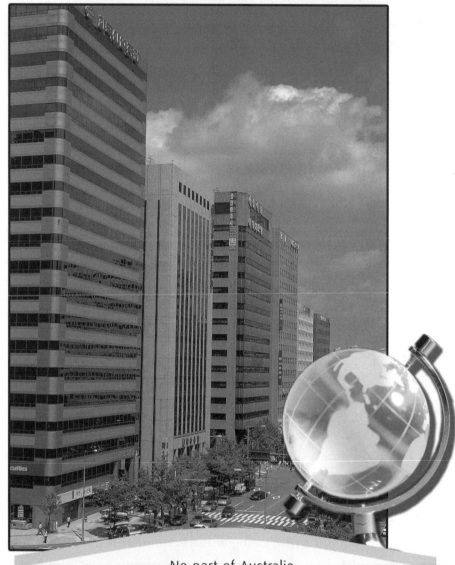

No part of Australia
is more than 1,000 km from the ocean.

Author Dedication & Biography

This book is dedicated to my two favorite traveling partners; my son Shanti, who spent the summer with me traveling through Europe and enjoying the Krakow Jazz Festival, and my wife Charoenkwan, who has shown me the wondrous mysteries of the Orient.

Dan Blacharski

Dan Blacharski is a writer and world traveler and has spent years traveling through both Europe and Asia, enjoying the cultures, unusual foods, and languages of other countries at every opportunity. He has eaten local delicacies such as rotten shark in Sweden; wild boar in Poland; and in Thailand, ant eggs, bird's nest soup, and fried fish bladders. He has ridden on the backs of elephants and water buffaloes, stood among wild monkeys, and once allowed a python to crawl around his neck. He moved to Bangkok simply because it was as far away from his hometown as he could possibly get without leaving the planet, and while there, he met his wife Charoenkwan, one of Thailand's best-known authors, who now travels the world by his side. A professional writer, Dan has written six books and ghost-written several others; he has produced thousands of online and print features, and has worked closely with many Internet and high-tech companies seeking to join the dotcom boom. His own blog, The Next Dotcom Boom (**http://www.thenextdotcomboom.com**) chronicles the next move in this direction.

Dan and his wife currently live in South Bend, Indiana, but having never gotten quite used to the frigid Midwest, they spend their winters in Bangkok and stay until the end of Songkran, the annual new year's celebration in April, when revelers and the occasional stray elephants have their faces painted and spray water on passers-by as part of what is the most fascinating week-long festival in the world.

Index

foundation in the search for laws and principles that can guide our thinking and improve our understanding of the physical and social forces inherent in natural hazards. Toward those ends, we emphasize the "general" through broad principles and lead to the "particular" by using examples of many kinds of natural hazard from different parts of the world.

This book was written with upper-level undergraduate students and graduate students in mind. It also should be useful to researchers in such fields as geography, anthropology, sociology, and environmental studies as well as to geologists, meteorologists, and engineers who have an academic fascination with natural hazards. It consists of eight chapters. Chapter 1 describes the problem of natural hazards and introduces key ideas and arguments; trends in losses due to natural disasters are discussed from both temporal and spatial perspectives and various characteristics of the toll taken by such extreme events are considered. Chapter 2 focuses on geophysical events, taking common physical traits they exhibit (such as magnitude, duration, and frequency) as the organizing framework. By understanding the physical characteristics shared by hazards, we may facilitate implementation of coping mechanisms. Owing to the range of parameters by which geophysical events can be measured and analyzed, this is the longest chapter in the book; however, it is not a substitute for natural science texts and articles focusing on the physical aspects of extreme geophysical events. In Chapter 3, we begin our examination of the human dimension of natural hazards with individual perceptions, considering how different and perhaps changing views can affect behavior and response. Chapter 4 pursues the human dimension by examining responses at the community rather than individual level; because individual responses do not translate readily into group and community responses, it is important to focus on community attitudes in order to develop a systematic analysis of community responses across hazards. Chapter 5 addresses the larger picture of policy and response to natural hazards, bringing out political and economic factors not addressed in previous chapters and providing numerous examples of planning-based responses to hazards management. Chapter 6 follows with an assessment of the economic factors associated with natural hazards and disasters, emphasizing broad principles related to the full range of economic impacts associated with events and alleviation measures rather than the accounting of direct economic losses. Chapter 7 addresses risk assessment as it applies to natural hazards and natural hazards management, incorporating economic as well as other types of risk. Finally, Chapter 8 pulls together the themes of the book in an integrative manner, providing a theoretical framework through which natural hazards might be examined and managed.

Preface

In both the physical and social sciences, natural hazards research has come a long way over the years. Our scientific endeavors have provided many answers, helping us to recognize that issues of natural hazards cannot be solved solely through study of the physical mechanisms of the natural world. It also has become patently clear that we cannot control every element of nature such that all risks are eliminated. Indeed, human factors are just as much a cause of natural hazards as are extreme geophysical processes. It is the recognition of human involvement that has enhanced hazards research and led many, including politicians, economists, and hazard managers, to advocate comprehensive development planning. Of course, there is still a long way to go, as evidenced by the tendency for societies to respond to disasters with aid that frequently serves to reinforce predisaster conditions, leading to a continual cycle of disaster–damage–repair–disaster. In many cases, little attention is paid to changing social conditions and the long-term needs of those affected by hazardous events. Thus, victims remain vulnerable to the same geophysical events time after time; to a large extent, it matters little what the nature of the hazard is.

In this book, we have adopted an integrated approach to natural hazards, incorporating facets of both the physical and social sciences to examine responses to extreme geophysical events. This has required a focus on general principles rather than detailed analyses of particular events or hazard categories. Many other texts have taken a hazard-by-hazard approach, which has added greatly to our understanding of natural hazards and specific disasters. We are indebted to these authors. However, it is time to build on their findings and apply an integrated approach to hazard assessment, and that is our goal for this book; in so doing, we have tried to synthesize many diverse theories and concepts. This book does not offer a panacea for all hazard-related problems, nor is it based on the premise that there is a single theoretical structure on which solutions might rest. Instead, it is founded on the belief that there is much to be learned from the varied disciplinary approaches of hazard researchers, providing a

Printed in the United States of America

This book is printed on acid-free paper.

Last digit is print number: 9   8   7   6   5   4   3   2

Library of Congress Cataloging-in-Publication Data

Tobin, Graham A.
    Natural hazards : explanation and integration / Graham A.
  Tobin and Burrell E. Montz.
        p.   cm.
    Includes bibliographical references and index.
    ISBN 1-57230-061-2. – ISBN 1-57230-062-0 (pbk.)
        1. Natural disasters.   I. Montz, Burrell Elizabeth.   II. Title.
  GB5014.T63   1997
  363.34–dc21   97-1191
                                                                   CIP

Natural
Hazards
Explanation and Integration

GRAHAM A. TOBIN
and **BURRELL E. MONTZ**

THE GUILFORD PRESS
New York   London

NATURAL HAZARDS

A central theme of this book was inspired by Germaine Greer, who wrote in *Sex and Destiny: The Politics of Human Fertility,* "Perhaps catastrophe is the natural human environment, and even though we spend a good deal of time away from it, we are programmed for survival amid catastrophe." Catastrophes, loosely defined as extreme disasters, are a common feature of human existence, something with which both societies and individuals must contend. The human spirit, therefore, directly challenges catastrophe, leaving us with an enduring saga of survival. "Survival amid catastrophe" is a story we have tried to relate.

Acknowledgments

This book could not have been written without the direct and indirect contributions of many people. We are grateful to all of them, especially the many hazard researchers upon whose work this book is founded and to whom we owe a great deal of thanks. We also would like to acknowledge the support of our departments, Geography at Duluth and Geography as well as Geology and Environmental Studies at Binghamton. And, of course, Peter Wissoker, our editor at Guilford, showed long-standing patience and gave continuous support during this project.

Several students, who have now left us for other places, provided invaluable research assistance, including Tracy Scharpe and Susan Bonfigt at Duluth and Chris Carangelo, Eric Nelsen, and Tim Worrall at Binghamton. As cartographer and draftsperson, Anne Hull endured many rounds of editorial changes and was able to translate many of our concepts into comprehensible diagrams. We also are indebted to Eve Gruntfest and John Cross, who reviewed an earlier draft of the manuscript. We cannot forget our colleagues who encouraged us as we developed this book, including Neil Ericksen, Jenny Dixon, Jim Sorauf, Jane Ollenburger, Paul Matthews, and the late John Webb. It was through discussions with them over the years that many of our ideas took shape. We also must recognize our old friends, Tom Newton, Rick Stusse, and Mary Bridge, each of whom provided moral support, in his or her own way, as this project progressed.

Finally, we owe a great debt to Gilbert F. White, whose influence is clear throughout the book. Most of all, we want to thank our partners, Liz Bird and Paul Covey, who put up with this project for far longer than any of us ever imagined.

Contents

1 Natural Hazards and Disasters

When Potential Becomes Reality

NATURAL DISASTERS: A NORMAL STATE OF AFFAIRS

In 1985, an earthquake in Mexico killed 20,000 people; a tropical cyclone killed 11,000 in Bangladesh, and one in Vietnam killed 670; 300 died from landslides in the Philippines; a volcano erupted in Colombia killing 25,000; a flood in China added 500 to the death toll; a storm in Algeria killed 26; cold waves were responsible for 290 deaths in India and 145 in the United States; a heat wave killed 103 in the United States; and 52 died in Egypt from a fire (Glickman, Golding, & Silverman, 1992). It was not an exceptional year. In 1995, over 700 people in Chicago died from heat stress during a period of abnormally high temperatures, and Hurricanes Luis and Marilyn both devastated the Caribbean; on the island of St. Thomas alone, Hurricane Marilyn damaged more than 70% of the approximately 18,000 homes. Thus, although disasters and deaths were slightly above average in 1985, comparison with selected events from 1995 suggests that in many ways, 1985 was a "normal" year.

The litany of disasters goes on with deaths and damages mounting almost daily. In just 1985, globally there were 47 disasters that killed 62,500 people, averaging 1,330 deaths per disaster (Glickman et al., 1992). In 1945 through 1986, the 42-year period studied by Glickman, Golding, and Silverman (1992), there were on average 30 disasters and 56,000 deaths per year; this means that 2.34 million people lost their lives to extreme natural events during this period. The cost to the global economy is in excess of $50 billion every year, two-thirds of which is direct loss and one-third of which is the cost of prevention and mitigation (Alexander, 1993).

Describing such deaths and losses as normal seems callous, but represents reality for many people around the world. People do live in high-risk areas, and many communities are particularly vulnerable to the vicissitudes of natural events. The situation is complicated further when the data are examined more closely. Glickman et al. (1992) showed that the number of disasters and the number of deaths had increased during the 42-year study period; however, some disasters, such as droughts, were not even included in their data set because of difficulties in accounting for losses from such events. This means that their statistics probably underestimate significantly the total impact of extreme natural events.

The available data raise many questions. One, at least for many Westerners, concerns the total number of disasters and deaths (Tables 1.1 and 1.2). How could it be that so many disasters have occurred and so many people around the world have died, but for the most part, we remain totally unaware of the tremendous dimensions of such problems? This lack of awareness involves issues of place and significance as well as concerns about media coverage of such events. Other questions arise as to why such disasters occur and whether anything can be done to mitigate them. Since we know a great deal scientifically about their physical properties, why do natural disasters continue to cause such losses? Who is at fault: is it the government or the victims themselves or is it more complicated than that? In addition, there are questions about the nature of disasters. Given that the incidence of events appears to be increasing and the number of deaths rising, is the environment becoming more hazardous or are there other explanations for this disturbing increase? What can we learn from the spatial patterns of disasters? Do some regions experience more events, do they have more hazardous conditions than others, and if so, why? Can spatial and temporal trends be explained by differences in the incidence of extreme geophysical events or are there human/structural factors that need to be addressed to explain hazard vulnerability? Finally, will answers to the above questions provide sufficient information to develop an explanatory model of disasters and disaster impact? In order to resolve these questions, we must be clear about terminology so that we understand the components that combine to create disasters in some places, but not others.

In this chapter, we begin by distinguishing natural "hazards" from "disasters." While one represents a potential or threat, the other presents a very real set of problems and losses. We then turn our attention to the natural versus human debate that has concerned so much of hazards research. Because it does not suffice to look at disasters per se, we also devote attention to understanding risk and vulnerability, assessing the potential for catastrophe, and identifying and defining the victims of disasters. This discussion is followed by detailed examination of the spatial

TABLE 1.1. Deaths from Natural Disasters, a Selection of Events Worldwide

Year	Event	Location	Approximate death toll
1900	Hurricane	United States	6,000
1902	Volcanic eruption	Martinique	29,000
1902	Volcanic eruption	Guatemala	6,000
1906	Typhoon	Hong Kong	10,000
1906	Earthquake	Taiwan	6,000
1906	Earthquake/fire	United States	2,500
1908	Earthquake	Italy	75,000
1911	Volcanic eruption	Philippines	1,300
1915	Earthquake	Italy	30,000
1916	Landslide	Italy, Austria	10,000
1919	Volcanic eruption	Indonesia	5,200
1920	Earthquake/landslide	China	200,000
1923	Earthquake/fire	Japan	143,000
1928	Hurricane/flood	United States	2,000
1930	Volcanic eruption	Indonesia	1,400
1932	Earthquake	China	70,000
1933	Tsunami	Japan	3,000
1935	Earthquake	India	60,000
1938	Hurricane	United States	600
1939	Hurricane/tsunami	Chile	30,000
1945	Floods/landslides	Japan	1,200
1946	Tsunami	Japan	1,400
1948	Earthquake	USSR	100,000
1949	Floods	China	57,000
1949	Earthquake/landslide	USSR	20,000
1951	Volcanic eruption	Papua New Guinea	2,900
1953	Floods	North Sea coast	1,800
1954	Floods	China	40,000
1954	Landslide	Austria	200
1959	Typhoon	Japan	4,600
1960	Earthquake	Morocco	12,000
1961	Typhoon	Hong Kong	400
1962	Landslide	Peru	5,000
1962	Earthquake	Iran	12,000
1963	Tropical cyclone	Bangladesh	22,000
1963	Volcanic eruption	Indonesia	1,200
1963	Landslide	Italy	2,000
1965	Tropical cyclone	Bangladesh	17,000
1965	Tropical cyclone	Bangladesh	30,000
1965	Tropical cyclone	Bangladesh	10,000
1968	Earthquake	Iran	12,000
1970	Earthquake/landslide	Peru	70,000
1970	Tropical cyclone	Bangladesh	300,000
1971	Tropical cyclone	India	25,000
1972	Earthquake	Nicaragua	6,000

continued

TABLE 1.1. (cont.)

Year	Event	Location	Approximate death toll
1976	Earthquake	China	300,000–700,000
1976	Earthquake	Guatemala	24,000
1976	Earthquake	Italy	900
1977	Tropical cyclone	India	20,000
1978	Earthquake	Iran	25,000
1980	Earthquake	Italy	1,300
1982	Volcanic eruption	Mexico	1,700
1985	Tropical cyclone	Bangladesh	10,000
1985	Earthquake	Mexico	10,000
1986	Volcanic eruption	Colombia	23,000
1988	Hurricane	Caribbean Islands	343
1988	Earthquake	USSR	25,000
1989	Hurricane	Caribbean Islands	56
1990	Earthquake	Iran	40,000
1991	Tropical cyclone	Bangladesh	138,000
1992	Earthquake	Turkey	547
1993	Floods	United States	50
1994	Earthquake	United States	57
1995	Earthquake	Japan	6,300

Note: Drought, famine, and certain meteorological events such as heat waves are excluded from this table. *Data sources:* Glickman, Golding, and Silverman (1992), International Federation of Red Cross and Red Crescent Societies (1993), Noji (1991), Office of U.S. Foreign Disaster Assistance (1990).

and temporal patterns of death and damage associated with disasters at the global scale. The significance of disasters to society and the individual also is addressed. Finally, we look to the future and discuss how a theoretical framework to model the hazardousness of a place (see, e.g., Hewitt & Burton, 1971) might improve our knowledge of natural hazards. Throughout this chapter, one basic theme prevails: natural hazards are normal

TABLE 1.2. Disaster Mortality by Type, 1960–1989

Disaster type	Deaths		
	1960–1969	1970–1979	1980–1989
Floods	28,700	46,800	38,598
Cyclones	107,500	343,600	14,482
Earthquakes	52,500	389,700	53,740
Hurricanes			1,263
Other disasters			1,011,777
Total			1,119,860

Data sources Glickman, Golding, and Silverman (1992), International Federation of Red Cross and Red Crescent Societies (1993), Noji (1991), Office of U.S. Foreign Disaster Assistance (1990).

events with which societies must deal. In some locations, they are frequent or perhaps even commonplace; in others they represent a relatively rare, once-in-a-lifetime event. To all societies, however, they present a challenge.

DEFINITIONS: NATURAL HAZARDS AND DISASTERS

The term "natural hazard" sometimes creates confusion. It has been used imprecisely and with different implicit meanings, but in addition, the term has evolved with understanding of the components that interact to comprise hazardousness.

A *natural hazard* represents the potential interaction between humans and extreme natural events (Figure 1.1). It represents the potential or likelihood of an event (it is not the event itself). By definition, then, natural haz-

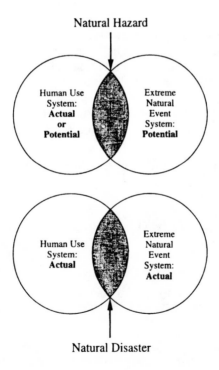

FIGURE 1.1. Natural hazards and natural disasters. In both cases, the overlap between human and physical systems is of concern; their difference relates to potential versus actual occurrences. Hazards represent potential events while disasters result from actual events. *Source:* Ericksen (n.d.). Reprinted by permission.

ards constitute a threat to society. Such a threat is ever-present, representing an intrinsic force with which all societies must cope in one way or another. The hazard exists because humans or their activities are constantly exposed to natural forces. For example, by locating property on flood plains, undertaking viticulture on the slopes of active volcanoes, or developing homes and resorts in hurricane-prone coastal zones, humans are exposed to natural hazards. It is possible, therefore, to identify some hazard-prone areas from a geographical perspective: for example, the Ring of Fire (along the margins of the Pacific Ocean) delimits an active zone of earthquakes, tsunami, and volcanoes. While not all hazards can be defined so accurately in space, it is possible to calculate levels of risk and the probability of dying from various events (Table 1.3). For instance, one's chance of being hit by a meteorite is 1 in 100 billion, a threat that is not high; however, if the earth were to find itself in the path of a succession of meteorites (such as those that crashed into Jupiter in 1994), clearly the risk would increase and the threat would be imminent. The same can be said of other hazards. The risk and ultimately the threat (often defined as *hazardousness*) change over time as human use and environmental processes change.

Only after an event occurs do we term it a "natural disaster." A *disaster* is usually defined as an event that has a large impact on society. Because our subject matter is limited to natural hazards, it is geophysical events that create disasters. Unfortunately, there are no definitive boundaries to determine exactly when a threshold has been reached such that we can categorically say, "*this* constitutes a disaster." A disaster might be defined qualitatively as a hazardous event that significantly disrupts the workings

TABLE 1.3. Risk of Death from Involuntary Hazards

Risk	Risk of death/person/year
Struck by automobile in the United States	1 in 20,000
Struck by automobile in the United Kingdom	1 in 16,600
Floods in the United States	1 in 455,000
Earthquake in California	1 in 588,000
Tornadoes in midwestern United States	1 in 455,000
Lightning in the United Kingdom	1 in 10,000,000
Falling aircraft in the United States	1 in 10,000,000
Falling aircraft in the United Kingdom	1 in 50,000,000
Release from atomic power station	
At site boundary in the United States	1 in 10,000,000
At 1 km in the United Kingdom	1 in 10,000,000
Flooding of dike in the Netherlands	1 in 10,000,000
Bites of venomous creatures in the United Kingdom	1 in 5,000,000
Leukemia	1 in 12,500
Influenza	1 in 5,000
Meteorite	1 in 100,000,000,000,000

Source: Dinman (1980). Copyright 1980 by American Medical Association. Reprinted by permission.

of society. It may or may not lead to deaths, but it usually has severe economic impacts. We might also search for quantitative measures and, by default, several researchers have categorized disasters by numbers of deaths or extent of damage. For instance, Sheehan and Hewitt (1969) defined disasters as those events leading to 100 deaths, 100 injuries, or $1 million in damages. Glickman et al. (1992) used 25 deaths as their threshold, a figure we shall use in this book.

In addition, there are *catastrophes*, which for a given society might be defined as an event leading to 500 deaths or $10 million in damages. These figures, however, are arbitrary since levels of impact mean different things to different people in different situations. Furthermore, we cannot ignore the element of scale. It would be a catastrophe for a small community if every building were totally destroyed by flooding (as occurred in 1993 in Valmeyer, Illinois), but at the global scale, it would be an insignificant event if only 350 houses were involved (Horton, 1994). Similarly, $10 million in damage to some communities, would be devastating to some communities, especially in less wealthy societies, but others would be able to cope relatively easily.

The preceding discussion has introduced several concepts relevant to hazards, including risk, threat, vulnerability, impact, disaster, and catastrophe, each of which means something slightly different. In this text, these concepts delineate different elements of natural hazards.

This chapter focuses on losses from disastrous events. These indicate not only the enormity of the problems, but also help in understanding why the components of natural hazards are so important. Because of the probabilistic nature of hazards, they tend to be ignored or considered low priorities. However, it is too late to protect against loss once the event has occurred; instead, we must take action prior to the event (or before the next one). Understanding natural hazards provides opportunities, too often missed or ignored, to alter hazardousness of an area so as to reduce loss. The figures, concepts, and themes presented in this chapter illustrate the magnitude and range of the problems as well as how our evolving knowledge of interacting variables has altered approaches to mitigation. In particular, hazards research has undergone an evolutionary process, best exemplified by changing perspectives on just what constitutes a natural hazard as well as which factors need to be addressed in order to avoid disasters.

THE COMPONENTS OF NATURAL HAZARD: AN EVOLUTION

The earth is indeed hazardous to your health. According to Atkisson, Petak, and Alesch (1984), there are 516 active volcanoes with an eruption every 15 days (on average) somewhere in the world; global monitors record approximately 2,000 earth tremors every day, and there are approximately 2 earthquakes per day of sufficient strength to cause damage to homes and buildings, with severe damage occurring 15 to 20 times per year; there are 1,800 thunderstorms at any given time across the earth's surface; lightning strikes 100 times every second; during the late summer there are something like 5 hurricanes developing at any one time; and tornadoes average 4 per day or 600 to 1,000 per year. As a physical environment, the earth is a risky place to live.

Physical Perspectives

The traditional view of natural hazards has ascribed all or almost all responsibility for them to the processes of the geophysical world. This approach has meant that the root cause of large-scale death and destruction has been attributed to the extremes of nature rather than encompassing the human world. Frequently, disaster victims have been viewed as unfortunates who could do little but react to physical processes. The physical world, then, has been seen as an external force, separate from human forces. For example, Burton and Kates (1964), defined natural hazards as those elements of the physical environment harmful to man and caused by forces extraneous to him.

This is not to say that early researchers did not concern themselves with the human dimension. Indeed, building on the work of White (1945, 1961, 1964), scholars such as Burton (1962), Kates (1962), Dynes, Haas, and Quarantelli (1964), Hewitt and Burton (1971), and many others established hazards research as a human-based discipline. Although their definitions and classifications were oriented toward physical extremes, through their contributions it was soon recognized that the human-use system could not be ignored and, in fact, probably played a profound role in determining hazardousness and eventually the extent and degree of death and destruction from natural disasters. Thus, by the midseventies, natural hazards were defined somewhat more liberally: "The concept of natural hazards is somewhat paradoxical; the elements of a natural geophysical event (e.g., wind and storm surge of a hurricane) are hazardous only when they prove detrimental to human activity systems" Baker (1976, p. 1).

Although still based on the physical perspective, this refinement places greater emphasis on the human use system. In essence, extreme geophysical events that do not affect human activities do not constitute a hazard. (The threats posed by an earthquake in a remote part of Alaska, a mud slide on an uninhabited slope in South America, or a tornado in the desert region of New Mexico are not natural hazards under this definition.) Although the actual events may be of great interest to the geologist, geomorphologist, and meteorologist and may substantially enhance our knowledge of physical processes, they are not hazards.

Human Dimensions

By the end of the 1970s, steps toward a more human explanation of natural hazards were being taken by many others. For instance, the American Geological Institute (1984) defined as a natural hazard "a naturally occurring or man-made geologic condition or phenomenon that presents a risk or is a potential danger to life or property." Since the 1980s and early 1990s, natural hazards have been seen as the product of an interaction of physical and human forces that, in combination, determine the significance and impact of disasters. Thus, Smith (1992) suggests that natural hazards result from a conflict between geophysical processes and people; that is, hazards lie at the interface between the natural-event and the human-use systems (as shown in Figure 1.1).

In some instances, physical processes are almost eliminated. In defining disasters, Taylor (1989, p. 10) focuses entirely on disruption to society rather than any underlying physical process, emphasizing "catastrophic events that (a) interfere severely with everyday life, disrupt communities, and often cause extensive loss of life and property, (b) overtax local resources, and (c) create problems that continue far longer than those that arise from the normal vicissitudes of life." However, in classifying disasters, he utilizes physical processes as one criterion by which to distinguish hazard type (Table 1.4). His classification of disasters is quite wide-ranging, encompassing natural, industrial, and human-induced events.

In any classification system, of course, some events do not fit easily into categories. In particular, the primary event may be compounded by secondary impacts that may actually trigger greater losses than the original event. How should such events be classified? A classic example is the San Francisco earthquake of 1906. The shaking lasted for only about 40 seconds whereas the fires, sparked by broken pipelines, raged for three days before being brought under control; an estimated 2,500 people died (Palm, 1995), of whom ten times more died from the fire than from the earthquake (Bolt, 1978; Thomas and Witts, 1971). Nor are the secondary impacts of earthquakes confined to fire hazards; they can trigger landslides on unstable slopes, snow and ice avalanches in mountainous terrain, soil liquefaction, land subsidence, dam failures and, when they occur underwater, tsunami. For example, the Alaskan earthquake of April 1964 caused $235 million in damage to public property, $77 million to real estate, and 131 deaths; these losses resulted from not only ground shaking but landslides and slumping that damaged many buildings, submarine slides and tsunami that

TABLE 1.4. A Disaster Classification

	Natural	Industrial	Human

Earth	Avalanches
	Earthquakes
	Erosion
	Eruptions
	Toxic mineral deposits
		Dam failures
		Ecological neglect
		Landslides
		Radioactive pollution
		Subsidence
		Toxic waste disposal
		Outer-space debris
		  fallout
			Road and train
			  accidents
			Ecological
			  irresponsibility

Air	Blizzards
	Cyclones
	Meteorite and
	  planetary activity
	Ice storms
	Tornadoes
	Thermal shifts
	Dust storms
		Acid rain
		Chemical pollution
		Underground
		  explosions
		Radioactive cloud and
		  soot
		Urban smog
			Aircraft accidents
			Hijackings
			Spacecraft accidents

Fire	Lightning	Boiling
		  liquid/expanding
		  vapor accidents
		Electrical fires
		Hazardous chemicals
		Spontaneous
		  combustion
			Fire setting

Water	Drought Floods Storms Tsunami	Effluent contamination Oil spills Waste disposal	Maritime accidents

People	Endemic disease
	Epidemics
	Famine
	Overpopulation
	Plague
		Construction accidents
		Design flaws
		Equipment problems
		Illicit drugmaking and
		  taking
		Plant accidents
			Civil strife
			Criminal extortion by
			  viruses and poisons
			Guerrilla warfare
			Hostage taking
			Sports crowd violence
			Terrorism
			Warfare

Source: Taylor (1989). Copyright 1989 by AMS Press, Inc. Reprinted by permission.

destroyed docks and ports, and liquefaction that undermined a residential neighborhood (Bolt, 1978).

There are numerous examples of secondary impacts. Paradoxically, floods can produce fire dangers. In the Texas floods of October 1994, a gas pipeline burst, spreading flames and burning buildings along the San Jacinto River; similarly, in Egypt in November 1994, flooding led to 250 deaths when a train of flammable liquids was overturned and burst into flames. During hurricanes, a host of secondary processes can spawn tornadoes, sea surges, heavy rain, flooding, and wind damage; combined with human-induced threats, these culminate in widespread destruction. The devastating surge of seawater often causes more damage and death than other aspects of hurricanes. Certainly, many of the 300,000 to 500,000 who died in Bangladesh in 1970 were victims of rising seawater rather than the tropical cyclone itself (Murty, 1988). (Note that different sources have different estimates of the number of deaths, illustrating the difficulties of using numbers alone to register the impacts of disasters.)

Many disasters in Taylor's classification can be triggered by human activities. To a degree, this is closely tied to human use of resources. On the one hand, floods are useful in that they deposit fertile sediment in agricultural areas; on the other, they represent destructive agents bringing hardship and death. A holistic view that includes the careful management of resources would seem pertinent. In Nepal, for example, the gathering of wood for fuel has resulted in deforestation of steep hillsides, altering the hydrological regimes of local streams and aggravating flooding and sedimentation downstream. In the United States, urbanization and the drainage of wetlands for agricultural purposes have greatly exacerbated the flood hazard in some areas; it has been suggested that some of the flooding in the Midwest in 1993 resulted from these human actions.

Another set of hazards includes those related to technological and industrial activities. Industrial accidents, chemical spills, radioactive fallout and the like can be triggered by geophysical processes, adding to the hazardousness of a place. For the most part, however, large-scale industrial accidents do not occur as frequently as natural events and they usually involve fewer deaths (Glickman et al., 1992). While this book does not deal directly with such events, any consideration of total risk at a place should not overlook these concerns, for they do create catastrophic losses.

Thus, focus on the physical world and geophysical processes yields an incomplete understanding of natural hazards. The human element is important not only because people are the victims when events occur, but also because humans define the very essence of a natural hazard. Natural hazards constitute a complex web of physical and environmental factors interacting with the social, economic, and political realities of society. In this sense, the risk from natural hazards is heightened by human use of resources.

Structural Constraints and Societal Controls

Hewitt (1983) castigated hazards researchers for the overwhelming attention devoted to geophysical processes and neglect of societal forces. He stressed three points. First, natural hazards are neither explained by nor uniquely dependent upon the geophysical processes that may initiate damage; this is not to say that geophysical processes do not play a role, but that too much causality has been attributed to them. Second, human awareness of and response to natural hazards are not dependent solely on geophysical conditions. Hewitt saw hazards as more dependent on the concerns, pressures, goals, and risks of society, not least the effectiveness of measures to reduce calamity; these factors, he said, reflect the values and institutions of the society. Third, the causes, features, and consequences of natural disasters are not explained by conditions or behavior peculiar to calamitous events; these can be explained by everyday forces. The important parameters are social order, its everyday relations to the habitat, and larger historical circumstances that help shape society. Thus, disasters result more from social than geophysical processes, and hazardousness varies as much (or more) as a result of social as of geophysical processes.

While not completely rejecting the role of the physical world, Blaikie, Cannon, Davis, and Wisner (1994) also place more responsibility on the structure of society. Natural hazards and hazard vulnerability can best be determined by understanding social processes that affect the choices of a society's members. Their description of natural hazards takes us even further along the physical-human continuum, placing much of the cause, blame, and responsibility squarely on the human-use system.

> . . . The relative contributions of geophysical and biological processes, on the one hand, and social, economic, and political processes on the other, vary from disaster to disaster. Furthermore, human activities can modify physical and biological events, sometimes at many kilometers' distance (e.g. deforestation contributing to flooding downstream) or many years later (the introduction by people of a new seed or animal, or the substitution of one form of architecture with other less safe ones). The time dimension is extremely important in another way. Social, economic, and political processes are themselves often modified by a disaster in ways that make some people more vulnerable to an extreme event in the future. The "natural" and the "human" are so inextricably bound together in almost all disaster situations, especially when viewed in an enlarged time and space framework, that disasters cannot be understood to be "natural" in any straightforward way. (Blaikie et al., 1994, pp. 5-6)

It is their focus on constraints and ties of the human-use system that makes Hewitt's (1983) and Blaikie et al.'s (1994) contributions to hazards research so valuable. Their work posits that natural hazards cannot be understood without close attention to the political, economic, and social structures that constitute a given society. Further, the relationships of that society to other societies also may affect hazard impact. Loss is not caused solely by the geophysical event (although it is certainly the agent of destruction and can affect the degree of loss), but also by the makeup of society and how it responds to physical processes.

To take one example, marginalized groups invariably suffer more than nonmarginalized groups during hazardous events. Kreimer (1980) pointed out that squatters and slumdwellers (who number 1.72 million in Calcutta, 1.13 million in Jakarta, and 0.81 million in Karachi) have little or no choice of location and few options in responding to floods, tropical cyclones, or earthquakes. In fact, such marginalized groups are excluded from many of the basic elements of life and are too poor to move elsewhere. Disasters compound their difficulties, exacerbating the poverty trap in which they exist. The inequities of shared resources are "normal" conditions for them (Anderson, 1991). Even strategies for recovery can aggravate their vulnerability. Crisis management, emergency relief, restoration policies, and reconstruction attempt to recreate "normal" conditions (Hewitt, 1983), but restoration of the status quo perpetuates the difficulties of marginalized groups and does nothing to change risk. Thus, social constraints can determine how a society responds not only to hazardous events, but to their long-term impacts.

A Time–Space Framework for Hazard Classification

In defining natural hazards, it is clear that both physical processes and the human-use system are important. In addition, we should incorporate spatial and temporal components of the hazard. The timing of events can affect the outcome of hazards. In 1994, an earthquake in Northridge, California occurred at 4:31 A.M. before roads were crowded with commuters; the death toll of 57 was probably much lower than it would have been hours later. Similarly, in the Loma Prieta earthquake of 1989, many deaths resulted from collapsed interstate highways and the crushing of cars on multilevel roadways; however, because a World Series game was due to start, many fewer people were commuting than would have been usual at that time of day. The importance of timing can be seen with other hazards; for example, flooding may benefit farmers by providing soil moisture and sediment at certain times of the year, but destruction may be high if crops are already growing or cannot be replanted.

In some hazard classifications, events are categorized in terms of a time scale. Natural hazards often are described in relation to the rate of onset of the geophysical process, distinguishing between rapid events and those of a more pervasive nature. Rapid onset events include those hazards that usually provide little time for warning or preparatory action such as tornadoes, flash floods, earthquakes, and windstorms. More creeping or pervasive hazards (such as droughts, heat waves, and cold waves) have no definite beginning or end, hence complicate how society might respond. In-between are those geophysical processes (such as floods, tropical cyclones, and tsunami) that can be more accurately delineated in time to permit some form of response. Other physical and temporal traits also can delimit natural hazards, including measures of intensity, duration, and frequency of occurrence. These characteristics are considered in detail in Chapter 2; it is sufficient here to note that the physical dimensions of hazards can help shape coping strategies. However, hazard response definitely is not based solely on reaction to physical processes.

A second criterion of classification is spatial location. Not all hazards occur in every part of the world. It is too cold for hurricanes in Arctic regions, blizzards do not occur in tropical areas (at least not at sea level), and tornadoes are relatively rare outside the United States and Australia. Again, physical parameters can help to define risk. It is very unlikely, for instance, that a volcano will erupt in Duluth, Minnesota, but a blizzard or ice storm is a distinct possibility most years. The distribution of natural hazards around the world should not be overlooked when considering their significance. In addition, geographic location helps to determine whether a particular event constitutes a hazard. A 15 cm snowstorm in Duluth would not be considered a particular problem, but in Atlanta, Georgia, such an event could effectively stop activities within the city; similarly, freezing is not a problem in temperate climates, but it can lead to considerable difficulties in subtropical areas, especially for citrus farmers.

Finally, some argue that the areal or spatial extent of a hazard should be considered. Some hazards are diffuse, extending over thousands of square miles but impacting relatively few people. For example, the Australian drought of 1994 created serious problems for farmers throughout many agricultural areas, but did not seriously affect the general population; if it had continued, however, more people would have been affected because food supplies would have been compromised. Similarly, heat and cold waves might be classified as diffuse hazards. At the other extreme, some hazards are highly concentrated in place and impact large numbers of people. Earthquakes and flash floods typically fit this category since even low-intensity events can cause major losses.

Nevertheless, it is not possible to classify hazards by a purely physical determination of spatial extent, since levels of concentration-diffuseness depend on land use and population density. It could be argued that any hazard occurring in Hong Kong or Singapore will be highly concentrated whereas hazards in the Sahara will be diffuse. Thus, classifying hazards by spatial extent poses its own problems. Over what area must an event extend in order to be considered a natural disaster? Is the destruction of a small hamlet by an avalanche in the Alps or Andes the same as an earthquake in San Francisco or Tokyo? Do distinctions based on spatial extent provide better insight into natural disasters? These questions are not easily answered. Clearly, an individual who loses his or her house has suffered some form of personal disaster, though in the global picture it may be insignificant; to a community destroyed by a tornado, the event is a catastrophe, but to the country at large it may go virtually unnoticed.

Thus, the question remains how many individuals must be affected before an event can be described as a disaster. As discussed above, a hazard can exist anywhere, but a disaster occurs only when a significant component of society is impacted. Where we set our limits and definitions can significantly alter our view of hazards.

THE SIGNIFICANCE OF LOSSES

As must be evident by now, natural hazards exact a heavy toll from society, killing large numbers of people, destroying homes, and damaging property throughout the world every year. The data presented at the beginning of this chapter for 1985 are merely a sample of the yearly devastation that occurs in one form or another in virtually all nation-states. However, in our attempts to evaluate the significance of natural hazards we are confronted with serious data limitations that prevent us from obtaining an accurate picture of the global costs of natural hazards. In many cases, data on death and damage are unavailable or have been poorly collected. In others, "closed" governments have not released details of disaster impacts for fear of losing control within their countries. This occurred in the Sahel in the 1970s when several African nations initially refused to acknowledge the famine resulting from severe drought in the area; national pride effectively blocked efforts to alleviate the problem.

Immediately after an event, when emergency relief and rehabilitation have priority, it is difficult to assess the number of deaths and total property loss and other damages. Data on impacts can be sketchy and often are exaggerated, especially during the early hours of the event. It is not unusual to find initial, unofficial estimates of deaths and damages to be considerably higher than final figures (Alexander, 1993), and, on many occasions, the media accept and repeat these wild estimates. The opposite scenario also can occur, further confounding hazards researchers. In some events, particularly earthquakes, the death toll gradually rises as bodies are discovered under the ruins of buildings, roads, and bridges. In Guatemala after the earthquake in 1976, the estimated death toll rose from 1,000 to over 20,000 during the first seven days as officials became aware of the size of the problem (Seaman, 1990). In 1995, initial reports from Kobe, Japan stated that several hundred had been killed, but the final death toll exceeded 6,000. By contrast, in Algeria in 1980, the opposite happened as estimates fell from 20,000 deaths on the first day to 4,000 later.

Estimates of economic loss also are vague and notoriously inaccurate. Often, losses are inflated immediately after the impact phase, especially if certain damage thresholds must be attained in order to receive aid from a central government agency. In the United States, a presidential declaration of disaster can result in federal financial support and other help becoming available to the affected communities. Thus, it is in victims' interests to exaggerate losses. To complicate matters, data are extremely difficult to gather even after the impact phase is over and reconstruction has begun. Should losses include only direct economic damage or indirect and intangible losses as well?

The plethora of estimates can vary widely from one person to the next and from one time to the next. When Riebsame, Diaz, Moses, and Price (1986) reviewed estimates for average annual losses from weather-related disasters in the United States, they discovered that estimates ranged from $1 to $3.18 billion for floods, from $800 million to $1.8 billion for hurricanes, $200 million to $2 billion for tornadoes, and from $800 million to $1.2 billion for droughts. Consequently, a reasonable assessment of impact is not always possible, and any economic statistics should be treated with extreme care.

In spite of these limitations, the data do provide an indication of the widespread and severe loss associated with hazards. Several different estimates are discussed below, illustrating differences among as well as allowing comparisons between data sets. Much of that difference is based on the use of different criteria by different researchers. However, even when the statistics are not identical, researchers have established common patterns and trends.

A number of studies have assessed global economic loss. Noji (1991) estimates that in a 20-year period, 3 million lives were lost and a further 800 million people were affected by the impacts of natural hazards, with an annual cost to the global economy of $23 billion. Alexander (1993) cites similar figures, but sets annual costs to the global economy at $50 billion, one-third of which represents the cost of predicting, preventing, and mitigating against disasters and the other two-thirds direct damages. Burton, Kates, and White (1993) give a similar breakdown of costs allocated to prevention and damage recovery, but put global losses at $40 billion. Their estimates are based on Sheehan and Hewitt's (1969) criteria for classifying disasters, to wit, events that kill over 100 people, injure 100, or cause $1 million damage. Although these thresholds are useful in defining which events should be included in the survey, they preclude significant hazardous events below given thresholds. For example, to some communities, $1 million does not represent too great a hardship; in San Francisco, for example, a minor earthquake would rapidly lead to such losses, but in California's overall economy they would not be significant. By contrast, a tropical cyclone in Bangladesh might destroy considerable food reserves which in themselves would not amount to $1 million, but their loss would have very serious repercussions for local communities. Again, place—or more precisely, the differences between wealthy and poor nations—helps to define the significance of hazardous events. To ignore such differences is to over- or underestimate the impact and severity of such events for the victims.

Probably the most comprehensive and valuable survey of disaster-related deaths since World War II was undertaken by Glickman et al. (1992). Their study examined disasters worldwide, using 25 deaths or more as the principal criterion by which to include events. Information was collected from over 20 data bases in Europe and the United States for the years 1945–1986. Two major drawbacks to this study must be mentioned. First, the failure to include "smaller" events (resulting in less than 25 deaths) means that some significant hazards may have been missed. Some events may have brought about catastrophic economic losses, especially in less wealthy nations, but because of few reported deaths would have been excluded. Second, several pervasive and long-term hazards such as droughts, epidemics, famines, and infestations were not included in the data set so that these statistics probably represent serious underestimates of the actual deaths accruing from hazards during this time period. Despite these limitations, it is a valuable study.

As mentioned at the beginning of this chapter, from 1945 to 1986, there were 1,267 disasters leading to 2,343,000 deaths, or an average of 56,000 deaths per year (Glickman et al., 1992). When the statistics are examined in greater detail, we find that three events accounted for a substantial number of the deaths. Two earthquakes, one in the former Soviet Union (1948) and one in China (1976), killed 110,000 and 700,000 people, respectively, and a tropical cyclone in Bangladesh (1970) killed 300,000 to 700,000 people. When those major catastrophes are removed from the data set, the number of deaths per year falls to 25,000. Obviously, large events can distort short-term trends. Further, deaths are not evenly spread among events; rather, a relatively small number of events accounts for a large proportion of deaths. For instance, Smith (1992) found that since 1949, there had been 17 major disasters, defined as those killing more than 10,000 people. In Glickman et al.'s (1992) data, 2.3% (or 29) disasters reached this threshold of 10,000 deaths, accounting for 81.2% of all deaths (1.9 million) from natural disasters. While hazards represent an ever-present risk, most deaths are caused by relatively few but very large events.

Temporal Trends

The data on disasters and deaths show distinct trends over time (Table 1.5). The number of disasters per 7-year period has risen gradually from 107 in 1945–1951, to 303 in 1980–1986; the number of deaths shows a similar pattern, at least until the last period. This temporal pattern of increasing deaths and disasters has been noted by others, including Berz (1988), Dworkin (1974), Hagman (1984), and Smith (1992). Economic losses also have increased dramatically, even when the data are controlled for inflation. According to Kreimer and Munasinghe (1991), global economic loss from weather-related disasters has increased approximately 90 times since 1950. Again, caution is advised with regard to temporal trends; inflation can alter the significance of these patterns as can better reporting of impacts. Nonetheless, the absolute amount of losses is staggering.

Spatial Trends

The spatial pattern of disaster deaths and damages is even more intriguing. According to Burton, Kates, and White (1993), 95% of all disaster-related deaths occur among the 67% of the world's population who live in the poorer nations, but 75% of economic loss is in the wealthy nations. Thus, an undue burden is borne by poorer nations in coping with natural hazards. That burden is even greater than revealed by statistical data, which do not reflect the relative impacts of economic loss in different countries. In real terms, losses sustained in poorer nations are actually greater than in wealthy nations. For example, a family in Bangladesh might lose all its possessions, but accrue only tens of dollars in damage; by comparison, when an automobile in California is destroyed by earthquake, the economic

TABLE 1.5. Temporal Trends in Natural Disasters, 1945–1986

Period	Number of disasters	Deaths (in thousands)	Deaths per disaster	Deaths per year
1945–1951	107 (106)	252 (142)	2,351 (1,336)	36,000 (20,286)
1952–1958	172	92	534	13,143
1959–1965	197	239	1,212	34,143
1966–1972	244 (243)	706 (206)	2,892 (847)	100,857 (29,428)
1973–1979	244 (243)	921 (221)	3,777 (912)	131,571 (31,571)
1980–1986	303	134	441	19,143
Total	1,267 (1,264)	2,343 (1,033)	1,849 (837)	55,786 (24,595)

Note: Numbers in parentheses exclude the three worst disasters. *Source:* Glickman, Goldman, and Silverman, 1992. Copyright 1992 by Resources for the Future. Reprinted by permission.

loss is high but the personal costs may be low. Some have suggested using a ratio of loss to income in order to determine the significance of loss for communities, at least in the aggregate.

Figure 1.2 and Table 1.6a illustrate the spatial impact of disasters around the world in terms of deaths and number of disaster events. These data show that East Asia and the Pacific region, southern Asia, Latin America, and the Caribbean have experienced the most deaths and disasters during the 42-year period studied by Glickman et al. (1992). However, controlling for size of resident populations shows the Middle East and North Africa joining Latin America and the Caribbean as the worst

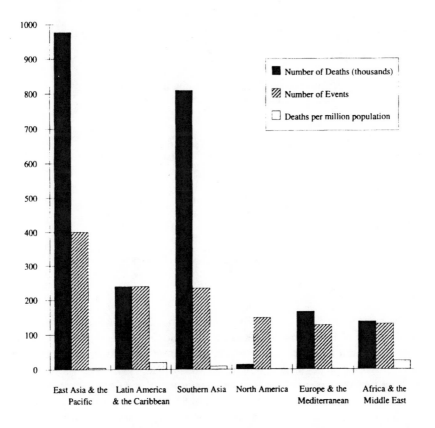

FIGURE 1.2. Spatial patterns of natural disasters, 1945–1986. The regional pattern of deaths and events reveals striking differences. Notably, the number of deaths is not closely related to the number of events, on either an absolute or relative basis. Other factors clearly come into play. *Source:* Glickman et al. (1992). Copyright 1992 by Resources for the Future. Reprinted by permission.

TABLE 1.6. Spatial Patterns of Natural Disasters

**(a) 1945–1986 (does not include droughts,
epidemics, famines, accidents, and civil strife)**

Region	Number of disasters	Deaths (in thousands)	Deaths per disaster	Deaths per million population
East Asia/Pacific	401 (400)	977 (277)	2,435 (691)	5
Latin America/Caribbean	239	239	999	21
Southern Asia	234 (233)	808 (308)	3,452 (1,321)	10
North America	149	15	104	2
Europe/USSR	127 (126)	166 (56)	1,308 (444)	2
Middle East/N. Africa	86	129	1,505	24
Sub-Saharan Africa	45	9	201	1
Total	1,267 (1,264)	2343 (1,033)	1,849 (837)	

Note: Numbers in parentheses exclude the three worst disasters. *Source:* Glickman et al. (1992). Copyright 1992 by Resources for the Future. Reprinted by permission.

**(b) 1967–1991 (includes droughts, epidemics,
famines, accidents, and civil strife)**

Region	Total number of disasters	Percentage of total	Total number of people killed by disasters	Percentage of total
Africa	1,168	15.0	3,631,804	49.7
The Americas	1,730	22.3	434,131	5.9
Asia	3,244	41.8	3,120,025	42.7
Europe	1,124	14.5	116,375	1.6
Oceania	500	6.4	5,223	0.7
Total	7,766		7,307,558	

Data source: International Federation of Red Cross and Red Crescent Societies (1993).

hit areas (Figure 1.2). Table 1.6b shows data produced by the International Federation of Red Cross and Red Crescent Societies (1993); although this survey relied on different disaster criteria than Glickman et al.'s and used slightly different regions, the patterns are similar.

When the data are broken down by country, different patterns emerge. Between 1900 and 1988, India experienced the most disasters, the USSR suffered the most disaster-related deaths per event, and Italy had the most economic loss (Office of US Foreign Disaster Assistance, 1988; 1993). These data, particularly for the USSR, must be interpreted with caution because the Office of Foreign Disaster Assistance includes in its tally civil strife, emergency, and displaced and expelled persons. Moreover, these statistics do not include the United States (see Table 1.7). Data for disasters between 1967 and 1991 indicate that the greatest number of deaths per year occurred in Ethiopia, Nigeria, Bangladesh, and Cambodia (Table 1.8), but these data include a diverse range of events such as droughts, epidemics, famines, accidents, and civil strife.

Clearly such patterns are to be expected. With greater wealth, more property is at risk so that losses can be high. Atkisson et al. (1984) estimated that just nine hazards in the United States accounted for $8 billion in damage in 1970, a figure they projected would rise to $17.78 billion in

TABLE 1.7. Rankings of Countries That Experienced 25 or More Disasters, 1900–1988

Number of disasters		Average number of disaster related deaths		Disaster damage per event (thousand U.S.$)	
Country	Total	County	Total	Country	Total
India	199	USSR	284,334	Italy	611,694
Philippines	134	China	80,812	Spain	374,686
Indonesia	110	India	44,379	Chile	121,505
Bangladesh	109	Bangladesh	26,981	USSR	90,645
Japan	91	Ethiopia	16,138	Argentina	84,758
China	89	Niger	7,826	Mexico	80,563
Brazil	68	Mozambique	7,262	Colombia	51,969
Mexico	60	Italy	2,949	Pakistan	39,370
Peru	55	Pakistan	2,061	China	39,296
Iran	53	Japan	2,005	Peru	32,498
Turkey	43	Peru	1,355	India	31,940
Colombia	39	Chile	1,107	Sri Lanka	31,734
Italy	39	Iran	1,103	Japan	30,416
Korea	38	Turkey	1,027	Bangladesh	26,831
Chile	37	Colombia	705	Korea	25,116
Burma	36	Haiti	429	Philippines	13,393
Pakistan	33	Vietnam	412	Haiti	10,460
Vietnam	32	Sri Lanka	317	Turkey	10,320
USSR	31	Mexico	287	Mozambique	9,588
Ecuador	30	Ecuador	261	Ecuador	8,830
Argentina	29	Indonesia	225	Brazil	6,964
Sri Lanka	29	Philippines	222	Indonesia	6,838
Niger	27	Argentina	202	Niger	4,322
Haiti	26	Burma	176	Burma	4,280
Ethiopia	25	Korea	107	Ethiopia	3,129
Mozambique	25	Spain	106	Vietnam	2,296
South Africa	25	Brazil	99	Iran	1,415
Spain	25	South Africa	73	South Africa	40

Note: The United States was excluded from these data. The types of disaster included are accidents and unusual phenomena; civil strife, emergencies, and displaced and expelled persons; cyclones, hurricanes, typhoons, and storms; drought, food shortage, and famine; epidemics; earthquakes and tsunami; fire; floods; infestation; landslides and avalanches; and volcanic eruptions. *Data source:* Office of U.S. Foreign Disaster Assistance (1988).

TABLE 1.8. Average Number of People Reported Killed by Disasters per Country per Year, 1967–1991

The 25 highest-ranking nations		Nations having 20 or fewer annual deaths on average	
Country	Deaths	Deaths	Countries
Ethiopia	48,517	20	Mauritania
Nigeria	40,720	19	Democratic People's Republic
Bangladesh	40,408		of Korea
Cambodia	40,005	17	Jamaica
Sudan	16,195	16	Albania, Ireland, Cuba,
Mozambique	15,763		Senegal, Morocco
China	12,755	14	Fiji
Pakistan	8,688	13	Ghana, Djibouti
Iran	5,801	12	Guinea, Tajikistan, Zimbabwe
India	5,044	11	Costa Rica, Hungary, Congo,
Nicaragua	4,619		Ivory Coast, Panama
Burundi	4,229	10	Denmark, Azores, Togo,
Peru	4,140		Sweden
Colombia	2,967	9	Oman, Maldives
Iraq	2,937	8	Czechoslovakia, Botswana
Lebanon	2,577	7	São Tomé and Príncipe,
Philippines	2,558		Mauritius, Surinam
Turkey	2,045	6	Switzerland, United Arab
Somalia	1,695		Emirates
Former U.S.S.R.	1,367	5	Libya, Laos, Bahrain, Bulgaria,
El Salvador	994		Comoros
Guatemala	981	4	Austria, Israel, Vanuatu,
Indonesia	856		Bahamas, Reunion, Trinidad
United States	837		and Tobago, Syria
Honduras	830	3	Grenada, Gabon, Guinea Bissau, Cape Verde, Martinique
		2	Lesotho, Georgia, St. Lucia, Mongolia, Finland, Dominica, New Zealand
		1	Anguilla, Central African Republic, Malta, Samoa, Singapore, Western Samoa, Kiribati, Netherlands, Iceland, Uruguay, Tonga, Equatorial Guinea

Data source: International Federation of Red Cross and Red Crescent Societies (1993).

2000; in fact, losses for 1992, 1993, and 1994 in the United States actually exceeded that projection. Yet, the more numerous disasters and catastrophes in poorer nations did not reach comparable levels of economic loss. In Latin America and the Caribbean for 1980–1987, Anderson (1991) figured $16.4 billion in loss, or approximately $2 billion damage per year due to earthquakes, floods, droughts, volcanic eruptions, and hurricanes (Table 1.9).

Nevertheless, if losses are examined as a proportion of total wealth, those in poorer nations are significantly higher than in wealthy nations. In five Central American countries (members of the Central American Common Market) during the years 1960–1974, natural disasters reduced growth of the gross domestic product (GDP) at an annual rate of 2.3% Anderson (1991). Similarly, hurricane losses in the Dominican Republic (1979) and in Haiti, St. Lucia, and St. Vincent (1980) amounted to 15% of the gross national product (GNP) (Smith, 1992). Thus, as a percentage of GNP, disaster losses can be much higher in poorer nations than in wealthy ones. Zupka (1988) argues that they can be up to 20 or 30 times higher, though Anderson (1991) suggests a more conservative figure of 20% higher. Nanjira (1991), who looked at several disasters in 10 East African countries, found that the costs to the GNP in 1980 ranged from 1.6% in Djibouti to 69% in Kenya (Table 1.10).

The scale of such losses means that many countries' development potential is undermined by recurring disasters and the human efforts and finance needed for recovery. Hazards also pose long-term difficulties. First, foreign investment may be curtailed and plans upset throughout the economy by a perceived high-risk environment. For example, Anderson (1991) states that in Bangladesh, the Philippines, and the Sudan, the business climate has been adversely affected by hazards.

Second, marginalized groups are often most severely affected (Kreimer, 1980). Those with the least economic and political power can do little to break the cycle of poverty and disaster, a cycle that Anderson (1991) suggests is perpetuated by disaster threat and failure to deal with its real causes. Thus, disaster and poverty are mutually reinforcing as vulnerability is exacerbated and events are repeated. It is interesting to consider how increasing populations will affect these statistics; although the number of

TABLE 1.9. Economic Losses from Natural Disasters in Latin America and the Caribbean, 1980–1987 (in U.S.$ million, 1987)

Losses and effects	Earthquakes	Eruptions	Hurricanes	Floods and droughts
Total Losses	9,679	224	2,485	3,970
Direct	7,671	154	1,975	1,311
Indirect	2,008	70	510	2,659
Secondary effects				
Public finances	4,286	—	1,132	n.a.
Exports/imports	12,567	—	1,076	621

Source: Anderson (1991). Copyright 1991 by The World Bank. Reprinted by permission.

TABLE 1.10. Disaster-Proneness and Economic Vulnerability of East Africa

Country	Area (mi^2)	Climate	Population (1987)	Population density (per km^2)	Per capita GNP (U.S.$)	Type of disaster	Cost of disaster (%GNP)
Burundi	10,747	Tropical, volcanic soils, irregular rains	5,001,000	179.7	240	Drought, famine, refugees, civil strife, displaced persons	8.50
Djibouti	8,958	Arid volcanic rocks, torrid, high tropical monsoons	483,000	20.8	460	Drought, pests, epidemics, civil strife, floods, refugees	1.60
Ethiopia	483,123	Tropical plateau, semidesert	46,184,000	36.9	130	Drought, famine, desertification, pests, civil strife, epidemics, earthquakes, floods, refugees	40.80
Kenya	224,960	Equatorial tropical forests	21,163,000	36.5	390	Drought, refugees, desertification, floods, pests, earthquakes, epidemics	69.30
Mozambique	308,641	Tropical forests, dry and hot	15,127,000	18.9	270	Floods, civil strife, cyclones, epidemics, famine	32.62
Rwanda	10,169	Tropical wet and dry, marshy	5,700,000	218.6	220	Drought, civil strife, famine	11.50
Somalia	246,201	Savannah, semidesert	6,860,000	10.8	260	Drought, pests, desertification, epidemics, oil spills, refugees, civil strife	11.10
Sudan	967,000	Flat tropical plains, semidesert	18,681,000	8.2	360	Desertification, famine, floods, epidemics, pests	67.40
Tanzania	364,900	Tropical rain forests, woodlands	23,217,000	23.8	270	Desertification, famine, drought, epidemics, floods	48.90
Uganda	93,104	Equatorial tropical	12,630,076	52.4	200	Epidemics, drought, floods, famine, civil strife, refugees, desertification	24.90

Source: Don Nanjira (1991). Copyright 1991 by The World Bank. Reprinted by permission.

deaths will quite likely increase, the proportions could decrease, masking the true impacts of hazards. Those patterns support the argument that the human component of natural hazards outweighs physical characteristics in defining hazardousness and vulnerability.

Geophysical Classifications

When losses are broken down by geophysical event, other patterns emerge (Table 1.11). Clearly, floods have been the most common event of the post–World War II period, accounting for over 30% of all disasters; hurricanes and tropical cyclones make up about 20%, earthquakes between 15% and 17%, and tornadoes approximately 12% (Glickman et al., 1992; Shah, 1983; Thompson, 1982). In terms of deaths, however, the pattern is different. Glickman et al.'s study (1992) indicates that earthquakes and hurricanes or tropical cyclones killed more people than any other event, with floods dropping to third; even after the three largest events are removed from their data, earthquakes and tropical cyclones account for more deaths than other hazards. The International Federation of Red Cross and Red Crescent Societies (1993) found similar patterns, but drought (a hazard excluded from Glickman et al.'s [1992] data) exceeded all other natural disasters.

TABLE 1.11. Global Loss of Life by Geophysical Event, 1945–1986

Type of disaster	Number of disasters	Deaths (in thousands)	Deaths per disaster
Meteorological			
Flood	395	244	618
Tropical cyclone	272 (271)	791 (291)	2,907 (1,072)
Other storm	212	28	131
Heat wave	23	5	223
Cold wave	15	4	275
Geological			
Earthquake	191 (189)	1,198 (388)	6,272 (2,053)
Volcanic eruption	27	40	1,494
Tsunami	7	3	271
Other			
Landslide	85	25	295
Fire	40	6	157
Total	1,267 (1,264)	2,343 (1,033)	1,849 (837)

Note: Numbers in parentheses exclude the three worst disasters. *Data source:* Glickman et al. (1992). Copyright 1992 by Resources for the Future. Reprinted by permission.

Like other hazard data these should be treated with care. It is not always possible to determine exactly how or why individuals died, and problems may follow from categorizing hazard victims too neatly. For example, a whole village was destroyed in 1970 when an earthquake triggered an avalanche that sent snow, ice, mud, and rocks careening down a valley, overrunning the village of Yungay, Peru. Should the 20,000 people who died be classified as earthquake, avalanche, or landslide victims? Similarly, in Bangladesh in 1970, at least 300,000 people died when a tropical cyclone surged up the Bay of Bengal; they died from varied causes, including drowning, windblown trauma, and exposure. As Murty (1988) points out, many deaths from tropical cyclones result from drowning due to the storm surge; Murty claims that the flooding of low-lying areas is responsible for up to 97% of deaths from tropical cyclones.

Other hazards present similar difficulties. For instance, deaths that result from extreme drought may be directly attributable to famine. More often than not, however, most victims succumb to disease before starvation because the body's immune system has been severely weakened by lack of food. Understanding how people died can help in assessing vulnerability and improving future responses.

Several researchers have focused on how people have died. Gruntfest (1977), who studied what people did during the Big Thompson flood in Colorado in 1976, found that many died trying to outrun the rushing wall of water, so were drowned in their cars. Work such as hers has led to the posting of warning signs that alert travelers to the potential for flash floods and recommended that they leave their cars and climb to safety. Similar findings (albeit from a limited data set) were reported by French, Ing, Von Allmen, and Wood (1983) in a study on flash floods in the United States; of 190 reported deaths, 93% were from drowning and 42% were car related.

Finally, spatial data also show that hazard threat is not equally distributed and that similar events can have significantly different impacts (Table 1.12). For instance, floods are ubiquitous, but deaths are highest in Asia and relatively low in North America. Noji (1991) reports that 70% of all flood deaths have occurred in India and Bangladesh.

Causes of Spatial and Temporal Trends

Various hypotheses have been proposed to explain the spatial and temporal trends in hazards data. For example, it is conceivable that geophysical events have been increasing in frequency during the past twenty or thirty years, which would account for the rise in number of hazards and the jump in number of deaths. Although this seems highly unlikely (at least on a

TABLE 1.12. Loss of Life by Disaster Type and by Continent and Region, 1947–1980

Disaster type	Asia	Oceania	Africa	Europe	South America	Caribbean and Central America	North America
Floods	170,664	77	3,891	11,199	4,386	2,575	1,633
Typhoons, hurricanes, and cyclones	478,574	290	864	250	–	16,541	1,997
Earthquakes	354,521	18	18,232	7,750	38,837	30,613	77
Tornadoes	4,308	–	548	39	–	26	2,727
Thunderstorms	20,210	–	–	120	60	310	244
Snowstorms	6,360	17	–	1,340	–	200	1,910
Heat waves	4,705	100	–	340	135	–	2,190
Cold waves	1,330	–	–	1,440	–	–	600
Volcanoes	2,805	4,000	–	2,000	440	151	34
Landslides	4,021	–	–	300	912	260	–
Rainstorms	1,648	–	5	26	145	–	49
Avalanches	335	–	–	340	4,350	–	–
Tsunami	4,459	–	–	–	–	–	60
Fog	–	–	–	3,550	–	–	–
Sand- and dust storms	150	–	–	–	–	–	10
Total	1,054,090	4,502	25,540	26,694	49,265	50,676	11,531

Source: Shah (1983). Copyright 1983 by Blackwell Publishers. Reprinted by permission.

global scale), there is some limited evidence that incidence has increased locally. For example, Changnon and Changnon (1992) examined temporal fluctuations in weather-related disasters from 1950 through 1989 in the United States. They found the incidence of disasters to be high in the 1950s and again in the 1980s, with relatively lower periods in-between; however, data on losses and extent of hazards did not follow the same pattern. Other studies of geophysical processes have proved inconclusive.

Current research on the impacts of global climate change suggests that increases in the frequency and magnitude of certain meteorological events are likely. Specifically, increasing global and regional temperatures may trigger more volatile weather conditions and some extremes in weather (Mitchell and Ericksen, 1992); for instance, warmer ocean waters in the tropics could mean longer and more intense hurricane seasons. This does not bode well for disaster losses since three-quarters of all disaster deaths and economic losses have been attributed to weather-related hazards (Kates, 1979). Population growth in vulnerable urban areas (especially in poorer countries) and a reluctance by governments to recognize the hazard potential of climate change aggravate the situation. As a result, the uneven spatial distribution of losses—with a larger proportion in poorer countries—will continue.

In some places, natural hazards may have increased because of human activities. Urbanization, deforestation, and the drainage of wetlands have changed hydrologic regimes, suggesting the possibility of more incidents of small-scale flooding. However, it is unlikely that humans have had any significant impact on the frequency of volcanic activity, major earthquakes, or tropical cyclones. Some large events, such as the 1993 floods in the midwestern United States, would have occurred with or without human intervention; floodplains would have been flooded and low areas inundated regardless whether levees had been constructed. The human factor only exacerbated the problem, placing large numbers of people at risk while increasing their vulnerability by ignoring the inadequacy of alleviation structures.

Another explanation for spatial and temporal trends might be increased population, which places more people in hazardous areas. Millions of people live in flood-prone environments, many of them vulnerable to the extremes of nature. In addition, increased urbanization puts more people in dense concentrations, often in high-risk areas. In the United States, for example, approximately 7% of the country can be defined as floodplain (94 million acres), which puts 9.6 million households and $390 billion worth of property at risk. Some 3.5–5.5 million acres have been urbanized, with over 6,000 communities (each with populations of 2,500 or more) exposed to flood hazard. Moreover, the rate of urban growth on floodplains (1.5 to 2.5% annually) has been twice that of the rest of the country (Federal Interagency Floodplain Management Task Force, 1992).

The situation is much the same or even more severe in other parts of the world. Cohen (1991) describes the tremendous increase in urban populations around the world. In 1960, Shanghai was the only city with a population over 10 million, but by 2000, there will be 17 such cities; with about 20 million people, Mexico City is believed to be the largest urban agglomeration. Such rapid growth has pushed many cities toward developing hazardous areas or developing them more densely, greatly increasing the number of people exposed to natural hazards. The problem is exacerbated by the fact that 25% of urban dwellers (an estimated 1.3 billion) live below the poverty line (Cohen, 1991). The interdependence of urbanized living further increases hazard problems, as their impacts are felt regionally.

Urbanization is complicated by unplanned development and the expansion of squatter settlements within urban agglomerations. According to Torry (1980), some of these communities have been growing at twice the rate of most urban areas. Further, squatters generally occupy some of the most risk-prone areas within a city, and their poorly constructed housing offers only minimal protection from extreme geophysical events. For example, squatters have settled the steep unstable hillsides of Lima, Rio de Janeiro, and Hong Kong; the flood-prone marshes and lowlands around Santiago, Karachi, and Davao; and ravines and verges in Caracas, Mexico City, Algiers, Delhi, and Manila (Havlick, 1986; Turner, 1969). People living in these areas are vulnerable to many different natural hazards.

Thus, increased population puts more people into risk areas; can promote changes in the physical environment, altering the risk potential; and can increase vulnerability by putting additional strains on a nation's resources so that remedial action is not possible. With concurrent marginalization of many groups, it is not surprising to find deaths and damages increasing. According to Cohen (1991), "Our phenomenal ignorance of the escalating economic, social, and political stakes of urban growth makes risk assessment exceedingly difficult. The likelihood of environmental disasters is probably increasing, and the prospects of their affecting cities is certainly increasing—because human activity is becoming densified" (Cohen 1991, p. 93).

A third possibility is that changing spatial and temporal patterns reflect better data collection and reporting. With new technological advances and improved global communications, famines and floods in formerly "remote" parts of the world can be flashed onto our television screens almost instantaneously. However, though greater media awareness may account for some of the apparent increase in deaths and hazards, it does not explain all of it. Media coverage is notoriously distorted, depending on which events are deemed both important and newsworthy relative to competing stories. Adams (1986), who examined 35 disasters between 1975 and 1985, found that U.S. television did not discern the severity of disaster events and was strongly biased toward Western nations; Adams suggested that in the eyes of the media, one Western European's death was equal to that of 3 Eastern Europeans, who were equal to 9 Latin Americans, who were equal to 11 Middle Easterners, who were equal to 12 Asians. Clearly, media coverage also adds to data difficulties, confounding our understanding of patterns and perhaps explaining why many people remain ignorant of deaths despite improved data collection and reporting.

HAZARDS: RISK, IMPACTS, AND VULNERABILITY

Examination of patterns of disaster losses is useful in documenting the magnitude of the hazard problem in its many variations, but it is essentially the easy part of data analysis. Other aspects of hazards are less visible and more probabilistic, but are critical to reducing losses. If we are to see any change in disaster losses, we must come to grips with both the hazard and the risk from a number of angles. Indeed, it is only through decreased loss that we will know if, when, and where we have made any progress in reducing hazardousness. Thus, we must consider all parts of the puzzle, and the issue is how to examine those parts.

A hazard scenario is described simply but coherently by Alexander (1993). Briefly, hazard risk becomes a threat once an event is impending, culminating eventually in impact and post-impact phases. The sequence of events passes through several stages, namely: hazard, risk, threat, disaster (impact), aftermath (postimpact). This sequence operates at different spatial scales and for all natural hazards (whether a small local flood, severe tropical cyclone, major earthquake, or some other type). However, the sequence does not account for variability in the time frame for each stage; for example, the threat phase depends on the speed of onset of the geophysical process, but direct impact is controlled, for the most part, by the duration of the event. Thus, a local flood event may pose a direct threat over several hours or even days, a tropical cyclone may be tracked for over a week before hitting the shoreline, and an earthquake may permit no warning time.

Once an event occurs and society has been disrupted, the impact phase has been reached. This would seem to be easily defined: a flash flood sweeps through a mountain community destroying property; an earthquake reduces buildings to rubble in seconds; a tornado tears apart houses to their very foundations. However, the start of some hazards (such as droughts, heat waves, and cold waves) cannot be pinpointed with such precision. Disaster impact and intensity may be moving targets, both spatially and temporally variable for individual societies. For example, the floods in the midwestern United States in 1993 did not affect all communities at the same time; flooding began in late spring and continued throughout the summer with different communities inundated at different times. Nor were these incidents clearly correlated with downstream flooding; in many instances, the same communities were flooded on several separate occasions, such as Chelsea and Tama, Iowa (Tobin and Montz, 1994b).

Usually, the aftermath phase is one of cleanup, relief management, and rehabilitation. It often focuses on how to respond to the next event (at least until this one is forgotten) and may be extended by more permanent alleviation strategies. The disaster and aftermath stages vary considerably between places and depend on both physical and human factors. Haas, Kates, and Bowden (1977) attempted to model this temporal scale for an urban disaster. Recovery was divided into four phases, emergency, restoration, reconstruction I, and reconstruction II, with each phase taking approximately ten times longer than the previous one to complete (Figure 1.3). However, we cannot expect their model to hold for all urban areas, particularly given the different social, political and economic conditions that prevail in cities around the world. In addition, in a catastrophe (which may be related more to socioeconomic components than to magnitude or duration of an event), some phases may be insignificant as the focus turns to long-term reconstruction. Indeed, a catastrophe not only disrupts society, but may cause a total breakdown in day-to-day functioning. One aspect of catastrophes, is that most community functions disappear; there is no immediate leadership, hospitals may be damaged or destroyed, and the damage may be so great and so extensive that survivors have nowhere to turn for help (Quarantelli, 1994). In disaster situations, it is not unusual for survivors to seek help from friends and neighbors, but this cannot happen in catastrophes. In a disaster, society continues to operate and it is common to see scheduled events continue; for example, after the Northridge earthquake in California in 1994, life went on for most Los Angeles residents. In contrast, the people who survived the 1988 earthquake in Spitak, Armenia, had nowhere to turn because the devastation was so extensive (Goenjian et al., 1994).

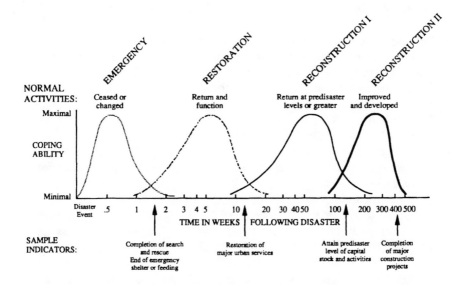

FIGURE 1.3. The postdisaster recovery period. Of the four distinct intervals, each lasts approximately 10 times longer than the previous one. Evidence to support this time frame, however, is sketchy. *Source:* Haas, Kates, & Bowden (1977). Copyright 1977 by MIT Press. Reprinted by permission.

All this points to the conclusion that if disasters are to be studied scientifically, a quantitative assessment of their impacts would be useful. So far, however, attempts to delineate disasters quantitatively (examples of which were described above) have met with limited success. Inevitably, it seems, different proposals omit important measures or incorporate others that are extremely difficult to assess with any degree of accuracy. Nevertheless, such problems should not deter researchers from attempting to model disaster impacts. For instance, de Boer (1990) proposes a ranking system based primarily on medical terms, notably injuries and deaths. He assigns scores to seven components of disasters: (1) effects on the surrounding community; (2) cause or agent, whether human or natural; (3) duration of the event; (4) radius of the disaster area; (5) number of casualties; (6) nature of injuries; and (7) the time required by rescue organizations to undertake initial primary treatment. Although this model may be medically valuable, it does not address other social, economic, and political aspects of disasters that are inextricably linked with spatial and temporal scales. Therefore, we leave the level of disaster and catastrophe somewhat vague and certainly qualitative in order to allow for factors of scale.

Vulnerability

Although it appears that communities are constantly faced with the risk of hazards, does this mean that all are equally vulnerable? Some societies have adopted stringent regulations and constructed major projects to ameliorate risk, reducing their vulnerability through positive action. It would be foolish to describe two communities located in similar risk areas as equally vulnerable if one has adopted strategies to cope with the hazard and one has not. However, changing social and economic conditions may increase vulnerability to catastrophe. When the design capacity of a mitigation measure is exceeded (such as when a dam fails), the resulting devastation can be catastrophic.

Nonetheless, measures of vulnerability should take into account those activities undertaken to mitigate hazard threats. Such activities can reduce the susceptibility of the society to the hazard event even if the risk (or probability of occurrence) is high, at least up to the design level. For example, the construction of earthquake-proof structures can significantly reduce the vulnerability of a community to earthquake losses; similarly, efforts to zone land for low-density use can minimize the impacts of flooding and tsunami. Thus, vulnerability represents a combination of risk and response. Either lack of response or an inability to respond to risk can heighten vulnerability; in contrast, well-laid plans aiding all segments of society can reduce vulnerability.

The International Federation of Red Cross and Red Crescent Societies

(1993) divided vulnerability into three components: material, organizational, and sociopsychological. Lack of resources and high levels of poverty (as in many poor nations) subject many societies to material vulnerability. By comparison, societies with more resources are less likely to be constrained in their responses; in short, it is easier to survive if one has more. Similarly, societies lacking in organization (whether at the state, local, or family level) are more vulnerable than those with more established structures. The final category, sociopsychological vulnerability, is less clearly defined, but an ability to cope with disasters and the associated stress is related to personality and psychological factors.

This is not to suggest that all human activities reduce the hazard risk. On the contrary, many land-use and societal decisions actually aggravate problems; for instance, the expansion of urban areas into floodplains, wetlands, and steep slopes heightens risk and vulnerability. The increasing complexity of society also may add to levels of vulnerability; our reliance on international trade and commerce, and our dependence on technological innovations (especially computers) put society at increasing risk whenever hazards threaten. Thus, human activities can exacerbate a population's vulnerability by increasing the sequence of events and minimizing the ability of individuals or society to respond (Blaikie et al., 1994).

Notwithstanding the human dimension, changes in the physical characteristics of natural hazards also may significantly change levels of vulnerability. For instance, global climate change models project that a warmer earth will result in stronger and more frequent hurricanes, more frequent droughts in many parts of the world, and much higher sea levels that would threaten many coastal cities (Mitchell and Ericksen, 1992). The picture is ominous for many communities that will have to face new, more frequent, or more devastating natural hazards.

PLAYERS AND DECISION MAKING IN THE HAZARD GAME

The physical-social complex that constitutes natural hazards means that people from all sectors of society are invariably involved at one stage or another. Town planners, developers, politicians, businesspersons, real-estate agents, social workers, hospital personnel, agency workers (both volunteers and official) and, of course, residents and workers all have stakes in the outcome of natural hazards. Individuals may be directly impacted by events, becoming hazard victims or survivors as well as statistics in the disaster toll. They may play the role of relief workers and participate in hazard prevention teams. For example, when floods are imminent, it is not unusual to see large numbers of people constructing sandbag levees to protect land and property; in the United States in 1993, thousands of people from around the country were transported by buses and cars to participate in such efforts.

The roles of these various actors will be addressed further in subsequent chapters, but it is important to note that everyone in society is affected by and at times must make a decision about a hazard threat. This may entail life-and-death decision making on where to go in the event of a tropical cyclone or merely how to get from home to work because floods have washed out a particular bridge. Nevertheless, decisions must be made and the workings of society (at least in disaster situations) usually go on. In this sense, hazards may be no more than part of the normal operating procedure of any society.

Living with Hazards

A primary concern of many has been to explain why people continue to reside and work in hazardous areas. For instance, residents of floodplains invariably return to flooded structures where they are constantly reacting to the hydrological processes of the riverine environment. Earthquake residents rebuild homes in tectonically active areas, and victims of volcanic eruptions continue to occupy risky slopes.

The gradual change in emphasis in hazards research from the physical process to the human-use system has provided a better understanding of why people continue to occupy hazardous areas. An example of the former approach can be found in Kates (1962). Like many others, he attempted to model individual response to flood hazard based on perceived hazard frequency. He noted that communities with a flood frequency of less than 1.44 floods in a ten-year period had a negative certainty about undertaking flood alleviation; that is, they were unlikely to adopt a mechanism by which to mitigate flood damage. Communities with flood records in excess of 4 floods per ten-year period had a positive certainty so that alleviation action was taken. In the middle there was a large area of uncertainty one way or the other. Obviously, Kates established a relationship between perception of flood frequency and action undertaken. However, his results do not show that flood frequency causes behavior patterns, only that it may play some role.

Hewitt and Burton (1971) developed these ideas, suggesting that community response to hazards is based on society's level of acceptance of geophysical variability. It is founded on the notion that society is oriented to accept some measure of variability in the physical environment, which might be described as the "norm." During extreme geophysical events, normal thresholds are exceeded, which may stimulate a response to the hazard. Thus, floods can represent a resource by depositing fertile sediment on the floodplain, but a hazard when certain thresholds are exceeded and damage outweighs benefits. Thus, locational advantages outweigh disadvantages—except when an event occurs. (The same can be said of volcanic slopes, earthquake zones, and so on.) Smith (1992) pursued the fine line between resources and hazards using the example of water under control (as in a reservoir) and out of control (as during a flood).

All these findings, particularly from the earlier period, suggest some degree of environmental determinism. That is, the environment (in this instance, the frequency of flooding) is said to "cause" the response to the hazard. In reality, of course, many social, political, and economic factors combine to affect decision making, although it would seem reasonable to suggest that physical events help to trigger some behavior patterns, such as when individuals or communities seek to reduce risk and minimize vulnerability. Adaptation to minor fluctuations is an important indication that the human use system may operate according to expected norms for a particular environment. Thus, the role of actors at each level is important.

The concern over how society copes with hazard threats also has been addressed sociologically. Dynes, DeMarchi, and Pelanda (1987) advocate consideration of social, economic, and psychological factors, but they continue to treat disasters as departures from the norm. As we have seen, Hewitt (1983) and Blaikie et al. (1994) provide a different framework. First, they suggest that disasters do not represent abnormal conditions because society is constantly locked in a battle to deal with the vagaries of the geophysical world. Second, they propose that vulnerability be the focus of hazards research. Although they recognize that physical processes may trigger mitigation efforts, they also criticize those who do not emphasize the human element sufficiently. In their view, response to hazards is constrained and determined not simply by physical processes, but also by the ever-present web of social, economic, and political factors. To paraphrase Blaikie et al. (1994), natural hazards are not caused by external geophysical factors per se nor are they necessarily abnormal; in fact, they are part of the natural and human-use systems.

Not all researchers agree entirely with this argument. For example, Quarantelli (1994) does not see disasters on a continuum of day-to-day affairs. A small daily emergency such as a traffic accident does not involve the many different facets of society that might be associated with a disaster; he suggests that a different order of problems comes to the fore in disasters and catastrophes that are not invoked by small-scale emergencies. However, whether part of the normal functioning of society or found only in exceptional circumstances, both scenarios emphasize the importance of human factors if hazards are to be fully understood. It is important to examine all facets of society and to look at all actors in order to determine the structural constraints on decision making. Who are the people in control, who makes decisions, and who has little or no influence on future activities?

In adopting this schema, some of the earlier notions of behavior in disaster environments must be revised. For example, it had been assumed that individuals would behave in ways that would optimize outcomes of hazardous events and minimize future problems; in other words, disaster victims would seek to protect themselves and attempt to save material goods. However, it also was recognized that their efforts would be constrained by their understanding of the hazard (that is, perception or awareness of it) and by prior knowledge of the response alternatives. The conceptual principle is called "bounded rationality," meaning that behavior is generally rational or logical, but is limited by perception and prior knowledge (Burton et al., 1993).

However, even after bounded rationality is taken into account, there remains the problem that a free, equal, or full choice of action is not available to everyone. There are other restrictions based on economic hardship, political bias, and sociological forces such as moral proscriptions. Consequently, individuals may be confined to a hazardous location for many reasons. For example, are the same economic opportunities available to residents of floodplains in Dhaka, Dusseldorf, and Des Moines? In addition, it may not be possible for many to relocate to safer areas. Bangladeshi peasants may have no option but to live on levees. Social norms may prohibit the use of some wells by certain castes in India even during severe droughts while religious beliefs may curtail the availability of certain foods, even during famine. Political action may force people to adopt strategies foreign to them or beyond their means; for instance, controlled flooding has forced some flood-recession farmers in Senegal to change their agricultural practices and others to leave agriculture altogether (Magistro, 1993).

Victims, Survivors, and Others

The dead, the injured, those who lost their homes—these are the primary victims of extreme physical events who often appear in news reports and as disaster statistics. But because the range of impacts is large and varied, there also are those who are indirectly affected, such as those with families and friends in the impact zone. There are those affected because they work in the impacted area, and there are those who neither live nor work there but who have links to the damaged community through business dealings.

Many of these victims may find themselves dealing with long-term problems such as economic hardship caused by the magnitude of direct or indirect losses or psychological stress resulting from the event itself. Still others may encounter difficulties in repair and reconstruction, which can extend disaster impacts for several years after physical evidence of the event is erased or cleaned up.

The real impact of events spreads well beyond the impact site. Through social networks, disasters encompass a broad spectrum of any community. Taylor (1990) describes disaster victims based on their experiences: (a) the extent to which they have suffered personal injury and sickness, bereavement, and property loss; (b) the particular emotional or task focus they have used for coping; and (c) the social chaos, disruption, and havoc they have experienced. His classification expands the concept of the disaster victim beyond media stereotypes, such as someone whose home has just been destroyed by wildfire or whose family has just been killed in an earthquake.

Understanding the different levels of impact is critical to responding to the aftermath of an event because different groups require different support strategies. Taylor (1989, pp. 17–18) identifies six levels of hazard victims (modified for exposition here). *Primary victims* have been exposed directly to the disaster and have experienced property damage, injury, death, or severe disruption. *Secondary victims* have family or friends in the impact zone; they may suffer from grief or feelings of guilt. *Tertiary victims* are persons whose occupations require them to assist in recovery and rehabilitation efforts. *Quaternary victims* live outside the disaster area, but express concern and may feel somewhat responsible; they may provide both appropriate and inappropriate goods and services to the hazard area. *Quinternary victims* have psychological problems such as exhibiting ghoulish behavior or participating in mob behavior. *Sesternary victims* comprise a miscellaneous group on the peripheries of the hazard; they may have considered themselves potential victims, encouraged others to go into the area, been close relatives of tertiary victims, or been associated in some distant way. Finally, Taylor identifies the possibility that clinicians and researchers also may suffer stress, ranging from a desire to help to feelings of guilt and emotional problems.

However, the problem lies in measuring such impacts quantitatively. Taylor's six levels relate to the post-impact stage, representing an important component that feeds directly into recovery activities. However, we cannot ignore the pre-event stage in which potential victims live with the hazard every day, knowingly or not. What sorts of stress result from living in a hazard-prone area? What are the emotional concerns of those vulnerable to the hazard and do their concerns translate into mitigative actions? How do all these pre-event experiences play out when the event occurs?

HEALTH CONCERNS

A great deal of attention has been focused recently on the health aspects of natural hazards. From a medical perspective, knowledge of the causes of death and of types of injury and illness can help greatly in distributing disaster aid (Beinin, 1985; Noji, 1991; Noji and Sivertson, 1987). Without such information, response to disasters can be seriously compromised; in Noji's estimation, "emergency health decisions are often based on insufficient, nonexisting, or even false information, resulting in inappropriate, insufficient, or unnecessary health aid, waste of health resources, and counter effective health measures" (Noji, 1991, p. 273). For example, Noji (1991) found that global relief efforts sometimes provided useless drugs and medical supplies. After the Guatemalan earthquake of 1976, 90% of the 120 tons of unsorted medicines were of no value, consisting of expired or already opened medications often labeled in foreign languages; after the Armenian earthquake of 1988 international relief operations provided over 5,000 tons of drugs, but only 30% could be used immediately and 20% had to be destroyed. Similarly, untrained personnel sometimes contribute to problems or skilled personnel arrive too late to be of any assistance.

Although the chaotic conditions associated with disasters often arouse health-related fears among survivors and emergency workers, most of these are unfounded. Epidemics are rare, and most communicable diseases result from contaminated water supplies rather than contagious diseases. Many events destroy water-purification plants and damage sewage-treatment works such that contamination from fecal matter can represent a serious threat to health. However, incidents of widespread disease are not common (Noji, 1991); only a few examples are found in the literature. Blake (1989) reports several cases, including Balantidiasis in Truk (1971) following a tropical cyclone, an outbreak of respiratory illnesses following flooding in the Marshall Islands (1978), and typhoid in Mauritius (1980) after a tropical cyclone. French and Holt (1989) report possible increases in disease following floods including a doubling of scarlet fever cases and increased dysentery in Moldavia in 1969, and higher levels of typhoid and paratyphoid in Russia in 1954 and again in 1969. However, Seaman (1990) suggests that most such examples cannot be so easily attributed to an event; he maintains that only one incident is associated unambiguously with a disaster, an increase in malaria in Haiti following a hurricane in 1963. Further, the presence of dead bodies has not led to problems to date; there appears to be only a negligible health risk involved, though decaying bodies clearly increase the potential for problems (Noji, 1991). Of course, it would seem appropriate to maintain a close watch over disease and other health hazards because disaster conditions certainly increase the potential for diseases to spread.

Health impacts can be divided into two categories: physical and mental. In the following sections, we look at some of these impacts across various events. Although there are distinct differences among hazards, there also are similarities.

Physical Problems

Natural disasters are associated with a whole range of physical impacts, from death to minor scrapes and bruises. Virtually all disasters have a large number of minor cases, many of which go unreported by individuals who fend for themselves. Nevertheless, when the incidence of reported deaths and injuries is examined, certain patterns are apparent. In a literature survey, Noji (1991) focused on the geophysical agent as a useful basis for analysis. Of course, there are physical impacts common to all geophysical agents, associated with the restoration and rebuilding stages: injuries and heart attacks during cleanup are examples.

Flooding

Flood disasters are often associated with a large number of deaths, but for the most part they include very few injuries requiring sophisticated medical attention (Seaman, 1990). In fact, only 0.2–2% of survivors need care for lacerations, ulcers, and skin complaints. The greatest problems came from water contamination and potential exposure to salmonella, hepatitis A, Norwalk virus, typhoid and paratyphoid, shigella, and *Escherichia coli*. In addition, arbovirus, malaria, and yellow fever can be spread under certain flood conditions. Nonetheless, Noji (1991) suggests that mass vaccinations are unnecessary, can lead to a false sense of security, and may encourage neglect of basic hygiene. In addition, they may divert resources and personnel away from critical areas (French and Holt, 1989). For example, although there has not been an outbreak of typhoid or tetanus after a flood in the United States, mass vaccinations often are carried out after floods (Noji, 1991), as was done in several communities in 1993 (Tobin and Montz, 1994b).

Tropical Cyclones

Tropical cyclones show other patterns of health impacts. As we have seen, most deaths from tropical cyclones (and there can be many) stem from sea surge rather than the winds and rain; nine out of ten victims die by drowning (French, 1989). The majority of survivors need treatment for lacerations and fractures which, without attention, can quickly become infected. Characteristic injuries, described as "cyclone syndrome," include severe abrasions of the chest, arms, and thighs incurred by holding on to trees for survival in fast-flowing water (Seaman, 1990). In addition, some survivors may have long-term health problems because of lack of food and potable water. Finally, the elderly, the very young, and women are over-represented in the death toll (Noji, 1991). In many cases, the lack of shelter under severe atmospheric and hydrologic conditions means that only the physically stronger individuals can survive.

Tornadoes

Tornado injuries are similar to those of tropical cyclones, in part because debris and glass become flying weapons. The major causes of death are craniocerebral trauma followed by crushing wounds (Noji, 1991). Tornadoes can lead to very high death and injury rates, especially in the United States. Between 1970 and 1980, 856 people were killed, 22,012 persons were injured, and 909,605 needed emergency care such as food, shelter, clothing, and medical supplies (Sanderson, 1989). Some injuries can become infected as all sorts of dirt can be attached to the debris (Seaman, 1990); consequently, sepsis is relatively common, ranging between one-half and two-thirds of all cases. Again, more deaths occur among the elderly than in other age groups.

Tectonic Activity

Tectonic activity brings a different set of health problems. During volcanic eruptions, falling tephra and ash, fast-flowing pyroclastic flows, and mud slides can lead to large numbers of dead and injured. For example, the 1985 eruption of Nevado del Ruiz in Colombia resulted in the deaths of 25,000 persons when a lahar buried the town of Armero. In addition, skin burns, asphyxiation, electrocution from lightning, drowning, dehydration, lung burns, and lacerations are common in volcanic situations (Baxter, Bernstein, Falk, French, and Ing, 1982; Noji, 1991). Eye problems, respiratory difficulties, bronchial obstructions, and acute pulmonary injuries also have been recorded. Thus, it is also important to protect emergency workers because gas releases can continue and lung irritants can be present for some time (Baxter et al., 1982).

Earthquakes also lead to many deaths and injuries, primarily from the collapse of buildings. As a rule of thumb, the ratio of deaths to injuries in earthquake disasters is 1:3 or higher (Seaman, 1990). It is often said that earthquakes do not kill people—buildings do. Most trauma results from crushing, although individuals trapped in collapsed buildings (sometimes for several days) suffer other problems, including dehydration, internal hemorrhaging, and asphyxia. Of the 6,308 casualties of the 1995 earthquake in Kobe, Japan, 75% died from crushing or suffocation under debris while 10% died from fire. Most of the rest died from shock, injuries sustained in the event, or stress-related illness. There were eight confirmed suicides, and the actual number is believed to be much higher (Natural Hazards Research and Applications Information Center, 1996).

As a result of earthquakes, some people die from hypothermia and hypovolemic shock (Noji, 1991) although, as Seaman (1990) points out, people are very effective at reestablishing "microenvironments" for shelter. In Guatemala, for example, 50,000 improvised dwellings were set up within 24 hours of the earthquake in 1986. However, many people also need to be treated for fractures, bruises, and crushing-related injuries. Of 4,832 patients admitted to hospitals after the 1988 earthquake in Armenia, Noji (1991) found that combinations of injuries constituted 40% of the cases, superficial trauma (such as lacerations and contusions) 25%, head injuries 22%, lower-extremity injuries 19%, crush syndrome 11%, and upper-extremity injuries 10%. These findings are consistent with patterns found elsewhere.

Other Hazards

Other hazards also have physical impacts. Heat waves can lead to increased death rates from cardiovascular diseases, respiratory problems, pneumonia, and cerebral hemorrhaging (Kilbourne, 1989a). In 1995, for example, more than 1,000 people died in the upper Midwestern United States from heat-related problems. Cardiovascular and cerebrovascular diseases appear to be the biggest killers, accounting for over 90% of heat-wave deaths in Italy in 1983 and 85% in Memphis, Tennessee, in 1980. Of course, heat waves also can lead to heat stroke, dehydration, and heat exhaustion. Cold waves produce other medical problems such as hypothermia, frostbite, and ulceration, which can have significant complications (Kilbourne, 1989b). Drought is associated with food shortages, hence famine and starvation. Without an adequate food supply, populations are subject to all sorts of disease, which can spread dramatically throughout the close confines of refugee camps. In fact, most famine-related deaths accrue from infections (such as measles) rather than hunger; for example, in the Sudan in 1985, the mortality from measles among the young rose to 50% (Toole and Foster, 1989). Similarly, diarrhea, pneumonia, cholera, and dysentery are usually found in populations who have inadequate food.

Mental Health

Natural disasters do not lead to a major breakdown in the mental health of victims, despite widespread media reports to the contrary (Baisden,

1979). If we were to believe such reports, then flooding, tornadoes, earthquakes, or any number of hazardous events would culminate in a total disintegration of the physical, mental, and moral climate of communities as inhabitants engaged in antisocial acts, irrational behavior, panic, and looting. Certainly, instances do occur, such as the racial disturbances and breakdown in law and order in St. Croix following Hurricane Hugo in 1989 and similar events in St. Thomas following Hurricane Marilyn in 1995. However, severe, incapacitating emotional breakdown from disasters is quite rare (Baisden, 1979).

Nevertheless, while sweeping mental illness may not be the norm, disaster victims and emergency workers do experience considerable stress and psychological trauma that may be prolonged and quite disabling to those involved (Wood and Cowan, 1991). However, understanding the relationship between natural hazards, natural disasters, and personal stress is not easy. The individual psyche is bound up in complex social, psychological, economic, geographic/environmental, and political concerns, all operating at various scales ranging from the individual to the community and beyond. The stress of living in a hazardous area varies among people based on many factors, including their knowledge of the hazard, level of preparedness, ability to deal with uncertainty, and previous experience with events. When an event occurs, stress may be severe (as when a family member or close friend is killed); in other instances, post-event stress may be little more than a minor annoyance (as when a journey to work or visit to the local shopping mall is disrupted). Similarly, some individuals deal well with the extreme conditions associated with natural hazards while others experience debilitating depression from an encounter with even a small event.

In addition, there is the temporal element of long-lasting socioeconomic impacts that can aggravate the stress of living in hazardous environments. Dynes (1974) associated disasters with eight stages, each requiring different responses from individuals and community leaders if stress was to be minimized. For example, short-term relief and rehabilitation require substantially different resources than do long-term, community recovery programs. The magnitude of the physical event also enters the equation; larger events are frequently more disruptive, both spatially and temporally, and can lead to greater levels of depression and stress throughout a community. Moreover, the scale of disaster in relation to community size may influence levels of stress; for example, a low-magnitude flood in a small community may lead to proportionally greater psychological problems than a major flood in a large community. Of course, some disasters generate only temporary anxiety which is soon forgotten as day-to-day concerns return to the top of people's agenda.

Research on stress has focused on four interrelated groups of variables:

personal and family traits, hazard experience and perceived locus of control, community and structural constraints, and prevailing health conditions (Tobin and Ollenburger, 1994). The first two categories are discussed in greater detail in Chapter 3, which focuses on perception of hazards and the factors influencing perception.

A substantial literature looks at relationships between various individual and family characteristics and impacts from hazards. Some studies suggest that the elderly, who generally are not wealthy, may experience more stress and relatively greater personal loss than younger persons during hazard events. In some events (such as tornadoes and tropical cyclones), the elderly also suffer more injury and death (Bolin and Klenow, 1982–83; Cutrona, Russell, and Rose, 1986; Huerta and Horton, 1978; Krause, 1987; Noji, 1991; Phifer, 1990; Phifer and Norris, 1989; Russell and Cutrona, 1991). Taylor (1989) has identified certain groups that should receive special attention; children, young mothers, the elderly, and the physically challenged might be classified as highly vulnerable to adverse stress-related effects.

Prior experience with disaster can play an important role in individuals' ability to cope with natural hazards (Burton et al., 1993; Norris and Murrell, 1988; Solomon, Regier, and Burke, 1989), and hence with stress and related health effects. The picture is complicated by the magnitude of hazards experienced, extent of loss sustained, and behavioral traits that influence how individuals perceive their ability to control their environment (i.e., the locus of control).

Beyond the family unit, research has focused on structural constraints, including community characteristics such as neighborhood homogeneity/heterogeneity, levels of social involvement following a disaster (Madakasira and O'Brien, 1987; Russell and Cutrona, 1991; Solomon, Smith, Robins, and Fischbach, 1987), social relationships (Coyne and DeLongis, 1986), and race and ethnicity (Bolin and Klenow, 1988). It has been found that individuals in homogeneous neighborhoods experience greater levels of support from the social network than those "isolated" in heterogeneous communities; thus, homogeneity may mitigate stress in individuals while heterogeneity may exacerbate it. Community homogeneity/heterogeneity also may play a significant role in influencing levels of formal and informal support. It is often argued that inefficient and ineffective relief efforts of leaders and government agencies account for some of the stress manifested by victims (Baisden, 1979). In addition, level of involvement (either in providing or receiving support) affects stress responses and probably has a direct bearing on organization and vulnerability.

Finally, physical and mental health conditions prevailing before the event may well have a significant effect on levels of stress manifested after the event (Canino, Bravo, Rubio-Stipec, and Woodbury, 1990). For example, individuals in poor health, and who have difficulty getting around, would be restricted in the actions they could take to mitigate hazard losses, which could lead to high stress levels. Tobin and Ollenburger (1994) found this to be true of individuals flooded in 1993 in the midwestern United States.

Mental health studies of disaster victims and emergency workers have shown a relatively high incidence of stress. Some have utilized the measure of posttraumatic stress disorder to evaluate disaster stress (Goenjian et al., 1994; Holen, 1991; McFarlane, 1988; Steinglass and Gerrity, 1990). These and other studies demonstrate that stress is a very real factor in hazards and that it can have long-term, sometimes debilitating impacts. To counter these impacts, some nations have adopted crisis intervention programs. However, their effectiveness needs to be addressed, with attention focused on what kinds of intervention are appropriate, what kinds of people might respond to intervention, and what kinds of disasters might be included (Flynn, 1994; Wood and Cowan, 1991).

Health effects illustrate the broad range of impacts associated with natural hazards and disasters. They are indicative in that they affect groups differently, they have pre- and post-event manifestations, and they can be both short- and long-term. Thus, they exemplify the complexities faced in all analyses of natural hazards.

AN INTEGRATIVE FRAMEWORK FOR ANALYZING HAZARDS

Too often hazards are analyzed separately—case by case, hazard by hazard. While such an approach provides in-depth examination of specific geophysical situations and human reactions to or interactions with it, opportunities for comparison are lost. Indeed, numerous issues transcend the nature of the events. Only through cross-hazard analysis can we investigate similarities and differences to obtain a more comprehensive view of hazards.

In addition, locations are subject to more than one hazard, which needs to be recognized in planning and management. While experiences from one hazard are not always applicable to another, many are. Certainly, the impact of losses is widely applicable, and the losses from disasters are extensive, as this chapter has shown. They force many countries to divert financial and other resources away from such areas as development planning and social welfare and into relief and reconstruction—only to experience loss again with the next event. For the most part, it does not much matter which type of event caused the loss. What is important is what we can learn from events that have already occurred and how to lessen loss from subsequent events. In what seems to be an increasing number of events, it is not the physical event that caused loss, but rather the socioeconomic context in which the event occurred.

Natural hazards present a probability of occurrence and a potential for action; if recognized, actions can be taken to reduce their impacts. Natural disasters, on the other hand, represent our failure to recognize the extent of the hazard and/or our inability to implement appropriate mitigation measures. Unfortunately, it seems that only after a disaster has occurred do we learn what could or should have been done. The accounting of absolute and relative losses and recognition of the spatial and temporal patterns of events can help in evaluating what has happened, to what extent, and by extension, what can happen in the next event. Yet, losses continue to grow despite our evolving understanding of the processes that create hazardous situations and our increasing technological ability to mitigate loss. The relative contribution of the human component to hazardousness and loss is critical. While there may be some variations from place to place and from hazard to hazard, the evidence is mounting that it is factors operating within society, rather than in the physical environment, that explain the increased losses that have been documented.

As our understanding increases, we see greater complexity in natural hazards. It is no longer merely a matter of building to specific standards or of disallowing development in hazardous areas. While these can be effective measures, they are not likely to have much impact until such problems as poverty, land and income distribution, and equity issues are resolved. Thus, when we see newspaper headlines like "Hurricane Kills 39" or "Earthquake Leaves Thousands Homeless," we cannot lay the blame on the physical event, which is only the agent. It is within the socioeconomic environment that we can usually find the causes of such disastrous effects. Rectifying these, of course, is more difficult.

The global spatial pattern of disasters is also indicative of a phenomenon exceeding purely physical causes. For example, Hagman (1984) has demonstrated very clearly that the poorest nations have the highest death rates per disaster, the most casualties per 10,000 people, and the largest number of people killed per 1,000 square miles. These death rates correspond to the density of population and are inversely related to the GNP of the particular countries concerned (Figure 1.4). In addition, mortality and morbidity associated with natural disasters has changed over time with greater impacts in poorer regions.

Deaths, injuries, and economic loss result from interactions between the physical and human environments, and it is the relative contributions of each that require our investigation. We have a reasonably good understanding of the physical events (Chapter 2), although it varies somewhat among events. For the most part, however, it is possible to forecast, warn,

FIGURE 1.4. Relationships among disaster mortality, population density, and the economy. Note that the graphs give only a rough picture as national economies vary within each income group. *Source:* Hagman (1984). Copyright 1984 by Swedish Red Cross. Reprinted by permission.

or build in order to avoid many of the adverse effects of a flood, hurricane, volcanic eruption, or earthquake. The more difficult task—where our ability to control the causes is less refined—is found within the political, economic, and social components of hazardousness. It is with these factors and variables that this book is concerned. We have moved away from blaming natural hazards and disasters on a Supreme Being or uncontrollable forces, and instead are beginning to see the role that humans play in contributing to hazardousness and, ultimately, disaster.

2 *Physical Dimensions of Natural Hazards*

In Chapter 1, we argued that natural hazards emanate from a complex interplay of physical processes and human activities that culminate in negative consequences for the human-use system. We stressed that losses accruing from such events are not simply the result of extreme geophysical forces, but often are caused, exacerbated, and perpetuated by human activities. Having said this we cannot ignore the fact that natural hazards do have a significant physical component that must be addressed if we are to understand natural hazards fully. In this chapter, we examine the physical dimensions of extreme geophysical events. Hazards outside the traditional geophysical realm (such as insect vectors that cause disease and the spread of locusts or grasshoppers that destroy crops and may cause famine) are not considered.

We begin the chapter with a discussion of the relative importance of the physical system, placing the traditional view in historical context. We introduce many different natural hazards and examine more closely some of their associated physical processes. We do not attempt to provide a complete, detailed explanation of all pertinent processes; that information is voluminous and there are many full-length texts devoted to each one. Nevertheless, we have tried to supply sufficient information to place physical characteristics within the broad picture of natural hazards (references are provided for those who wish to follow up particular points in greater detail). This section is followed by a look at commonalities among natural hazards that could provide a framework around which to address physical processes. The chapter concludes by integrating physical dimensions across events in order to develop a theoretical base for analyzing dynamics of natural hazards more holistically.

THE TRADITIONAL APPROACH

The traditional approach to the study of natural hazards has taken a physical science perspective. When problems associated with natural hazards were seen as consequences of extreme geophysical events, most attention came from natural scientists, principally those concerned with meteorology, geology, and hydrology. Research tended to be narrowly focused and response to events limited.

Indeed, the ability and perceived need to control extreme physical forces dominated scientific thought. Engineering "solutions" became symbols of modernity and human dominion over nature (McPhee, 1989) and, as a result, engineers often assumed the role of managers. In the United States, this era is typified by the years between World Wars I and II, when many dams were constructed to retard flood flows in upstream reaches and extensive levee systems were erected to restrict floodwater on controlled sections of floodplains. Billions of dollars were spent on such projects, yet flood losses continued to rise. It was the failure of such approaches to reduce loss, combined with increasing expenditures on flood alleviation projects, that eventually led to criticisms of the engineering viewpoint and subsequently to the integration of human and physical considerations (White et al., 1958). Although there are fewer examples of structural control of other events such as tornadoes or earthquakes—because engineering measures cannot control these physical processes so easily—engineering solutions have been sought for many hazards. This is discussed in later chapters when we look at perceptions, responses, and adjustments.

While many criticisms have been leveled at early hazards research, there is a danger that physically based studies will be regarded as passé. It is true that we understand more about the physical processes associated with hazards than about the human-use system. Modeling streamflow during floods or meteorological events that determine thunderstorm activity would seem far less complex than predicting human behavior before, during, or after a particular event. However, research into the physical dimensions of natural hazards is not complete, and considerable work must be undertaken if we are to appreciate fully how such events originate. Further, while not necessarily determined by the geophysical events themselves, our responses to natural hazards must take into account the physical processes operating in locales if they are to be successful. To take an obvious example, it is highly unlikely that exactly the same mitigation measures will be implemented to ameliorate the impact of ice storms and heat waves. On the other hand, knowledge of physical processes may provide learning experiences across hazards such that lessons from one event may be applicable to another; there may be important similarities between ice storms and heat waves such that adjustments to one could instigate improved response to the other.

Geophysical Processes

Traditionally, natural hazards have been classified by underlying physical process, usually grouped into meteorological, hydrological, geological, and sometimes extraterrestrial events (Table 2.1). The last category, involving meteorites and other major planetary changes, would seem unlikely and certainly of low risk, but nevertheless should be included; In 1994 the crashing of meteorites into Jupiter raised the possibility of similar occurrences on Earth. In the short term, however, the probability of a person being killed in a given year by a meteorite is extremely low, estimated by Dinman (1980) at only 1 in 100 billion. It is not worth spending too much time, then, on concern that "the sky is falling," and we do not explore extraterrestrial events any further in this book. However, just to affirm the

TABLE 2.1. Natural Hazards Classified by Physical Processes

Category of hazard	Types of event
Meteorological	Tropical cyclones/hurricanes
	Thunderstorms
	Tornadoes
	Lightning
	Hailstorms
	Windstorms
	Ice storms
	Snowstorms
	Blizzards
	Cold waves
	Heat waves
	Avalanches
	Fog
	Frost
Geological	Earthquakes
	Volcanoes
	Tsunami
	Landslides
	Subsidence
	Mudflows
	Sinkholes
Hydrological	Floods
	Droughts
	Wildfire
Extraterrestrial	Meteorites

rule of probabilities, Elizabeth Hodges of Sylacauga, Alabama, was alleg-
edly hit by an 8.5-lb meteorite on November 30, 1954 (she was not killed,
but did suffer severe bruising). Many scientists believe that the extinction
of the dinosaurs was caused by the collision of an asteroid with Earth at
the end of the Cretaceous, which ejected sufficient debris into the atmosphere to cause global cooling.

Clearly, the presence of any one of these physical events does not
constitute a hazard in itself. The destruction of a forest by wildfire,
devastation of a coastline by a tsunami, or creation or obliteration of a
mountain by volcanic eruption may change the physical landscape and
destroy wildlife habitat, but it is not defined as a hazard unless human
activities are impacted. (Such events may be detrimental to wildlife and
may even lead to species extinctions in particular areas, but they still are
not classified as hazards by our definition.) However, classifying natural
hazards by physical process does provide a useful framework for detecting
similarities among and making generalizations about hazardous events.
The physical framework enhances our explanatory models of natural
hazards and, ultimately, promotes sound management practices.

In Table 2.1, the list of natural hazards covers a range of events
stemming from many different physical processes. The evidence suggests
that virtually any meteorological, hydrological, or geological process can
nurture a potential hazard. Because of their diversity, it would seem that
hazards are physically unique and that few similarities can be identified
among them. Avalanches, for example, would appear to have little in
common with earthquakes or tornadoes while volcanic eruptions and
hurricanes would seem to be completely dissimilar. If local environmental
conditions that enhance or diminish physical impacts are added to the mix,
making generalizations across events looks foolhardy indeed.

The traditional strategy has been to adopt a hazard-by-hazard approach (sometimes even case study by case study) based on physical criteria
and with only minimal integration between hazards. This strategy has
yielded important information and should continue if we are to strengthen
our knowledge of physical processes. Nevertheless, it is not the only way to
study natural hazards and associated physical processes. If we examine the
physical processes themselves, many similarities can be seen. For instance,
it was suggested above that avalanches, earthquakes, and tornadoes have
nothing in common beyond causing death and destruction. However, a
cursory examination indicates that each of these events usually occurs with
little warning and lasts for only a short period, but can be very powerful.
Similarly, volcanoes and hurricanes are fairly predictable in a spatial
context, although at present, only hurricanes can be forecast in a temporal
context. Similarities among processes are worth exploring, especially the
following physical parameters of natural hazards:

1. Physical mechanism: magnitude, duration, spatial extent
2. Temporal distribution: frequency, seasonality, diurnal patterns
3. Spatial distribution: geographic location
4. Countdown interval: rapidity of commencement, preparation time, speed of onset

PHYSICAL MECHANISMS

Following any natural hazard, we are inevitably caught up in questions of size: how large was the volcanic eruption, how deep was the floodwater, how strong were the winds in the tornado or tropical cyclone? There is a desire by media and public alike to know how powerful the physical processes were that could cause such damage. Hence, evening news programs incorporate the seemingly inevitable photographs of houses in ruins, people suffering, and emergency teams at work while a voice-over describes the tornado as having winds "in excess of 322 kilometers per hour (200 mph)" or the earthquake as registering "7.5 on the Richter scale."

Such measures also interest scientists. By measuring magnitude, we can refine our knowledge of nature and possibly improve our response to hazards. However, it should be stressed that any simple measure of magnitude (e.g., the explosive power of a volcanic eruption, depth of floodwater, or high wind speed in a tornado or tropical cyclone) represents a force of nature, *not* a determination of the vulnerability of a particular population. Such measures are nothing more than appraisals of the physical parameters of the geophysical events. For instance, a small, relatively weak earthquake in one area could precipitate greater damage and have more severe impacts than a significantly larger event in another place. Why is this? Two hypotheses come to mind: (1) variable impacts are associated with physical processes, and (2) the vulnerability of human populations is not simply a product of magnitude. In Chapter 1, we discussed how human factors (including social, economic, and political forces) affect communities and individuals' response to geophysical events. Vulnerability is tied to a web of human activities that confine and restrict behavior, ultimately defining hazard impacts. At the same time, variable impacts are associated with physical parameters and (as in the earthquake example above) can lead to differential levels of destruction even when all other factors remain constant.

Magnitude

Magnitudes of geophysical events rest primarily on scientifically based measures of the strength of physical processes. Appraisals of wind speeds in thunderstorms, assessments of flow velocities in floods, or estimations of energy released by earthquakes follow standard procedures, thereby permitting comparisons between similar events (wind speed in one thunderstorm might be 80 km/h [50 mph] compared to 70 km/h [44 mph] in another). Measures of magnitude, then, do not concern impacts of an event on the human-use system (that is, intensity), but rather the physical processes involved. In some hazards, because direct measurements are difficult, assessments are made based on "intensity" estimates that incorporate human variables as indices of destruction; for instance, both the Fujita scale for tornadoes and the modified Mercalli scale for earthquakes include building damage among indicators of magnitude (see below). In this section, we review some of the dynamics of events and the measures used to determine magnitude.

Meteorological Events

All the usual measures of wind speed, wind direction, humidity, temperature, and radiation are pertinent to understanding severe meteorological events. In fact, it is these measures that first attract our attention to particular hazards; descriptions of them can be found in standard texts such as Ahrens (1994) and Barry and Chorley (1992). Nevertheless, their usefulness for classifying natural hazards is somewhat limited. For instance, in many cases it is difficult to define exactly when an extreme geophysical event has started, or even whether particular events are in progress, based on physical criteria.

For example, take heat and cold waves. Although we can attribute deaths to them and even identify periods of more intense activity (see Riebsame, Diaz, Moses, and Price, 1986; Ruffner and Bair, 1984; Shah, 1983), we have great difficulty determining when individual events begin. Are they declared when temperatures reach some certain point, perhaps when abnormally high death rates are recorded for a particular season? Do we look for temperature change over time?

A simple measure of temperature is not satisfactory because temperature means different things to different people. In 1930, 1934, and 1936, record temperatures occurred throughout the Dust Bowl. Temperatures of 49.4C (121°F) or higher were recorded in North Dakota and Kansas; 48.9C (120F) in South Dakota, Oklahoma, Arkansas, and Texas; and 42.8C (109F) in Indiana, Louisiana, Maryland, Michigan, Minnesota, Nebraska, New Jersey, Pennsylvania, West Virginia, and Wisconsin (Ruffner and Bair, 1984). However, temperature alone does not suffice. Some assessment of humidity, especially during heat waves, also is important. Thus, we often use a temperature–humidity index to indicate comfort (Table 2.2). At 26C (79F) and only 10% relative humidity, few people feel uncomfortable, but

TABLE 2.2. Temperature–Humidity Index

RELATIVE HUMIDITY

	10%	20%	30%	40%	50%	60%	70%	80%	90%	100%	
19°C	16.5	17	17	17	18	18	18.5	18.5	19	19	
20°	17	18	18	18.5	18.5	19	19.5	19.5	20	20	few people feel uncomfortable
21°	18	18.5	18.5	19	19.5	19.5	20	20.5	20.5	21	
22°	18.5	19	19.5	19.5	20	20.5	21	21.5	21.5	22	
23°	19	19.5	20	20.5	21	21.5	21.5	22	22.5	23	about 1/2 of all people feel uncomfortable
24°	19.5	19.5	20	20.5	21	21.5	22	23	23.5	24	
25°	20	20.5	21	21.5	22	23	23.5	24	24.5	25	nearly everyone feels uncomfortable
26°	20.5	21	21.5	22	23	23.5	24.5	25	25.5	26	
27°	21	21.5	22	23	24	24.5	25	25.5	26.5	27	
28°	21	22	23	23.5	24	25	25.5	26	27	28	rapidly decreasing work efficiency
29°	21.5	23	23.5	24	25	25.5	26	27	28.5	29	
30°	22	23.5	24	25	25.5	26.5	27	28.5	29	30	
31°	23	24	24.5	25.5	26.5	27	28.5	29.5	30	31	
32°	23.5	24.5	25	26	27	28.5	29.5	30.5	31	32	extreme danger
33°	24	24.5	25.5	26.5	28	29	29.5	30.5	31.5	33	
34°	24.5	25.5	26.5	27	28.5	29.5	30.5	31.5	33	34	
35°	25	26	27	28.5	29.5	30.5	31.5	33	34	35	
36°	25.5	26.5	28	29	30	31	32	34	35		
37°	26	27	28.5	29.5	31	32	35				
38°	26	28	29	30	31.5	33	34				
39°	26.5	28.5	29.5	31	32	33.5	35				
40°	27	29	30	31.5	33	34.5	35.5				
41°	28	29.5	30.5	32	34	35.5					
42°	28.5	30	31.5	33.5	35						
43°	29	30.5	31.5	31.5	35.5						

Source: Ruffner and Blair (1984). Copyright 1994 by Gale Research Company. Reprinted by permission.

as relative humidity increases, so too does the level of discomfort until at 90–100% humidity, everyone is affected. At 34C (93F) and 80% relative humidity, there is extreme danger that deaths can occur (see the bottom right section of the table). A separate index for cattle is called the livestock weather safety index.

Similar measurement difficulties arise with cold waves. Rather than cold temperatures, it is a combination of extreme cold and high wind that creates most problems for humans. The wind chill index combines these two variables in a "temperature" chart to show how cold it actually feels. In extreme conditions, severe problems can result from even minimal exposure of bare skin (Table 2.3); minus 29C (-20F) is cold but bearable in calm conditions, but in a 32-km/h (20-mph) wind, conditions become intolerable.

Thus, the intensity of a hazard cannot be expressed independently of human factors. Some people are more prone to cold or heat than others. For example, the elderly often suffer more than younger people because of their physiological makeup. Consequently, demographics may be indicative of intensity and serve as a surrogate measure of magnitude. Besides age, wealth plays a leading role in determining intensity of such hazards. Inevitably, the poor suffer most as they are unable to pay for heating or cooling during extreme conditions. In some cities, temperatures can be extremely high during the summer months, contributing to the stress on

TABLE 2.3 Wind Chill

Dry bulb temperature (°C)	WIND SPEED (km/h)										
	6	10	20	30	40	50	60	70	80	90	100
	Equivalent temperature (wind chill)										
20	20	18	16	14	13	13	12	12	12	12	12
16	16	14	11	9	7	7	6	6	5	5	5
12	12	9	5	3	1	0	0	-1	-1	-1	-1
8	8	5	0	-3	-5	-6	-7	-7	-8	-8	-8
4	4	0	-5	-8	-11	-12	-13	-14	-14	-14	-14
0	0	-4	-10	-14	-17	-18	-19	-20	-21	-21	-21
-4	-4	-8	-15	-20	-23	-25	-26	-27	-27	-27	-27
-8	-8	-13	-21	-25	-29	-31	-32	-33	-34	-34	-34
-12	-12	-17	-26	-31	-35	-37	-39	-40	-40	-40	-40
-16	-16	-22	-31	-37	-41	-43	-45	-46	-47	-47	-47
-20	-20	-26	-36	-43	-47	-49	-51	-52	-53	-53	-53
-24	-24	-31	-42	-48	-53	-56	-58	-59	-60	-60	-60
-28	-28	-35	-47	-54	-59	-62	-64	-65	-66	-66	-66
-32	-32	-40	-52	-60	-65	-68	-70	-72	-73	-73	-73
-36	-36	-44	-57	-65	-71	-74	-77	-78	-79	-79	-79
-40	-40	-49	-63	-71	-77	-84	-86	-86	-86	-86	-86
-44	-44	-53	-68	-77	-85	-87	-91	-92	-94	-94	-94
-48	-48	-58	-73	-82	-89	-93	-96	-98	-99	-99	-99
-52	-52	-62	-78	-88	-95	-99	-102	-104	-105	-105	-105
-56	-56	-67	-84	-94	-101	-105	-109	-111	-112	-112	-112
-60	-60	-71	-89	-99	-107	-112	-115	-117	-118	-118	-118

Danger from freezing of exposed flesh

Little   Considerable

Slight   Great

the urban poor. Thus, the magnitude of the physical component combines with the material vulnerability of the human population to produce disaster statistics.

Most extreme meteorological events have received considerable scientific attention. Snowstorms and blizzards, for instance, are frequently classified as a function of snow depth, drifting potential, and wind speeds that might lead to whiteout. Visibility is emphasized during foggy weather, and thunderstorms are tracked because of the possibility of hail, heavy rain, gusty winds, and tornadoes, all of which present problems for the human-use system. In the United States, alone, there are approximately 100,000 individual thunderstorm cells that develop every year (Grazulis, 1993); of these, about 10,000 become severe and 1,000 spawn tornadoes. Every year, tornadoes kill 92 people on average, lightning approximately 100, and flash floods originating from thunderstorms another 100. Some attention to the physical processes underlying these events may help reduce death and damage. For instance, it has been found that lightning is usually concentrated in a few thunderstorm cells at any given time (Pierce, 1977) while others have documented the movement of lightning through the atmosphere (Salanave, 1980); the study of such phenomena as ribbon lightning, beaded lightning, and other loop-the-loop patterns may lead ultimately to better protection strategies. Similarly, an understanding of hail development may eventually reduce the huge agricultural losses accruing from hailstorms; descriptions of hail size (pea, golf ball, baseball, grapefruit) are certainly colorful and useful in alerting people to the possibility of damage, but they do not replace more careful measures of size, time, and intensity that would allow meaningful comparisons of events. Difficulties arise in defining events by such physical criteria, as illustrated below by tornadoes and tropical cyclones.

Tornadoes. Grazulis (1993, p. 6) describes a tornado as a "violently rotating column of air, a vortex, spawned by a thunderstorm, in contact with both the thundercloud and the ground, often accompanied by a funnel shaped cloud, progressing over the land in a narrow path." The pressure in the center is relatively low, and the whole system moves across the ground at up to 112 km/h (70 mph). Most winds within tornadoes are less than 240 km/h (150 mph), but the combination of variable wind speed and changing direction can add to damage totals (Figure 2.1). Although higher wind speeds are sometimes reported, the extreme velocities above 485 km/h have not been confirmed and are highly unlikely. In large tornadoes there may be multiple suction vortices that have lower pressures and even higher wind speeds; these may account for some of the more extreme damages that have been recorded. A tornado typically cuts a path 400 m (1/4 mi) wide and 26 km (16 mi) long (Ruffner and Bair, 1984), although there are considerable variations on these averages. However, the

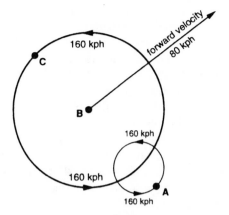

FIGURE 2.1. Tornado velocity. Measurements relate to the rotational speed of all vortices (A and C) the speed of forward movement along the ground (B). At the outer edge, the forward velocity (80 km/h, kph, or 50 mph) combines with the rotational velocity of the tornado (160 km/h, or 100 mph) and the rotational velocity of the suction vortex (160 km/h) to produce winds of 400 km/h (250 mph) for a few seconds at A. *Source:* Grazulis (1993). Copyright 1993 by The Tornado Project. Reprinted by permission.

famous 1917 tornado that reputedly lasted 7 and 20 min, moving at 64 km/h (40 mph) and staying on the ground for 471 km (293 mi) as it crossed Illinois and Indiana, was probably a series of tornadoes that formed and decayed in sequence rather than a single event. Similarly, the Tri-State tornado of March 18, 1925, which traveled 352 km (219 mi), from Missouri through Illinois and into Indiana, also consisted of a series of tornadoes (Grazulis, 1993).

There are two major difficulties with tornadoes. First, while we know the conditions under which they occur, we do not understand fully what causes them to form. Second, tornadoes are very difficult to measure directly since weather gauges are invariably destroyed by the events. It is known that wind speeds are very high and that internal pressures can fall to low levels. Several studies have been undertaken to measure the strength of tornadoes by placing special instruments in their paths. For instance, a team of scientists developed a device known as TOTO (named after Dorothy's dog in the Wizard of Oz) that was placed in the potential path of oncoming tornadoes in the hope of recording pressure and wind speed (Bluestein and Golden, 1993); unfortunately, no direct hits were recorded with this equipment. Until reliable statistics are available, we must rely on surrogate assessments of a tornado's magnitude from damage evidence; thus, tornadoes usually are classified according to the Fujita scale, the

Pearson path length scale, and the Pearson path width scale. Attributing damage to physical processes is not foolproof and, of course, interpretation of damage is subjective. Nevertheless, used with care, these standards allow researchers to compare tornado characteristics across the United States.

The Fujita scale (Table 2.4; F scale) runs from F0, characterized by light damage and wind speeds of 64-115 km/h (40-72 mph), to F5, characterized by incredible damage and winds of 420-512 km/h (261-318 mph). The F6 tornado, with wind speeds of 513-610 km/h (319-379 mph), is classified as "inconceivable." Pictures depicting typical damage within each category serve as referents for observers. Unfortunately, tornado strength still may be over- or underestimated because buildings are made of different materials having different strengths and weaknesses; for example, a sloping roof may survive higher wind velocities than a flat roof. Further, the F scale focuses on conditions that may have existed for only a small time in one part of the tornado; higher or lower wind speeds probably occurred in other places, but were not "recorded" because the tornado did not hit a structure at that time.

The Pearson scales are not used as often as the Fujita, but they provide some interesting statistics (Table 2.5). It appears that tornado path lengths

TABLE 2.4. Tornado Intensity on the Fujita Scale

Level	Description	Damage
F0	Gale tornado	64-116 km/h (40-72 mph) winds; light damage; some damage to chimneys; branches broken off trees
F1	Moderate tornado	117-180 km/h (73-112 mph) winds; moderate damage; surface peeled off roofs; mobile homes pushed off foundations or overturned
F2	Significant tornado	181-253 km/h (113-157 mph) winds; roofs torn off frame houses; mobile homes demolished; trees snapped or uprooted
F3	Severe tornado	254-331 km/h (158-206 mph) winds; severe damage; roof and some walls torn off; most trees uprooted; heavy cars lifted off ground
F4	Devastating tornado	332-418 km/h (207-260 mph) winds; devastating; well-constructed houses leveled; cars thrown and large missiles generated
F5	Incredible tornado	419-512 km/h (261-318 mph) winds; incredible damage; strong frame houses lifted off foundations and carried considerable distances; steel-reinforced concrete structures badly damaged
F6	Inconceivable tornado	513-610 km/h (319-379 mph) winds; very unlikely; damage probably not recognizable along with damage from F4 and F5 winds surrounding F6

Source: Grazulis (1993). Copyright 1993 by The Tornado Project. Reprinted by permission.

TABLE 2.5. Fujita–Pearson FPP Scales

Scale	Fujita wind speed in, in km/h (mph)	Pearson path length, in km/h (mi)	Pearson path width, in m (yd)
	0–64 (0–40)	<0.5 (0.3)	<5.5 (<6)
0	65–116 (41–72)	0.5–1.5 (0.3–0.9)	5.5–15.5 (6–17)
1	117–180 (73–112)	1.6–5.0 (1.0–3.1)	16–50 (18–55)
2	181–253 (113–157)	5.1–16 (3.2–9.9)	51–160 (56–175)
3	254–331 (158–206)	16.1–50 (10–31)	161–518 (176–566)
4	332–418 (207–260)	51–159 (32–99)	0.5–1.5 km (0.3–0.9 mi)
5	419–512 (261–318)	160–507 (100–315)	1.6–5 km (1.0–3.1 mi)

Source: Grazulis (1993). Copyright 1993 by The Tornado Project. Reprinted by permission.

and widths are positively correlated with tornado magnitude as measured by wind speed. Path lengths range from less than 0.5 km (0.3 mi) to over 482 km (300 mi); widths range from less than 5.5 m (18 ft) to over 5 km (3 mi) (Grazulis, 1993).

Extensive research by Grazulis (1993) demonstrates how valuable these scales can be. Using the F scale to analyze tornado data for 1880–1989, he found that among tornadoes leading to deaths, 32% were classified as strong (F2–F3), 67% as violent (F4–F5), and only 1% as weak (F0–F1). In terms of frequency, however, strong tornadoes made up 30% of the total, violent only 2%, and weak accounted for 68% of all tornadoes. Similarly, he undertook an extensive review of the data on tornadoes' path length and width. Building on the work of Fujita in particular, and Pearson, Grazulis has revealed important new information about the hazardousness of tornadoes. From his and other research, we can make meaningful generalizations and model the tornado hazard, placing research on a sound scientific footing.

Tropical Cyclones/Hurricanes. Hurricanes are large cyclonic storms originating over tropical oceans (Pielke, 1990). They have been described, somewhat poetically, as

> tropical children, the offspring of ocean and atmosphere, powered by heat from the sea, driven by the easterly trades and temperate westerlies, the high planetary winds, and their own fierce energy. In their cloudy arms and around their tranquil core, winds blow with lethal velocity, the ocean develops an inundating surge, and as they move toward land, tornadoes now and then flutter down from the advancing wall of thunderclouds.
>
> Compared to the great cyclonic storms of the temperate zone they are of moderate size, and their worst winds do not approach tornado velocities. Still, their broad spiral base may dominate weather over thousands of square miles, and from the earth's surface into the lower stratosphere. Their winds may reach 220 miles per hour (354 kilometers per hour), and their lifespan is measured in days or weeks, not minutes or hours. (Ruffner and Bair, 1984, p. 42)

In some ways, measures of magnitude are more problematic for hurricanes than tornadoes. The term "hurricane" itself is defined by a physical parameter: cyclonic winds above 120 km/h (74 mph). Yet, hurricanes are associated not only with high cyclonics winds and low atmospheric pressures but also with heavy rain that may lead to flash floods, sea surges that can inundate huge coastal areas, and large waves that can scour and batter the coastlines. Hurricanes also may propagate thunderstorms, lightning, and tornadoes. Nonetheless, classification has focused on wind speeds and pressures (Table 2.6). The Saffir–Simpson damage potential scale categorizes hurricanes into five classes, extending from level 1, where pressure exceeds 980 mb and winds are 119–153 km/h (74–95 mph), to level 5, where pressures might fall below 920 mb and winds exceed 250 km/h (155 mph).

Hurricanes develop from tropical disturbances that originate over warm ocean areas. For example, between 1931 and 1983, there were 516 tropical cyclones in the North Atlantic with 181 reaching the coast of the United States; of these, 295 were classified as hurricanes, of which only 87 reached the United States (Ruffner and Bair, 1984). As summarized by Pielke (1990), studies have shown that several conditions are necessary for hurricanes to develop:

1. An ocean area with a surface temperature above 26C (79F)
2. Small wind speed and direction changes between the lower and upper troposphere of less than 15 kn
3. The presence of a preexisting region of lower tropospheric horizontal wind convergence, or what some have termed a tropical wave disturbance
4. A distribution of temperature with height which will overturn when saturated, resulting in cumulonimbus clouds
5. A location at least 4 or 5 degrees away from the equator

Pielke (1990) suggests that these conditions are met globally where oceanic regions have sufficiently warm surfaces, at least during the sum-

TABLE 2.6. The Saffir/Simpson Damage Potential Scale for Hurricanes

Level	Central pressure, in millibars (inches of mercury in barometer)	Maximum sustained winds in knots (km/h)
1	>980(28.94)	64–83(119–153)
2	965–979(28.50–28.94)	84–95(154–175)
3	945–964(27.91–28.49)	96–113(177–209)
4	920–944(27.17–27.90)	114–35(210–249)
5	<920(<27.17)	>135(>249)

Source: Pielke (1990). Copyright 1990 by Routledge. Reprinted by permission.

mer, and are significantly toward the equator side of the polar front. Although measurement of magnitude includes many of these weather variables, wind speed, pressure, and cyclonic action are usually used to define hurricane-type activity (as in the Saffir–Simpson scale); a cross section of a hurricane shows how these variables might be related (Figure 2.2). Given specific water depth and fetch conditions, wind velocity also is related to wave heights and lengths (Table 2.7); for example, a wind speed of 19 kn can produce waves of 3.7 m (12 ft), but a 51-kn wind can generate waves of 15.5 m (51 ft) (Kotsch, 1977). Between 1900 and 1995 in the United States, there were 20 category 4 hurricanes and 4 category 5 events, the most intense of which (in terms of pressure gradient) was the Labor Day hurricane over the Florida Keys in 1935.

The naming of hurricanes has an interesting history. According to Ruffner and Bair (1984), tropical cyclones originally were named after the saint's days on which they occurred (at least in Central America and the

FIGURE 2.2. Relationships among surface pressure, wind, and wave height in a hurricane. As barometric pressure decreases, wind speeds and wave heights increase. Yet, within the hurricane, when pressure is at its lowest, wind speed decreases, only to rise sharply again with increasing pressure. The decrease in wave heights lags slightly behind the decrease in pressure. *Source:* Pielke (1990). Copyright 1990 by Routledge. Reprinted by permission.

TABLE 2.7. Relationship between Wind and Wave Characteristics
(a) Probable Maximum Wave Heights with Various Wind Speeds

Wind speed (kn)	Wave height (m)
8	1.0
12	1.5
16	2.4
19	3.7
27	6.1
31	7.6
35	9.0
39	11.0
43	12.0
47	15.5

(b) Average Wave Length Compared To Wind Speed

Average wave length (m)	Wind speed (kn)
16	11
38	20
80	30
117	42
252	56

Source: Kotsch (1977). Copyright 1977 by The Naval Institute Press. Reprinted by permission.

Caribbean). Gradually a more refined system developed using latitude and longitude to identify storms for shipping. However, because the storms are always moving and several can exist at the same time, this system gave way to using the alphabet (able, baker, charlie . . .). Many other systems were and have been tried, usually involving some alphabetized sequence such as animals, things, and even descriptions (annoying, blustery, churning . . .). The current system, which came into vogue in 1953 (and has been modified since then), uses names in alphabetical order. These are set for a 6-year cycle and, since 1979, have alternated male and female names. If some name is associated with a particularly severe event, it is removed from further use. In addition, there are different sets of names for the different oceans so that the general location of storms can be identified immediately. Because of the high frequency of events in the western Pacific, names are based on four sets that rotate regardless of year.

Hydrological Events

The measurement of hydrological factors presents different problems. Some measures are relatively simple, such as water quantity, quality, and velocity in floods and droughts. Once again, it is not possible to determine the hazardous nature of these events without close attention to the human-use system. However, hydrological texts (such as Chow, 1964; Dunne and Leopold, 1978; Ward, 1978) provide valuable background information of which the hazards researcher must be aware in assessing community vulnerability or modeling hazard behavior.

Floods. Under natural conditions, the hydrological processes of the fluvial system have created and maintained stream channels, with channel maintenance maximized during bankfull discharges (Petts and Foster, 1985). A river is said to be in flood whenever flows exceed bankfull capacity and spill over onto the adjacent floodplain. This occurs at some time or another on all rivers; consequently, it can be said that all rivers flood. Therefore, any human activities along a river's edge are subject to inundation, or flood hazard. For instance, in 1992, unprecedented rainfall in northern Pakistan resulted in extensive flooding that caused over $1 billion in damage; more than 1,000 people were killed, 6 million displaced, 176,000 houses destroyed or damaged, and 122,000 livestock killed (United Nations, 1993).

The magnitude of floods is usually viewed in terms of recurrence intervals. For instance, a 100-year flood refers to a particular discharge that is expected to occur once every hundred years; more accurately, this discharge or larger has a 0.01 probability, or 1% chance, of occurring in any given year. The greater the discharge, the larger the flood flow (see section below on frequency for details). This "measure" of magnitude is appropriate and useful for engineers and floodplain managers, but it does not describe potential discharges with sufficient accuracy for many activities. Although velocity and depth of floodwater are implicit in this approach, the probability does not include energy and intensity.

While all rivers flood, not all flood in the same way, and adjustments to flood hazard must take account of their differences. The response of the hydrological system to a given level of precipitation determines the characteristics of a flood. In particular, hydrological characteristics of a drainage basin depend on geology, vegetation, soils, meteorology, and size of the drainage basin. If the peak discharge rises rapidly (usually with considerable force) the drainage basin is said to be flashy, and flash floods can be devastating; they provide little warning and often hit with tremendous velocity. At the other end of the hydrological continuum are sluggish drainage basins that may take several weeks to flood, providing ample time for warning (Figure 2.3).

In July 1976, a deadly flash flood hit Big Thompson Canyon in Colorado. Heavy rain occurred in the central portion of the watershed for more than 4 hours, and over 12 in fell in the western part. Both the amount and the intensity of precipitation in the narrow canyon contributed to the flash flood that killed at least 139 people and destroyed more than 400 homes and businesses. Discharge was officially recorded as 883 cubic

FIGURE 2.3. Flood hydrographs. Stream discharge (Q) varies over time (T) as a function of precipitation, runoff, and drainage basin characteristics. Although both flows represent the same volume of water, on the flashy stream, the peak flow occurs higher and sooner; on the sluggish stream, discharge is spread more evenly over time. Flashy streams are characteristic of steep canyons and urbanized areas. Bankfull discharge or flood stage (dashed line) shows the difference in duration of the two events.

meters per second (m^3/s) (or 31,200 cubic feet per second, cfs), but the extent of loss suggests that the flows were even higher (Gruntfest, 1977, 1987). Four years earlier, Rapid City, South Dakota, had experienced even greater losses following torrential rains that culminated in a dam breaking and a flash flood sweeping through the community (Haas, Kates, and Bowden, 1977). Again, high-velocity flows contributed to the high level of destruction. Traditional measures of flood magnitude, then, provide a consistent means of comparing flows within a stream, but they hardly reflect the intensity of those flows.

By comparison, the 1993 floods in the United States were generally of the extensive type, causing over $12 billion damage but leading to only 48 deaths (Tobin and Montz, 1994b). The immediate cause was exceptionally heavy precipitation during the months preceding the floods. Many weather stations throughout the upper Midwest received record totals for January–July 1993; the spatial distribution of precipitation for the first seven months of 1993 is shown in Figure 2.4. In Iowa, the *Des Moines Register* reported the wettest conditions on record for July; for 2 months (June–July); and for 3 months (May–July) (Fuson, 1993). The 12-, 24-, 36-, and 48-month totals also were the wettest on record. These rainfall totals were combined with the preceding 15 months of below-average sunshine, temperature, and evaporation. The National Weather Service reported that Bismarck, South Dakota, and Cedar Rapids, Iowa, had the second wettest May–July period

FIGURE 2.4. Spatial pattern of rainfall in the upper midwestern United States, January–July 1993. Isohyets show the total rainfall (in mm) for the 7-month period. *Source:* Wahl, Vining, and Wiche (1993).

in 104 years of record keeping (Wahl, Vining, and Wiche, 1993). In addition, July was one of the 3 wettest months ever in 8 of the 9 states affected (National Weather Service, 1993). However, whether these were the highest or second highest totals on record does not really matter; the result was a saturated drainage basin that was unable to store further precipitation. The rain continued and when added to the antecedent soil moisture conditions, led to unprecedented flooding in many places.

It is interesting to examine these rainfall totals and compare them with averages for the same 7-month period. Table 2.8 shows the 1993 precipitation data as percentages of the 30-year averages for ten midwestern weather stations. Minneapolis, Minnesota, received the least rain for the 7-month period (121%) while Bismarck, North Dakota, received the most (221%); none of the stations received less than its average. Closer examination shows that rainfall was particularly heavy in July throughout the upper Midwest. Again, Minneapolis got off relatively lightly at 158%, as did St. Louis, Missouri (131%), but other weather stations received well over 200%; Bismarck, in fact, reported 643% and Manhattan, Kansas, received 535%. Relatively high rainfall variability is not unusual during summer months because of the changing patterns and occurrences of thunderstorms. However, these percentages were much higher than in preceding years and, when combined with the extremely wet conditions and cooler temperatures that persisted prior to July, they served to exacerbate the flood hazard in the upper Midwest. Figure 2.5 shows similar trends, using monthly precipitation patterns for the same weather stations in 1993.

Although these patterns help us to understand the pre-flood situation, they do not tell the whole story. The flood problem was compounded by a series of very severe storms that tracked across the upper Midwest, dumping huge quantities of water over local areas. For instance, southeastern Iowa

TABLE 2.8. January–June 1993 Precipitation as a Percentage of 30-Year Averages, for Selected Stations

	Bis-marck, N. Dak.	Sioux Falls, S. Dak.	Minnea-polis, Minn.	Des Moines, Iowa	Cedar Rapids, Iowa	Mad-ison, Wisc.	Peoria, Ill.	Man-hattan, Kans.	Kansas City, Mo.	St. Louis, Mo.
Jan.	64	137	132	166	117	150	235	121	180	196
Feb.	77	127	44	137	109	109	118	120	116	130
Mar.	49	124	64	138	156	152	140	147	88	92
Apr.	75	104	82	88	121	186	130	64	179	176
May	108	273	119	205	146	121	88	241	145	99
June	168	189	155	172	193	182	143	123	120	191
July	643	293	158	258	414	276	242	535	249	131
7-month period	221	199	121	174	206	180	155	210	159	141

Source: Parrett, Melcher, and James (1993).

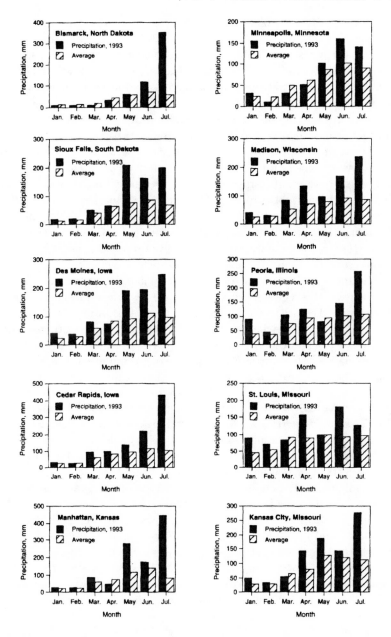

FIGURE 2.5. Monthly precipitation at selected midwestern stations, January–July 1993. Comparisons with 30-year averages show that extreme conditions were experienced. While the stations exhibit different patterns, the overall trend of extremely wet conditions is identical. *Source:* Wahl et al. (1993).

received over 10 cm (4 in) of rain on July 4–5. On July 15–16, a storm moved through North Dakota dropping 13.4 cm (5.27 in) on Bismarck and 18.4 cm (7.25 in) on Jamestown (Wahl, Vining, and Wiche, 1993). July 22–24, 33 cm (13 in) of rain fell in parts of Nebraska, Kansas, Missouri, Iowa, and Illinois (Parrett, Melcher, and James, 1993). In combination, these storms and the accumulating precipitation that year led to flood incidents in many communities; cumulative totals for three communities are shown in Figure 2.6. Such patterns were repeated throughout the river basins.

In terms of both absolute and relative magnitudes, the precipitation data serve to explain the widespread and often devastating flooding that occurred. However, magnitudes were not the same throughout the Mississippi River basin; subbasins with different hydrological and physical characteristics experienced different levels of flooding (Table 2.9). Not only does the table illustrate differences in magnitude as defined by recurrence interval (see below), but it also shows the variations in discharge that define given recurrence levels. The extent of flooding in the upper Midwest during the summer of 1993 can be explained best perhaps by the fact that virtually all streams in this region reached flood stage sometime during this period; many reached record levels. The comparative data showing 1993 levels and previous maximum discharges clearly illustrate the severity of these floods.

It is evident from the data that the flooding took various forms. Occasionally, flash floods swept rapidly through communities, aggravating damage because of the high velocities of flows; other floods were slower to develop and steadily inched higher toward towns and across farmland, destroying virtually everything in their paths. The latter were often larger floods that inundated houses to considerable depths and remained for long periods, adding to the misery of victims who could not return to their homes for several weeks. Some floods resulted from the failure of alleviation projects, especially levees; in such cases, water that had been held back by an embankment often surged through narrow openings at great velocity, then remained on the floodplain for weeks at a time. Thus, it is not always the magnitude of flooding that defines the extent of the hazard, but also frequency and duration of flooding. Alleviation structures, which were breached, also played a significant role in the severity of impacts.

Droughts. Conceptually, droughts can be distinguished from other natural hazards in several ways. Although the word conjures scenes of dry fields, barren cropland, and emaciated cattle, the truth is that drought can be these and much more. It is not usually possible to identify or forecast a drought prior to the event; it is said to creep up so insidiously that it is impossible to say exactly when one has started until the drought is well under way. Unlike many other events, droughts can persist for considerable lengths of time. Again, it is important to comprehend the physical processes that contribute toward drought conditions if we are to develop

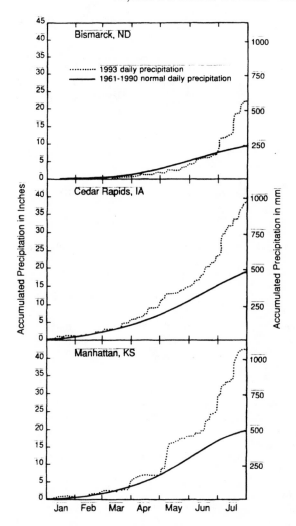

FIGURE 2.6. Accumulated daily precipitation at selected midwestern stations, January–July 1993. Comparisons with 30-year averages illustrate the extreme conditions, characteristics of patterns throughout the upper Midwest. *Source:* Wahl et al. (1993).

effective management strategies and avoid unscientific behavior. Perhaps more than any other natural hazard, droughts evoke the superstitious and bizarre. Folklore is replete with attempts to modify drought conditions through rain dances, sacrifices, and pseudoscientific practices. Weems (1977) points out that the Connecticut Weather Control Board was estab-

TABLE 2.9. Peak Stages and Discharges at Selected Streamflow Gauging Stations

Station name	Drainage age (km²)	Flood of 1993 Peak stage (m)	Peak discharge (m³/s)	Date	Recurrence interval range (yr)	Previous maximum discharge Peak stage (m)	Maximum discharge (m³)	Date
Little Minnesota River near Peveer, S.Dak.	1,158	3.32	96	7/25	>100	4.07	122	4/1952
Little Cottonwood River near Court-land, Minn.	596	3.19	100	6/20	>100	2.73	38	3/1985
Black River near Galesville, Wisc.	53,872	5.07	1,811	6/21	>100	4.71	1,854	4/1967
Mississippi River at Clinton, Iowa	221,704	7.00	6,934	7/05	10–50	7.5	8,688	4/1965
Iowa River at Mar-shalltown, Iowa	4,051	6.26	543	7/10	10–50	6.24	1,189	6/1918
Iowa River at Lone Tree, Iowa	11,119	7.00	1,613	7/07	>100	6.18	1,010	5/1974
Iowa River at Wapello, Iowa	32,372	9.00	3,028	7/07	>100	8.81	2,660	6/1947
Mississippi River at Keokuk, Iowa	308,210	8.28	12,311	7/10	>100	6.40	10,188	6/1851
Des Moines River at Humboldt, Iowa	5,843	4.63	538	7/13	>100	4.69	5,094	4/1969
Middle Raccoon River near Bayard, Iowa	971	8.75	679	7/09	>100	7.53	413	7/1973
South Raccoon River at Redfield, Iowa	2,574	8.22	1,245	7/10	>100	8.85	991	7/1958
Des Moines River near Tracy, Iowa	32,321	7.36	3,085	7/12	>100	8.08[a] 5.52[b]	4,387 1,206	6/1947 6/1984
Des Moines River at Ottumwa, Iowa	34,639	6.74	3,170	7/12	>100	6.71[a] 4.46[b]	3,962 1,353	5/1903 6/1984
Des Moines River at Keosauqua, Iowa	36,358	9.95	3,056	7/13	>100	8.49[a] 8.54[b]	4,132 2,043	6/1903 4/1973
James River near Manfred, N.Dak.	655	2.86	74	7/23	>100	2.80	57	4/1979
James River at Jamestown, N.Dak.	7,304	4.14	37	7/23	>100	4.82[a] 2.67[b]	181 28	5/1950 6/1983
James River near Scotland, S.Dak.	53,491	6.08	546	7/06	(>50)–100	6.23	832	6/1984
Platte River at Ash-land, Nebr.	218,078	6.53	3,226	7/25	>100	5.86	3,679	3/1993
Missouri River at Rulo, Nebr.	1,074,591	7.73	8,688	7/24	>100	7.8[a] 7.44[b]	10,131 6,849	4/1952 6/1984
Missouri River at St. Joseph, Mo.	1,088,577	9.77	9,481	7/26	>100	8.17[a] 7.26[b]	11,235 5,858	4/1952 5/1987

TABLE 2.9. *(cont.)*

| Station name | Drainage age (km²) | Flood of 1993 | | | | Previous maximum discharge | | |
		Peak stage (m)	Peak discharge (m³/s)	Date	Recurrence interval range (yr)	Peak stage (m)	Maximum discharge (m³)	Date
Kansas River at Wamego, Kans.	143,175	8.33	4,839	7/26	(>50)–100	9.31[a] 5.70[b]	11,320 2.063	7/1951 10/1973
Missouri River at Kansas City, Mo.	1,256,668	14.90	15,310	7/27	>100	11.58[a] 8.80[b]	17,688 8,858	6/1844 9/1973
Missouri River at Boonville, Mo.	1,299,403	11.27	21,225	7/29	>100	10.00[a] 9.71[b]	20,093 9,452	6/1844 10/1986
Missouri River at Hermann, Mo.	1,357,678	11.27	21,225	7/31	>100	10.82[a] 10.91[b]	25,244 15,537	6/1844 10/1986
Mississippi River at St. Louis, Mo.	1,805,230	15.11	30,564	8/01	>100	13.18	36,790	6/1844
Mississippi River at Thebes, Ill.	1,847,188	13.87	27,593	8/07	10–50	13.76	38,913	7/1844

[a] Before regulation.
[b] After regulation.
Source: Parrett, Melcher, and James (1993).

lished in the 1860s to "control rain-producing activities and make sure people didn't hurt anybody with rain dances and cloud seeding and that sort of thing."

While droughts are associated with moisture deficits, the real problem comes from the integration of society with the natural environment. A society becomes dependent on a water budget for an area, with additional water perhaps imported from other regions. When these supplies fail, that society suffers a drought and, to some extent, the degree of water shortage determines the level of impact. Thus, water supply—loosely interpreted as the needs and wants of a society—determines the presence or absence of drought. For example, a humid area could have ample water for crops and human needs, but experience water shortages if some usual needs cannot be met. In arid areas of the world, communities have learned to exist on very small amounts of water; although the failure of the water supply can be devastating, aridity alone does not constitute a drought. Further, moisture deficit is not just a product of diminished rainfall or snowfall; it also can result from reduced runoff conditions, heightened evapotranspiration, or increased infiltration. Thus, hot, dry summers add to drought conditions, but do not in themselves imply drought. Droughts cannot be defined in absolute terms of either precipitation or water budgets (Wilhite and Glantz, 1987).

The extent of the drought hazard also is variable, again distinguishing it from many other natural hazards that are more easily defined in space.

To quote the World Meteorological Organization (1975), "Drought is undoubtedly one of man's worst natural enemies. Its beginning is subtle, its progress insidious and its effects devastating. Drought may start at any time, last indefinitely and attain many degrees of severity. It can also occur in any region of the world with an impact ranging from slight personal inconvenience to endangered nationhood."

There is an additional temporal aspect to droughts. Although rainfall may be plentiful, drought conditions may prevail if it does not fall at the right time for particular crops. For instance, storms in August in the midwestern United States may be ideal for soybeans, but too late to save a corn crop. The spatial impact, therefore, depends on local rainfall conditions and characteristics of the agricultural system. In this instance, the soybean farmer may be less affected than the corn farmer.

Attempts to classify drought based on quantitative criteria have been relatively unsuccessful. Proposed thresholds have included critical rainfall levels, number of days without rain, evapotranspiration levels, and even temperature measures (Wilhite and Glantz, 1987). In some countries, drought is defined as a sustained period without rain. Examples are Bali, with 6 days without rain (Wilhite and Glantz, 1987); Britain, 15 days with less than 1 mm of rain; India, seasonal precipitation at less than 2 mean deviations from the norm; Libya, annual rainfall of less than 180 mm; and the United States, more than 14 days with less than 1 mm of rain (Alexander, 1993).

Perhaps the most common standard in the United States is the Palmer drought severity index (PDSI), a uniform measure of physical variables that combine to make dry conditions. According to Wilhite and Glantz (1987), the PDSI has been employed in many other countries, including China, South Africa, and Australia. It includes measures of precipitation, potential evapotranspiration, antecedent soil moisture, soil moisture estimates, and runoff. The value of a drought severity index is that comparisons can be made of areas' potential dryness. However, no account is taken of the human-use system, including different agricultural land use, modes of adjustment to the hazard, or socioeconomic effects, all of which are important factors in considering severity.

What it comes down to is that droughts mean different things to different people. Subrahmanyan (1967) described six types: meteorological, climatological, atmospheric, hydrological, agricultural, and water management. A farmer may see drought conditions as soon as crops do not grow at average rates, a hydrologist may be concerned about river levels or ground water recharge, a meteorologist may focus on lack of rainfall, and an economist may define drought as something that negatively affects the economy of an area. Ultimately, droughts are complex hazards for which the physical considerations tell only a small part of the overall story. Two examples, drawn from England and Africa, illustrate this complexity.

Although generally considered a humid climate, southern England experienced quite a severe drought in 1976. It was the worst recorded in 200 years, and Parliament eventually created a Minister for Drought to address the problem (echoes of Connecticut). Needless to say, it began raining shortly after this action was taken, showing that if you wait long enough, drought will break. Although precipitation had been below average for months extending back to 1975, when the drought began cannot be stated with exactitude. It was only when shortages were apparent that action finally was taken. Dry conditions gradually led to falling reservoir levels and inadequate water supplies; water shortages accumulated and parts of the country were placed on rationing, which for some residents meant that all domestic water supplies were shut off and public standby pipes turned on for just a few hours a day. For 7 to 11 weeks in 1976, supplies in southeast Wales were cut off for up to 17 hours every day (Perry, 1981).

The late response was typical of wet environments, where everyone expects the drought to break at any time. As it continued to rain in some parts of the country, newspapers published photographs of residents, umbrellas in hand, standing in the rain collecting water from outdoor pipes that served as the communal water supply. Moreover, there remained enough water in Britain to supply all its inhabitants; the problem was getting the right amount of water to the right place at the right time. There also were political and economic ramifications of the drought. While residents of south Wales collected water rations, those of Birmingham and some other English cities faced no shortages at all; the irony was that these cities were supplied with water from reservoirs located in the Welsh mountains. In addition, many commercial enterprises were adversely affected by limited water supplies.

Countries in various parts of Africa have experienced numerous droughts over the years. Periodic droughts have affected the Sahel at least four times since the turn of the century as dry conditions have spread across the continent. Between 1968 and 1974, a particularly severe event killed up to 150,000 people, destroyed cattle, and left many people homeless. In 1974, 200,000 people in Niger were entirely dependent on food aid, and another 250,000 in Mauritania and Mali had moved to urban centers where many were destitute (Wijkman and Timberlake, 1984). At its peak, the drought directly affected Mali, Niger, Chad, Sudan, Mauritania, Senegal, and Burkina Faso. In 1984, similar problems left over 150 million people in 24 countries on the brink of starvation because of famine associated with drought (Wijkman and Timberlake, 1984). In that case, the ravages of the physical processes were inextricably tied to political maneuvers. Ethiopia,

Mozambique, and the Sudan were all involved in civil wars that compromised organizational structures and greatly exacerbated vulnerability; many of those struggles, both political and natural, continued into the 1990s. In 1992, ten southern African countries were severely hit by drought (Angola, Botswana, Lesotho, Malawi, Mozambique, Namibia, Swaziland, United Republic of Tanzania, Zambia, and Zimbabwe), affecting 18 million people as water shortages increased and food supplies declined. In places, 70–90% of the maize crop failed (United Nations, 1993).

The response of the global community has changed somewhat over the years. During droughts of the early 1970s, the wealthy nations were criticized for not responding promptly. To some degree, suffering nations hid or redistributed the impacts of drought (Burton, Kates, and White, 1993). It has been suggested that in newly independent nations, pride may have prevented officials from admitting or recognizing that the problems were too large to address without outside help. Food aid is now more timely and often distributed more effectively. However, distribution problems persist, often due to the whims of local officials or warring factions.

Geological Events

The magnitude of geological events is usually expressed in energy released (as in earthquakes and volcanic eruptions) or wave height (as in tsunami and seiches). Other geological events practices (such as landslides, mudflows, and sinkholes) are measured by standard geomorphological practices.

Earthquakes. The cause of earthquakes has been attributed to many mythical figures and explained in religious terms. Traditionally, Japanese culture described a giant underground catfish whose movements caused the ground to shake. Chinese myths say that the constant struggle between yin and yang sets the earth in motion. Many cultures have ascribed responsibility to god(s), from Vulcan stoking his smith's fire to an angry god bringing death and destruction to an errant people (Bullard, 1979; Palm, 1995). For instance, the Lisbon, Portugal, earthquake of 1775, which virtually destroyed the city, was viewed by some Europeans as representing the wrath of God and they urged punishment of "sinners." (The mordant savant, Voltaire, reputedly commented that "the sight of several persons being slowly burned in great ceremony is an infallible secret for preventing earthquakes.")

More recently, some religious leaders attributed the 1994 Northridge earthquake in California to God's punishment of immoral behavior (Palm, 1995). This particular earthquake killed 60 people, injured 5,000, left 25,000 homeless, and caused an estimated $20 billion in damage. Understanding physical processes must surely promote more positive action than that based on myth and superstition. Nevertheless, strongly held beliefs may help to explain some seemingly irrational behavior in the face of a disaster.

Aristotle was one of the first who tried to explain earthquakes scientifically. He theorized that the earth was composed of caverns and filled with hot air that periodically broke free under pressure, causing volcanic eruptions and earthquakes. While this is not true (the earth is definitely not hollow), it was a reasonable hypothesis for the time. Since then, considerable advances have been made in understanding tectonic activity, but it was not until the latter half of the twentieth century, with the development of the theory and concepts of plate tectonics, that a more satisfactory explanation of earthquakes and volcanoes became possible (Figure 2.7). The earth is a dynamic system whose lithospheric plates move at varying speeds; to some extent, their relative movements around the globe determine the location and degree of tectonic activity (Gabler, Sager, Brazier, and Wise, 1987). For example, the North American plate is moving away from the European plate at the rate of approximately 2–3 cm (1 in) per year. Los Angeles could be contiguous with San Francisco in just 12 million years (Ingwerson, 1986).

Plate activity can be very fast or exceedingly slow, but all processes bring about change, especially along the boundaries of the plates. Thus, the locations of earthquakes and volcanoes are inextricably tied to the boundaries of the lithospheric plates: where plates are sliding past one another (as along the Californian coast); where continental plates undergoing obduction (such as the Indian subcontinent moving into Asia); where ocean floors are tearing apart along the midoceanic ridge; and where oceanic material is being subducted into the earth's interior (such as along the Japanese and Aleutian Island arcs). Other tectonic activity can be identified at hot spots (such as Hawaii). The Pacific Northwest of the United States and southwestern Canada illustrates some of the complexity of this activity. The region is composed of several small, independent plates moving at different rates, with both subduction and obduction taking place; not surprisingly, there has been considerable tectonic activity. A hot spot located under Idaho and Wyoming is associated with geothermal activity at Yellowstone National Park. Indeed, a recent earthquake disrupted the cavernous structure of the park such that Old Faithful is no longer so regular in its eruptions as it used to be.

Earthquakes do not occur only at or near plate boundaries, although these are generally the most frequent location. There also are intraplate earthquakes that occur in what have been described as "comparatively aseismic regions" (Whittow, 1979). Probably the best known examples occurred in December 1811, January 1812, and February 1812 around New Madrid, Missouri (Penick, 1976). It is said that the earthquake was so intense

FIGURE 2.7. Tectonic plates and their directions of movement. This map corresponds closely with the location of earthquakes and volcanoes. *Source:* Gabler, Sager, Brazier, and Wise (1987). Copyright 1987 by Saunders College Publishing. Reprinted by permission.

that the flow of the Mississippi River was reversed for some time; while this may be an exaggeration, the magnitude of the earthquake was clearly exceptional and certainly disrupted the river's flow and altered its course (Ritchie, 1988). Another intraplate earthquake occurred in Charleston, South Carolina, in 1886; although not so large as the New Madrid event, it was felt over an extensive area. Our understanding of these intraplate earthquakes is incomplete, although several hypotheses have been put forward to explain them. It is possible that fault lines from old plate boundaries deep within the earth's crust may move periodically, inducing seismic activity. Accumulation of sediment from riverine or marine deposits may exert sufficient stress to cause geological failure. In some instances, it seems that earthquakes have been induced by the construction of reservoirs; the weight of water can cause the crust to flex while percolation of water into the ground may reduce frictional resistance, thus contributing to earth movements. For example, the filling of Hoover Dam was associated with increased seismic activity (as happened at other dams in the United States and India).

Earth movements associated with earthquakes produce several forms of activity. First, pressure can build up along fault lines and eventually slip, sending shock waves radiating from the focus of the movement. This seismic activity generally expands from the epicenter as body waves (traveling through the earth) and surface waves (Ruffner and Bair, 1984). The body waves consist of two main types; primary or "P" waves, and secondary or "S" waves (Bolt, 1978, 1982). The P wave is fast-moving, traveling approximately 6.5 km/s (4 mi/sec) at the earth's surface and more than twice as fast below the Mohorovicic discontinuity (Ruffner and Bair, 1984). It can travel through most media, although it is affected to some extent by different geological material. The P wave is propagated by longitudinal forces that displace particles by extension and compression (Figure 2.8); Ruffner and Bair (1984) describe the motion as like being on a long train that is just setting off. The S waves move at about half the P waves; velocity and can travel only through rigid media. Consequently, they do not move into the liquid interior and are not transmitted through water. They also are called shear or transverse waves because they displace particles at right angles to the wave direction (Ritchie, 1988). They have greater wave amplitude than P waves (up to 4 in), which can result in greater damage than P waves alone.

Surface waves—called Love and Rayleigh waves—last much longer than body waves, are much slower at 4 km/s (2.5 mi/sec), and can cause shaking for 30 seconds or more. Love waves propagate a shearing motion in a horizontal plain whereas Rayleigh waves generate an elliptical action in a vertical plain (Ruffner and Bair, 1984). It is the combination of these waves that causes the damage associated with ground shaking (see Figure 2.8).

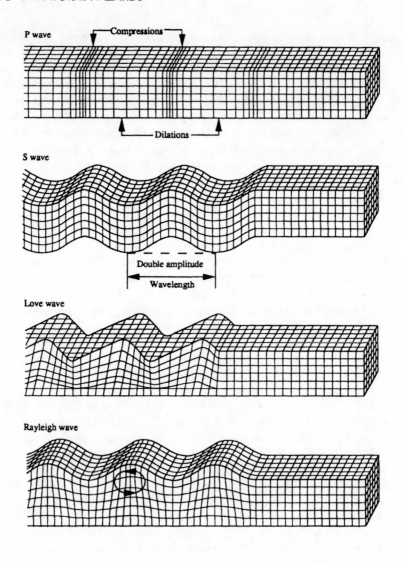

FIGURE 2.8. Types of wave caused by earthquakes. P and S waves are body waves while Love and Rayleigh waves are surface waves. *Source:* Whittow (1979). Copyright 1979 by Penguin Books, Ltd. Reprinted by permission.

However, the picture is complicated because waves respond in different ways to changing geological material. Figure 2.9 shows the idealized effect on these waves as they move from a solid to a loosely consolidated medium. Note that as wave velocity decreases, wave amplitude increases; it is the slower shaking associated with the larger amplitude waves that usually causes the greatest damage to structures. In addition, waves are refracted as they move from one rock type to another. Thus, any measure of earthquake magnitude would be inadequate without some account of the local geology.

Movements within the earth's crust can have various effects on the landscape, including ground shaking, surface rupture, ground failure, and generation of tsunami. Ground shaking is by far the most common and is responsible for most of the damage associated with earthquakes; it is a direct result of the movement of energy through the earth's crust by the waves described above. Surface rupture is quite uncommon by comparison, although some notable events have been recorded in the United States (Table 2.10). Clearly, any structure straddling a surface fault has little chance of survival if the motion is rapid. Some fault lines show a slower movement, termed creep, that is less catastrophic but eventually leads to failure. The shaking of certain materials also can initiate slope failure, landslides, mud slides, and ground failure as the soils liquefy. Finally, tsunami can be generated when earth movements occur underwater; the energy transmitted through water can have severe consequences for coastal communities (see below).

Magnitude of earthquakes is usually measured on the Richter scale, first developed in 1935 by Charles Richter. It is based on calculations incorporating wave amplitude, wave period, and focal depth, among other

FIGURE 2.9. Speed and amplitude of ground waves. Different subsurface materials respond differently to seismic waves. Thus, lower amplitudes are found in more compacted materials where they cause less damage. *Source:* Griggs and Gilchrist (1977). Copyright 1977 by Wadsworth Publishing. Reprinted by permission.

TABLE 2.10. Surface Faulting Associated with California Earthquakes

Year	Fault	Richter magnitude	Surface effects
1836	Hayward	7.0*	Ground breakage
1838	San Andreas	7.0*	Ground breakage
1852	Big Pine	—	Ground breakage (questionable)
1857	San Andreas	8.0*	Right lateral slip (30 ft)
1861	Calaveras	—	Ground breakage
1868	San Andreas	—	Long fissure in earth
1868	Hayward	7.0*	Strike slip
1872	Owens Valley fault zone	8.3*	Right lateral slip (16–20 ft), left lateral movement (?), vertical slip (23 ft)
1890	San Andreas	—	Fissures in fault zone, railroad tracks moved
1899	San Jacinto	6.6*	Surface evidence (questionable)
1901	San Andreas	6.3*	Ground breakage
1906	San Andreas	8.3	Right lateral slip (21 ft)
1922	San Andreas	6.5	Ground breakage
1934	San Andreas	6.0	Ground breakage
1934	San Jacinto fault zone	7.1	Distinct fault trace
1940	Imperial	7.1	Right lateral slip (19 ft)
1947	Manix	6.4	Left lateral slip (3 in)
1950	Unnamed fault	5.6	Left lateral slip (5–8 in)
1951	Superstition Hills	5.6	Right lateral slip (slight)
1952	White Wolf	7.7	Left lateral slip (2 ft), upthrown (2 ft)
1956	San Miguel	6.8	Right lateral slip (3 ft), vertical slip (3 ft)
1966	Imperial	3.6	Right lateral slip (0.5 in)
1966	San Andreas	5.5	Right lateral slip (several inches)
1968	Coyote Creek, Superstition Hills, Imperial, and San Andreas	6.4	Right lateral slip (15 inches) on Coyote Creek, slight right lateral slip on others
1971	San Fernando	6.6	Left lateral slip (5 ft), thrusting up to 3 ft

*Estimated magnitude.
Source: Ruffner and Bair (1984). Copyright 1984 by Gale Research Company. Reprinted by permission.

variables. Seismographs record earth movements, then data are standardized by computing statistics for 100 km from the epicenter. The scale produced by Richter is logarithmic and open-ended (although measures usually range up to 9); an increase of 1 unit (e.g., from 5 to 6 on the Richter scale) represents a tenfold increase in the measured wave amplitude, which translates into a 30–60 times increase in energy involved (Alexander, 1993). That is, the wave amplitude of an 8.2 earthquake is not twice as large as a 4.1 event; it is 10,000 times larger and the corresponding increase in energy released is perhaps up to 10 million times greater. A magnitude 7 earthquake releases 1 billion times more energy than a magnitude 1 earthquake (Ruffner and Bair, 1984). Each year, the energy released by earthquakes is between 10^{25} and 10^{26} ergs (Bolt, 1978); to put this energy into other terms, an earthquake of magnitude 5.5 releases about 10^{20} ergs which may be compared to the 1946 nuclear explosion of 10^{19} ergs at Bikini. Table 2.11 shows the estimated energy in weights of TNT released from earthquakes of various size.

Richter's original scale has been modified considerably to adapt to different conditions around the world and to different seismographs. For instance, deep-seated earthquakes give slightly different results than shallow-seated ones. According to Bolt (1978), the 1964 Alaskan earthquake had a body wave magnitude of 6.5 as measured by the P wave, but a surface wave magnitude of 8.6. In this case, the latter measurement gives a better estimate of earthquake magnitude. It is not surprising, then, that different magnitudes often are given for a particular earthquake.

However, the Richter scale obscures some of the impacts of earthquakes. Table 2.12 shows that deaths associated with selected major earthquakes are not well correlated with magnitude. The Armenian earthquake of 1988 measured 6.8 on the Richter scale and killed over 25,000 people

TABLE 2.11. Energies of Earthquakes

Richter magnitude	Approximate energy, relative to TNT
1.0	170 g
1.5	907 g
2.0	5.9 kg
2.5	28.6 kg
3.0	180 kg
3.5	903 kg
4.0	5443 kg
4.5	29,030 kg
5.0	180,533 kg
5.5	907,200 kg
6.0	5,688,144 kg
6.5	28,622,160 kg
7.0	90,720,000 kg
7.5	907,200,000 kg
8.0	5.7×10^9 kg
8.5	2.8×10^{10} kg
9.0	1.8×10^{11} kg

Source: Petak and Atkisson (1982). Copyright 1982 by Springer-Verlag. Reprinted by permission.

TABLE 2.12. Major Disastrous Earthquakes

Year	Richter magnitude	Location	Approximate number of deaths
365	unknown	Crete and Greece	50,000
526	unknown	Syria Area	250,000
893	unknown	India	180,000
1138	unknown	Syria	100,000
1293	unknown	Japan	30,000
1455	unknown	Italy	40,000
1556	unknown	Shaanxi, China	830,000
1667	unknown	Caucasia	80,000
1693	unknown	Sicily	60,000
1737	unknown	Calcutta, India	300,000
1755	8.7	Lisbon, Portugal	60,000
1783	unknown	Italy	50,000
1797	unknown	Ecuador	41,000
1868	unknown	Ecuador and Colombia	70,000
1908	7.5	Southern Italy	58,000
1915	7.5	Central Italy	32,000
1920	8.6	Kansou, China	200,000
1923	8.3	Yokohama, Japan	103,000
1927	8.3	Nan-shan, China	200,000
1932	7.6	Kansou, China	70,000
1935	7.5	Northern India	60,000
1939	7.8	Chile	40,000
1939	7.9	Erzincan, Turkey	23,000
1949	7.3	Pelileo, Ecuador	6,000
1960	5.8	Agadir, Morocco	12,000
1970	7.7	Chimbote, Peru	67,000
1976	8.0	Tangshan, China	242,000
1976	7.5	Guatemala	23,000
1978	7.7	Northeast Iran	25,000
1985	8.1	Mexico City, Mexico	9,500
1988	6.8	Armenia	25,000
1990	7.7	Northwest Iran	40,000
1995	6.9	Kobe, Japan	6,000

Source: Nuhfer et al., 1993; updated for 1995. Copyright 1993, American Institute of Professional Geologists. Used with permission.

(Nuhfer, Proctor, and Moser, 1993); the Friuli, Italy, earthquake measured 6.4 and killed less than 1,000 (Geipel, 1982); and the Loma Prieta earthquake of 1989 in the San Francisco Bay Area measured 7.1 and killed 62.

Thus, intensity of seismic activity is difficult to measure. There is no absolute scale of intensity; instead, estimates reflect size of the seismic wave, distance from the epicenter, geological structure, and social factors such as land use and type of construction. The most commonly used measure of intensity is the modified Mercalli scale which ranges from 1, which is not even noticeable, to 12, which represents almost total destruction with large rock masses displaced and objects thrown into the air (Table 2.13). This scale has been adapted many times to different situations. Essentially, it is similar to the Fujita scale for tornadoes in that it is based on damage rather than objective measures of physical processes; an earthquake that causes doors to swing, displaces small unstable objects, awakens sleepers, and is felt outdoors would be classified as category 5, whereas earthquakes that bend rails and put underground lines out of service would be category 11. The Mercalli scale is useful for comparing impacts of earthquakes, but does not provide a measure of magnitude. It also varies from place to place because of different construction materials and possessions whereas, in theory the Richter scale remains the same.

TABLE 2.13. Modified Mercalli Scale

Scale	Description
I	Not felt. Marginal and long-period effects of large earthquakes.
II	Felt by persons at rest, on upper floors.
III	Felt indoors. Hanging objects swing; vibration like the passing of light trucks; may not be recognized as an earthquake.
IV	Hanging objects swing; vibration like the passing of heavy trucks, or sensation of a jolt like a heavy ball striking the walls; standing cars rock; windows, dishes, and doors rattle; glasses clink. In the upper range, wooden walls and frames creak.
V	Felt outdoors. Sleepers awakened; small, unstable objects displaced or upset; doors swing, close, and open; pendulum clocks stop, start, and change rate.
VI	Felt by all. Many frightened; persons walk unsteadily; windows, dishes, and glassware broken; pictures fall off walls; furniture moved or overturned; weak plaster cracked; trees shaken.
VII	Difficult to stand; noticed by drivers of cars; furniture broken; weak chimneys broken at roof line; falling plaster, stones, tiles, and loose bricks; waves on ponds.
VIII	Steering of cars affected; damage to some reinforced masonry; fall of stucco and some masonry walls; fall of chimneys, smokestacks, monuments, towers, elevated tanks; frame houses move off foundations if not bolted; cracks in wet ground and on steep slopes; changes in flow or temperature of wells and springs.
IX	General panic; most masonry seriously damaged; frame structures shifted off foundations, if not bolted; serious damage to reservoirs; underground pipes broken; conspicuous cracks in ground.
X	Most masonry and frame structures destroyed; some well-built wooden structures and bridges destroyed; serious damage to dams, dikes, and embankments; large landslides; rails bent slightly.
XI	Rails bent greatly; underground pipelines completely out of service.
XII	Damage nearly total; large rock masses displaced; lines of sight distorted; objects thrown into the air.

Source: Wood and Neumann (1931). Copyright 1931 by Seismological Society of America. Reprinted by permission.

TABLE 2.14. Volcanic Explosivity Index (VEI)

Category	Description	Volume of ejecta (m^3)	Column height (km)	Classification	Tropospheric injection	Stratospheric injection	Historic eruptions
0	Nonexplosive	$<10^4$	<0.1	Hawaiian	Negligible	None	Nyiragongo, Tanzania (1977)
1	Small	10^4-10^6	$0.1-1$	Hawaiian/ Strombolian	Minor	None	
2	Moderate	10^6-10^7	$1-5$	Strombolian/ Vulcanian	Moderate	None	Unzen, Japan (1792; 1991)
3	Moderate–large	10^7-10^8	$3-15$	Vulcanian	Substantial	Possible	Nevado del Ruiz, Colombia (1985)
4	Large	10^8-10^9	$10-25$	Vulcanian/ Plinian	Substantial	Definite	Pelée, Martinique (1902); El Chicón, Mexico (1982)
5	Very large	10^9-10^{10}	>25	Plinian/ Ultraplinian	Substantial	Significant	Mount St. Helens, U.S. (1980); Pinatubo, Philippines (1991)
6	Very large	$10^{10}-10^{11}$	>25	Ultraplinian	Substantial	Significant	Krakatoa, Indonesia (1883)
7	Very large	$10^{11}-10^{12}$	>25	Ultraplinian	Substantial	Significant	Tambora, Indonesia (1815)
8	Very large	$>10^{12}$	>25	Ultraplinian	Substantial	Significant	

Source: Newhall and Self (1982). Copyright 1982 by American Geophysical Union. Reprinted by permission.

Volcanic Eruptions. Volcanoes are associated with the ripping apart of the earth's crust around the midoceanic ridges and at subduction zones where oceanic material is subducted into the earth's molten asthenosphere as lighter, less dense material moves to the surface (Figure 2.7). Fujiyama in Japan and Mount Katmai in Alaska are examples of active volcanoes at subduction zones while volcanoes at oceanic ridges include those on the island of Iceland. In addition, some volcanoes are formed over hot spots such as the of Hawaiian Islands. During the last 10,000 years, 1,300 volcanoes have erupted, half of which are still considered active (Alexander, 1993).

The magnitude of eruptions often is measured in terms of explosive capacity, so that the released energy is said to be equivalent to "x" nuclear weapons or "y" tons of TNT. Nuhfer, Proctor, and Moser (1993) describe the volcanic explosivity index (VEI) devised by Newhall, which defines eight classes of volcanic eruptions from 1 (relatively small) to 8 (catastrophic) (Table 2.14); it is based on measures of the ejecta, height of the cloud column, and other observations. The Krakatoa eruption of 1883 probably was a 6, while the Mount St. Helens eruption in 1980 was only a 5. The largest in the current era was probably that of Mount Tambora in 1815, which rated a 7; its eruption in which 150 km^3 (36 mi^3) of material was ejected, was equivalent to 10,000 atomic bombs. By comparison, less than 4 km^3 (1 mi^3) was ejected in the Mount St. Helens eruption (Lipman and Mullineaux, 1982). Because the figures are so large they are often incomprehensible, but having a scale for comparison is valuable.

Several recent volcanic eruptions have received considerable scientific attention. Paricutín, which erupted in 1943, built a cone fairly rapidly. On the first day, the cone reached 6 m (20 ft), after 6 days 167 m (548 ft), and by the end of the year 325 m (1,066 ft); after intermittent ash eruptions over the next few years, the cone reached 410 m (1,345 ft) by 1952 (Rees, 1979). (For a detailed discussion of Paricutín, see Luhr and Simkin, 1993.) Similarly, Mount St. Helens has been the focus of intensive scientific research since 1980, and Mount Pinatubo's eruption in 1991 generated considerable interest because of its global repercussions. It is believed that like Krakatoa 100 years earlier, Pinatubo led to cooler temperatures around the earth as the sulfur dioxide injected into the upper atmosphere combined with water vapor, forming microscopic sulfuric acid droplets that effectively reflected some incoming solar radiation back to space. Along with dust injected into the atmosphere, this had a significant impact on surface temperatures.

The composition of lava gives a good indication of the explosive potential of a volcano. Basaltic magma is low in silica content and high in iron and magnesium (and other metallic gases); it is generally rather fluid, hence has low explosive potential. Rhyolitic magma has high levels of silica;

it is acidic, more viscous, and has high explosive potential (Sheets and Grayson, 1979). Thus, the magma in Hawaiian volcanoes is generally basic and free flowing whereas that in the Aleutian chain is acidic and explosive. Andesite has an intermediate composition. Thus, simple measures of silicon dioxide content of magmas provide a good indication of explosive potential. More than 66% are acidic (rhyolitic); 52–66% are intermediate (andesitic); and less than 52% are basic (basaltic) (Bullard, 1984). However, not all eruptions from the same volcano have the same silica content. The Taupo volcanic zone on New Zealand's North Island has experienced more than 15 rhyolitic eruptions in the past 20,000 years as well as several basaltic eruptions (Johnston and Nairn, 1993); the largest had a volume of approximately 110 km^3 (26.4 miles^3) (Houghton, 1982).

Temperature is another important factor because it affects the speed of lava movement. Typically, lava reaches the surface at 800–1,200C (1,472–2,192F) (Sheets and Grayson, 1979). It moves rapidly near the source, then slows down as it cools. Thus, lava on Mauna Loa, Hawaii, reaches a peak at 16–40 km/h (10–25 mph), but usually flows at just meters per hour. The velocity of lava flows at Paricutín reached more than 15 m/min (50 ft/min) at the source, but 1.6 km (1 mi) away slowed to 15 m/h (50 ft/hr). Of course, since slope angle also determines the velocity of lava flows, the steeper the slope, the greater the velocity of flow.

Associated events also must be examined, including lava flows, mudflows (lahars), tephra deposits, and pyroclastic flows. Lava flows are usually identified with the less explosive sort of volcanic activity; magma erupts and issues from fissures, moving down slope relatively slowly and often following preexisting landforms. While lava flows may constitute a significant hazard for property, they rarely kill people because there often is ample warning of oncoming flows. Major lava flows are found in Hawaii and Iceland.

Mudflows or lahars are a more destructive volcanic phenomenon. As an eruption occurs, the heat may melt ice and snow which, along with heavy rain, can lead to substantial mudflows; some are "hot," being associated with pyroclastic activity (see below) while others are "cold," being rain-induced (Verstappen, 1992). These flows usually follow existing drainage patterns, but if large enough, can override small ridges and significantly alter the topography. Furthermore, they can move very rapidly, covering wide areas and posing serious problems for humans in the vicinity. In 79 C.E., Herculaneum was overrun by a lahar. The Mount St. Helens eruption involved several lahars, one of which completely filled Spirit Lake and moved down the North Fork of the Toutle River; the event caused major damage to aquatic and biological systems and destroyed 200 homes (Nuhfer et al., 1993). Deaths from lahars can be very high. In 1985, for instance, 25,000 people died as ice melted from the Nevado del Ruiz volcano in

Colombia and three lahars traveled over 60 km before destroying the village of Armero (Verstappen, 1992). This was the worst recorded event in Armero, but it was certainly not the first, for poor siting of the town at the mouth of the Lagunilla Canyon had subjected it to many floods and small lahars in the past (Hall, 1992).

Tephra deposits refer to any material that is transported through the air and falls to land, rather than that flowing across the surface. It is composed primarily of ash deposits, but also includes boulders and rock fragments known as "volcanic bombs," pumice, and small rounded particles or lapilli (Bullard, 1979). The fallout pattern often is determined by wind direction rather than eruption, but gradually declines with increasing distance from the eruption. Larger events, however, force dust particles and gas high into the atmosphere where they circle the globe and can affect the weather significantly. This ash in the atmosphere also heightens other hazards; in 1989, an airplane flying over Alaska near Redoubt experienced trouble after the volcano's eruption when ash entered the jet engine.

Ash falls are particularly problematic for humans and the environment. Like snow, ash collects on all surfaces, but unlike snow it does not melt and, when combined with water, exerts heavy pressure on structures beneath it. Roofs can collapse and other structures can sustain damage unless the weight is removed, but ash removal can take on momentous proportions. For example, 540,000 m³ (706,250 yd³) were removed from the highways in Washington State after the eruption of Mount St. Helens (Nuhfer et al., 1993). Ash deposits on Heimay after the Hekla eruption in 1973 were greater than 1 m (3 ft) deep, burying over 100 houses and crushing several buildings (Thorarinsson, 1979).

Even small layers can create difficulties (Sheets and Grayson, 1979). Ash in the atmosphere often leads to respiratory problems if it is inhaled; people and animals can die of asphyxia because mucus in the lungs reacts with the ash to plug airways. Of the 23 bodies found after the eruption of Mount St. Helens, 18 had been asphyxiated (Manni, Magalini, and Proiett, 1988). Even persons living some distance from an eruption can develop bronchial and respiratory difficulties. Similarly, as ash enters fluvial and biological systems, fish gills become clogged and the photosynthetic processes of plants are compromised. However, these immediate difficulties can be balanced against long-term advantages of ash deposits. Ash can have a beneficial effect on soils, improving subsequent crop harvests; for example, farmers in Indonesia exploit rich volcanic ash in growing coffee.

Pyroclastic flows comprise incandescent hot gas releases, incorporating ash and boulders, that move rapidly outward from the volcano. The pyroclastic mixture is lethal, often composed of toxic gases at temperatures of 600–1,000C (1,112–1,832F) (Sheets and Grayson, 1979). The flows can move at speeds in excess of 320 km/h (200 mph) and can cover hundreds of square kilometers. Initially, they emanate from the center of the eruption, destroying virtually everything in their paths; in the immediate vicinity of the volcano, buildings, plants and other surficial features are obliterated. As the pyroclastic flows move outward and lose strength, the destruction weakens and surficial features become recognizable. Several kilometers from Mount St. Helens, large trees were blown down in the direction of the flows. On Martinique in 1902, a pyroclastic flow from Mount Pelee's eruption killed 30,000 people (Nuhfer et al., 1993).

Gas emissions from volcanic activity also generate hazardous conditions. Thorarinsson (1979) reports that carbon dioxide releases can collect in gas "ponds"; he found evidence of one pond extending over 6,000 m² (7,176 yd²) that had led to a number of animal deaths. Thorarinsson (1979) also examined the toxicity of fluorine gases on Vestmannaeyjar in Iceland. Fluorosis, from fluorine gas, was a problem when associated with low levels of ash deposits. Higher levels of ash were identifiable so that farmers took precautions to exclude animals from contaminated areas, but when ash was not so visible, sheep and cattle consumed fluorine and often died. Although this problem was identified at least as early as 1693, it continues to pose problems because it is frequently an invisible threat. Other gas releases have been noted as well. In Cameroon, carbon dioxide releases from crater lakes killed 37 people around Lake Monoun in 1984 (Sigurdsson, 1988) and 1,746 people around Lake Nyos in 1986 (Kling et al., 1987; Walker, Redmayne, and Browitt, 1992); however, these events have not been linked definitively with tectonic activity and are more likely a function of tropical lake processes.

Tsunami. Tsunami are ocean waves generated by underwater earthquakes, volcanic eruptions, and landslides. Abrupt seafloor movements displace water, thereby generating disturbances that become waves. As energy is transmitted through the water, the ocean rises and falls by only 1 m (3 ft) or more, but when the energy comes close to shore, the tsunami crests because of wave dynamics, rises, then plunges onto the shoreline. Depending on water depth, the wave may move at more than 965 km/h (600 mph) before cresting and pounding the shoreline. The actual wave size and degree of destruction depend on topographical features, which can significantly influence wave dynamics, and on building location and construction that can exacerbate damage. The direction of wave travel in relation to coastal features also influences tsunami characteristics. Tsunami can occur as a series of waves with a wavelength up to several hundred kilometers (Ruffner and Bair, 1984).

While any seismic activity under the oceans and shorelines may generate some form of tsunami, it is usually the larger events, perhaps in excess of 7 on the Richter scale, that lead to serious problems (Ayre, Mileti, and Trainer, 1975). Ayre et al. (1975) describe four measures relevant to tsunami magnitude: wave height above current stage of tide; wave height above wave trough; wave amplitude (i.e., vertical distance between wave crest and trough); and wave height above some arbitrary datum point. The first two are most commonly used. Research has shown that tsunami migrate across the oceans at fairly regular periods. For tsunami generated in Chile, there may be a warning time of 15 hours for Honolulu. Indeed, a warning system (sited on Honolulu) provides time lines for wave movement across the whole of the Pacific (Figure 2.10), based on time of travel from the tsunami's generation point to the location of a community. However, when earthquakes occur near coastal communities (as happened in 1964 in Alaska), warning time is minimal at best. Most attention has been focused on the Pacific because of the higher incidence of events around the Pacific rim; during the first three-quarters of this century, there were 181 tsunami in the Pacific (Ayre et al., 1975).

The devastation and destruction from tsunami can be almost total for the area hit (Table 2.15). The Krakatoa eruption (1883) generated waves over 30 m (100 ft) high and killed 36,000 people while the Tambora tsunami (1815) killed 10,000, all in Indonesia. Similarly, the Lisbon earthquake of

FIGURE 2.10. Predicting tsunami. Because we cannot predict earthquakes, tsunami are difficult to forecast. The Pacific Tsunami Warning System is based on seismic and tide stations located throughout the Pacific Ocean. This map shows tsunamis' known travel times to and from Hawaii, the center of the system. *Source:* Office of Emergency Preparedness (1972).

TABLE 2.15. World's Major Tsunami

Year	Origin	Location of losses and damage	Drownings
1628 B.C.E.	Santorini eruption	Crete, Greece, Egypt	unknown
1640 C.E.	Komagatake eruption	Hokkaido, Japan	1,470
1707	West Pacific earthquake	Osaka Bay, Japan	4,900
1755	Offshore Lisbon earthquake	Western Europe, North Africa	20,000–30,000
1792	Unzen volcano, avalanche into sea	Kyushu, Japan	10,000 (landslide killed 5,000 more)
1815	Tambora eruption	Indonesia	10,000
1883	Krakatau eruption	Indonesia	36,000
1896	Honshu earthquake	Sanriku, Japan	27,000
1905	Norway, landslide into fjord	Loen Lake, Norway	61
1918	Landslide	Philippines	100
1933	Honshu earthquake	Sanriku, Japan	3,000
1936	Norway, landslide into fjord	Loen Lake, Norway	73
1946	Aleutian Islands earthquake	Wainaku, Hawaii	173
1960	Chile, earthquake	Chile, Hawaii, Japan	1,900
1964	Alaska, earthquake	Crescent City, California Valdez area, Alaska	126
1971	Peru avalanche into lake in Andes	Chungar, Peru	600
1976	West Pacific earthquake	Philippines	3,000
1979	Werung volcano with avalanche into sea	Lomblem Island, Indonesia	539
1992	Southwest Nicaragua, earthquake	Masachapa, Nicaragua, and 200 miles of coast and nearby islands	116

Source: Nuhfer et al. (1993). Copyright 1993 by American Institute of Professional Geologists. Reprinted by permission.

1755 led to between 20,000 and 30,000 deaths from tsunami (Ayre et al., 1975). Japan has been particularly vulnerable to tsunami, suffering 150 major events since the seventh century. For instance, 27,000 people died in Sanriku in 1896; more recently, a western Pacific earthquake in 1976 produced tsunami that killed 3,000. In 1946, 173 died in Hawaii and, in 1964, 126 died in Alaska as a result of tsunami (Petak and Atkisson, 1982). In 1992, a powerful earthquake (7.2 on the Richter scale) generated a tsunami over 14 m (45 ft) high on the Pacific coast of Nicaragua; reports stated that 150 people had been killed, 500 were missing, and immediate material losses were valued at $25 million. However, there are many smaller tsunami that do not have much impact.

In enclosed bodies of water, activity similar to that causing tsunami creates comparable wave effects, termed seiches. These are not on the same destructive scale and can have different causes. Seiches can be triggered by various phenomena, including tectonic activity and strong winds that cause bodies of water (such as bays, harbors, bayous, canals, and lakes) to rock back and forth. Korgen (1995) discusses several of these, including seismic waves associated with the 1964 Alaskan earthquake that reached the Gulf Coast of Texas and Louisiana; in 14 min these eventually created seiches of up to 2 m (6 ft). Similarly, severe storms on Lakes Michigan and Erie have caused various seiches, resulting in property damage around the lakes.

Summary

The events discussed above illustrate both the difficulties and importance of measuring magnitudes of occurrence and resulting impacts. The examples serve to emphasize three points. First, no matter how scientific measures of magnitude vary from place to place and from time to time, in part because of variations in local sites and situations but also because of the dynamic nature of events. Second, measures of magnitude may permit comparisons of geophysical events, but they do not necessarily allow direct comparison of impacts. Third, measures of magnitude are neither simple in themselves nor should they necessarily stand alone. Other characteristics must enter the analysis of physical dimensions in order to have a complete understanding of extreme geophysical events.

Duration

A second feature of physical mechanisms is duration of a geophysical event. In many ways it is relatively easy to measure, making possible a degree of generalization across hazards. Human impacts are not included in this discussion although they can last for many years (see Chapter 7). The duration scale for physical events ranges from a few seconds (as in most earthquakes and tornadoes) to years (as in many droughts). In addition, ongoing processes—such as soil creep, land slipping, subsidence, and swelling soils—are responsible for substantial damage to buildings and transportation facilities. Preparation and planning for different hazards may show many similarities based on duration.

At one end of the duration scale for any given place, the physical processes are short-lived. Although an event might last for an hour or so, such as a tornado moving across the landscape, its duration is shorter for a given place. In a few seconds, a tornado can cut a swath through a residential neighborhood, destroying all property in its path; in minutes a flash flood can sweep down a canyon, overturning cars and sweeping away houses; or a tsunami might invade a coastline, killing hundreds of people within minutes. Perhaps earthquakes are the quintessential short-duration geophysical event; huge areas of cities can be completely leveled in seconds as shock waves move through the earth's crust. For example, the initial shock of the San Francisco earthquake of 1906 lasted only 40 sec, followed by a secondary wave of approximately 30 sec (Ritchie, 1988), yet it killed 2,500 (Palm, 1995). The Friuli earthquake in 1976 lasted approximately 1 min, killed 939 people and, following a series of aftershocks, left a total of 157,000 people homeless (Geipel, 1982). The Armenian earthquake in 1988 lasted only seconds but resulted in 25,000 deaths (Nuhfer et al., 1993), and the Kobe, Japan, earthquake in 1995 lasted just 22 sec, but caused over 6,000 fatalities and billions of dollars in damage.

At the other end of the duration scale are droughts. Many have lasted for years, particularly those in the Great Plains (U.S.) and the Sahel. For example, one Sahelian drought lasted from the late 1960s until 1974; East Africa experienced droughts in 1972–1974, and again in 1982–1984 (Giorgis, 1987; May, 1987). The Great Plains suffered droughts in 1930–1936, in the mid-1950s and again in the late 1980s. Each drought generated research on appropriate adjustments so that the impacts of the next were generally less severe. However, changing trends in population growth, availability of resources, and international dependencies have exacerbated adverse consequences (Warrick, 1975). The long duration of drought offers opportunities for adjustments even as drought conditions are occurring, but the longer the duration, the wider the range of impacts and more intense the severity.

Other events fall somewhere in-between on this time scale. Floods can be fairly lengthy or brief like the Big Thompson event. During the 1993 midwestern floods in the United States, several communities were under water for several weeks; residents in neighborhoods of Ottumwa and Davenport, Iowa, could not return to their homes for nearly 6 weeks, while for others the water rose and retreated in the same day (Tobin and Montz, 1994b). Such events therefore present different problems for the individuals and communities involved. In many instances, events of longer duration require semipermanent shelters (some people in the Midwest were still in temporary housing 18 months after the flood occurred) and support facilities to meet the needs of a displaced population. Short-duration events may require immediate emergency action; work on cleaning up can begin quite rapidly. Of course, magnitude and intensity of the event must also be taken into account, especially if short-duration events are also of high magnitude.

Volcanic activity can vary in duration. In Hawaii, where volcanic eruptions are relatively slow-moving, the current activity has continued for a number of years. The outpouring of lava has encroached gradually on property, burning or burying many houses. In some ways this activity is akin to droughts because the physical processes begin almost imperceptibly, creeping ever onward and affecting more and more people. By contrast, explosive volcanic eruptions may be preceded by the venting of gases and release of ash, but the final, cataclysmic eruption generally lasts for only a few hours. Clearly, the destruction can be profound and the landscape changed dramatically. The type of eruption, then, is important, and vulcanologists provide important data on these events.

Hurricanes provide another example where duration is variable and, as a result, problematic. They generally begin as tropical storms that sometimes intensify by amassing thermal energy until cyclonic winds exceed 119 km/h (74 mph). Once developed, they continue to move and may eventually make landfall. This process can take many days during which time the evolving storms can be tracked by meteorologists; in the United States, this job is undertaken by the National Hurricane Center in Miami, Florida. Once hurricanes reach land, they lose their contact with the warm ocean surface and immediately begin to diminish in strength. Nevertheless, this does not prevent them from creating considerable devastation along coastal and near-coastal communities in places like Bangladesh, the Caribbean, and the United States.

Landslides and debris flows represent events that usually have a very short duration of impact, but a long period of onset. These events generally follow periods of heavy precipitation so that hill- and mountainsides become sufficiently saturated and unstable to initiate a landslide (DeGraff, 1994). While the slide itself may take only minutes, the antecedent conditions may have been building for hours, days, weeks, or even years, as in the case of soil creep. The situation is complicated by other factors that reduce slope stability, including deforestation, construction, and earthquakes. Furthermore, many landslides can be disruptive and expensive to remedy because they can block transportation routes and clog stream channels. Individual events generally do not cause significant losses because they are spatially limited in extent, but their cumulative effect amounts to millions of dollars in damages every year. This is not say that landslides have not caused major problems. In 1979, one in Dunedin, New Zealand, destroyed or removed 62 houses in a matter of 30 minutes; however, this occurred after 70 days of slow movement around the margins of the main slide area (Coombs and Norris, 1981). In another New Zealand example, heavy rain and deforestation caused a landslide in Te Aroha that deposited 40,000 m³ (52,318 yd³) of debris in the town center, killed 3 people, and resulted in several million dollars in damage (Montz, 1992).

These examples illustrate the importance of understanding the duration of geophysical events. Although events of longer duration tend to generate greater losses than those of shorter duration, when duration is combined with magnitude, losses can increase significantly. In some events (such as drought and soil erosion), longer durations make possible the adoption of adjustments even while the event is occurring. Thus, it is appropriate and necessary to evaluate duration among the physical characteristics of extreme geophysical events.

Spatial Extent

Any discussion of physical dimensions of events must take into account the area over which they occur. We might be tempted to generalize that at least for some events, the larger the magnitude, the larger the area affected. For floods, greater discharge means more water and consequently a larger area of inundation. Similarly, a high-magnitude earthquake means greater shaking that resonates farther from the epicenter than in low-magnitude events. However, this generalization does not always hold.

As a general rule, extreme geophysical events can be defined by spatial extent, to wit, (1) high-magnitude processes are generally more spatially confined than low-magnitude events, and (2) low-probability events with a long return period are spatially extensive. At first, this might seem contradictory since low-probability events are generally of higher magnitude, but the corollary pertains to measurements at a given place. At the epicenter of an earthquake, for instance, tectonic processes may be intense, especially in a high-magnitude event, but the intensity declines with increasing distance. There is a two-way relationship between magnitude and spatial extent.

In the first case, it is not possible to have very intense physical processes extending over wide areas. Very heavy storms, for example, tend to be spatially confined; unfortunately, very little work has been done to establish relationships between depth, area, and duration of precipitation (Dunne and Leopold, 1978). However, a study of the eastern United States showed a relationship between point rainfall of a specified duration and average rainfall up to 1,036 km^2 (400 mi^2) (U.S. Weather Bureau, 1957), which held regardless of storm frequency. For example, using figures from the graph in Figure 2.11, an area of 518 km^2 with a 1-hour storm could expect approximately 67% of the point rainfall totals; this figure would rise to nearly 80% in a 3-hour storm and 92% in a 24-hour storm. Similarly, the intensity of precipitation is constrained by time (Table 2.16); the most intense rainfall (rainfall total divided by time interval) occurs during shorter intervals and cannot continue at the same rate for lengthy periods.

Other natural processes follow similar patterns. For example, large-magnitude earthquakes are felt over a wide area, but the strength of the

FIGURE 2.11. t.k. Depth–area curves for the continental United States. These make it possible to represent average depth–area relationships as a percentage of single gauge observations. *Source:* Hershfield (1961).

seismic wave diminishes fairly rapidly away from the epicenter (depending, of course, on geological conditions). For example, the New Madrid, Missouri, earthquakes of 1811–1812 were felt as far away as Minnesota and New England, but smaller events generally have more local impact. Likewise, the spatial extent of other meteorological events is related to magnitude. For a tornado, magnitude is related to area; the broader the funnel and longer the path length, the stronger the winds. This is not to say that there is a simple relationship between the two, since some very large funnels have lower wind speeds.

In the second case, we find that large events affect large areas, hence increase the probability that more people will be impacted. For example, the droughts of the Sahel in the early 1970s and Ethiopia in the mid-1980s covered thousands of square miles and eventually involved the world community. These were genuinely intense events, but the severity of the droughts was not uniform across the continent; depending on the magnitude of processes at a particular location, specific areas were affected to a greater or lesser degree. Regions at the centers of the problems were those where the meteorological events were most intense. The same can be said of earthquakes; to some extent, the magnitude of ground shaking decays in relation to distance from the epicenter. Therefore, small-magnitude events impact smaller areas than large-magnitude events, but the most intense processes occur over smaller areas. It is the relationship between magnitude and intensity in space that must be understood in hazards research.

TABLE 2.16. Maximum Rainfall Intensities for Various Durations of a Storm

Duration (min)	Maximum rainfall during interval (mm)	Maximum rainfall intensity for duration (mm/h)
20	19.2	57.6
40	26.1	39.2
60	31.2	31.2
100	48.8	29.3
120	52.6	26.3
160	57.7	21.6

Source: Dunne and Leopold (1978). Copyright 1978 by W.H. Freeman and Company. Reprinted by permission.

TEMPORAL DISTRIBUTION

Understanding the temporal distribution of extreme geophysical events provides further insight into natural hazards. In planning and mitigation, it is useful to know how an event is expected to occur at any given place and when it might take place. Such information would transform seemingly random geophysical events, and enhance hazard response. We now examine the frequency and return periods (recurrence intervals) of specific events, and further ascertain whether particular physical processes can be delineated within a temporal framework, such as at seasonal or diurnal scales.

Frequency

The traditional approach to measuring event frequency and return periods has been to analyze historical records by calculating the average number of occurrences over a specific time period. If there have been 20 blizzards in a 60-year record, the event is said to have a recurrence interval of once every 3 years; that is the blizzard has a 0.33 probability or a 33% chance of occurring in any given year. Similarly, 40 floods, earthquakes, or tornadoes in a 100-year record would suggest a return period of once every 2.5 years, and a 0.04 probability of occurring in a given year. The level of analysis can be further refined by size of the event. We know, for example, that large-magnitude events have a lower probability of occurrence than small-magnitude events. In practical terms, there are thousands of low-magnitude earthquakes occurring around the world every day, but only a few that measure 7.0 or more on the Richter scale in any year. Consequently, there is an inverse relationship between frequency, or probability of occurrence, and magnitude; any data set of natural phenomena will contain more small events than large ones. Analysis of frequency must account for magnitude of the event and incorporate a measure of the expected return period.

Table 2.17 (a)-(b) shows how such calculations are made using a hydrological data set from the Raccoon River at Van Meter, Iowa. In the first instance, the annual maximum series (AMS) is used to determine flood probability (Dunne and Leopold, 1978). The highest flow for each year is recorded, then ranked (from high to low) according to other years. The return period is solved using the formula: $Tr = (n + 1) / m$, where Tr represents the return period, m represents the rank number of the event, and n is the number of years of record.

In the data set (Table 2.17a), the flood of 860 m³/s (30,400 cfs) is ranked 10th and can be expected to recur once every 8.1 years. The probability of this event occurring in any given year is the reciprocal, that is, 0.124. The return period, however, does not give any indication when an event might occur; it only states the frequency at which we can expect such events to be equaled or exceeded. In other words, the 8-year event could occur two or three times in 1 year and not again for 50.

For a long data set and for events in excess of a 10-year recurrence interval, this simple equation gives fairly accurate results and is a useful guide to flood frequency. What it does not capture so well is the frequency of smaller events, for which a different data set might be used. The problem with the AMS is that many high flows are ignored in favor of the single high annual event. In some years, there may be several floods, of which only one is included in the analysis, while in "dry" years, a relatively low flow will be included that is not a flood. For example, the Raccoon River data show that in 1941 and 1942, no flows exceeded bankfull discharge, but they still are included in the analysis of flood flows. In many other years (such as 1947, 1965, 1974, 1984, and 1993), there were several peaks representing independent flood events that perhaps should be included in the data set.

To overcome this problem, the partial duration series (PDS) is used. This data set includes all flows above some arbitrary point (e.g., bankfull discharge). Analyzing the same drainage basin but using the PDS with a discharge cutoff of 283 m³/s (10,000 cfs), the flood of 860 m³/s (30,400 cfs) is now ranked 13th and has a recurrence interval of 6.2 years, or a 0.16 probability of occurring in any given year. For smaller events, the difference is even greater. For instance, a discharge of 566 m³/s (20,000 cfs) has a return period of 3 years (0.33 probability) according to the AMS, but only 2 years (0.5 probability) according to the PDS.

This level of analysis provides some useful information about hazards that can be related to risk (see Chapter 7), but the technique has limitations. It does not transfer easily from one location to another; it does not permit

TABLE 2.17. Discharge Records for the Racoon River at Van Meter, Iowa

	(a) Annual Maximum Series				(b) Partial Duration Series		
Date	Ranked discharge (m^3/s)	Return period	Probability	Date	Ranked discharge (m^3/s)	Return period	Probability
7/10/1993	1,984	81.00	0.01	7/01/1993	1,984	81.00	0.01
6/13/1947	1,166	40.50	0.02	6/13/1947	1,166	40.50	0.02
7/01/1986	1,138	27.00	0.04	7/01/1986	1,138	27.00	0.04
7/04/1973	1,007	20.25	0.05	6/25/1947	1,075	20.25	0.05
6/07/1917	996	16.20	0.06	7/04/1973	1,007	16.20	0.06
7/03/1958	996	13.50	0.07	6/07/1917	996	12.54	0.08
6/16/1990	979	11.57	0.09	7/03/1958	996	12.54	0.08
4/02/1960	914	10.13	0.10	6/16/1990	979	10.13	0.10
9/20/1926	906	9.00	0.11	8/29/1993	959	9.00	0.11
5/19/1974	860	8.10	0.12	7/01/1973	917	8.10	0.12
3/19/1979	846	7.36	0.14	4/02/1960	914	7.36	0.14
4/30/1984	807	6.75	0.15	9/20/1926	906	6.75	0.15
3/31/1951	784	6.23	0.16	5/19/1974	860	6.23	0.16
3/19/1948	756	5.79	0.17	3/19/1979	846	5.79	0.17
6/11/1953	736	5.40	0.19	4/30/1984	807	5.40	0.18
7/03/1983	722	5.06	0.20	3/31/1951	784	5.06	0.20
5/21/1944	662	4.76	0.21	3/19/1948	756	4.76	0.21
2/20/1971	651	4.50	0.22	6/10/1917	736	4.38	0.23
3/25/1969	645	4.26	0.23	6/11/1953	736	4.38	0.23
4/06/1965	631	4.05	0.25	6/05/1947	724	4.05	0.25
5/29/1915	611	3.86	0.26	7/03/1983	722	3.86	0.26
4/19/1991	594	3.68	0.27	6/17/1984	679	3.68	0.27
6/25/1954	589	3.45	0.29	5/21/1944	662	3.52	0.28
3/22/1978	589	3.45	0.29	2/20/1971	651	3.38	0.30
6/12/1967	583	3.24	0.31	3/25/1969	645	3.24	0.31
6/25/1924	569	3.12	0.32	4/16/1973	642	3.12	0.32
6/18/1957	566	3.00	0.33	6/23/1984	634	3.00	0.33
3/31/1962	558	2.89	0.35	4/06/1965	631	2.89	0.35
3/14/1929	549	2.79	0.36	5/29/1915	611	2.75	0.36
6/20/1950	498	2.70	0.37	5/08/1973	611	2.75	0.36
6/23/1964	492	2.61	0.38	4/19/1991	594	2.61	0.38
6/13/1966	458	2.53	0.40	9/06/1958	591	2.53	0.40
3/06/1949	450	2.45	0.41	6/25/1954	589	2.42	0.41
11/24/1932	430	2.38	0.42	3/22/1978	589	2.42	0.41
6/06/1945	427	2.22	0.45	6/12/1967	583	2.28	0.44
9/09/1946	427	2.22	0.45	6/07/1991	583	2.28	0.44
4/01/1952	427	2.22	0.45	6/08/1951	572	2.19	0.46
6/15/1982	472	2.22	0.45	6/25/1924	569	2.11	0.48
6/06/1918	419	2.06	0.49	3/17/1965	569	2.11	0.48
5/14/1970	419	2.06	0.49	6/18/1957	566	2.03	0.49
8/26/1987	416	1.98	0.51	5/02/1951	563	1.98	0.51
6/18/1975	405	1.93	0.52	6/03/1947	560	1.93	0.52
3/12/1939	396	1.88	0.53	3/31/1962	558	1.88	0.53
6/04/1959	382	1.84	0.54	3/14/1929	549	1.82	0.55
8/16/1943	354	1.80	0.56	6/30/1983	549	1.82	0.55

TABLE 2.17. (cont.)

(a) Annual Maximum Series				(b) Partial Duration Series			
Date	Ranked discharge (m^3/s)	Return period	Probability	Date	Ranked discharge (m^3/s)	Return period	Probability
4/24/1919	351	1.74	0.57	6/14/1991	535	1.76	0.57
9/16/1992	351	1.74	0.57	6/10/1967	529	1.72	0.58
3/05/1936	345	1.65	0.60	6/16/1944	507	1.69	0.59
3/04/1937	345	1.65	0.60	6/20/1950	498	1.65	0.60
3/11/1963	345	1.65	0.60	6/23/1964	492	1.61	0.62
8/28/1977	337	1.59	0.63	8/12/1993	492	1.61	0.62
4/11/1922	323	1.56	0.64	6/23/1947	490	1.56	0.64
3/27/1961	320	1.53	0.65	10/1/1974	484	1.53	0.65
3/05/1935	311	1.50	0.67	4/02/1979	481	1.49	0.67
3/27/1923	291	1.47	0.68	5/17/1986	481	1.49	0.67
3/14/1920	283	1.45	0.69	6/13/1966	458	1.45	0.69
2/22/1985	280	1.42	0.70	3/14/1973	453	1.42	0.70
9/11/1972	275	1.40	0.72	3/06/1949	450	1.40	0.72
9/20/1921	274	1.37	0.73	7/04/1951	447	1.37	0.73
5/11/1942	249	1.35	0.74	7/18/1993	442	1.35	0.74
3/07/1994	248	1.33	0.75	3/30/1993	439	1.33	0.75
4/24/1976	246	1.31	0.77	3/14/1971	436	1.31	0.77
4/24/1955	244	1.29	0.78	8/18/1978	433	1.29	0.78
6/02/1938	243	1.27	0.79	11/24/1932	430	1.27	0.79
8/27/1928	240	1.25	0.80	6/06/1945	427	1.22	0.82
8/07/1925	228	1.23	0.81	9/09/1946	427	1.22	0.82
9/06/1989	216	1.21	0.83	4/01/1952	427	1.22	0.82
4/05/1933	214	1.19	0.84	6/15/1982	427	1.22	0.82
6/02/1941	196	1.17	0.85	4/29/1974	425	1.17	0.86
2/08/1927	195	1.16	0.86	4/29/1991	425	1.17	0.86
5/13/1930	166	1.14	0.88	6/27/1952	422	1.13	0.89
3/16/1916	165	1.13	0.89	5/29/1962	422	1.13	0.89
7/31/1940	163	1.11	0.90	3/11/1983	422	1.13	0.89
6/21/1931	149	1.09	0.91	6/06/1918	419	1.09	0.92
6/15/1980	138	1.08	0.93	5/07/1960	419	1.09	0.92
6/27/1968	123	1.07	0.94	5/14/1970	419	1.07	0.94
9/04/1956	118	1.05	0.95	7/10/1969	416	1.05	0.96
7/04/1981	117	1.04	0.96	8/26/1987	416	1.05	0.96
6/09/1988	111	1.03	0.98	4/17/1983	410	1.03	0.98
4/07/1934	57	1.01	0.99	6/18/1951	408	1.00	1.00
				4/02/1983	408	1.00	1.00

Note: The data set contains another 100 observations that exceed bankfull discharge of 283 m^3/s.

easy extrapolation of the data set to a longer time frame; and it does not perform well for all geophysical events. First, the return period calculations are specific to the location from which the data were collected (in the example, the Van Meter gauging station of the Raccoon River drainage basin in Iowa). Ideally, we could use the data set to estimate return periods in similar, possibly adjacent but ungauged basins. There are ways of extending the record to comparable drainage basins, but the error factor associated with the return periods will increase (Dunne and Leopold, 1978).

Second, temporal extrapolations are similarly constrained. For example, the data set in Table 2.17 can be used to project the 200-year event by plotting points on a graph. There are various ways to do this, but essentially different forms of probability paper have been used to determine flood frequency (Figure 2.12 a–b). Inevitably, the extrapolation of data beyond the station record significantly increases potential error, as shown by the dotted line. On the other hand, plotting the data also shows that some events do not fall on the distribution curve; in the example, the 1993 event is way above the line, suggesting that this flood was actually of longer return period than 80 years. Extending the 1993 point left of the curve gives a truer indication of the flood return period. Thus, a graphic analysis may provide a different perspective on the data (although selection of the appropriate type of graph paper is somewhat subjective).

The 1993 flooding in the upper Midwest of the United States demonstrates some of the problems with this approach since flows surpassed many previously recorded peak discharges. For example, flows at 154 river gauging stations exceeded the 10-year flood, and 42 were above all known records (Parrett, Melcher, and James, 1993). Another 14 exceeded previous maximum regulated discharges, that is, flow conditions for rivers altered by river control projects or reservoirs. In terms of flood recurrence intervals, 46 stations had estimated return periods in excess of 100 years, and it is probable that many streams experienced floods with a recurrence interval in excess of 200 years. Figure 2.13 shows the historic peak discharges for several of these stations and the 100-year return period for each. Clearly, the 1993 flooding was significantly higher than peak discharges for most previous years of record and consequently inundated much larger areas than earlier events.

Many meteorological and hydrological events fit into this statistical distribution of randomness over the long term. Tropical cyclones, tornadoes, windstorms, heat waves, cold waves, blizzards, and droughts all occur with a given probability for a place for a given year. Figure 2.14 shows the probability of a hurricane hitting a particular section along the coastline of the United States. These data are based on records of hurricanes since the last century. However, we might ask ourselves whether the same patterns of storms are occurring now as occurred earlier this century, and whether the same probability patterns remain. Further, there is a tendency toward persistence in weather patterns such that particular systems come to dominate for a time. This does not affect long-term variability and

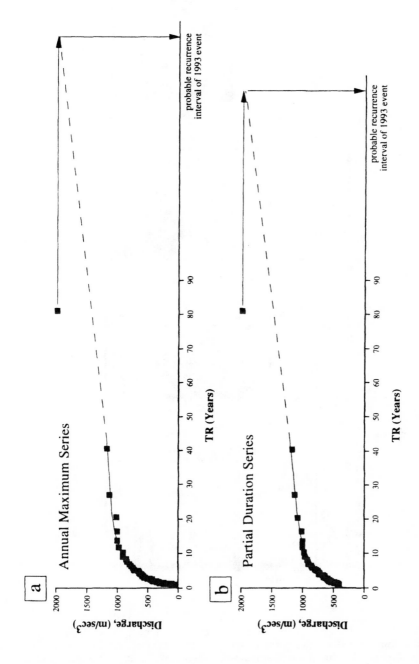

FIGURE 2.12. Predicting floods. The curves show the way in which annual maximum series and partial duration series data can be used to project flood events. They also illustrate the difficulty of extrapolating beyond a station record.

FIGURE 2.13. Differences in flood experience. Four gauging stations in the Mississippi River Basin had distinct patterns of frequency of flooding. For all but St. Louis, the 1993 flood was the greatest on record. *Source:* Parrett, Melcher, and James (1993).

FIGURE 2.14. Probability of occurrence of hurricanes along the Atlantic and Gulf Coasts of the United States. Based on the historical record of events, the numbers represent the probability (in percent) that a hurricane will make landfall along a 50-mile stretch of coast in a given year. *Source:* Office of Emergency Preparedness (1972).

probability estimates, but it does mean that conditions for specific events may be higher at particular times. For example, during the summer of 1993, the jet stream stalled over the midwestern United States, allowing storm tracks to follow the same paths repeatedly. Consequently, the upper Midwest experienced extremely heavy rainfall that culminated in widespread flooding (Figure 2.15).

Persistence in weather patterns can be observed at various temporal scales. For example, weather does not change drastically from day to day, but alters as new air masses or synoptic systems enter the area. In fact, there is a 60% chance that tomorrow's weather will be the same as today's. Thus,

FIGURE 2.15. Area affected by flooding in the Mississippi River Basin, 1993. The jet stream remained stationary for some time, allowing storms to track across the Midwest. Some communities had several floods during the spring and summer, others experienced the worst flood on record, and others had neither frequent nor particularly large flooding. *Source:* Adapted from Parrett at al. (1993).

Within the figure:

Unseasonably
cool, dry air

Warm, moist air

Approximate position of jet streams

Area of flooding streams

Boundary of Mississippi River Basin

400 Miles

400 Kilometers

days or weeks of above-average heat, cold, rainfall, or dryness are not very unusual. The same trends can be identified in longer data sets; climatological records show distinct groupings of wet and dry years and of warm and cold years. To some extent, floods and droughts, are concentrated in particular years. For example, the early 1930s and the mid-1950s were warmer than usual for large parts of the United States. Drier and warmer years were recorded throughout much of northern Europe and North America in the late 1980s, and southern Britain experienced droughts in the mid-1970s and early 1990s. Graves and Bresnock (1985) also found temporal patterns on an annual basis for tornado activity; years with high tornado activity are usually followed by another year with higher than normal frequency.

Thus, although frequency distributions assume that patterns remain the same over the long term, variations at different scales can influence the pattern of extreme geophysical events. The apparent grouping of events is described statistically as a Poisson distribution. A further complication exists if the climate is changing. Any global warming or cooling will change global atmospheric circulation, hence the distribution of storms and precipitation around the world. Our records, therefore, are compromised because even in this century, we can identify periods of warmer and cooler conditions.

An examination of the incidence of tornadoes demonstrates another statistical problem. Figure 2.16 shows the annual number of tornadoes, in the United States for 1916–1991. It appears that their frequency has increased dramatically during this period. However, as depicted by the lower line of the graph, the actual number of days per year on which tornadoes occurred did not change. This suggests that the apparent increase in number of tornadoes can be explained by other factors, notably improved tornado spotting techniques and better data collection and recording. A simple analysis of tornado frequency over time, therefore, can provide misleading information. In an average year (using 1953–1991 as the base), the United States experiences 768 tornadoes which occur on 169 different days (Grazulis, 1993). However, if we look at tornado frequency for the years 1970–1991, the average number of tornadoes per year is 865 (Table 2.18). An even shorter time frame could produce other results. Some years have much higher incidences of tornadoes than others, and it is not unusual to find major outbreaks occurring in other years; for instance, on April 3–4, 1974, 148 tornadoes passed through 13 states (Tierney and Baisden, 1979).

Similar statistics have been found for avalanche hazards. For example, there has been a remarkable rise in the number of deaths in many countries, primarily because of the increase in alpine recreation rather than increased avalanches (Prowse, Owens, and McGregor, 1981). In New Zealand, for

FIGURE 2.16. Officially counted tornadoes per annum in the United States, 1916–1991. The number of officially counted tornadoes reflects increasing sophistication in spotting techniques as well as an expanded spotters' network. However, because the number of tornado days has not increased (as the lower line on the graph indicates), the actual number of tornadoes probably has not risen. *Source:* Grazulis (1993). Copyright 1993 by The Tornado Project. Reprinted by permission.

example, no avalanche deaths were recorded between 1900 and 1909, but by 1970, 16 people had died. Using different data sets, therefore, can alter the apparent risk involved with different geophysical events. Consequently, care must be taken in interpreting frequency data.

While it is possible to determine a simple frequency of occurrence for all hazards by dividing the number of years of record by the number of incidents, this would not necessarily give a true picture of the temporal distribution for all processes. Thus, our third concern is that the technique is not universally applicable to all geophysical events: not all geophysical events have this element of randomness. Some exhibit periodicity that stems directly from buildup and release of pressure. For example, if there have been seven earthquakes of similar size in 100 years at a particular place, then according to the flood formula, the frequency of occurrence would be once every 14.3 years or a 0.07 probability of occurring in a given year. This figure is useful in defining long-term risk, but masks the true danger involved. Since many earthquakes are associated with the buildup of strain on fault lines, the risk of fault failure increases accordingly; thus, the probability of occurrence changes from year to year. The probability of a major earthquake along the San Andreas fault in San Francisco has increased significantly since the 1906 event, so that scientists predict that another major event is imminent. Indeed, a major earthquake (Richter magnitude 7.0 or greater) on the Hayward fault east of San Francisco Bay has a 0.67 probability of occurring between 1990 and 2020 (Ward and Page, 1990); some estimate the probability of the same magnitude earthquake on the northern San Andreas fault at 50% for the period 1990–2020. Similarly, the chance of a magnitude 7 or greater earthquake in Los Angeles by the year 2024 is 86%, raising earlier estimates of a 60% chance before 2018 (Working Group on California Earthquake Probabilities, 1988, 1995).

Explosive volcanoes may follow a similar pattern. A historical analysis of Mount St. Helens, for example, shows that it is the most active volcano in the Pacific Northwest (Crandell, Mullineaux, and Miller, 1979). Four major periods of activity have been identified, each lasting several hundred years; the periodicity is similar to that of earthquakes. It is interesting to note that Crandell et al. (1979) predicted activity on Mount St. Helens before the 1980 eruption, including explosive eruptions, lava domes, pyroclastic flows, and lahars. Less explosive volcanoes also have periods of activity and dormancy. Hodge, Sharp, and Marts (1979) traced the eruptions of Mauna Loa and Kilauea; Mauna Loa erupted on average every 3.5

TABLE 2.18. Tornadoes in the United States: Occurrences and Deaths, 1970–1991

Year	Number of tornadoes	Number of days with tornadoes	Number of deaths	Number of "killer" tornadoes
1970	653	171	72	18
1971	888	192	156	22
1972	741	194	27	11
1973	1102	206	87	37
1974	947	184	361	71
1975	920	204	60	29
1976	835	169	44	29
1977	852	189	43	15
1978	788	173	53	17
1979	852	186	84	20
1980	866	176	28	18
1981	783	175	24	14
1982	1046	182	64	30
1983	931	190	34	27
1984	907	166	122	34
1985	684	168	94	20
1986	764	168	15	11
1987	656	151	59	14
1988	702	156	32	18
1989	856	160	50	12
1990	1133	181	53	18
1991	1132	179	39	15

Source: Grazulis (1993). Copyright 1993 by The Tornado Project. Reprinted by permission.

years whereas Kilauea was almost constantly active between 1823 and 1924. Between 1924 and 1965, Kilauea erupted 21 times. However, no distinct pattern can be discerned from these data that would allow meaningful predictions of impending eruptions to be made. The same can be said of tsunami. For example, Japan has experienced 65 destructive tsunami since 684 C.E. (Pararas-Carayannis, 1986); a reported 1,000 people were killed in 869, another 1,700 in 1361, 500 in 1498, and 27,122 in 1896. However, a regular pattern is not discernible. Predicting temporal trends will probably always lie beyond our capabilities.

Frequency of occurrence is important, then, because of what it tells us about potential magnitude in some instances, and about recurrence intervals and planning needs in others. However, it is clear that frequency of occurrence may not be sufficiently indicative of any trend.

Seasonality

Another temporal feature of some natural hazards is seasonality. Some events occur more frequently at certain times of the year than others. Snowstorms, cold waves, and blizzards are clearly winter phenomena in temperate latitudes, while heat waves and tropical cyclones are typically summer events. Thus, planning for such events can be targeted for particular times. Northern European and North American cities must be prepared for snow clearing and severe cold conditions during January and February. In contrast, heat waves might be expected during summer months. On the other hand, a heat wave (relatively speaking) is not necessarily a hazard if it occurs during winter months; in fact, the event is often welcomed. Not everyone will be happy, however, since ski resort operators, winter sports enthusiasts, and ice fishing resorts can be negatively impacted. In such instances, hazard is relative to the interests of the persons concerned; the lack of snow can be catastrophic for ski businesses, but reduce snow-clearing costs and automobile accidents. In heat waves, most deaths occur during the early stages rather than later on (see Chapter 1); some people gradually become acclimated, but those susceptible to heat, particularly the very young and the very old, do not survive the initial impacts (Ruffner and Bair, 1984).

Although we would expect many meteorological and hydrological events to be seasonal to some degree, this pattern can be obscured by inadequate consideration of spatial location. Figure 2.17 shows the global incidence of flood disasters from 1900 to 1992, broken down by month (Office of US Foreign Disaster Assistance, 1993). While slightly more floods are recorded for August, the overall variation throughout the year

FIGURE 2.17. Seasonality of floods, worldwide. Globally, flood disasters show a seasonal pattern, with greater numbers occurring in July, August, and September. However, variation throughout the year appears small because of aggregation of global data. *Data source:* U.S. Office of Foreign Disaster Assistance (1993).

is small. This is not surprising. With the exception of the United States, the data were collected from around the world, thus crossing many climatic zones and the equator. Therefore, any seasonality exhibited by the data is hidden. A similar graph depicting snowstorms around the world also would produce an even distribution because the Northern and Southern Hemispheres would effectively cancel out each other. There is a danger, then, in data that are aggregated at too large a level; in the flood graph,for example, if the analysis were undertaken for specific locations we might expect to find a seasonal component. Figure 2.18 shows flood frequency at the Raccoon River drainage basin using data from Table 2.17(b); we can see that floods occurred more often in spring and summer than winter months. The data also show that 9 of the 10 largest floods occurred in June and July.

Tornadoes have a distinctive seasonal component in the United States. As winter gives way to spring in the Northern Hemisphere, the jet stream migrates north, bringing with it severe weather systems that move along the boundary between cold and warm air masses. For southern states, thunderstorms begin early in the year and their progression north can be tracked; by May or June, they have reached Minnesota. Along with these storms come lightning, hail, and tornadoes. Although thunderstorms occur in many areas, the very intense events are associated with this jet stream movement.

Figure 2.19 shows the monthly incidence of tornadoes in the conti-

FIGURE 2.18. Seasonality of floods in the Raccoon River Basin at Van Meter, Iowa.

nental United States for 1950–1989. While tornadoes develop throughout the year, the majority occur from April through June, with the modal month being May. At a different spatial scale, however, other patterns emerge. Grazulis (1993) showed that North Dakota and Mississippi have distinctly different temporal distributions, reflecting the progression of storm tracks to the north (Figure 2.20). Mississippi has a bimodal distribution with peaks in April and November, while tornadoes peak in North Dakota in June. Thus, the seasonality of tornadoes is spatially defined and communities from south to north hold tornado awareness weeks at different times as the spring progresses.

Hurricanes are another seasonal meteorological event. The buildup of heat is essential for their formation (as discussed above), hence, it is during the late summer, when heat has been stored in the oceans, that the majority occur. As with other events, however, the pattern is not consistent around the world. Gray (1975) found that tropical cyclones are most common in the western North Pacific from October through December, in the western Atlantic between July and September, and from January through March in the southern Indian Ocean (Table 2.19). The western North Pacific accounts for most of the tropical storms in any given year. In 1993, there were 38 significant tropical cyclones there (the highest since 1967 when there were 41); of these, 21 became typhoons, 9 were tropical storms, and 8 were tropical depressions. This higher incidence of tropical cyclones has been linked to an El Niño–Southern Oscillation event, especially when the Southern Oscillation index remains negative until late in the year (Pielke, 1990). William Gray of Colorado State University uses these weather traits to predict hurricane activity in the North Atlantic each year; in 1995, his projections for a particularly active hurricane season proved quite accurate. If these relationships hold in future years, a predictive model of tropical cyclones in the western North Pacific may become possible.

Nonmeteorological and nonhydrological events cannot be defined in a temporal context, and consequently are less easy to predict. Earthquakes and volcanoes are unaffected by seasonality and occur at any time. This is notwithstanding Aristotelian ideas on tectonic activity. Aristotle believed that in hot weather air migrates underground where it is heated further by underground fires; under escalating pressure, the air eventu-

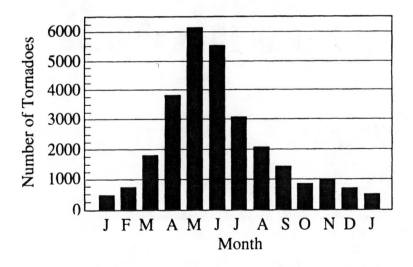

FIGURE 2.19. Monthly incidence of tornadoes in the United States, 1950–1989. More tornadoes occur in the spring than in any other season. *Source:* Grazulis (1993). Copyright 1993 by The Tornado Project. Reprinted by permission.

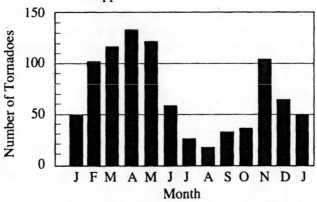

FIGURE 2.20. Monthly incidence of tornadoes in North Dakota and Mississippi, 1950–1989. *Source:* Grazulis (1993). Copyright 1993 by The Tornado Project. Reprinted by permission.

ally breaks through the ground surface, causing earthquakes and volcanoes. According to Pliny the Elder, "Tremors of the earth . . . occur [only] when the seas are calm and the air so still that birds cannot fly" (quoted in Ritchie, 1988).

One legacy of these views is that for some time, people believed that earthquakes occurred only during hot weather. Clearly, this is not true. Figure 2.21 shows the monthly incidence of earthquakes and volcanoes around the world for 1900–1992 (Office of US Foreign Disaster Assistance, 1993). December is higher than other months for earthquakes and April for volcanoes, but these are statistical anomalies rather than reflections of any causal mechanism. Even if we examine earthquakes and volcanoes at a specific place, there is no definable temporal distribution. Mud slides may be the only type of geomorphological event retaining an element of seasonality. They often are associated with wetter conditions that can initiate the movement of unstable slopes. For example, Southern California experiences many mud slides during winter months.

Diurnal Factors

As the temporal scale is reduced, other patterns can be identified. Thunderstorms occur most frequently during the late afternoon in response to heat buildup during the day, especially in summer or in tropical climates. The presence of atmospheric instability can then promote thunderstorm activity; lightning, hail, and tornadoes also are more common at this time. The probability of thunderstorm activity at the Kennedy Space Center in Florida at 4:00 P.M. is approximately 0.23 in July, 0.04 in April, and only

TABLE 2.19. Origin of Hurricanes by Season

Months	Genesis region	Average number of occurrences
January–March	South Indian Ocean	8.5
	South Pacific	6.3
	North and west Australia	5.8
	Western North Pacific	1.4
April–June	Western North Pacific	4.0
	North Indian Ocean	3.3
	South Pacific	2.3
	East Pacific	1.9
	South Indian Ocean	1.2
July–September	Western North Pacific	11.7
	Western Atlantic	8.4
	East Pacific	7.7
	North Indian Ocean	5.8
October–December	Western North Pacific	12.4
	North Indian Ocean	4.9
	South Indian Ocean	3.9
	West Atlantic	2.7
	South Pacific	2.0
	East Pacific	1.8

Source: Pielke (1990). Copyright 1990 by Routledge. Reprinted by permission.

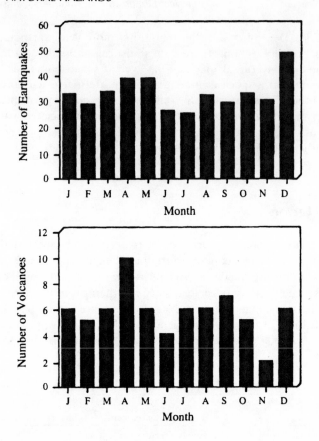

FIGURE 2.21. Monthly incidence of earthquakes and volcanic eruptions world-wide, 1990–1992. As would be expected, there is no seasonal pattern. *Source:* U.S. Office of Foreign Disaster Assistance (1993).

0.02 in October. In October, the probability of thunderstorms is fractionally higher at 6:00 P.M. than at other times of the day (Pierce, 1977). We see these patterns again with tornadoes; Figure 2.22 shows that they occur most frequently at 3–6 P.M., which would be expected from the temporal distribution of thunderstorms.

Diurnal variation is not associated with other geophysical events. Floods, droughts, and geological events can occur at any time. This is not to say that the timing of such events is not crucial. If the next major earthquake in San Francisco occurs during a rush hour, more people will die than if the event occurs at night. Unfortunately, predicting the precise timing of tectonic activity is not possible.

SPATIAL DISTRIBUTION

A third feature of natural hazards is geographic location. Since most extreme geophysical events can be spatially defined, their mapping enhances our knowledge of risk. However, the occurrence of the extreme geophysical *events* does not mean that the *hazard* has the same spatial pattern, although the correlation between the two is obviously very strong. It also should be remembered that impacts often extend well beyond the boundaries of a physical process.

In looking at geographic distribution, it is important to consider spatial scale. At the global or regional scale different patterns can emerge. For instance, temperate latitudes experience by far the greatest number of snowstorms, cold waves, ice storms, and so on. A simple climate map can indicate the potential for such events around the world. Tropical cyclones, on the other hand are a feature of lower latitudes where the heat and moisture of warm oceans help to spawn them. Floods are generally confined to floodplains and coastal areas, mudflows to steeper slopes, and sinkholes and subsidence to particular geologic conditions.

The mapping of geological processes shows that earthquakes and volcanoes are most frequent along plate boundaries (Figure 2.7). The broad picture shows the Ring of Fire around the Pacific rim where there are numerous active volcanoes and earthquakes, but it does not help in planning for such events. The cartographic scale is too small to define the conditions prevailing at a place. Because of the many different processes associated with tectonic activity, other, more localized information is

FIGURE 2.22. Incidence of tornadoes in the United States by time of day. Their diurnal variation reflects the temporal pattern of thunderstorms. *Source:* Grazulis (1993). Copyright 1993 by The Tornado Project. Reprinted by permission.

required to make effective plans. Nevertheless, the global pattern does help to confirm the theory of plate tectonics as an explanation of earthquakes and volcanoes.

Mapping hazards at a larger spatial scale reveals other important information. Thunderstorms, for example, occur most frequently over Florida in the United States (Figure 2.23). Lightning cells tend to be well separated spatially and are apparently unorganized. However, Pierce (1977) has shown that the average distance between strikes is 2–6 km (1.24–3.7 mi), and that if lightning strikes 2 km (1.24 mi) from an airfield, there is a 10% chance that the airfield will be struck within 30 sec. There is, then, a spatial pattern to lightning and thunderstorm activity.

Tornadoes show a distinct regional distribution and a map of their incidence helps to define the risk for any given place. Figure 2.24, for example, is based on over 20,000 tornado occurrences in 1953–1980, and depicts tornado frequency across the United States. However, even this mapping has drawbacks that can affect the overall pattern. In this case, tornadoes have been mapped using incidence per 25,900 km^2 (10,000 mi^2), producing an annual average for each area. The pattern reflects the high occurrences in the Great Plains, especially in the southern reaches. However, the high levels in Florida are somewhat misleading since they include many "weak" tornadoes that generally present fewer problems than those further west (Grazulis, 1993).

Grazulis illustrated the complexity of spatial data by analyzing how states can be ranked for tornado hazards. Table 2.20 shows the top 10 states, based on different criteria. For example, Texas has the highest number of tornadoes, but given its size and location, that is not surprising; when area is held constant, Florida has most. Similarly, most deaths from tornadoes are recorded in Texas, but when area or population is included in the analysis, then Massachusetts and Mississippi, respectively, appear deadliest.

At a small scale, the geographic location and movement of hurricanes are fairly well defined, as shown in Figure 2.25. The storms may eventually stray from these areas, but as they do so, they weaken and are modified by other meteorological conditions. In 1972, Tropical Storm Agnes was very damaging to the United States, but quickly lost intensity as it moved inland. When these spatial patterns are combined with temporal distribution (see Table 2.19), a much clearer picture of hurricane risk emerges. Complications arise when we try to forecast exactly how a specific hurricane will move. Since 1954, the 24-hour forecast of landfall for hurricanes affecting North America has improved by only 15 nautical miles, and the average error is still over 100 nautical miles. Nonetheless, satellite tracking of hurricanes has advanced our understanding of the spatial and temporal patterns of storm tracks.

FIGURE 2.23. Spatial distribution of thunderstorms. Based on the annual frequency of thunderstorm days, this pattern shows distinct regional differences. *Source:* National Oceanic and Atmospheric Administration, as cited in Eagleman (1983).

FIGURE 2.24. Spatial distribution of tornadoes, 1953–1980. This distribution is based on the number of tornadoes per 10,000 mi^2. *Source*: Grazulis (1993). Copyright 1993 by The Tornado Project. Reprinted by permission.

TABLE 2.20. Top Ten States, Based on Tornado Occurrence Criteria

Rank	Total number of tornadoes	Annual incidence per 25,000 km^2	Total deaths	Deaths per 25,000 km^2	Deaths per million population
1	Texas	Florida	Texas	Massachusetts	Mississippi
2	Oklahoma	Oklahoma	Mississippi	Mississippi	Arkansas
3	Florida	Indiana	Alabama	Indiana	Kansas
4	Kansas	Iowa	Michigan	Alabama	Oklahoma
5	Nebraska	Kansas	Indiana	Ohio	Alabama
6	Iowa	Delaware	Oklahoma	Michigan	Indiana
7	Missouri	Louisiana	Kansas	Arkansas	Texas
8	Illinois	Mississippi	Illinois	Illinois	North Dakota
9	South Dakota	Nebraska	Ohio	Oklahoma	Nebraska
10	Louisiana	Texas	Arkansas	Kentucky	Kentucky

Source: Grazulis (1993). Copyright 1993 by The Tornado Project. Reprinted by permission.

When the record of hurricanes is analyzed for a number of years, their spatial pattern in the North Atlantic shows seasonal variation. The area of formation shifts eastward gradually as the season progresses, with early season storms generally confined to the western Caribbean and the Gulf of Mexico. From mid-August to mid-September, the area of formation is at its easternmost boundary near the Cape Verde Islands, after which there is a westward movement of the formative area (Neumann, Jarvinen, and Pike, 1987).

The movement of hurricanes is dependent on (1) regional air patterns that act as steering currents, (2) the internal forces of the storm (Holland 1983), and (3) outflow jets from the storm. It is thought that regional air speed governs the overall speed of a hurricane; the faster the regional air, the faster a storm tends to move. Hurricane Hugo (1989), for example, got caught up in a southwesterly jet stream that accelerated it inland (Pielke, 1990). Unfortunately, rapid changes in the regional air flow and other synoptic systems produce different responses. Hurricane Betsy (1965) was moving north off the shore of the United States, a common path, when it migrated suddenly south and west, clipping southern Florida and eventually hitting land at New Orleans (Pielke, 1990). Two weather systems appeared to be responsible: a subtropical high ridge that developed near Bermuda, and winds that essentially forced the hurricane south.

Internal processes also affect forward direction, which explains some of the "unpredictable" routes followed by hurricanes. According to Pielke (1990), when the steering winds are cyclonic on the right, hurricanes move toward the right and slow down. Conversely, when the steering winds are more anticyclonic toward the right, hurricanes move to the left. As the storm moves away from the equator, the Coriolis effect strengthens and helps to move the storm polewards. In essence, the steering winds and

FIGURE 2.25. Incidence of hurricanes worldwide, 1952–1971. The number of hurricanes per region of genesis and the percentage each area contributes to the global total illustrate the lack of dominance of any one region. *Source:* Gray (1975). Reprinted by permission of author.

internal mechanisms of the storm are interrelated. More accurate forecasts of landfall may become possible when our understanding of these relationships grows more refined.

Finally, the outflow of winds from the storm also can affect forward motion. Winds in the upper troposphere transport air far away from the storm and descend some distance away. Pielke (1990) points out that if this action were uniform, there would be no effect, but since the storms are asymmetric, outflow winds tend to concentrate activity in particular areas. If this descending air accumulates in the front right quadrant, the storm will tend to move toward the left; if the accumulation is to the right rear, then movement will be to the right in the Northern Hemisphere. It was this asymmetry that pushed Hurricane Gilbert into northeast Mexico rather than Texas. An analysis of hurricane characteristics in the United States is shown in Table 2.21.

Summary

Many natural hazards can be identified with particular locations. Some (such as floods and earthquakes) are more defined geographically than others (such as droughts and blizzards). This is not to say that extreme geophysical events do not occur in many places. Indeed, every state in the United States experiences tornadoes, but clearly they are more prevalent in the Midwest than in Alaska. Similarly, heat waves are not common in the northern part of western Europe, but they do occur.

Thus, it is possible to make probability statements about the likelihood of occurrence of events at specific locations. In addition, the scale of analysis affects the spatial patterns of different extreme geophysical events. A small-scale map shows less detail, but covers a wider area than a large-scale map. If the object is to determine risk at a particular location, a large-scale map is more appropriate while for a more general level of analysis, the smaller scale suffices. Consequently, scale may be important to planners concerned with ameliorating the impacts of natural hazards.

COUNTDOWN INTERVAL

One characteristic of natural hazards that has bearing on the success or failure of remedial action is the speed of onset of an event. In general, the more rapid the occurrence of an event, the more unexpected, allowing less time to undertake remedial action.

For example, flash floods, tornadoes, and earthquakes permit only minimal warning and last-minute preparation. Consequently, mitigation

TABLE 2.21. Ranking of U.S. Hurricanes Based on Losses, Deaths, and Intensities, 1900–1992

	Costliest				Deadliest		
Hurricane	Year	Category	Damage ($)	Hurricane	Year	Category	Deaths
Andrew[a]	1992	5	25,000,000,000	Tex. (Galveston)	1900	4	6,000+
Hugo	1989	5	7,000,000,000	Fla.	1928	4	1,836
Frederic	1979	3	2,300,000,000	Fla./south Tex.	1919	4	600+
Agnes	1972	1	2,100,000,000	New England	1938	3	600
Alicia	1983	3	2,000,000,000	Fla. Keys	1935	5	408
Bob	1991	2	1,500,000,000	Audrey	1957	4	390
Juan	1985	1	1,500,000,000	Northeast U.S.	1944	3	390
Camille	1969	5	1,420,700,000	La.	1909	4	350
Betsy	1965	3	1,420,500,000	La.	1915	4	275
Elena	1985	3	1,250,000,000	Tex. (Galveston)	1915	4	275
Gloria	1985	3	900,000,000	Camille	1969	5	256
Diane	1955	1	831,700,000	Fla.	1926	4	243
Allison	1989	T.S.[b]	500,000,000	Diane	1955	1	184
Eloise	1975	3	490,000,000	Southeast Fla.	1906	2	164
Carol	1954	3	461,000,000	Miss./Ala./Fla.	1906	3	134
Celia	1970	3	453,000,000	Agnes	1972	1	122
Carla	1961	4	408,000,000	Hazel	1954	4	95
Claudette	1979	T.S.	400,000,000	Betsy	1965	3	75
Donna	1960	4	387,000,000	Carol	1954	3	60
David	1979	2	320,000,000	Southeast Fla./La./Miss.	1947	4	51

	Most Intense			
Hurricane	Year	Category	Millibars	Inches
Fla.	1935	5	892	26.35
Camille	1969	5	909	26.84
Hugo	1989	5	918	27.16
Andrew	1992	5	922	27.23
Fla./South Tex.	1919	4	927	27.37
Fla.	1928	4	929	27.43
Donna	1960	4	930	27.46
Tex. (Galveston)	1900	4	931	27.49
La.	1909	4	931	27.49
La.	1915	4	931	27.49
Carla	1961	4	931	27.49
Fla.	1926	4	935	27.61
Hazel	1954	4	938	27.70
Southeast Fl./La./Miss.	1947	4	940	27.76
North Tex.	1932	4	941	27.79
Gloria	1985	3	942	27.82
Audrey	1957	4[c]	945	27.91

TABLE 2.21. *(cont.)*

Most Intense

Hurricane	Year	Category	Millibars	Inches
Tex. (Galveston)	1915	4[c]	945	27.91
Celia	1970	3	945	27.91
Allen	1980	3	945	27.91

[a] Areas affected by named hurricanes are as follows:

Agnes	Northeast U.S.	Claudette	North Tex.
Alicia	North Tex.	David	Fla./Eastern U.S.
Allen	South Tex.	Diane	Northeast U.S.
Allison	North Tex.	Donna	Fla./Eastern U.S.
Andrew	Fla.	Elena	Miss./Ala./Northwest Fla.
Audrey	Southwest La.	Eloise	Northwest Fla.
Betsy	Southeast Fla./southeast La.	Frederic	Ala./Miss.
Bob	N.C./northeast U.S.	Gloria	Eastern U.S.
Camille	Miss./Ala.	Hazel	S.C./N.C.
Carla	Tex.	Hugo	S.C.
Carol	Northeast U.S.	Juan	La.
Celia	Southeast Tex.		

[b] Of only tropical storm intensity, but included because of high damage figures.
[c] Classified 4 because of extreme tides.
Source: NOAA (1995)

measures should be in place all the time. By examining the spatial and temporal occurrences of such events, we can identify their risk of incidence at a place and adjust management plans accordingly. By contrast, events that take a long time to develop (like droughts and extensive flooding) can permit the implementation of fairly complex strategies to deal with the impending problem. For example, during the 1993 floods in the midwestern United States, some communities had several weeks' warning of impending floods. Fifty miles south of St. Louis, St. Genevieve took these warnings seriously and added a major embankment of sandbags to the levee system. Similarly at Hannibal, Missouri, the residents raised the existing levees 1 m (3 ft) beyond the 500-year flood safety level. The flood finally crested 0.3 m (1 ft) above the permanent levee (Tobin and Montz, 1994b). Not all such actions were successful, but the communities involved were actively engaged in "fighting" the flood.

Most tectonic events cannot be forecast with any degree of accuracy at the present time, but it is feasible to predict where they will occur. It is astonishing, therefore, to find that earthquakes and volcanoes often come as a surprise to inhabitants of these high-risk locations. Scientists have tried for some time to identify precursors of such events, but have had little success. Attempts to forecast earthquakes have been based on such variables as land deformation, tilt and strain measures, number and frequency of foreshocks, seismic velocity, earth currents, and electrical resistivity changes (Kitizawa, 1986). In the United States, for example, researchers have been studying the Palmdale bulge in California, which between 1959 and 1974 showed ominous signs of increased strain. In all, there was a 0.5 m (1.5 ft) rise over a land area of 842,000 km^2 (325,098 mi^2). In Niigata, Japan, there was a similar uplift prior to the 1975 earthquake (Keller, 1988). Russian, Japanese, Chinese, and U.S. scientists among others, have been monitoring an array of physical variables in hope of identifying those that signal an oncoming earthquake (Figure 2.26). Based on the frequency of foreshocks, the Chinese made a successful forecast of the Haicheng earthquake in 1975, which measured 7.3 on the Richter scale; authorities evacuated a large number of people prior to the event, which undoubtedly saved many lives (Keller, 1988). However, subsequent warnings have proved inaccurate and have kept people from their homes unnecessarily. Furthermore, and presumably because no foreshocks were noticed, no warning was given of the T'anshon earthquake (7.8 magnitude) in 1976, that killed 240,000 people (Kitizawa, 1986).

Others have suggested that anomalous animal behavior can predict an earthquake. While research continues, it is difficult to see how unusual behavior before an earthquake (if it exists) can be distinguished from other unusual behavior. Dogs often bark for no apparent reason, chickens fly up to perches, cows lie down, and other animals scurry around. Doom-and-gloom prophets also have become part of the earthquake prediction act. In London in 1761, following two small earthquakes, a man by the name of Bell forecast a third on April 5; reports state that much of the city emptied that day as people expected to see buildings fall (Ritchie, 1988). Recently, a similar event occurred when Iben Browning projected a 50% chance of an earthquake within the New Madrid seismic zone on or around December 3, 1990; nothing happened, but public officials found themselves working nonstop responding to requests for advice and answering questions from the media (Showalter, 1991). Society should not depend on warnings from such sources until a scientific model can be developed, we remain most skeptical.

The onset of volcanic activity would seem to be similar to that of earthquakes, but there are indications that a degree of warning may be possible. Prior to some explosive events, for example, there are series of earthquakes (termed harmonic tremors) that increase in magnitude and intensity as magma migrates up through the earth's crust. At the same time, there may be surface displacement as slopes bulge because of changes within the earth (Nuhfer et al., 1993). Such signs preceded the Mount St. Helens eruption. Indeed, the final explosion came when the bulge completely destabilized the north face of the volcano, precipitating a major landslide and releasing the pent-up pressure inside; the eruption then

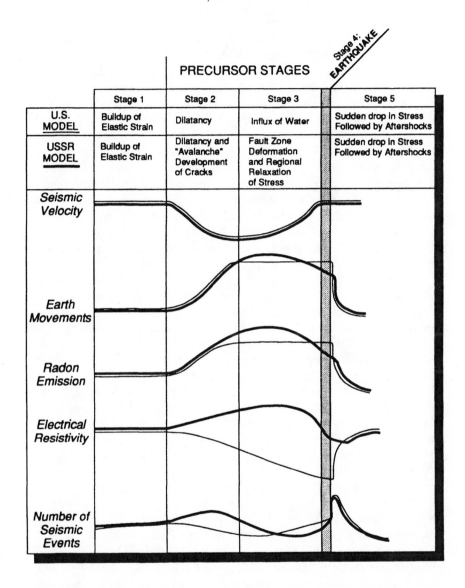

	Stage 1	Stage 2	Stage 3	Stage 5
U.S. MODEL	Buildup of Elastic Strain	Dilatancy	Influx of Water	Sudden drop in Stress Followed by Aftershocks
USSR MODEL	Buildup of Elastic Strain	Dilatancy and "Avalanche" Development of Cracks	Fault Zone Deformation and Regional Relaxation of Stress	Sudden drop in Stress Followed by Aftershocks

PRECURSOR STAGES

Stage 4: EARTHQUAKE

Seismic Velocity

Earth Movements

Radon Emission

Electrical Resistivity

Number of Seismic Events

FIGURE 2.26. Predicting earthquakes. Two models, dilatancy–diffusion (thinner line) and dilatancy–instability (thicker line), give different precursory signals at different stages, of earthquake development. *Source:* Press (1975). Copyright 1975 by Scientific American, Inc. Reprinted by permission.

spewed out toward the north. In Hawaii, scientists have tried to predict the timing, severity, and location of eruptions, but have had only limited success (Hodge, Sharp, and Marts, 1979). As research continues, therefore, the models must be tested to see whether such events precede all volcanic activity or if different models are necessary for different areas.

Tsunami are somewhat different because there often is warning of oncoming disaster. Once tectonic movements have been detected, coastal locations can be placed on a seismic wave alert. The real problem comes when an earthquake causes a tsunami in the immediate vicinity, which leaves no warning time, as occurred in Alaska in 1964. Tsunami fall into four categories based on their speed of onset (Ayre et al., 1975). Categories 1 through 3 include those that provide less than 30 min warning; though earthquake damage may offer physical clues of an impending event, very little organized evacuation can be accomplished in this short time. In the final category (between 45 min and 12 hr), effective action can be taken. Figure 2.10 shows some of the warning times across the Pacific basin.

Drought is another meteorological event that can be predicted in probability terms, but generally cannot be forecast as to time and place. However, two developments show promising signs of success. First, recent insights into the workings of the global atmospheric system have greatly advanced our understanding of weather processes. It appears that some extreme conditions are related to the El Niño-Southern Oscillation (ENSO) events that occur off the coast of South America (Rasmussen, 1987). Drought in Africa and wetter conditions in North America seem to be highly correlated with these ocean temperature changes. Hastenrath (1987) and Nicholls (1987) found that ENSO events preceded drier weather in northeastern Brazil and Australia/Indonesia, respectively. (Bryant [1991] has reviewed the significance of ENSO in combination with other global changes that potentially affect weather conditions.)

Second, more sophisticated technology has made the monitoring of weather systems less problematic. Even some events with short countdown intervals now can be forecast. For instance, Doppler radar has enhanced the ability to "see" into thunderstorms and identify conditions that might lead to tornadoes, flash floods, heavy hail, and lightning. Through satellite observations, major storms and hurricanes can be tracked accurately and drought conditions identified around the world. It is possible that as the global models become more sophisticated, we will have even better ability to forecast these meteorological events.

THEORETICAL INTEGRATION

The foregoing discussion has provided a foundation based on extreme geophysical events, on which analysis and integration of natural hazards can build. Development of our knowledge regarding the mechanisms that produce these extreme conditions—at different magnitudes, frequencies, and durations—aids in predicting, protecting, and responding. Each of these physical parameters has been the focus of research. For instance, it was thought that frequency of occurrence of extreme geophysical events held the key to understanding behavior in natural hazards; that is, the expected return period for floods, earthquakes, or tornadoes was presented as a significant variable in determining the response to hazards at the community or individual level. It was argued that if the recurrence interval was sufficiently short (that is high frequency–high probability), a community would respond by undertaking measures to mitigate the problem; longer return periods (low-probability events) would elicit few or no responses because the threat would be seen as remote. At first sight, this scenario appears sound: those communities with more incidence of extreme events do tend to take remedial action in one form or another. The certainty–uncertainty model developed by Kates (1962) is based on these arguments (Figure 2.27).

However, these arguments should not be carried too far. A model based on geophysical event–response mechanism is essentially deterministic. It does not allow other variables that influence decision making to be considered, but subsequent research has demonstrated that all facets of a geophysical event may have bearing on response. In the extreme, such a model is environmentally deterministic and, as such, is unsound. Associated with the work of Huntington (1919) and Semple (1911), environmental determinism hypothesized that physical features, such as landscape

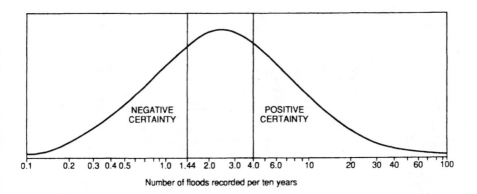

FIGURE 2.27. Event–response patterns. As illustrated by this study of floods, some researchers suggest that frequent events generate mitigative responses. Conversely, infrequent events generate a lack of action because of belief that the problem already has passed. *Source:* Kates (1962). Copyright 1962 by University of Chicago Press. Reprinted by permission.

and geographic location, significantly affected the intelligence and behavior of people in given areas; in its strictest form, environmental determinism served to "explain" why some nations were poor and others wealthy. The racial implications were, of course, horrendous. Fortunately, such arguments fell into disfavor, although we should not reject completely the idea of environmental factors playing a role in decision making. Indeed, environmental conditions can and do influence behavior, in a relationship that interested early hazards researchers.

It is helpful, then, to consider an integrative framework around physical dimensions so that lessons from specific events or occurrences can be applied to other places. Certainly, there are physical attributes that are unique to particular places and events, but comparison and generalization still are possible. The integration of physical characteristics, both within and across types of geophysical events, is useful in analyzing what to expect as well as how to prepare and respond. Integration within an event type is shown in Table 2.22, where a number of traits associated with earthquakes are considered in relation to each other. Thus, magnitude, intensity, frequency, and spatial extent are interrelated and, together they provide a comprehensive view of the physical dimensions of the geophysical event. Similar comparisons can be made for other natural events.

A second level of analysis is integration across different geophysical events. For any given spatial scale and place, it is possible to organize geophysical events according to the levels of magnitude, duration, spatial extent, frequency, seasonality, and countdown interval. Figure 2.28 shows how such events might be evaluated. Such a system allows analysis of natural events and related hazards from a broader conceptual perspective, avoiding the bias toward geophysical processes and hazard-specific concepts. Certainly, the different measures and scales present problems, especially if one looks to apply them to specific locations, but they enable

TABLE 2.22. Earthquake Magnitude in Relation to Other Variables

Magnitude	Modified Mercalli intensity	Expected annual incidence	Felt area (km^2)	Felt distance (km)
3.0–3.9	~II–III	49,000	1,940	25
4.0–4.9	~IV–V	6,200	8,850	50
5.0–5.9	~VI–VII	800	38,850	110
6.0–6.9	~VII–VIII	120	165,350	200
7.0–7.9	~IX–X	18	518,000	400
8.0–8.9	~XI–XII	1	2,072,000	725

Note: These data show average conditions. Due to depth of focus and local geology, any given earthquake will probably have effects felt over distances and areas somewhat different from those shown here. *Source:* Costa and Baker (1981.) Copyright 1981 by John Wiley and Sons, Inc. Reprinted by permission.

FIGURE 2.28. Comparison of hazard characteristics. Different hazards "score" differently on this dimensionless scale. It is not the position of a given hazard on any one characteristic that is important, but rather the combination of positions that affects perception and the ability to respond to different hazards. Even for the same event, different places score differently because of spatial differences in occurrence. Thus, while this figure is useful for generalizing about various characteristics through which events can be compared, it cannot be generalized over space. *Source:* Adapted from Burton, Kates, and White (1993).

lessons learned from one type of event to be transposed to others more easily.

Although an integrative framework can develop into broader analysis of physical processes, particularly by managers and planners, generalization also brings problems. For example, Figure 2.23 would look different at varied spatial scales; the distribution of tornadoes is not identical at global, continental, and regional scales. In addition, some characteristics, such as speed of onset and duration, are virtually universal, but others are locationally dependent, such as frequency and temporal spacing. Finally, events of different intensity should result in different configurations in the diagram. Thus, scale of generalization must be considered in any attempt to apply integrative concepts to a given situation.

Another concern is that hazards rarely are caused by a single physical process (except perhaps being hit by a meteorite). Invariably, other aspects of the natural system must be examined to understand fully the impact of an event. For instance, the seismic activity associated with earthquakes can be documented; different waves moving through the earth's crust can cause distinct problems, as can different physical and geological conditions. As a result, the data in Table 2.22 are not necessarily applicable to a particular event at a particular location. Further, the degree of impact also is determined by rock type and geologic structure. Loosely consolidated material is subject to considerably more shaking than solid material, and young slopes are more prone to land slippage and failure than older, established slopes. The Vaiont Dam earthquake of 1963 is a classic example of multiple events; a small earth movement on saturated dipping rock strata triggered a major landslide of 238,000 km^3 (57,129 mi^3) of debris on Mount Toc that fell into Lake Vaiont. The slide was so catastrophic that the resulting air blast removed roofs and damaged trees. The rock fall also created a tsunami in the reservoir, with waves over 90 m (295 ft) high, which spilled over the dam and continued down the valley, still at 70 m (230 ft) in height. An estimated 2,600 people died (Keller, 1988). Hurricanes, too, often are associated with tornadoes, strong winds, tidal surges, heavy rain, and floods. Similar comparisons can be made for volcanic activity and the volcanic explosivity index. In a different capacity, the physical aftermath of volcanic activity can be longstanding. Changes in topography, removal of vegetation, and hydrological alterations may make communities more prone to disaster. For example, Mount Pinatubo continued to present problems in the Philippines in 1992; bare slopes and heavily silted rivers, resulting from the 1991 eruption, led to mud slides and flash floods when several typhoons deposited large amounts of precipitation. Damage exceeded $74 million (United Nations, 1993).

So far, our discussion of natural hazards and the physical dimensions of events has focused on the "extreme" aspects of physical processes. By extreme, it is meant that processes outside the norm can lead to problems. This argument assumes that human activities have adjusted to some "normal" range of physical processes (a point taken up in Chapter 5). If we accept this premise for the time being, we can see that "normal" constitutes a certain range of events defined by measurable physical attributes. Even droughts are declared in relation to normal conditions for a community, so that it is departures from the norm that constitute a hazard. The same is true of floods. A community located on the banks of a river may come to expect frequent flooding, and it may adapt by zoning flood-prone neighborhoods as open space or raising structures on stilts; although the flood problem still exists, the community has adapted to the magnitude of the physical event up to some threshold. In the case of floodplain zoning, this threshold might be defined by the boundary between open space and urban development. A flood event that exceeds the flood threshold will inevitably lead to losses as water inundates residential or commercial properties adjacent to the open space. Thus, response appears to be related to magnitude and frequency, suggesting that physical characteristics engender certain adjustments even if they do not prescribe mitigation policies.

Is the concept of adjustment a reasonable premise on which to base judgments concerning natural hazards? There is little argument that the physical characteristics of extreme events are important. Indeed, size, scale, and temporal features help to define hazardousness of place and shed light on response frameworks. Yet, while physical processes are necessary components of any assessment of natural hazards, they are not sufficient. For example, take two flash floods of similar magnitudes. These floods could have very different impacts on society. One might completely devastate a community, washing away homes and killing many people, a situation not unlike the Big Thompson flood of 1976 or the Rapid City, South Dakota, flood of 1972. If communities have implemented mitigation measures (such as rigorous building codes) or curtailed all development within a flood-prone area, such flooding might lead to minimal disruption. Indeed, since 1976, Rapid City has taken precautions to move dwellings and businesses off the floodplain, reducing its susceptibility to damage from flash floods. Still, a flood exceeding the design standard would be destructive.

Similarly, a drought in one place might not be considered a drought in another. A great deal depends on community dependence on water, which in turn is related to physical conditions. In humid parts of the United States, economic activities are based on the availability of large quantities of water; when supplies are restricted, a drought condition exists. By contrast, in arid areas of Africa, many societies have adjusted to a norm of low levels of water supply. Although the variation in dependence is significant in its own right, droughts occur in either community when water supply is reduced. Needless to say, in arid areas, a water shortage can mean a life-and-death struggle. However, an examination of rainfall patterns alone would not provide a meaningful assessment of the drought hazard, but would merely indicate variations from average conditions.

Finally, events of similar scale can have significantly different results. As discussed by Burton et al. (1993), Tropical Storm Agnes, which struck the United States in 1972, and the typhoon that devastated Bangladesh in 1970 were of comparable magnitude. However, at least 300,000 people died in Bangladesh as compared to 118 in the United States. Their social, political, and economic conditions accounted for many of these differences. Consequently, an assessment of natural hazards based purely on measures of physical process is not complete. This is not say that the physical event is not important, only that other factors play important roles in determining the size and extent of the natural hazard.

3 *Perception Studies*
The Individual in Natural Hazards

INTRODUCTION

In the first two chapters we examined some of the extreme physical processes that, in interaction with the human-use system, can create natural hazards and precipitate catastrophes. It is this combination of natural and human processes that must form the basis of our research into natural hazards if we are to comprehend the real and underlying causes of disasters. Natural hazards are clearly neither solely of physical origin nor simply the result of ill-contrived human activities. As pointed out in Chapter 2, the extremes of nature are recurring events to which we can often assign levels of probability. At the same time, the human-use system needs to be considered in detail if we are to answer such questions as: why do people live in hazardous areas? Why do they relocate in hazardous areas even after experiencing an extremely traumatic event? Why do some individuals behave "irrationally" during and after a hazardous event? Why have some communities but not others implemented policies to mitigate the impact of natural hazards? How have decision makers selected particular alleviation projects over other proposals?

The number of questions is almost infinite and certainly beyond the scope of a single text. Nevertheless, we can identify a number of factors as well as different levels of responsibility that might be used to explain individual and community behavior in hazardous environments. At the larger level, national and regional (state or provincial) governments usually are empowered to undertake measures to protect citizens; community-level institutions also are legitimately concerned with public safety. The concerns of society involve all components of the human-use system, and it is as important to examine the social, political, and economic factors as it is to examine the physical processes at work in a particular vicinity. The significance of social structure is addressed in subsequent chapters. At the base of the hierarchy is the individual, which is the focus of this chapter.

This chapter looks at individual perception of natural hazards as both a contributor to and detractor from effective response. Taylor, Stewart, and Downton (1988) define perception as "the range of judgments, beliefs and attitudes" that individuals hold. We can assume that behavior and actions result from these judgments, beliefs, and attitudes—a grouping we refer to as cognition. Yet, the expected, so-called "rational" response is not always result, an irony that requires us to examine more closely the factors on which actions are based. Since we know that behavior depends on the perceived environment, we must incorporate psychological and sociological research in our analysis.

Although we may be tempted to assume that individual perceptions and resulting actions after or in anticipation an event are directed toward reducing individual and community vulnerability, frequently such action does not occur. In part, this may stem from the fact that the individual may feel relatively powerless when faced with the extremes of the natural environment and when confronted with the realities of major political, social, and economic structures that serve to constrain individual activity. In addition, the actions of some individuals may not be considered appropriate by others. Faced with expected flooding, some individuals may sandbag their homes while others pray for divine intervention to stop the floodwaters. Others, when warned that mobile homes are inappropriate housing within a coastal flood or hurricane zone, may respond that mobile homes are cheaper and quicker to replace than more permanent structures. (In fact, these potential victims do have less wealth at risk than those in permanent buildings.)

Other, perhaps external factors may affect action. Civil unrest sometimes promotes action leading to the implementation of protective measures and, in many societies, individuals may turn to the ballot box to promote change, but such activities require coordination and leadership from major "actors." Although there have been cases where natural hazards instigated political change, these occurred where the government already was perceived as weak. The ability of an individual to instigate major change is severely limited and much depends on how individuals, separately and in groups, view the hazard and their relationship to it.

In part, this chapter focuses on the role and perception of the individual within the context of society. Sometimes, however, individuals seem to be little more than pawns who may be sacrificed to protect institutional power and preserve the status quo. For example, the supply of aid, while often important to the immediate relief and survival of hazard victims, can be patronizing and frequently does little more than perpetuate the difficulties of those living in hazardous areas (Smith, 1992). There have been instances of wrong foodstuffs being sent to some countries and too many blankets to another (see Chapter 5). Such decision making would appear to be poor and certainly not oriented to individual needs.

Invariably, our responses to disasters do not address their root causes, merely patching over problems with "feel good" policies. How do we explain the continued suffering of marginalized groups (discussed in Chapter 1) in light of the billions of dollars in foreign and domestic aid? Squatter settlements around the world continue to be located in extremely hazardous environments with inhabitants exposed to considerable risks. In Turkey, survivors of earthquakes regularly rebuild "dwellings" in exactly the same place and constructed of the same materials (Torry, 1980). The aid that does find its way to such individuals often relieves their immediate need for sustenance, but does nothing to counter the long-term threat of hazards.

Thus, it is important to recognize that individuals frequently are ignored when natural hazards, events, and disasters are analyzed or addressed. Indeed, the aggregate statistics of global problems and aid presented in Chapter 1 do not portray the individual tragedies associated with hazards though individuals are their victims, often through no fault of their own. While this chapter deals specifically with understanding individual perception, that focus does not exhaust our interest in or concern for the individual.

UNDERLYING HYPOTHESES, OR EASY QUESTIONS/COMPLEX ANSWERS

The behavior of individuals before, during, and after hazardous events has been the focus of many studies by researchers from a range of academic disciplines, including geographers, sociologists, anthropologists, psychologists, and medical professionals. Some of this work has been primarily descriptive, providing important background information on the relationship between perception and action; for example, we know that an individual is not likely to take action to prevent loss from hazards unless that person is aware of the nature of the hazard and of personal consequences that may result from it. Some behavior may seem irrational, especially to an outside observer, unless that observer looks into the range of factors influencing such behavior. Three scenarios serve to illustrate this. First, why do individuals frequently relocate in areas of known problems? For example, residents along the Trinity River in Texas were flooded four times in three years in the early 1990s. Why do they remain on the floodplain? The fact that some people experience the same natural hazard time after time, suggests that other factors must be influencing individuals' locational decisions.

Second, why do many individuals risk their lives by remaining in their homes in spite of repeated warnings to evacuate at the onset of an event? For example, in 1992, many Florida residents decided to "ride out the storm" as Hurricane Andrew moved through the southern part of the state; at the time, it was the most expensive natural disaster ever to hit the United States, with over 80,000 homes damaged or destroyed. In 1969, in Pass Christian, Mississippi, 25 individuals attended a hurricane watch party to celebrate the forthcoming event; the apartment complex in which the party was held was completely obliterated and all but one partygoer was killed. Why do individuals expose themselves and their property to such risks?

The third scenario addresses behavior of individuals prior to and during hazardous events which can appear bizarre. For instance, some remedial activities are of minimal value or may even be counterproductive (such as sandbagging front doors against flooding but ignoring other openings in the building). These waste time needed for more effective measures. Even more confusing, what might the impartial observer conclude upon witnessing individuals leaving their automobiles in hazard-prone areas or relocating valuable possessions (such as sophisticated stereo equipment) to lower levels in the face of imminent flooding?

There may be perfectly logical explanations for the risk taking exhibited by individuals in the first two scenarios and their implicit vandalism in the latter. Some are rooted in people's perceptions of the nature of the hazard and their own vulnerability. Thus, we must combine psychological variables with attitudinal characteristics in a set we term *cognitive factors*. Cognitive factors include personality characteristics that influence one's view of nature, hence one's propensity to take or avoid risk. A set of *situational factors* complicates an individual's range of choices. Situational factors include a person's physical location in relation to a hazardous area as well as income, age, and social system factors that may affect a person's ability to undertake specific actions.

Figure 3.1 depicts relationships between these two sets. Cognitive and situational factors may work individually, in combination, or even in sequence to influence response (Frazier, Harvey, and Montz, 1983, 1986). The dashed lines suggest direct relationships between each set and response; the dotted line shows that different elements of each set can interact to lead to response. There also are linkages between cognitive and situational factors. The solid arrows suggest feedback between the sets of factors and, while visually simple, represent a complex interplay of internal and external variables. Situational factors affect cognition, and the reverse can be true. Further, these interactions persist, change, and influence

FIGURE 3.1. Relationships between perception and response. Hazard perception derives from sets of situational and cognitive factors, which interact within and between sets to define responses. That is, the factors work separately and in combination.

response over time. Thus, the sets must be considered both separately and in various combinations.

The complexity of the relationship between cognitive and situational variables can be illustrated by returning to our three scenarios. First is continued residence in a hazardous area. For instance, people continue to live in San Francisco in spite of repeated warnings about impending earthquakes, including media attention to the high probability of the "big one" occurring before the end of the twentieth century. The incidence of "small" events (such as Loma Prieta in October 1989) may have influenced some individuals to leave the area, but most residents have remained. Obviously, many other factors—some situational and some cognitive—enter into residents' locational decisions besides the threat of earthquakes. This is the pattern for all hazards. They are viewed as anything from an element of excitement to a minor irritation to a traumatic event, with some factors controllable by the individual and others not.

The literature on location and decision making typically relates residential location to such factors as family ties, employment opportunities and other social, political, and economic factors (Grether and Mieszkowski, 1974). The San Francisco resident or the Texas floodplain dweller may have pressing reasons to remain in the area. For some, spatial scale may constrain locational choices. In the case of San Francisco residents, relocation is usually not a feasible option because new, "safe" areas would be far from the San Francisco Bay area; the nearest alternative is many kilometers away. In contrast, the Texas floodplain dweller would have to move only a short distance to get off the floodplain. Similar spatial differences can be seen with other natural hazards (such as hurricanes, tornadoes, and blizzards or avalanches, cliff erosion, and landslides). In addition, some people deliberately choose to live in a hazardous location because of its other amenities. For example, many people residing in the Florida Keys are well aware of the hurricane risk, but have the attitude that other advantages of the location far outweigh the rare occurrence of a hurricane. Thus, residential location involves more than individual choice; in fact, choice may be curtailed by circumstances beyond the control of the individual—our situational set.

Residents also may find it extremely difficult to move away from hazardous areas because of considerable investment in property and housing. This circumstance compounds situational dependence because events outside their control may dictate individuals' decision making. For instance, house values usually fall after a hazardous event because of direct damage or sheer association with the hazard (Montz and Tobin, 1988, 1990; Tobin and Montz, 1988, 1990). Table 3.1 shows temporal changes in house values following flooding; immediately after a flood in central California, some houses sold at less than 50% of their pre-flood values. Consequently, hazard victims often are faced with the prospect of damaged valuables and major losses in investments (see Chapter 6). Similarly, many farmers and residents throughout the midwestern United States, which flooded in 1993,

TABLE 3.1. Housing Values Following Flooding in Linda and Olivehurst, California

	List price ($)	Selling price ($)
Before flood		
Median	53,000	50,000
Average	52,768	49,871
Range: High	59,900	59,900
Low	46,000	37,000
6 months after flood		
(Sold property)		
Median	41,000	37,800
Average	43,816	40,600
Range: High	62,500	60,000
Low	32,000	27,500
12 months after flood		
(Sold property)		
Median	52,000	47,000
Average	50,567	46,250
Range: High	65,950	63,000
Low	35,000	20,000
(Unsold property)		
Median	53,500	
Average	49,875	
Range: High	72,500	
Low	25,000	

Source: Tobin and Montz (1988). Copyright 1988 by JAI Press. Reprinted by permission.

have remained in the area because their alternatives are constrained. Another residential group, squatters, has virtually no choice but to stay in hazardous environments because their situational set—lack of assets—provides little or no capacity to escape. In the final analysis, individuals may have few options but to remain in hazardous environments.

In the second scenario, individuals may have a number of reasons not to evacuate during a hazardous event. For instance, they may have only minimal awareness of the hazard. White and Haas (1975) pointed to cases where residents were either completely unaware of a local hazard or perceived it as nondamaging. There are many examples of residents potentially underestimating the threat: Saarinen (1966) found perception of the drought hazard on the Great Plains of the United States to be low; Jackson (1974) indicated that farmers in Utah usually underestimated the probability of frost events; and Murton and Shimabukuro (1974) discovered that Hawaiian residents rarely perceived the volcano threat as life-threatening. Further, it is not unusual to find that individuals perceive a problem as "solved" once mitigation measures have been adopted; official response to a hazard invariably generates a false sense of security because the hazard is regarded as eliminated (in other words, the "government" or "agency" has stopped all risk). Gradually, this perception can lead to increased development in the "safe" area and, ultimately, to catastrophic losses when the alleviation system fails; this is known as the levee effect (discussed in more detail in Chapter 4). Thus, both the physical threat and society's actions influence individual perception and ultimately behavior.

The fact that hurricanes, earthquakes, tornadoes, and other hazards may not be perceived as serious threats is part of our cognitive set. Why would anyone take remedial action if he or she does not believe that an event is of high risk? The same attitudes operate for hazard warning systems. Unless a threat is perceived as imminent, the warning may be ignored. If previous warnings resulted in "unnecessary" evacuation because nothing untoward occurred, residents may decide to sit the next one out. This outcome reflects a combination of cognitive and situational variables. Consequently, we may find automobiles left exposed to a flood hazard.

This last aspect of hazard warning systems is known as the "crying wolf syndrome" (Breznitz, 1984). Warnings issued when no disaster follows can lead to unsatisfactory behavior in subsequent events. Individuals may become blasé about warnings and indifferent to the hazard, and therefore they may take little or no remedial action. This can be true even among hazard experts; a fire alarm during a workshop of hazards researchers and practitioners elicited no response at first and eventually a reluctant evacuation. Clearly, to the extent possible, managers of warning systems must take precautions to ensure that warnings generate an appropriate response by potential victims. Successful warnings (where warnings have been issued, action taken, and a hazardous event did ensue) invariably generate, at least for a time, a greater and more accurate awareness of the hazard. In China, for instance, several successes have been reported in earthquake forecasting, such as in Haicheng in 1975, where warnings and subsequent movement of people to tents in open areas probably saved many lives (Kitizawa, 1986). On the other hand, there are many examples of unnecessary warnings in China and elsewhere, including Lima, Peru, and New Madrid, Missouri (Showalter, 1991). (The latter was not really a warning because it was made by a nontechnical person, but it did stir up all sorts of sociological issues.)

It is interesting to examine what makes some people respond to warnings while others ignore them. Research on warning systems and responses has led to the identification of individual traits as well as characteristics of the warnings themselves that influence response (Gruntfest, 1987). Such individual traits as age, education, gender, and previous experience with a hazard have been considered to have an influence, although the direction of that influence (toward an appropriate response or away from one) may vary under different circumstances. In addition, characteristics of the warnings themselves (such as credibility of the source, specificity and lack of ambiguity in the message, and provision of clear instructions) are seen to increase the potential for effective response. In this regard, it is pertinent to examine individuals' perceptions of hazards and their proposed responses. Other factors also can affect an individual's ability to undertake remedial action in the face of an impending disaster. Situational factors (such as physical health, mental health, and age) may well constrain the range of options open to individuals (Bolin and Klenow, 1982–1983, 1988; Huerta and Horton, 1978; Norris and Murrell, 1988; Phifer and Norris, 1989).

At the same time, cognitive factors play a role. Those residents of southern Florida who were reluctant to evacuate in the face of Hurricane Andrew often were making rational decisions based on previous experience. In the past (that is, the residents' lifetimes), hurricanes rarely had reached inland areas and, certainly, few had been as strong as Andrew. The power of Hurricane Andrew was unexpected (and has forced a reassessment of warning and evacuation policies). The behavior of the partygoers in Pass Christian was different; they were effectively challenging the storm and, in this instance, a lack of understanding of the force of hurricanes contributed to their deaths.

Evacuation also involves other questions. For instance, is a particular hazard threat greater than the perceived risk of leaving one's house empty and exposed to looting? Looting of evacuated property may not be common, but in the United States, the National Guard frequently is mobilized to protect against it. Any decisions by individuals must balance the perceived benefits of taking evasive action against the perceived risks of looting. While this may not be a major issue during small events that are generally nonlife-threatening, the problem is greater with low-probability, high-risk hazards; for example, in the buildup to Hurricane Andrew (1992) and prior to flooding in the Midwest (1993), some residents had to be coerced to evacuate their dwellings. The anxiety and reality of looting are certainly not recent. Johnson (1889) described scenes of looting in Johnstown after the 1889 flood and resultant fires which killed an estimated 3,000 to 15,000 people; several years later, Halstead (1900) attested to comparable events during the Galveston hurricane which killed approximately 6,000 people. In light of the huge numbers who died in these and similar events, decisions to remain in hazardous environments appear illogical, but those decisions are made on the basis of incomplete information and biased by perceptions of the individuals concerned. Furthermore, the incidents of looting are invariably magnified by the media looking for a different perspective on the hazard.

In the third scenario, involving the apparently bizarre actions of individuals, it must be recognized that their behavior may be contingent upon knowledge of the hazard, the perceived effectiveness of past actions, and understanding of other support structures. For example, if preventive strategies have worked in the past, an individual is quite likely to employ similar techniques again. It was found that many floodplain residents in Carlisle, England, were convinced of the effectiveness of sandbagging because it had been helpful in previous floods. However, such success had been attained only by those residents flooded to minimal depths; if flood water were to exceed six inches, sandbagging would prove less effective, requiring additional preventive measures (Tobin, 1978). In Carlisle, many residents undertook completely inappropriate action prior to the 1968 flood. Many sandbagged front doors and forgot to protect air vents and even back doors; problems also occurred with communal foundations in row houses, where floodwater seeped up through floorboards of sand-bagged houses. Working under stress—having no prior experience with and little knowledge of proven measures—clearly presents difficulties. Small events can be a learning experience for many, and they can have positive effects on future behavior by stimulating appropriate responses. However, a slight change in hazard characteristics (in this case, deeper floodwater) also can affect the applicability of learned techniques, rendering later efforts worthless or ineffective.

It also may appear that some individuals display particularly bizarre behavior by placing or intentionally leaving valuables in hazardous environments. For example, stereo systems may be moved to lower levels prior to floods, television sets may be placed adjacent to windows during windstorms, or delicate electronic equipment may be left on shelves where it "falls off" during earthquakes. Though morally reprehensible, this wanton destruction of property may not be the action of deranged individuals. If they have insurance policies that cover replacement of personal possessions or if they expect disaster aid in the form of financial relief, then the intentional destruction of old or obsolete property may be internally logical and consistent. That is, such behavior is perfectly rational to the individual concerned because insurance covers the costs of new equipment, but appears irrational to the outside observer. Further research is needed to test these hypotheses regarding the impacts of insurance and aid to assess the balance between the needy and the greedy.

In each case, therefore, our research must go beyond the descriptive and seek out explanations of the apparently foolish behavior of individuals. Inevitably, behavior is tied to *perception* of our environment (rather than governed by imperatives of the real world). Yet, it is also constrained or enhanced by situational variables. Even panic may lead to some inappropriate activities. This interplay of environment and perception is constantly changing as more information is received and processed, further complicating our attempts to uncover decision-making processes.

PHILOSOPHICAL APPROACHES

A number of approaches have been taken to try and understand the behavior of individuals before, during, and after hazardous events. They all attempt to make generalizations about the relationship between perception and response and, in one way or another, they all address those elements pertinent to human decision making, including economic, social, and political forces. These approaches are neither independent nor mutually exclusive, but involve a group of interacting forces which may be viewed and evaluated differently. Discussion of the various approaches will help us to understand the complexities of the interactions involved. Equally important, these approaches illustrate the evolution of thought within hazards research on the importance and role of individual perception. Following description of these approaches, we turn to specific variables within the cognitive and situational sets that have been identified as important.

Behavioral Approaches

To examine the role and behavior of individuals in the context of natural hazards, scientists have frequently adopted a logical positivist approach, leaning heavily on hypothesis formulation and deductive argument. Much of the early work, borrowed from the economics literature, included simple assumptions regarding economically rational behavior. The fundamental premise was that individuals in hazardous environments would behave in an economically rational way to maximize all outcomes (see utility models below). Recognizing that such simplistic cost–benefit analyses could not explain the locational decisions of most individuals and certainly could not account for many of the seemingly odd activities found within hazard zones, Gilbert F. White and the Chicago School took hazards research in a slightly different direction (Johnston, 1991). This new focus incorporated more social analysis in the research questions, based on the premise that social factors were instrumental in governing behavior. In part, White also adopted a more inductive approach, as when studying human adjustment to flood hazard (White, 1945).

The change in strategy was important. In prior work, hazards had been "solved" by implementing structural engineering projects, usually at great expense and with little concern for the human element. Such approaches did not prove conducive to reducing hazard losses. Indeed, White frequently pointed out that investment in mitigation measures was rising every year, yet hazard losses were increasing. It was this apparent paradox that stimulated the new research directions. One example will serve to illustrate the need for sociological considerations. Theiler (1969) showed that farmers on Coon Creek in Wisconsin did not respond as expected to implementation of a structural flood alleviation measure. It had been assumed that agricultural land use would intensify after the construction of levees as farmers would change from cattle to crop farming. Unfortunately, local farmers were not consulted. Agricultural patterns did not change because (1) the volatile agricultural economy reduced the demand for crops, (2) there was an increase in the number of retired and semiretired farmers, and (3) some farmers felt they already had sufficient cropland. In other words, the situational set of the individual farmers gave no incentive for change.

Others soon incorporated similar strategies in their research paradigms, and it was not long before the Chicago School was turning out a series of hazard papers (for example, see Kates, 1962; Roder, 1961; White et al., 1958; White, 1961). By the early 1970s, White's influence had become international and gradually, his research strategy was adopted in a variety of hazard studies. For instance, White (1974) pulled together over thirty hazard studies from around the world, many following similar methodologies, especially surveys using questionnaires to elicit opinions from people located in hazardous areas. These permitted comparisons between hazards as well as cross-culturally. For example, residents of Opotiki, New Zealand, were asked about their expectations of future flooding and perceptions of protective works. The results showed that while residents generally expected new levees to decrease future flooding, most who remained in the floodplain did so because of socioeconomic constraints (that is, situational factors) rather than faith in flood control (Ericksen, 1974). Locational choices were perceived differently by other groups. Ramachandran and Thakur (1974) studied residents of several villages on the Ganga floodplain in India. Residents tended to perceive flooding as inevitable (that is, the result of divine forces) and almost universally, depended on government action for remedial relief; indeed, opinion was equally divided on the relative advantages and disadvantages of the floodplain location and, consequently, only a small number expected to relocate. A similar dependence on state action was found by Kirkby (1974) in Oaxaca, Mexico, where adjustments to rainfall variability were initiated and provided by the government. Again, a relationship between situational and cognitive factors is evident, through the results of these questionnaires.

Behavioral work is sometimes criticized as normative and lacking in sound theoretical structure. This is not entirely valid as a theoretical structure was established through which comparisons could be made across social and ethnic boundaries. Indeed, it was the logical positivist approach, with a focus on the human element, that really helped to establish the significance of social and perceptual concerns.

Preference Models

Others have attempted to understand individual behavior through "revealed" and "expressed" preferences, that is, to document what people would do under given hazard conditions (as in the Opotiki example above). In essence, researchers have adopted two broad themes. First, in the revealed preferences approach, disaster victims have been surveyed to establish what they actually did before, during, and after an extreme geophysical event. For example, earthquake victims may have undertaken long-term preventive techniques to protect their own homes, taken appropriate action during the event, and adopted further protective measures for future events based on knowledge learned from experience. In contrast, others may have had no preparedness program and behaved irrationally during the event. In this way, preferences are revealed. We can determine exactly how individuals responded to specific threats and, in turn, their behavior can be related to the cognitive and situational models.

The expressed preferences approach has taken a different track. Since it is not always possible to interview potential disaster victims directly (in part because the event has yet to occur), some studies have looked at what individuals perceive as acceptable behavior: potential victims are asked to state a preferred choice of action under a detailed or synthetic hazard situation. One problem with the expressed preference approach is that individuals do not always do what they say they would (words do not always translate into action). For example, researchers may ask, "What would you do in the event of an earthquake?" The answer does not necessarily reveal what an individual really would do in that stressful time. While such an approach may show what people are thinking with regard to hazards, researchers must be cautious about what is termed "cognitive dissonance" (Festinger, 1957). In this situation, individuals who must remain in a hazardous area for socioeconomic reasons may well hide their perceptions and expectations of the hazard from an interviewer so that their responses or behaviors appear to be consistent. Further, in the reality of a disaster, stressful conditions may promote a set of actions completely different from those planned or considered prior to the event.

Utility Models

Burton, Kates, and White (1993) discuss individual behavior as a choice process from among a limited number of alternatives. However, they also point out that our understanding of these choice processes (and hence, behavior) is far from clear and might depend on economic or social factors. For example, from an economic perspective, it is argued that individuals assess all potential outcomes in a set order to determine the maximum outcome, termed the "expected utility model." The assumption is that individuals have sufficient information and the capability to make "rational" decisions that will result in optimal outcomes; all decisions are, thus, maximized. A companion model, the "subjective utility model," argues that individuals make choices based on subjective views of probable outcomes; that is, decision making is constrained by personal views or subjectivity of the individual. In the former model, economic considerations prevail; in the latter, other factors may influence decision making. In a hazardous situation, it is argued that individuals will consider a series of options and select those actions perceived to provide the greatest benefits; prior to a hurricane, therefore, one might opt to board up windows and ride out the storm, rather than remove all valuables and evacuate, if the perceived advantages of the former course of action outweigh the latter. The combination of economic and social factors can confuse the reality of the hazardous environment.

The assumption of the "rational individual" is central to economic theory and has been adopted by many social scientists. The model assumes an economically rational person who is in command of all the facts and capable of making a logical decision. As a normative model, this approach has merit in that it permits comparisons between real-world decision making and a theoretical ideal, but as a tool for explaining behavior, it fails.

Most individuals neither have nor could assimilate all the necessary information on which to base a decision. Indeed, it is clear that for most, lack of information limits any ability to maximize or to make sound, purely objective decisions.

Even if people did have all the information necessary to make "rational" decisions, not only would they have difficulty processing it but they might well have other goals than maximizing expected utility. Burton et al. (1993) term this "bounded rationality," pointing out that there are bounds of rational choice that restrict or inhibit the individual and lead to decisions that are less than optimal. Some of these bounds may be societal; others are reflected in what the individual sees as important. Burton et al. (1993) give the example of a fisherman deciding whether to evacuate in the face of a hurricane threat. While from an economic perspective the optimal choice (to minimize cost) might be to remain in place and protect his or her equipment and possessions, the fisherman instead considers the costs associated with evacuating versus staying, up to a point; instead of maximizing benefits or minimizing costs, the fisherman may have other priorities or goals such as staying out of debt. Thus, the bounds on this individual's decision are defined by the extent to which he or she views evacuation as affecting personal debt. In other words, behavior in this instance may be rational based on the level of financial commitment a decision is viewed to have. Clearly, many people could adopt sophisticated remedial action plans to mitigate the adverse impacts of events, but those choices would have to be weighed against the opportunity costs of other choices.

As a result, actions of individuals that appear irrational to the impartial observer may be internally consistent and logical in the context of the individual. In a small village in England, for example, one elderly resident with a lifetime experience of flooding had cemented a board across her doorway to prevent further damage. This strategy worked for small events but was totally inadequate for larger ones and, in fact, aggravated the problem by keeping water in the house during one particular flood; in addition, the day-to-day difficulties of climbing over this board were perceived as unimportant compared to the ever-present threat of flooding (Smith and Tobin, 1979). At the other extreme, many residents in the same community took no precautions and expressed a reliance on a fictitious flood forecasting and warning system. Part of their confusion arose from a river-level gauge located in the town which was actually part of the warning system for a community many miles downstream. The bounded rationality of these residents exemplifies why it is important to examine underlying causes of any behavior. The local context may provide important clues to potential behavior.

Behavior-oriented studies have addressed various attributes of the individual. Kates (1962) and later Burton et al. (1993) developed a schema for decision making at the individual level: "People who have been observed coping with risk and uncertainty in the environment appear to take account of likely economic outcomes . . . but the resulting behavior rarely conforms to what should be optimum by their standards of utility" (Burton et al., 1993, p. 100). As they explain, people cannot appraise the magnitude and frequency of extreme events with any accuracy, are rarely aware of all the alternatives open to them, and differ greatly in the way they judge the consequences of their actions. As will be seen in Chapter 7, perceptions of risk vary from group to group, from person to person, and from time to time (Slovic, 1987). Thus, people in hazardous areas have difficulty evaluating the risk involved in the location, but will take remedial action given some threshold of awareness; however, the potential for action is restricted because individuals usually have limited knowledge of possible options. It is further suggested that individuals assess each alternative in turn rather than all at once and that choice is based on perceived outcomes (Burton et al., 1993). The simplified model therefore suggests that individuals proceed in decision making to satisfy demands rather than maximize outcomes. This would mean that decision making is hierarchical and the choices structured (Slovic, Kunreuther, and White, 1974).

Factors external to individuals also come into play in influencing behavior and response. Palm (1990) identified macro- and mesoscale factors that constrain individual decisions and activity, both positively and negatively. At the macro scale are cultural norms and traditions as well as the political–economic context in which an individual functions (discussed in the next section and throughout the book). At the mesoscale are the "gatekeepers" or "urban managers" whose actions and decisions affect the range of choices available to the individual. For example, mortgage lenders, real-estate agents, and insurance brokers in Liberty County, Texas, were found to view flood-prone residences as of lower value than nonflood-prone houses, and many admitted a reluctance to handle such properties (Montz and Tobin, 1992). Although many factors likely contributed to that distinction, the fact that these three groups recognize it limits the range of choices open to residents of the floodplain because the gatekeepers control access to opportunities for change and mitigation. Thus, when evaluating the perceptions and subsequent decisions of individuals, it is crucial to consider the constraints of the context in which decisions are made.

Marxist Approaches

In attempting to explain individual behavior, some researchers have taken a more radical approach by looking at the political and social structures that affect human lives. Some have adopted a Marxist approach to hazards, pointing out that the political economy overwhelms many other factors (for an example of such arguments, see Susman, O'Keefe, and Wisner, 1983). One tenet of this approach is that people remain in hazardous areas because society does little or nothing to help them. Further, it is argued that hazards affect marginalized groups more than others. To paraphrase Susman et al. (1983), vulnerability can be defined as the degree to which a society is at risk both in terms of the probability of occurrence of an extreme physical event and the degree to which a community can absorb the effects and recover. In looking at levels of vulnerability, it is clear that not all individuals are equally vulnerable, but rather that different classes, groups, or even countries experience different degrees of risk. The poor are generally more vulnerable than the rich; their behavior is likely to be different and recovery for this group is usually very different (International Federation of Red Cross and Red Crescent Societies, 1993). The Marxist approach, then, focuses on situational factors which, for some people, may override the influence of cognitive factors. For instance, relief may not reach certain groups for a variety of reasons; in central California, some individuals neither received nor requested federal money following flooding because the regulations and paperwork were too complex (Montz and Tobin, 1988). Failure to reach certain groups also can be attributed to outright discrimination; in India Harijans (the "untouchables") are usually suppressed when loans are distributed after floods (Ramachandran and Thakur, 1974).

With a situational focus, differences in vulnerability can be explained in part through the marginalization of groups and individuals within society. Specifically, complex socioeconomic relationships produce a system in which some are less powerful than others; while still a part of the cultural weave of society, a group or individual is marginalized within the prevailing power structure. At a larger scale, the world economy may perpetuate technological dependency and unequal exchange between nations, thereby maintaining the impoverishment of underdeveloped countries and increasing the constraints of an individual's situation. Similar relationships can operate at a smaller scale within countries. The poor and less powerful frequently occupy the most risk-prone areas and are usually more vulnerable to hazards (Burton et al., 1993; Torry, 1980); consequently, relief reinforces the status quo by perpetuating dependency on the current system, which can lead to greater vulnerability of the poor as populations expand and resources are controlled by a smaller minority.

For example, Bangladesh illustrates many elements that support the notion of marginalization and increasing vulnerability. First, the hazards themselves have adversely affected the ability of the country to pull out of the poverty trap in which it finds itself (that is, they are combined with a sluggish economy that leaves few choices or options for most people). Throughout its history, Bangladesh has been faced with natural hazards and/or war and has had little time to concentrate on economic vitality (Hossain, 1987). During 1988–1989, 45% of its development budget went to pay for damages accruing from floods in the late 1980s (Brammer, 1990a, 1990b). Second, there is a progressive imbalance between the population and the available land. The proportion of landless households is now 30%, having increased 3.1% per year since the early 1960s. This figure is rising at a rate faster than population increase (Choudhury, 1987). In rural areas, 30–50% of the households are landless, hence are forced to occupy the most hazardous areas of the floodplain (Brammer, 1990a, 1990b). Thus, landlessness has pushed people into poverty and, because the landless are increasingly marginalized and vulnerable to hazards, situational factors may dominate the relationship between perception and response. This is not to suggest that perception of hazards is unimportant or that perception does not influence individual behavior prior to an impending event. Instead, we must recognize that the range of responses for a Bangladeshi farmer is severely restricted by external, situational factors, and any analysis of perception and response must be cognizant of that context. The same, of course, applies to persons vulnerable to disaster around the world.

Summary

The various approaches that have been taken to evaluate influences on responses to hazards reflect different theoretical leanings and different views of the interactions among groups of factors. Early research provided important building blocks for the studies that followed, in terms of both the range of factors seen to be at play and the complexity of interactions among them. All studies have been based on the recognized need to develop a framework to analyze the contributions of the many factors found to influence individual response. Unfortunately, development of such a framework has been difficult because of the variable influence that different factors have in different places at different times, complicating their analysis and sometimes clouding our understanding. Nonetheless, we do know that certain factors are at work, and it is to this range of factors that our attention now turns.

PERCEPTION OF HAZARDS: CONTRIBUTING FACTORS

A prevailing issue in natural hazards research has been to understand those factors that influence individual perception of natural hazards, with an eye toward understanding and perhaps influencing or guiding responses and actions. As we have seen, the physical environment, the socioeconomic climate, and psychological characteristics all influence the perceived world in one way or another.

We have considered two groups of variables: situational factors and cognitive factors. However, we can break these groupings down even further. White (1974) proposed that variation in perception and estimation of danger can be accounted for by the magnitude and frequency of an event, recency and frequency of personal experience, significance of the hazard to income interest, and personality traits. Thus, situational factors can be subdivided into those associated with the physical environment and those associated with the socioeconomic environment in which individuals operate. Cognitive factors can be subdivided into personality or psychological factors that influence perception and attitudinal factors that influence expectations (obviously, these are not mutually exclusive as personality characteristics also influence attitudes). Figure 3.2 depicts how these sets of factors interact between and within sets.

As discussed below, situational variables include those related to physical characteristics of a hazard, over which the individual tends to have little control; socioeconomic factors that may define the circumstances, and sometimes the boundaries in which an individual finds him- or herself; and some combination of the above. These factors work separately and in combination with each other and with the cognitive set to influence response and actions toward hazards. It is an oversimplification to assume that one or another variable or combination of variables will necessarily produce some desired result because of the mutable roles these factors play among people in different places. For instance, the Bangladeshi example above suggests that socioeconomic conditions make adaptive responses difficult if not impossible, regardless of experience or

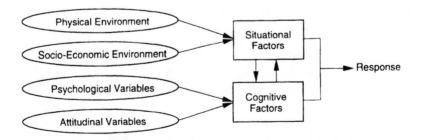

FIGURE 3.2. Influences on perception. Different variables influence the sets of situational and cognitive factors. While these variables are frequently subsumed to their respective sets, they also have individually distinct effects on the nature of each set of factors.

vulnerability; there, a specific subset of situational variables prevails over all others.

Situational Factors: The Physical Environment

Several physical characteristics of events have been shown to influence responses and actions, though not in isolation from other contributing factors. The magnitude and frequency with which events occur are important, as is their duration. These factors help to define an individual's experience with hazards and shape perception. Although our ultimate concern is with experience, it cannot be studied generically but must be evaluated with regard to its type, frequency, and magnitude. Consequently, it is essential to consider the effects of the physical environment in some detail.

At the community level, physical parameters of events have been linked to decision making (as discussed in the next chapter). For instance, the threat of hurricanes and associated flooding from storm surges was undoubtedly instrumental in the decision by Galveston, Texas, to construct a major defensive wall along the shoreline. Hurricanes in 1875 and 1886 were followed by a major hurricane in 1900, when 6,000 people lost their lives and damage exceeded $20 million in 1900 dollars (Halstead, 1900). After dealing with the dead, the immediate response of the city was reconstruction. Recommendations were made to raise housing and provide firmer foundations, and discussions were held on the effectiveness of a proposed sea wall. The question of hurricane frequency also was addressed. "Without question, Galveston is in the track of a certain abnormal but not infrequent West Indian hurricane which fails to be deflected from the Georgia and Florida coasts" (Halstead, 1900, p. 263). "Galveston's destruction and that of other towns similarly situated had been predicted" (Halstead, 1900, p. 264). Although the cause of hurricanes was not fully appreciated, the community did have an idea that hurricanes could and would continue to affect Galveston.

Although studies of natural hazards are replete with examples of postdisaster responses by communities (see Chapter 4), it is not possible to disaggregate the community-level response to the individual. Individuals respond differently to the processes of natural hazards, and the extent to which physical attributes of a hazard are responsible for their decision making is debatable. As Burton et al., (1993, p. 100) suggest, "They cannot appraise the magnitude and frequency of extreme events—states of nature—with accuracy." (As we saw in Chapter 2, scientists themselves sometimes have problems determining and estimating recurrence intervals for hazards.) Some combination of physical characteristics—including frequency, suddenness of onset, magnitude, and duration—influences individual perceptions, but that influence is not always consistent or direct, depending on how other factors come into play. It has been suggested, for instance, that unless floods are experienced relatively often and damages are comparatively high (determined by frequency and magnitude of events), appropriate responses may not be undertaken (Burton et al., 1993; Kates, 1971; Kunreuther, 1976).

Nevertheless, there is some evidence that certain individuals do make decisions based in part on perceived magnitude and frequency of events. Many have adopted lifestyles that directly confront hazards, living in some ways "on the edge." Some occupy swamplands and riverbanks while others live on steep slopes or unstable land, perceiving the hazard as a minor inconvenience rather than a threat to their lives. Research has shown that most of these individuals are knowledgeable about hazard frequency and have undertaken remedial actions to alleviate impacts. In many instances, they have taken preventive action to prepare for floods, hurricanes, tornadoes, and earthquakes. In a study of coastal erosion, for example, Rowntree (1974) found that an awareness of high risk translated into better understanding of the hazard and a larger expenditure on preventive measures, whereas residents of medium-risk areas knew little about the magnitude, frequency, and predictability of the problem. Residents of low-risk areas were aware of problems, but had difficulty articulating their thoughts and feelings about the hazard. However, the picture is complicated because awareness of hazard is dynamic. For example, Bernknopf, Brookshire, and Thayer (1990) showed that the perception of risk increases as a disaster approaches but soon decreases to pre-event levels. Cross (1990) also found increases and decreases over time among residents in hurricane-prone communities.

However, lack of awareness of a hazard should not automatically be attributed to or blamed on the individual. Clearly, the daily concerns of living are often more pressing than low-probability events over which individuals have little control (Drabek, 1986), and people are generally less sensitive to hazards that they cannot control (Lacayo, 1989). Land degradation is a good example; farmers feel they can do very little about it and usually have the more pressing demands of farm costs and income to concern them (Heathcote, 1983). Similarly, Cross (1985) demonstrated that while coastal residents expressed anxiety about high-probability flooding, they were more concerned about daily events.

Nonetheless, many individuals in hazardous environments do take remedial action. For example, some people living along the Trinity River in Texas have taken the initiative and raised their structures above perceived flood levels. In some cases, such action is enforced or encouraged by hazard-reduction programs; in others, it is not. Some people enjoy the lifestyle and seek out such locations, expressing excitement about their living environment and finding floods a "positive" distraction from the usual routine of life. There is no question that many people make a conscious decision to locate in these environments, recognizing the hazard to which they are subject. Their efforts to alleviate the effects of flooding are based on perceived risk and personal experience of the hazard, taking into account both frequency and intensity.

Researchers have leaned heavily on the premise that knowledge of the physical world will translate into more accurate perceptions of the real world, hence a better understanding of hazards. For example, Roder (1961) found that farmers (that is, those who had a close association with the land) had a more accurate perception of flood hazard than comparable urban dwellers. Others have demonstrated that direct personal experience with a particular hazard often leads to more accurate assessments of it. For example, Geipel (1982) studied earthquake victims in Friuli, measuring their expectations of future events. Results showed that 70% of the residents either expected a recurrence after more than 100 years or did not know what to expect. Geipel distinguished between different communities and found higher levels of expectation in the more isolated community (see Table 3.2). Individuals were classified in relation to technical estimates; the pessimists had a higher expectation of recurrence (less than 50 years), realists were closest to reality, and optimists felt overly safe.

However, not all studies have shown similar patterns. As Tazieff (1986) argues, "people know from experience that the probability of a cataclysm occurring twice in the same part of the world during their lifetime is, if not nil, at least negligible." For the most part, this is true. For instance, take the 50% probability of a major earthquake (magnitude 7.0) occurring in San Francisco between 1990 and 2020; to many, this is a remote possibility that can be ignored (although, given the temporal distribution of earthquakes, there is significant cause for concern). At the other extreme are the frequently recurring events in Bangladesh, where flooding is perceived fairly accurately by residents (although those in "safe" areas tend to perceive a lower frequency of events than those who experience regular flooding).

TABLE 3.2. Expectation Patterns of Earthquake Recurrence

Location	People (in %) expecting recurrence within:			
	50 yr (overestimates)	50–100 yr (realists)	More than 100 yr (underestimates)	Don't know (uncertain)
Portis	13.4	10.4	38.8	37.3
Braulins[a]	20.1	14.7	45.3	20.0

[a]Isolated, with fewer contacts than in Portis.
Source: Geipel (1982). Copyright 1982 by Routledge. Reprinted by permission.

The issues of frequency of hazard and probability of disaster recurrence are difficult to assess accurately. Geophysical events are changing as environmental conditions change. Some natural events can remain dormant for some time, then occur with devastating results. The incidence of volcanic action may fit this scenario; Mount St. Helens had been relatively inactive for over 100 years when it erupted in 1980, killing over 50 people. Human modifications of the physical environment also can affect the incidence of natural hazards. The increasing urbanization of rural landscapes, draining of wetlands, and modification of river systems to control flow have exacerbated the recurrence interval for smaller flood events. Consequently, even "accurate" perceptions of the risk will not hold true over time.

The effects of individual, personal experience are more complex than the examples above may indicate. Some veterans of hazards see themselves as capable of withstanding any event; they believe they already have made it through the "big one." While experience is related directly to the physical environment, it also relates to individual attitudes toward a hazard. Further, the physical environment (including hazards) does not necessarily determine behavior, but many researchers work on the premise that knowledge and curiosity may be increased through experience, which may lead to improved appreciation of a hazard. For example, Burton (1961) demonstrated the significance of personal experience in appreciating flood frequency. However, other studies have shown that such experience leads to less accurate perceptions. We will return to consideration of this point in our discussion of the other subset of situational factors relating to socioeconomic characteristics and constraints.

Situational Factors: The Socioeconomic Environment

Socioeconomic factors include a range of characteristics that also may be seen as cultural, such as education, employment, income, religion, and family ties. These factors do not always play the same role (in one case, high income may be associated with accurate perception of a hazard, but in another, it may not). It is cause for concern that such factors may be part of a larger psychological set-up, as posited below. In a study of two communities astride the Rio Grande River which flooded in 1954, Clifford (1956) demonstrated their different responses although both communities relied on family groups, religious connections, and social links for support. Establishing a predictive or explanatory model based on socioeconomic criteria, thus, poses many problems.

It should not surprise us that some studies have found better educated and wealthier people responding to the threat of hazards by adopting mitigation measures such as flood-proofing and purchasing earthquake insurance. However, this is not always the case nor does it necessarily indicate that high-income people have more "rational" perceptions of hazards than lower-income people. Such measures may merely reflect financial ability to do something. In one study, de Man and Simpson-Housley (1988) found that expectations of flooding and perception of severe damage were positively related to expectation of flood damage but unrelated to education; they concluded that education was not a major factor in influencing perceived risk.

Clearly, other factors must explain why some individuals adopt mitigation measures while others do not. For example, Laska (1985) reported that homeowners perceive a personal benefit–cost ratio that often works against adopting their own remedial measures; she argued that this benefit–cost ratio caused homeowners to reject mitigation measures, especially flood-proofing and other nonstructural adjustments. In a later study, Laska (1990) found that people are more willing to pay high taxes for community adjustments if they perceive that government action will provide a solution. Her evidence suggests that individuals are more willing to support societal action than to undertake direct personal investment, but whether this relates to personal benefit–cost evaluations or to perceived effectiveness and benefit remains to be seen.

Other socioeconomic variables include age, gender, household size, and experience with a hazard, which also relates to physical characteristics. In some cases, the relationship between a single characteristic and response has been in one direction; for example, family groups have been found to take adaptive or mitigating action (Mileti, Drabek, and Haas, 1975). However, it is not so clear-cut when other variables are considered, separately or in combination. Take age, for example. Some studies have shown that the elderly are less likely to evacuate their homes (Gruntfest, 1977; Moore, Bates, Layman, and Parenton, 1963) while others have found no association or a different relationship (Baker, 1976). It also has been hypothesized that age may elicit a curvilinear relationship with stress; middle-aged groups (aged 45–60) may experience higher levels of stress following disasters because they have greater responsibilities, such as looking after children and the elderly. Further research is needed to test these relationships.

Certainly, a variety of factors affect how people react to disaster stress and respond to hazards. A useful model might be developed incorporating three different scales: the individual, group, and community. At the individual or primary level, behavior is influenced by individual traits such as age (Huerta and Horton, 1978; Phifer, 1990), gender (Krause, 1987), race and ethnicity (Perry and Mushkatel, 1984), and family structure. Personal mobility, health, functional behavior, and personal experience also can have impacts on hazard victims (Solomon, Smith, Robins, and

Fischbach, 1987). Because personal characteristics can either constrain or enhance individual activity, they must be added to our situational set, suggesting the complexity of attempting to determine specific causes of variation in perception among individuals. For example, the elderly and infirm are generally less able to undertake any major moving of possessions in the event of a flood or hurricane. Race, ethnicity, and gender affect different patterns of behavior, which while not necessarily causal, nevertheless point to other concerns. In analyzing individual response to natural hazards, attention must be given to such variables.

At the secondary or group level, other factors come into play (and affect the individual level). Some variables include neighborhood characteristics, levels of social support, and social involvement following disaster (Madakasira and O'Brien, 1987; Russell and Cutrona, 1991; Solomon et al., 1987). For example, it has been suggested that inhabitants of homogeneous neighborhoods have greater levels of social support within the social network than those "isolated" in heterogeneous neighborhoods (Gubrium, 1973; Ollenburger, 1981). Thus, homogeneity should mitigate stress while heterogeneity should exacerbate problems. However, Ward, LaGory and Sherman (1985) found that age-concentrated neighborhoods (or those having age-homogeneous social networks) do not necessarily contribute to individual well-being; although their study does not specifically examine disaster conditions, it suggests that other variables must be considered when examining stress from natural hazards.

The level of neighborhood involvement also affects stress and behavior. Madakasira and O'Brien (1987) found that inadequate social support was correlated with higher levels of posttraumatic stress disorder among tornado victims, suggesting the importance of informal social networks. Indeed, high levels of social support appear to have a buffering effect that reduces the negative impact of stress on mental health (Cutrona, Russell, and Rose, 1986). On the other hand, social networks also can elicit higher levels of stress among disaster victims. Solomon et al. (1987) discovered that those network members on whom others relied heavily were more likely to experience stress following a disaster; in particular, they found that very strong social ties could be especially burdensome to women. Coyne and DeLongis (1986) argue that the context of any close relationship (such as spousal or family ties) affects the level of stress experienced. Further, age and organizational vulnerability, which are associated with social support, have combined effects on the mental health of disaster victims. There are, then, both positive and negative consequences of social involvement.

Finally, at the tertiary or community level, the model incorporates urban–rural distinctions. Various authors have explored the sociology of community and differences in the urban–rural continuum, including Craven and Wellman (1973), Haggerty (1982), Simpson (1965), and Wirth

(1938). In light of the findings described above, social support systems and networks within communities would appear to have an important bearing on aspects of hazard behavior. It is often argued that rural communities are more close-knit than their urban counterparts, but this may not be true since social networks may operate in urban neighborhoods much the same as throughout rural communities. In disaster situations, different relationships and leaders also emerge that can significantly change social patterns.

Although social contact is important, other factors also affect the response to natural hazards. For instance, differences in the level and timing of aid, the distribution of that aid, and other levels of support can directly affect how individuals later respond to other events. It is reasonable to hypothesize, for example, that aid may not get disseminated to rural communities as rapidly as to concentrated urban areas. These structural factors alone may have implications for stress and behavior.

Such arguments may not hold for megacities, defined by the United Nations as cities with populations exceeding 8 million people. Mitchell (1993) argues that megacities are becoming more exposed to natural hazards and that disaster losses are increasing as a result. Management of these cities is becoming increasingly difficult as size increases rapidly. Along with their growth comes an increase in concentrations of wealth and poverty. Social systems are changing rapidly and structural factors may work against the support systems seen in smaller settlements. As a result, size of urban area is a situational factor in its own right.

Culture

Hazard problems and perceptions do not stem only from the interaction of physical processes and the social system. Cultural and religious elements of society also can play important roles (Bryant, 1991). For example, fatalistic attitudes that accept disasters as divine acts or religious strictures against intermixing of social groups may exacerbate hazardous events and may not stimulate appropriate responses. Thus, vulnerability can be increased by cultural norms, sometimes by delineating appropriate or inappropriate courses of action, sometimes by marginalizing particular groups. Many variations may follow from cultural differences; for example, urban inhabitants often feel more hazard-free than rural inhabitants (Burton et al., 1993).

Experience

Personal experience with a hazard is a factor that relates to both socioeconomic circumstances and the physical environment. Being in the wrong place at the wrong time may make one a victim of a natural event, and the reasons for being there are usually situational (such as owning a house on a floodplain or being at work when a tornado tears through the business district). However, physical characteristics of an event (including magnitude, duration, frequency, and suddenness of onset) shape the nature of that experience, hence its influence on perception and response.

One might assume that greater personal experience with hazards correlates with more accurate assessments of the threat and more favorable responses. It has been found that the best predictor of flood response is often personal experience. Laska (1990) showed that personal flood frequency was related positively to the adoption of flood-proofing measures, and Tobin (1978) found that residents who had experienced flooding in two communities in Britain had a more realistic view of floods and more accurate perception of their own capabilities than did newcomers.

Another characteristic of personal experience is its recency. Hazardous events remain the focal point of peoples' lives for a long time; indeed, all other life events may be dated by that experience (that is, whether they occurred before or after the earthquake or flood). Inevitably, however, memories of the experience wane, to be replaced by a collection of thoughts that may or may not reflect the reality of the personal experience. Consequently, we can expect that the more recent the experience, the greater the awareness of the hazard, but 15 years after an event, knowledge of its particulars will be minimal and no longer directly impact decision making. Baumann and Sims (1978) refer to this pattern as motivational decay, noting that it may take only six months for the motivational effects associated with an event to decrease. However, major traumatic events become "life markers" for many individuals; the devastation caused by Tropical Storm Agnes in Wilkes-Barre, Pennsylvania, in 1972, is still remembered as a pivotal date by many residents just as the flooding of the Midwest in 1993 will become a significant temporal reference point to many.

The magnitude and temporal occurrence of an experience also affect how people perceive hazards. It has been suggested that if recent events are especially severe, earlier events will become less clear (Kirkby, 1974; Burton et al., 1993), but a first experience also may be a major event in an individual's life. In his study of drought, Saarinen (1966) found that individuals usually remembered their most recent drought except when recalling their first experience, which he termed a "factor of primacy." It is easy to see that Hurricane Andrew of 1992 will become for many Floridians the baseline event to which all subsequent hurricanes will be compared. This play between the recent and the first experience is important and can influence behavior. Indeed, Burton et al. (1993) suggest that the more memorable the experience, and in many ways the more severe, the better the perception of risk from that hazard.

The effect of experience also can be limiting. Individuals make choices from a range of alternatives, but the alternatives considered depend on their knowledge. Individuals may be prisoners of their own experiences because those experiences may set boundaries around their knowledge (Kates, 1962). In essence, individuals at risk are not cognizant of all options and strategies for dealing with a hazard and therefore cannot always act appropriately. If particular remedial activities worked in the past, the usual pattern is to repeat that behavior. If the experience was catastrophic, the individual may have little knowledge of what to do in subsequent disasters.

In addition, we must consider how individuals deal with the perceived risk of hazards (discussed in greater detail in Chapter 7). Since most natural hazards are low-probability events, it is not unusual to find them virtually ignored on a day-to-day basis. Given the other pressing concerns of life, this might be expected, but there are times when preparation for disastrous events would be common sense, especially for those in high-risk zones. Unfortunately, there is a tendency toward what is known as "gambler's fallacy"; that is, once an event has occurred, many individuals believe that it will not occur again within their lifetimes or, at least, not in the near future (Slovic, Kunreuther, and White, 1974). For instance, told that a flood was a 100-year event, a common response would be, "I'll be dead before the next one," based on the expectation that such events occur regularly. Misperceptions occur because hazards are viewed as cyclical or systematic, based on a simple law of averages, or the result of freak incidents (Slovic et al., 1974). While some events can be attributed to accidents (such as most dam failures), the general basis for assessing hazard threats is grossly inadequate. Again, experience does not have the expected or desired effect of leading to "rational" responses and actions.

Cognitive Factors

Cognitive factors include both psychological and attitudinal variables that, along with situational factors, influence behavior and responses. In general, cognition can be seen as expectations that an individual develops over time and that relate to beliefs about the threat of a hazard and personal vulnerability. In complex interplay, psychological characteristics and availability of information influence the processing of information about a threat. Furthermore, an individual's belief systems and attitudes (toward risk, the type of information available, and its source, to name a few) also come into play. As a result, consideration of both psychological and attitudinal factors is important because they are intricately tied to one another.

Psychological Factors

One assumption of environmental psychology is that situational factors play a significant role in affecting individual behavior. However, this relationship is quite complex and cannot be addressed fully without due consideration of additional cognitive factors. As Holahan (1982) says of the environment–behavior equation, the individual's role is active and vital; situational factors do not lead simply to behavior in a deterministic way (Figure 3.3a). Holahan (1982) suggests that the simplistic relationships in the determinist model imply that negative environmental conditions (like a damaging flood) promote negative behavior, which we know is not the case.

Psychologists instead assume a three-part relationship involving "an external environment or situation, an input from the environment to [the individual] actor initializing a psychological process, and an output from actor to environment termed action" (Garling and Golledge, 1993, p. 4). This relationship is represented in Figure 3.3b, which shows adaptive psychological processes interacting with situational conditions (the environment) to produce behavior (an action). The relationship is to some extent reciprocal and behavior is moderated by adaptive processes. Returning to the flood example above, Figure 3.3b might be interpreted to suggest that all negative situational factors can be overcome by coping strategies or that a failure to do so results from inadequate preparation by victims. As we have seen, marginalized groups (such as squatters in Turkey or

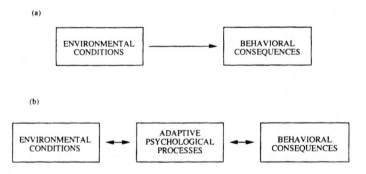

FIGURE 3.3. The environment–behavior relationship. Although Environmental conditions clearly have behavioral consequences (a), psychological processes intervene to alter and moderate behavior (b). These processes may influence subsequent environmental conditions. *Source:* Holahan (1982). Copyright 1982 by McGraw-Hill, Inc. Reprinted by permission.

floodplain residents in Bangladesh) have very little opportunity to mitigate disaster impacts because they have only minimal control over the situation. Similarly, those with greater control over their situations still may pursue nonadaptive actions because of other factors, including the lack of social support structures, a perceived inability to respond, or stress brought about by the event. Disaster victims are not necessarily to blame for any apparent inaction; instead, adaptive processes may be directed toward coping with a situation rather than changing or mitigating it.

Perception studies also must take account of cultural differences. As psychologists have pointed out, individual perception is affected by cultural background; people from different backgrounds have different world-views. Rather than focusing on isolated psychological processes, it has been suggested that we should take a more holistic view. Holahan (1982) argues for considering behavior patterns through a personality paradigm that incorporates the concept of the "total person," a theme in interactional psychology (Magnusson and Torestad, 1992). In other words, a person functions as a totally integrated and dynamically organized entity in the environment–behavior equation, purposefully and actively interacting with the environment in a multidimensional way. Consequently, it is difficult to study single actions or characteristics in isolation from the total functioning of the individual. Again, relationships between factors are interactive and reciprocal; there is not a simple deterministic model such that situational factors affect psychological processes (i.e., perception, cognition, and attitudes) which eventually influence behavior (Figure 3.4). Instead, Holahan suggests that psychological factors are processed simultaneously, then are translated into overt behavior. Behavior toward hazards (that is, a choice to take or not to take mitigative action) results from a continuous interactive process between the individual and the situation (the hazard or event) with psychological processes serving as a filtering system. Cognitive factors certainly influence this behavior, as does the psychological meaning that individuals attribute to the situation (Endler and Magnusson, 1976).

A number of psychological factors have been related to individual perception and behavior. One is the sense or locus of control exhibited by individuals (Burton et al., 1993). Persons who perceive the environment to be controllable by their own efforts are said to have an internal locus of control, while those who consider events beyond their control have an external locus of control (Baumann and Sims, 1974). It has been argued that individuals with external loci have higher levels of anxiety and expect hazards more frequently than those with internal loci (Simpson-Housley and Bradshaw, 1978).

The determining factor is whether an individual has perceived control of his or her fate or destiny (that is, is the environment

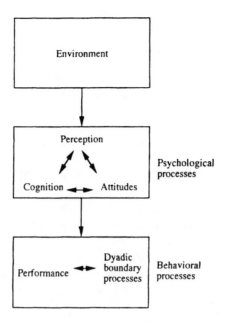

FIGURE 3.4. Determinants of behavior. Behavioral and psychological processes are more complex than indicated in Figure 3.3. The "total person" is determined by a nexus of psychological processes, consisting of perception, cognition, and attitudes. *Source:* Holahan (1982). Copyright 1982 by McGraw-Hill, Inc. Reprinted by permission.

controllable [internal] or beyond control, hence external?). As Smith (1992, p. 61) points out, "In order to reduce the stress associated with uncertainty, hazard perceivers tend to adopt recognizable models of risk perception with which they are more comfortable." He groups these models into three basic types.

Determinate perception. Because most people find the randomness of hazardous events unacceptable, they try to make hazards knowable or more ordered. For example, the hazard may be assigned orderliness in the minds of individuals such that floods occur on a regular basis. Kirkby (1974) found that 25% of her respondents perceived a drought hazard to be cyclical, often saying that "we get these every seven years." (While some hazards like earthquakes can be "regular," the pattern of most hazards is not.) In many ways, the technological fix also can generate a determinate perception (Penning-Rowsell, 1986); the false sense of security that results from many structural measures can promote the view that a hazard has been eliminated, which of course can have severe consequences if and when the project fails. Like the gambler's fallacy, the hazard becomes determinate and knowable to the individual in a way that may not reflect the reality of the risk.

Dissonant perception. Denial is a fairly common response to hazards, which can take various forms. First, there may be complete denial of the existence of a hazard. Second, people may relegate the threat to the past (saying that it no longer occurs); this frequently results from remedial measures that may bring about such statements as "they have stopped all that." Third, a threat may be accorded only minor importance (which may be reasonable in the context of other "threats" to daily life). For example, Saarinen (1966) noted that farmers were eternally optimistic, usually underestimating the frequency of drought and overestimating average crop yields. Others perceive certain hazards to be little more than minor disruptions, such as the snow hazard in upper Michigan (Earney and Knowles, 1974). Quite simply, some people tend to deny risk (Drabek, 1986) or to see themselves as completely unaffected by it (Lacayo, 1989).

Probabilistic perception. With this type of perception, an individual recognizes and accepts the occurrence and randomness of disasters, attaining a level of accuracy not found in the other types of perception. However, probabilistic perception may be combined with a fatalistic attitude and the transfer of responsibility to some higher authority (Smith, 1992). When hazards become "acts of gods," individuals feel no personal responsibility to respond to them (for example, see Murton and Shimabakur, 1974; Ramachandran and Thakur, 1974). Others blame government agencies and organizations. U.S. residents held the Federal Emergency Management Agency responsible for acting slowly to mitigate damage before Hurricane Hugo even though its role was to provide support after the event (Lacayo, 1989).

It is clear that individuals make decisions based on the perceived environment rather than the real world. In an extreme case, Harry Truman refused to move from his house near Spirit Lake on Mount St. Helens in 1980. As he explained to journalists, because the mountain had been stable all his life he did not believe it was about to erupt. His perception persisted in spite of physical evidence to the contrary, including small tremblers, a bulging mountainside, and the venting of gas. Exhibiting dissonance, Truman perceived his environment as safe and refused to relocate. He was, of course, killed when the volcano did erupt on May 18. Another way of viewing this case is that Harry Truman was forced into the situation by an ever-inquisitive media. They interviewed him fairly regularly before the real evidence became apparent and, having made many public statements, he would have been forced to back down from them. His stubbornness and bravado may have prevented him from doing so.

Attitudinal Factors

Attitudes probably represent a compendium of all the factors already discussed. They are both a product of some of the other variables and a significant contributor to response and action. Thus, attitudinal factors cannot be considered without incorporating psychological and situational factors.

Cognitive behavior theory assumes that reactions are perceived and interpreted based on an individual's own experience. Any event can provoke a great variety of outcomes because each individual has unique experiences on which to base decision making. Tuan (1974) argues further that the human mind thinks in terms of opposites (such as life and death or light and dark) that permeate individual behavior. This combination of situational and cognitive/attitudinal factors has been examined by de Man and Simpson-Housley (1988). In a study of earthquakes and floods, they concluded that (1) length of residence best measured perception of future earthquakes; (2) perception of reliable official support best measured potential damage; (3) individual trait anxiety best measured emotional reaction to earthquake prediction; and (4) damage expectation best measured the feeling of anxiety versus calmness after hearing flood forecasts. A mixed bag, admittedly, but these findings illuminate some of the complexities of victims' perception and behavior.

This set of interactions can be described best through examples. In a study of behavior related to the flood hazard, cognition was analyzed separately and in combination with situational factors as they influenced evacuation behavior (Frazier, Harvey, and Montz, 1983). These cognitions were defined by a number of beliefs floodplain residents held, including attitudes toward flooding as a threat to life and property, the flood hazard as a stress, and the belief that they had adjusted to the threat. Research disclosed that those who did not evacuate but remained "on watch" were not stressed by the flooding problem. They tended to be people who had experienced previous financial loss from flooding (that is, who had personal experience) and they tended to have larger family size (a situational characteristic). Clearly, belief in control over one's fate affects those attitudes, but it was not evaluated in the study.

Anxiety traits of individuals also can affect perception. High hazard anxiety may translate into greater preparation (de Man and Simpson-Housley, 1988). In a study of Chilean volcanic and earthquake areas, Larrain and Simpson-Housley (1990) found a correlation between high anxiety and perceived greater damages. Yet, that was not the only relationship at work; it was the combination of variables that made this study of attitudes useful, particularly because separate analysis of the contributing variables (such as experience or family size) would have missed their important combined effects.

It also must be remembered that attitudes can be influenced by information provision, though not always in the desired direction (as seen earlier in our discussion of warnings). Thus, attitudes cannot be looked at as separate factors, nor can they be ignored. More important, it may be easier to take account of attitudes than personality factors.

SUMMARY

As we have shown, decision making is tied inextricably to perception and is constrained by situational and cognitive variables. Nevertheless, it appears that decision making follows four major steps (Burton et al., 1993). First, individuals appraise the probability and magnitude of the hazard occurring; second, they consider a range of possible actions (remembering that these are limited by knowledge and experience); third, they evaluate the consequences of any selected actions; and fourth, a choice of one or a combination of actions is made. Burton et al. (1993) argue that these are dealt with in order and not simultaneously, particularly as different options are evaluated; an ordered choice process is supported empirically by Laska (1990).

In the end, we are left with a whole range of variables that affect human behavior at the individual level. Their modeling has proved somewhat elusive in that particular variables can sometimes have different effects under different situations, but despite such complexity, there are models and theoretical frameworks that can guide the hazards researcher. The division into situational and cognitive sets is well founded. What researchers must pursue now in greater detail is explanatory models for behavior at the individual level.

4 Behavioral Studies

Community Attitudes
and Adjustment

INTRODUCTION

Community attitudes and adjustments to natural hazards are not simple aggregations of individuals' perceptions and actions. It has long been recognized that individual perceptions do not translate readily into group and community attitudes (Burton, Kates, and White, 1993; Mitchell, 1984), but our understanding of the dynamics of collective perception, behavior, and decision making lags behind our knowledge of individual perception. Attitudes encourage or discourage the adoption of hazard mitigation strategies and are particularly important at the community level where there is a wider range of adjustments to be made. Costs are allocated, economies of scale can be realized, and the benefits of loss reduction accrue to everyone in the community, directly or indirectly, regardless whether individuals have acted to protect themselves personally.

Because community attitudes and collective adjustments are more complex than individual perception, greater variation should be expected (Burton et al., 1993). Many studies that purport to assess community attitudes do so by compiling individual responses to questionnaires in order to describe prevailing attitudes. For example, White (1974) coordinated a comprehensive, cross-cultural survey of hazard victims. Although these studies assessed the attitudes and responses of particular groups (such as flood victims), they did not necessarily describe communitywide attitudes. For example, the studies did not include groups that were not affected directly by an event (such as those who suffered financially from lost production). In fact, most studies have concentrated on understanding victims' perceptions and actions rather than collective perception, attitude, or response.

Here, community is defined as an organizational entity with an established means of collective decision making, though not necessarily having recognized geographic boundaries. It has, or should have, a different view of the probability of a hazardous event than would individuals, who may reasonably expect little chance of occurrence during their own tenure at a given place. For a community, occurrence of disasters is virtually inevitable (Burton et al., 1993), which poses different problems in appraising the significance of natural hazards.

Community decision-making processes yield a different understanding of the relationships between attitude and response. Identifying community attitudes requires exploring internal and external group dynamics, the vested interests of disaster subcultures, economic concerns, social constraints and values, and hazard characteristics, as they operate and change over time.

CHOICES FOR COMMUNITIES

Any community is presented with a number of natural hazards against which it can take some mitigative action. Although no action will be taken unless a hazard is recognized as a threat, there is no simple path from recognition of the threat to action or to implementation of mitigation measures. We might assume or wish that a community would recognize a threat, evaluate probable losses and their spatial and social variation across the community, consider alternative measures, and implement the most efficacious strategy, but that is not how it usually works. First, something must set the process in motion. It may be an event in the community that spurs action; for example, following a devastating tornado in 1982, decision makers in Paris, Texas, revised the city's emergency response capabilities, enhanced the warning system, and instituted a public awareness campaign (Rubin, Saperstein, and Barbee, 1985). It may be an event elsewhere that brings about action; after the 1976 Big Thompson Canyon flash flood alerted officials in Boulder, Colorado, to their city's vulnerability to a similar event, they adopted a comprehensive flash flood warning system (Montz and Gruntfest, 1986). Pressure for community action may come from external forces in the form of government directives and land use controls; in the United States, the National Flood Insurance Program requires participating communities to adhere to a zoning policy based on the 100-year floodplain. Or, local decision makers may simply recognize that their community is vulnerable and action should be taken.

Alternatively, the occurrence of an event, either locally or elsewhere, may have little impact. Political, social, or economic characteristics of the community or outside influences may work against the adoption of mitigation measures. For example, outside provision of relief after an event may ameliorate some of its impacts, effectively removing incentives for community action. Some external influences discourage local autonomy or decision making, such as court rulings against some land use restrictions, while others fail to support it by allocating inadequate funds to shift infrastructure and development out of harm's way. Finally, local decision makers are saddled with a number of problems, of which natural hazards represent just one set (and a probabilistic one at that); other problems and issues may take priority.

However, communities do have choices. They can choose to pursue or not to pursue mitigation. If they choose to take a course of action, then they must make other choices as well. For example, they may prefer measures directed toward warning and evacuation, those designed to minimize loss of life or property when an event occurs, those designed to control the event itself, or some combination of measures. Of course, a community's range of choices is constrained by the nature of the hazards to which it is subject and is either constrained or facilitated by characteristics of the community itself. In addition, timing is critical. If it is an event that sparks the choice process, there is a window of opportunity during which choices are made more easily, but if that window is passed, the hazard may well lose priority to other issues.

COMMUNITY ATTITUDES: A COMPLEX OF VARIABLES

It is difficult to generalize about communities because of the different social, economic, political, and cultural contexts that, separately and in combination, provide both opportunities and constraints for response and adjustment. However, it is possible to detect some common threads in attitude and response in different communities' experiences.

Social theory can help us to understand some of these commonalities. A community represents a functional entity that serves to satisfy needs while providing a cultural milieu for social integration and participation (Wenger, 1978). It exercises "control" over the members of a society through norms that permit that society to function as a system. A community consists of many individuals and groups with interests and concerns that vary from person to person, from group to group, and over time. (Documenting and measuring those differences and changes is a difficult task in itself even under "normal" conditions, let alone following disasters.) Community or collective issues arise and are discussed, conflicts develop that may or may not be resolved, and decisions are made that may generate additional issues and conflicts. At any given time, individuals and groups may or may not have a personal interest or stake in particular issues.

Further, the interrelationships among individuals, groups, and community decision makers change over time, in part as a function of the issues at hand. Communities also must respond to emergencies and crises that surpass the immediate concerns of individuals. Thus, a community is a dynamic system that "is constantly in the process of reallocating and reintegrating" (Thompson and Hawkes, 1962, p. 271).

Given an abrupt change to this functional system in the form of a natural disaster, a community responds by setting new social priorities, reallocating resources, and possibly establishing new social organizations. However, such responses often are a transient aspect of various postdisaster phases. Problems and issues evolve and roles change as emphasis moves from emergency response to rehabilitation, reconstruction, and mitigation. Resource availability also changes. This temporal dimension of disasters engenders diverse community attitudes at different points in time, and should be considered when analyzing community responses. Unfortunately, for hazard mitigation to be most effective (that is, to save the most lives and property), it must be implemented in the pre-event stage when awareness and salience of a hazard are low; after an event, community awareness is high, but it is too late to mitigate most impacts of the event.

Community Values, Norms, and Beliefs

Embedded within community systems are sets of cultural values and norms through which society functions (Wenger, 1978). Norms govern attitudes toward such issues as law and order, nationalism, patriotism, life and property, materialism, safety, justice, and humanitarianism, which constitute the collective views of a community. However, not all groups hold exactly the same views or adhere to the same critical values, and divergent community norms often lead to internal conflict. (One has only to consider such issues as smoking, gun control, and abortion in the United States during the 1980s and 1990s to see how divisive such norms can be.) In every community, there is a mix of groups with strong economic, political, cultural, religious, or other concerns. An individual may be a member of several such groups at the same time and hold various attitudes toward critical values. (Likewise, groups can have very different interests, that may be competing or mutually supportive.) Discriminating among these viewpoints is one challenge in understanding how communities respond to natural hazards.

Together, groups and organizations within communities help define critical values and norms. For instance, it may be agreed by a majority that mitigating the threat of hazards is a worthwhile task to which the community should direct itself. However, conflict may increase through discussion about how this should be achieved (Rowntree, 1974). For example, groups with economic interests (such as realtors and bankers) probably would be interested in quick recovery from a disaster and support adjustments to avoid its recurrence, but might prove unenthusiastic about mitigation measures that restricted economic activity such as land use controls, zoning, or disclosure rules (Montz, 1992; Palm, 1981; Rubin, Saperstein, and Barbee, 1985). If the decision makers or gatekeepers do not live in high-risk areas, they may not be interested in sharing the costs of alleviating a hazard that affects the poor or a politically marginal population. The disparity between awareness of a hazard and exposure to risk can translate into either action or inaction by a community. Thus, it is imperative to recognize the existence and degree of influence of various groups, and to understand their particular interests and interactions with each other and with community decision makers.

Belief systems held by a community also influence how it responds to natural hazards. If disasters are regarded as divine acts (perhaps preordained or punishing a sinful population), the community may respond somewhat passively. More active responses may be expected in communities where natural hazards are perceived as geophysical phenomena that can be controlled or modified. In fact, there are many belief systems between these two extremes, and all will influence behavior. In eastern Kenya, communities that suffered periodic drought gradually developed more active responses through involvement in local and community projects that had not existed under the previous, more centralized colonial system (Wisner and Mbithi, 1974).

Size and Interorganizational Relationships

Interrelationships among groups and organizations are critical to the smooth functioning of a community. They may take many forms, both internal to the community and external to other organizations and communities. Some subsystems are founded on relationships entirely within a community; for example, community businesses have links to local suppliers, financiers, and organizations that together constitute an economic unit with local significance. Beyond the local community, external links add to the complexity of community systems (Figure 4.1); these might be horizontal relationships of professional, business or social linkages to similar organizations or groups, or vertical relationships through a hierarchical system such as links to business headquarters or between regional plants (Wenger, 1978). Such interorganizational relationships vastly extend the sphere of influence of a community as well as the level of control that can be exerted over it. They also determine the magnitude and significance of the extracommunity impact of a natural disaster.

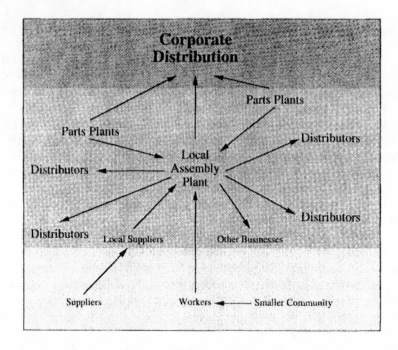

FIGURE 4.1. Geographic relationship in industry. Because of horizontal and vertical integration, a local assembly plant receives inputs from smaller communities (bottom layer) and its own community (middle layer) and produces outputs to all levels, including the corporate hierarchy (top layer). A disastrous event at any one layer can have far-reaching effects.

As the size of a community increases, there is a tendency for local relationships to diminish in importance as social and economic complexity increase. Scale may be a significant factor influencing community responses. Larger communities have a larger resource base compared to smaller units, but demands on those resources are many and varied. The distribution of impacts of a natural event is usually uneven in all but the most homogeneous communities, and that unevenness tends to increase with size and diversity of a community. In the Whittier Narrows earthquake of 1987, for example, damage in Los Angeles was proportionally higher in those neighborhoods having a greater number of ethnic minorities (Bolin, 1994; Rubin and Palm, 1987). (This pattern can also be understood as consistent with the socioeconomic marginalization theories introduced in Chapter 1.)

Depending on hierarchical organizational structures, we can expect the locus of power to change for particular organizations. If alternative locations exist, a large commercial enterprise may demonstrate little concern for the interests of a local population following a disaster, and may simply relocate to a "safer" area. For example, when a mall in Linda, California, was flooded in 1986, most of the stores chose not to reopen at that site, including the anchor stores needed to ensure the viability of a commercial center; ten years after the flood, the structure stands virtually empty. In contrast, businesses that benefit from close ties with a community may make considerable efforts to support disaster victims. Predicting how organizations will respond requires more in-depth research than currently exists and a greater understanding of the organizational structure and local commitment of commercial enterprises.

Increased community size also may exacerbate problems during times of emergency or crisis because there is no single entity in control. Smaller communities generally have closer internal ties, hence can be rapidly organized for action. In pluralistic societies, where power generally is not concentrated in a particular individual or agency, disaster can result in a period of confusion as new organizational structures come to the fore. This problem may be balanced by the broader resource base available to large communities. Thus, the scale of a community can affect response in varied ways, depending on a number of factors.

Influences on Decision Making and Decision Makers

Community decision makers do not make decisions in a vacuum, nor are individuals and groups within a community unaware of outside influences, especially restrictions or financial assistance emanating from state and national governments. Restrictions include state, provincial, and central government mandates or regulations, such as those regulating land use in hazardous areas or issuing standards for building materials and design (see Chapter 5 for details). The availability of financial aid also influences community attitudes toward a hazard and adjustments to it. The relief provided by national and international governments and organizations allows a community to get through the early days of an event, assisting it in getting organized so that the long-term recovery process can begin. However, the knowledge that relief is forthcoming also can foster community inaction. That relief is necessary is not at issue; where and how it conditions a do nothing attitude is. White (1974) noted that opinion differs on the extent to which this occurs. Almost two decades later, Burby et al. (1991) found that among communities they studied in the United States, those that received federal disaster relief following flooding tended to be more prepared for subsequent events than flooded communities that had not received relief, which they attributed to the conditions attached to relief. We do not yet know whether unconditional relief leads to what Burby et al. (1991) term "inefficient behavior."

Some adjustments to hazards (floods in particular) are very capital-intensive, and national governments frequently have assisted in financing such projects. In the United States flood-control works were almost entirely funded by the federal government until the 1960s, when more stringent cost-sharing regulations were implemented. By the 1990s, at least five federal agencies were involved in the funding, design, construction, and repair of levees, each with its own cost-sharing formula applied to specific phases of projects; their involvement resulted in some 25,000 miles of levees, flood walls, embankments, and dikes, and had a major impact on land use on floodplains (Tobin, 1995). In New Zealand, before 1987, the federal/local cost-sharing ratio was 3:1, which encouraged heavy reliance on flood-control works and emphasized a federal role in "adjusting" to the hazard; a community's attitude (and a very cost-effective one for a local area) was typically one of inaction, except as required to initiate federal interest in the problem. More recently, flood control has come to be viewed as a local government function, although the effects of this change remain to be seen. Similar examples would include federal or other extracommunity funding of irrigation projects to prevent drought-related crop losses.

Economics plays a significant role in community attitudes toward hazards and adjustments, but measuring and monitoring its effects can be a difficult undertaking indeed. Its influence may be as obvious as federal funding of a project or as subtle as local unwillingness to commit money to what is seen to be an improbable event. A community also may believe that external money (particularly in the form of disaster relief) will be made available if and when needed.

Finally, in delineating the problematic areas surrounding our understanding of community attitudes and decision making, it is important to recognize the influence of the event itself. Physical characteristics of a hazard elicit different perceptions, leading to more or less active responses. We know, for example, that frequency has a strong influence on both individuals and communities; communities with hazard experience have been found to respond more effectively (Mileti, Drabek, and Haas, 1975; Wenger, 1978). Other characteristics also influence community attitudes. For instance, if we adopt the pervasive–intensive continuum (Burton et al., 1993), we can posit that a pervasive hazard elicits different attitudes and evokes different responses than does an intensive hazard. With a pervasive hazard (such as a drought), a community can respond even as the hazard continues, which is not the case with intensive hazards (such as earthquakes, tornadoes, or flash floods) where adjustments must precede the event or the community will find itself in a crisis rather than an emergency (Wenger, 1978).

Horizontal and Vertical Integration

As discussed above, the extent to which communities are integrated horizontally and vertically affects both the level of impact of an event and the ability of a community to recover. Low horizontal integration reflects "a weakly knit social network" (Berke, Kartez, and Wenger, 1993, p. 100) and affects the ability of the community to converge on the problem at hand. Vertical integration means that a community has ties with other levels of government and with other institutions (see Figure 4.1). In general, the greater the vertical integration, the more readily resources can be directed to the community when needed.

Communities experience different levels of horizontal and vertical integration, resulting from historical, social, and economic factors that many times cannot be neatly defined or determined. As developed by Berke, Kartez, and Wenger (1993), the potential relationships between horizontal and vertical integration can be shown graphically (Figure 4.2). A Type 1 community exhibits the greatest potential for effective disaster recovery because it has "well-developed ties to external resources and programs" and "a viable horizontal network that will allow it to exert power and influence in the recovery process" (Berke et al., 1993, p. 101). At the other extreme, a Type 4 community has the least potential for recovery as it has difficulty marshaling either internal or external resources. Examples of each can be found throughout the world. Berke et al. (1993) cite Montserrat, West Indies as a Type 3 community that changed to a Type 1

Vertical / Horizontal	Strong	Weak
Strong	Type 1	Type 2
Weak	Type 3	Type 4

FIGURE 4.2. Types of communities. The degree of horizontal and vertical integration can serve to identify more and less resilient communities. In general and all other things being equal, the greater the linkages (especially vertical), the greater the potential for recovery from disaster. *Source:* Berke, Kartez, and Wenger (1993). Copyright 1993 by Basil Blackwell, Ltd. Reprinted by permission.

following Hurricane Hugo. International nongovernmental organizations (NGOs) worked with a local NGO to train local citizens to rebuild structures and implement local development projects (such as in agriculture); this work strengthened horizontal integration through local empowerment and strengthened preexisting vertical ties because of increased understanding of institutional capacity. By contrast, Type 4 communities represent those marginalized within the larger political system, exerting little or no influence over the larger societal network. Saragosa, Texas, is one example. Prior to a devastating tornado in 1987, Saragosa was a small, unincorporated community of Mexican-Americans; following the tornado, disaster recovery activities, which were administered from the county seat more than 20 miles away, served to aggravate the lack of horizontal and vertical integration. This result was evidenced in citizens' general belief that they were worse off two years after the disaster because they had had no say in the recovery process.

Several points are important with regard to levels of integration. First, whatever "type" a community is, is partly an artifact of historical processes. Thus, it helps to define the context in which a community is situated. While linkages or lack of linkages can change during the immediate postdisaster period, there is a tendency for communities to revert to their pre-disaster "type" when the emergency has passed (Wenger, 1978). Any effort to alter the range of response and mitigation alternatives available to a community probably should include alteration of its linkages. Second, size of a community affects both horizontal and vertical linkages. As size increases, horizontal integration tends to decrease unless there is a well-developed organizational framework through which local organizations can interact. The need for comprehensive, pre-disaster planning is critical, including consideration of local capacity to recover from a disastrous event.

TEMPORAL PATTERNS IN COMMUNITY RESPONSE

As discussed in Chapter 1, recovery from disasters follows fairly distinct temporal patterns (see Figure 1.3). An initial emergency phase corresponds to the period immediately after impact; the community focus is on search and rescue operations and the supply of emergency relief. This is followed by a restoration and rehabilitation period when the community returns to or approaches "normal" functioning (with the caveat that communities are dynamic systems that will never replicate preexisting conditions completely). In turn, this phase gives way to reconstruction, involving both short-term rebuilding and long-term development (Alexander, 1993; Burton et al., 1993; White and Haas, 1975). These phases last for varying times that may be overlapping for individual groups and organizations. Further, restoration may represent a "honeymoon" period when community activities are focused primarily on social and humanitarian goals, only to be replaced by a period of disillusionment as the reality of long-term consequences of the disaster sets in (Raphael, 1988). Other phases, dependent on victims' psychological well-being, may be defined by different periods of stress (Taylor, 1989). In addition, there may be a pre-disaster phase in communities warned of impending disasters; warnings permit time for remedial activities, but also may increase stress as the onset of an event approaches.

Immediate Postdisaster Response and Recovery

The hazards literature on community attitudes and responses is weighted heavily toward the immediate postimpact period, with a significant proportion of the contributions coming from sociologists. While this work has been criticized as having limited value for our understanding of community responses and adjustments at other times, including before and in the long term following an event (Mitchell, 1984), it does give us a good sense of the development of community attitudes in relation to natural events.

There are three major theses on the immediate postimpact period: (1) there is decreased conflict in the community as all members focus their energies and efforts on the problem at hand; (2) there is a convergence of social values centering on cleanup and recovery, that overrides pre-disaster differences; and (3) there is a new, but temporary organizational structure that emerges in the community, characterized by a high level of consensus on community priorities. These changes facilitate local recovery, but begin to fade as the community returns to pre-disaster conditions. Priorities also change during the pre-disaster period; social participation and individual goals are replaced by the role of good citizens and community service as volunteers help victims and neighbors help neighbors. For example, during the summer of 1993 in the upper Midwest of the United States, thousands of volunteers converged on flood-prone communities to help shore up levee systems, filling and distributing over 26.5 million sandbags (Tobin and Montz, 1994b); in a typical response pattern, their help continued during the emergency phase, but gradually declined as the flood receded. Some researchers suggest that crime and other antisocial behavior are reduced during the immediate postimpact period, and it does appear that claims of widespread looting often are exaggerated by the media looking for another angle on disaster (Wenger, 1978). This is not to say that crime or looting does not exist, but it may be that minor crimes go unpunished at these times as society deals with the more pressing crisis (Bryant, 1991).

During the immediate postimpact phase, a different structure emerges that exists for the express purpose of providing relief. In essence, a natural disaster creates a problem for a community around which critical values converge, maximizing community consensus. Following a disastrous event, the usually divergent values of a community become secondary. Individual and group interests tend to disappear or are subsumed to the community good (Wenger, 1978). A community's priorities become singularly focused on the protection of life and property, often at the expense of other norms; for instance, freedom may be curtailed as large areas are cordoned off for reconstruction. At times, this consensus may affect even external relationships within the vertical and hierarchical structures; thus, outsiders and other levels of government may be inculcated into the hazard experience (Quarantelli and Dynes, 1976). By strengthening organizational ties, disasters may create communities rather than destroy them (Wenger, 1978).

With the convergence of values comes decreased community conflict, at least compared to the pre-disaster situation. The extent and longevity of this convergence vary from community to community, depending on several factors. Quarantelli and Dynes (1976, pp. 141–143) associate seven factors with the absence of community conflict following an event:

1. "Natural disasters involve an external threat."
2. "In almost all natural disasters, the disaster agent can generally be perceived and specified."
3. "There is high consensus on priorities in natural disaster situations."
4. "Natural disasters almost by definition create community-wide problems that need to be quickly solved."
5. "Disasters lead to a focusing of attention on the present."
6. "There is a leveling of social distinctions in natural disaster situations."
7. "Disasters strengthen community identification."

Communities face a particular set of problems during disasters which they must develop the means to address. Typically, a new organizational structure of centralized decision making emerges as prior organizational relationships dissipate (Dynes and Quarantelli, 1972; Wenger, 1978). The emergent structure, sometimes termed a "synthetic community," is temporary, disappearing as the need for it becomes less apparent or as internal conflict develops. This usually occurs as the process of recovery moves toward a long-term focus; once the focus on the present has passed and attention turns to long-range reconstruction, individual and group interests reemerge. The postimpact structure serves an important purpose in directing energies and resources to immediate tasks and, once those needs pass, so does the emergent, synthetic community.

The preceding may suggest that every community experiences the same convergence of values, minimization of conflict, and emergent structure. Indeed, that is not the case, nor do all communities experience similar magnitudes or intensities of change for the same amount of time. Disasters create a set of new problems for communities. Some successfully rally around these problems, but some conflict is inherent in the process. Interorganizational conflict can occur horizontally in the community or vertically with other levels of government or authorities; in the former case, difficulties in meeting the immediate needs of different sectors of the population create problems. For example, following the Loma Prieta, California, earthquake in 1989, response to needs for temporary housing and relocation was complicated by conflicts that, in large part, reflected preexisting social inequalities (Bolin and Stanford, 1990). There are numerous examples of conflict between levels of government, based partly on delayed assistance (or what is perceived to be delayed assistance). Similarly, a lack of adequate coordination among individuals or organizations may contribute to the breakdown of the immediate postdisaster structure (Barton, 1969).

THE IMPORTANT COMPONENTS OF RECOVERY

Experience

Direct experience with natural hazards has an important bearing on how communities respond to them. Indeed, an extensive literature suggests that the effectiveness of mitigation strategies varies among communities in part because they have different experiences (Mileti et al., 1975) with repercussions throughout the community structure. Communities with prior disaster experience generally fare better in the recovery process. First, they tend to develop disaster plans and organizational arrangements for community response (Mileti et al., 1975; Wenger, 1978). These communities have acted already on their collective recognition of a hazard to implement locally initiated projects. However, experience alone does not generate appropriate responses in a community since, as we have seen with individuals, not all hazardous events bring about mitigative responses (Burton et al., 1993); the relationship is more complex, and other important factors will be considered later in the text.

Second, although geophysical attributes alone cannot provide a guide to community response, certain conditions may encourage particular forms of response and affect types of experience. Hazards that are predictable, perceived as controllable, and have definite onset times may elicit specific activities by communities (Wenger, 1978). However, predicting precise behavior is not possible because of the many other social consequences of disasters.

For example, take the degree of disaster. The more extreme the geophysical event, the greater the social consequences of disaster and the higher the probability of crisis; a less intense event might constitute only an emergency situation with which society could cope. A disaster must be defined by impact (that is, intensity and scope). A severe or intense geophysical event may disrupt many traditional social structures and have repercussions at higher levels of society, government, and the economy. Under extreme conditions of catastrophic proportions, social infrastructure could be temporarily, if not permanently damaged; for instance, hospitals, police, and other services could be obliterated within seconds in an earthquake, significantly reducing the ability to respond. In the 1995 earthquake in Neftegorsk, Russia, damage was so severe that the town was not rebuilt and, as refugees sought other accommodation, impacts were felt regionally and nationally. By contrast, a small event may have limited scope, perhaps affecting primarily family units and neighborhoods. The midwestern floods in the United States in 1993 had both levels of impacts, and in some cases, community response to the widespread destruction was a decision to relocate. For example, the community of Valmeyer, Illinois, which suffered extensive damage, decided to move completely off the floodplain; in other communities, flood problems have been replaced already by other challenges as structures have been rebuilt and life has returned to normal.

Thus, there is a question about the relationships among size of community, extent of disaster experience, and coping or response strategies. Very often, small communities have closer internal ties and can deal effectively and immediately with disaster needs, but they usually have fewer resources with which to launch effective countermeasures against future disasters. With a population of 1,000, Valmeyer must rely heavily on federal relocation assistance.

A third characteristic of disaster experience is the development of a subculture with its own set of norms and beliefs pertaining to a hazard (Moore, Bates, Layman, and Parenton, 1963). In order for such groups to emerge, the community must experience repetitive disasters by the same geophysical agent, some degree of warning must be possible to allow time for remedial and other action, and some damage must have been experienced (Wenger, 1978). Subcultures often deal with disasters on a routine basis as emergencies rather than crises (as would be seen elsewhere). That is, experience is translated into action for what might be termed "the expected." The behavior of these groups may be most effective against a particular type of threat and under similar intensity to that which occurred in the past. In such cases, the values and traditional behavior patterns of the subculture may be useful in mitigating losses. Some residents of the Trinity River Valley in Texas have adapted to the frequent floods that inundate subdivisions along the river to such an extent that they find them exciting.

Unfortunately, the subculture also may be overspecialized so that variation from the expected intensity or type of geophysical event can lead to inappropriate behavior. For instance, in Carlisle, England, flood victims who had employed sandbags successfully during a past flood continued to rely on them, but upon closer examination, it turned out that they had experienced only minimal flood depths; in a flood above those levels, merely sandbagging would undoubtedly lead to large losses. Indeed, the disaster subculture can prove deadly; behavior that might be reasonable during a minor storm, such as throwing a hurricane party, can be completely inappropriate during a major event, as was the case when Hurricane Camille struck Pass Christian, Mississippi (see Chapter 3 for details). Where the disaster subculture is unprepared for either the magnitude or the type of event, it may be replaced by an emergent structure (Bardo, 1978). In other cases, more extreme conditions may be required to generate appropriate responses. For example, there have been many occasions when subcultures have denied a threat because of past experiences (as shown in Chapter 3). Experience is not a sufficient condition for the development of an appropriate and efficacious community response to an event, though it remains unknown whether it is a necessary condition.

Leadership

One theme that comes through many case studies of community recovery is the importance of effective leadership (D'Souza, 1982; Kartez, 1984; Quarantelli and Dynes, 1976; Rubin et al., 1985). Evacuation plans and other large-scale activities (such as sandbagging) often require centralized coordination to be truly effective; it is not unusual to see the army, police, or other strong enforcement units brought in to disaster areas to maintain control. This is not to suggest that strong authority with centralized decision-making power is the only answer; indeed, it is hardly likely to be effective throughout the recovery period. In most societies, power is vested in many groups, organizations, and individuals and it is frequently associated with wealth and social standing (Wenger, 1978), but these attributes have little bearing during disasters when there tends to be a concentration of power among only a few leaders whose authority invariably derives from demonstrating expertise or controlling aid. It is well recognized that the "new," emergent leadership does not come from a particular position and need not be centered on one person (Dynes and Quarantelli, 1972; Rubin et al., 1985). Instead, it may be earned through action and shared among several individuals with different responsibilities. In fact, many leaders can emerge, usually temporarily. Although examples are drawn largely from U.S. experiences and case studies, they are applicable to other communities in other parts of the world (Clifford, 1956; Dynes, 1975).

Leadership alone is not sufficient to meet the tasks facing a community. As Rubin et al. (1985) point out, it must be combined with the administrative and technical resources to act and with knowledge of what to do and how, when, and where to get outside assistance (which supports the need for a degree of local control over recovery). After an event, some resources can be obtained from outside the community, but delays adversely affect recovery. In an analysis of disaster response networks in six natural disasters and one search and rescue activity in the United States, Drabek (1983) noted the importance of local control by local managers, even when outside resources were needed. Indeed, whether organizations can respond depends on their knowledge of what has happened as well as a full understanding of what is needed (Raphael, 1986), knowledge that is largely locally based. For example, the distribution of some forms of flood relief may utilize local resources, as was the case in 1992 in Liberty County, Texas, where a local resident delivered hot meals to those stranded in flooded homes. However, this can be complicated by the involvement of external hierarchical structures can either promote or inhibit relief.

Interorganizational and Intergovernmental Coordination

Although a postimpact community experiences decreased internal conflict as recovery gets under way, there may be increased conflict with outsiders (Quarantelli and Dynes, 1976). External ties are weakened during the emergency phase and the community may even become completely isolated for a short time; effectively, it turns inward, strengthening internal relationships (Wenger, 1978). The local view of what can and should occur takes precedence and is quite likely to differ, perhaps significantly, from the views of other levels of government (Rubin et al., 1985). Since these levels of government (particularly national) usually control the allocation of relief and rehabilitation funds, different views become very important and can lead to serious conflicts that impede recovery. However, this pattern is gradually reversed as the original hierarchical structure is reimposed. Central headquarters or national centers take over relief operations and local activities become subordinated to larger units. Because local entities do not always have the financial, technological, or material resources to deal with major crises, they have little choice but to rely on outside support (Wenger, 1978).

An example of uncoordinated response can be found following the 1993 midwestern floods in the United States. Given the number of levee failures, a report issued by the Office of Management and Budget called for careful review of any proposals to reconstruct them (Myers and White, 1993). Political reaction was adverse, to say the least (Koenig, 1993), and reconstruction of many levees proceeded rather quickly. Thus, local community interests prevailed in many cases over what might be considered broader national policies.

Without predisaster planning, lack of coordination among organizational structures can compromise postdisaster recovery, but even the best-laid plans can fall apart in a disaster emergency, especially if the community infrastructure is destroyed. While communities tend to expect immediate assistance, usually from other levels of government, they also reject anything that might be seen as interference in their activities. This may be particularly true as the lag time between the event and the appearance of organizations increases (Gillespie and Perry, 1976). The numerous examples of insufficient, inappropriate, or excessive supplies being sent to a disaster area illustrate the consequences of a lack of planning and coordination.

Bardo (1978) suggested that organizations may be changed or modified to cope with crisis conditions. For example, new organizations might be created specifically to deal with disasters, such as food distribution or search and rescue teams. Existing organizations, like the Red Cross or Red Crescent Societies, also may be expanded to meet new demands, while others might be adapted to respond to the specific emergency. For instance, local councils could form relief committees and set up temporary housing and food centers for victims. Finally, some organizations may be temporarily suspended so as not to interfere with postdisaster activities.

Whether real or perceived, a lack of coordination probably is not found in all communities. Where it is found, it is not necessarily to the same degree. As societal organizations and their relationships become more complex, there is increasing probability of lack of coordination and, ultimately, conflict (Dynes, 1975). For instance, the experiences following several earthquakes in California suggest that pre-disaster social trends are intensified by natural disasters. There, conflicts between marginalized groups and government agencies became more pronounced (Bolin and Stanford, 1991); specifically, the 1983 Coalinga earthquake, 1987 Whittier Narrows earthquake, and 1989 Loma Prieta earthquake all aggravated preexisting social conflicts among groups within the affected areas, albeit to different degrees. In addition, the magnitude and significance of national involvement in disaster recovery differs from place to place, which affects interorganizational relationships (Quarantelli and Dynes, 1976); in California, coordination of emergency federal housing, Red Cross shelters, and housing reconstruction was made possible only through coalition-building that incorporated local citizen groups.

Preparation for the Immediate Postimpact Period

Preparedness is key, to communities' ability to respond to natural disasters. In the short term, experience is important because it provides knowledge of what is needed, by whom, and when. Yet, communities do not prepare for events, which would involve ranking hazards as higher priority and organizational change to put in place appropriate structures for response at the local level.

In an attempt to deal with this issue internationally, the European Community Humanitarian Office has developed a program to promote community-based preparedness structures. It encourages locally appropriate and locally based mechanisms designed to strengthen local self-reliance. For example, work in Bangladesh includes setting up radio-based early warning systems, training for relief management, and establishing facilities for storage of emergency supplies. All activities are directed to the most vulnerable people in the population. In Bangladesh, they include the landless; in other places, the focus is on women and children. No matter where the hazard, the program's emphasis is on building local capacity.

Summary

In this section, we have presented the development and consequences of community attitudes in the immediate postimpact period. Although the literature that served as the basis for this section draws on a variety of hazards, most would be termed intensive; it suggests that communities tend to react and respond in similar fashion to any type of intensive natural disaster. Different responses would occur with pervasive hazards. In addition, we would expect differences in reaction, response, and attitudes toward technological disasters, in large part because blame can be allocated and energies directed in a different direction.

Although there is a preponderance of U.S. case studies, few differences in community response have been noted in non-Western studies. Those differences that have been seen among communities in different cultural contexts tend to be differences in degree rather than in fact. That is, variations among societies may create different relationships, attitudes, and organizational structures; in some countries, kin relationships may supersede community ties while in others, the relative roles of national, regional, and local governments change and promote different priorities during disasters. These differences are manifested in contrasting attitudes, particularly in the form of different patterns of response and adjustment.

Finally, while a number of complicating factors have been elaborated in this section, that of time has been ignored because of our emphasis on the immediate postimpact period. However, the element of time adds significantly to the complications. The emergent community attitude and response do not last as the recovery period moves into rehabilitation and beyond; as time passes, old conflicts emerge and new ones may develop. The immediate postdisaster situation does not persist when the community returns to "normal."

COMMUNITY ATTITUDES AND RESPONSES AFTER THE CLEANUP

The emergent structure lasts until the initial tasks are complete, a time that is marked by what has been called "white shirt day" (Quarantelli and Dynes, 1976). This is the time when, to the extent possible, people return to their pre-disaster routines. Internal conflict begins to reemerge, sometimes similar to the conflict that predated the disaster; the convergence of values also begins to end as individuals and groups grow concerned again with their own interests. The disaster is not forgotten; in fact, there may well remain visible indications of it as some structures may not be repaired yet and some members of the community may not be back in their homes. However, with the immediate tasks of search, rescue, and cleanup completed, attention can turn to "getting back to normal."

It is at this point, too, that communities begin to consider avoiding a recurrence of the crisis through planning, adjustments, or perhaps deciding collectively to do nothing. It is more difficult to sustain a singular attitude toward the hazard and much more difficult to obtain consensus on priorities. Thus, understanding community attitudes is probably more important after the initial recovery than before.

Return to Normalcy

There has been much debate among hazards researchers about the long-term effects of disasters. The central question is: Do communities, in fact, return to normal, that is, to their pre-disaster situation? The results of numerous research projects and case studies are quite disparate, with some researchers remarking on the extraordinary resilience of communities as compared to individuals (Wright, Rossi, Wright, and Weber-Burdin, 1979), others noting economic gains made by communities (Dacy and Kunreuther, 1969), and still others suggesting a return to pre-disaster trends, whether positive or negative (Haas, Kates, and Bowden, 1977). The common thread is that for the most part, communities do indeed appear to return to normal, but "normal" may not mean that they are better off as a collective than they were before the event.

The reemergence of conflict is regarded as a sign that a community is back to normal (Quarantelli and Dynes, 1976), though the subject of conflict may have changed. In the United States, where large amounts of aid are made available by state and federal governments, conflicts arise over its use and allocation. Vested interests take the place of convergent values (though some of those interests may be of an altruistic nature). The point is that the agreements on priorities that characterized the immediate postdisaster period give way to individual and group differences on needs, goals, and vision. As a result, it becomes more difficult to gain consensus on what must, should, and can be done with regard to hazard adjustment in the community.

In the United States, a noted exception to the return to normalcy is Buffalo Creek, West Virginia. Erickson (1976) gave two reasons in his case study of its flood. First was the extent of the destruction, which he characterized as communitywide. When the community was put back together, reconstruction was not in keeping with previous patterns of housing or neighborhoods. Second, the ties to kin and neighbors that had been so important in the community before the disaster broke down after the flood. That is, its horizontal integration was lost. It also might be pointed out that this community was economically depressed when disaster struck. A similar experience was had by residents of Saragosa, Texas, who believed themselves to be much worse off two years after a tornado struck (Pereau, 1990).

Clearly, the model of the normalized community does not always apply. Just as with the immediate postimpact period, there are some factors that, over the long term, influence community attitudes, response, and adjustment for better or worse. These include different interests in the community, the role of governments, and economics.

Community Response Options

Just as individuals have choices, communities have a number of options they can exercise to lessen collective vulnerability. The range of choice is greater at the community level where there is a larger resource base and, thus, a wider range of choices (planning and policy options are detailed in Chapter 5); however, the collective choice process is more complicated because of the range of community perceptions and interests. As described by Smith (1992), there are three broad categories of response into which a wide variety of adjustments fall. At this juncture, we intend to assess only the influential factors in community response, not the adjustments themselves. Those options are discussed in detail in the Chapter 5 and, when reading that chapter, it would be useful to keep the three broad categories in mind. Neither the categories nor the specific adjustments are mutually exclusive; indeed, it is rare for only one adjustment to be implemented, as they achieve different ends.

Modify the Loss Burden

This category of adjustments is centered on spreading financial losses, based on the recognition that some losses cannot be avoided. Typical adjustments that fall into this category include hazard insurance and relief; the former requires planning on the part of the community, but the latter may not.

Hazard insurance losses in the United States are considerable. For example, insurance companies paid $18 billion in Florida and Louisiana after Hurricane Andrew, or more than four times the insurance cost of Hurricane Hugo and seven times the insured losses from the 1994 Los Angeles earthquake (Sheets, 1995). Over 725,000 claims were filed in Florida (Bailey, 1995). One insurance company recorded "losses" of $500 million for Hugo and $3.6 billion for Andrew. After Andrew, 11 insurance companies failed and 40 more considered pulling out or curtailing insurance of property in Florida (Sheets, 1995); as Sheets points out, insured property located in hurricane-prone coastal counties in the United States has risen to $1.86 trillion, or 64% above 1980 figures.

In the United States, federal policy is aimed at encouraging local governments to insure public property against loss from hazardous events. Burby et al. (1991) made an extensive study of this subject and concluded that though perception of risk by local public officials helps to explain insurance adoption, so does political pressure. However, there is strong sentiment among many local officials that the risk is too low to warrant insurance. In fact, though the availability of federal disaster relief may work against adoption of insurance, even communities denied aid are no more likely to purchase insurance against future events (Burby et al., 1991).

Individual hazard insurance also can have a significant impact at the community level. Large numbers of uninsured victims translate into greater demands on community resources, including financial and other services. Some communities have lowered residential property taxes in proportion to decreased property values resulting from flood damage, raising the tax rates as repairs are made. Obviously, it is in the best interests of a community as a whole to bring property back to at least pre-event values as soon as possible, a process that insurance facilitates.

The National Flood Insurance Program (NFIP) requires residents of designated flood-prone communities to carry flood insurance or face sanctions in the event of a flood. Under the circumstances, insurance may be considered a community-level response. Coverage has repercussions throughout a community, as would the failure of any insurance companies.

Other means include loss-sharing projects. Many communities contribute to mutual pools or other funds from which individual communities can draw following an event. In most cases, these arrangements do not reduce losses over the long term unless tied to other measures designed for that purpose. The floodplain management requirements of the NFIP (described below) combine loss sharing and loss reduction; this set of adjustments requires forethought and planning, as investments must be made in anticipation of a possible event.

Modify the Event

In general, this category focuses on structural responses that control either events or damage from them. We are best able to control flooding, through the construction of dams, levees, and flood walls and by channelization. This has had the effect of keeping water away from the people up to the design stage of the structure. Historically, the availability of cost sharing between levels of government has made this a popular adjustment, but the capital costs are high and attitudes as well as cost-sharing formula are changing, which may make such projects less attractive. It is difficult to modify other events, at least at present. Research on weather modification may have implications for meteorological hazards, but we do not yet have sufficient knowledge to make this an effective adjustment, and modification of most geological hazards is beyond our reach. Among the exceptions, attempts have been made to slow down and direct lava flows in Iceland, with some limited success.

Modify Vulnerability

This category represents efforts to modify the human-use rather than natural system in order to reduce loss. There are many ways this can be achieved. Forecasting and warning systems, in conjunction with emergency preparedness plans, allow for the implementation of a variety of measures such as evacuation and the protection of property before an event occurs. Public awareness programs that provide residents with information on what to expect, what to do, and when to take action also serve to reduce vulnerability to loss (whether of life, property, or both). In addition, we can modify our vulnerability by identifying hazard zones and restricting or regulating their development. Modifying vulnerability requires planning and preparedness and, although capital requirements vary, they generally are lower than those associated with modifying the event.

Other means of modifying vulnerability include hazard-resistant design. By designing buildings to withstand earth movements or tolerate high winds, we are able to protect property from damage. Such action allows human use of high-risk areas because that use is hazard-proofed by technology; hazard-proofed buildings allow high-risk land to be developed. They tend to be high in capital requirements, but the investment probably is less than repair and replacement costs. Although individual developers or builders are responsible for making the adjustments, communities play a significant role by instituting and enforcing building codes that include hazard-proofing. Interestingly, hazard-proofing the public infrastructure varies among communities and hazards, relating in part to differences in perceived vulnerability to damage, particularly by floods (Burby et al., 1991).

Do Nothing

It would be foolish to suggest that community choices always fall into one of the three categories above. In fact, doing nothing is a very popular "adjustment." Because the other categories require recognizing vulnerability, planning, and investing capital, doing nothing is chosen frequently. If hazards are not on the political agenda, if other priorities take precedence, and if there is little interest in investing time, personnel, and money in a probabilistic venture, then doing nothing makes sense (at least until an event occurs). As we have seen, there is usually great interest in mitigation immediately after an event, but if momentum is lost, as the community returns to the normal, nothing is done.

Actors and Factors in Long-Term Community Response

Several factors have been found to influence community attitudes and adjustment. The role of experience has been discussed already, but rarely does it prevail over all other constraints, such as when capital commitments are required for or result from particular courses of action. In addition, individual and collective attitudes toward the efficacy of specific actions must be considered, as must the context in which decisions are made. Indeed, all these enter into the process, with one or another coming to the fore at any given time.

Economic Interests

It is useful to combine economics and interest groups in a single category. Interests frequently are economic in nature, which should not imply that purely individual, economic interests prevail; communities as a whole also have a vested interest in economic well-being. However, while there may be general agreement on community goals, there are quite likely to be differences of opinion on how to achieve them. In addition, outside influences present both opportunities and constraints, and opinion probably will differ on the merits of those influences.

Any community decision, particularly with regard to low-probability events, is very complex. As Burby et al. (1988, p. 85) comment, "The floodplain land conversion process results from a complex of decisions and actions by individuals and groups, each guided by their own incentives," an observation that can be extended to other hazardous situations. The choice of protective measures depends on the views and interests of the individuals involved. Recalling that economic interests may work against the adoption of certain mitigation measures (Rubin et al., 1985), their importance in community decision making cannot be ignored.

However, the community decision-making context, is much more involved than that. Community leaders and decision makers are guided partly by economic interests, but also by personal interests. Personal risktaking characteristics of decision makers, as well as their own views on the seriousness of a natural threat (Palm, Marston, Kellner, Smith, and Budetti, 1983), influence both public and private decisions. We can neither assume that economic interests always prevail over community needs for loss reduction nor ignore the influence of individual and group economic interests. In one extreme example, an NGO project to site four flood shelters in char (new land) areas in Bangladesh was influenced heavily by local elites; in the end, the shelters were located where they would secure the local landowners' physical and social positions rather than serve the landless majority (Khan, 1991).

Extracommunity Governmental Influences

Regional, provincial, state, and national governments can influence communities toward or away from specific courses of action (which also involves influencing community attitudes toward various adjustments). This extracommunity influence can be subtle, as in the long-term effects of postdisaster assistance that sometimes encourages occupancy of hazardous areas. It also can be overt, as in regulating building and land uses in hazardous zones (such as seismically active areas or floodplains). In effect, central governments can both prescribe and proscribe mitigation at the community level.

With regard to subtle impacts, it has long been recognized that while postimpact disaster relief is essential to recovery, it may also discourage individuals and communities from acting on their own. Similarly, government provision of flood-control structures (with minimal or no cost-sharing requirements) has convinced communities that they no longer have a problem or that the government will take care of it; consequently, people come to rely on emergency and long-term aid in a "dependency syndrome" from which it may be difficult to break (Bryant, 1991). Therefore, it is generally recommended that emphasis be placed on reestablishing normal patterns as soon as possible with appropriate and constructive aid that allows affected areas to ensure their own self-reliance (de Goyet, 1993). For example, sending kitchen supplies rather than simply food may be useful.

More recently, these extracommunity forces have been using their resources overtly to encourage community action. Their resources include regulation, financial assistance, and technical assistance. In the United States, the Federal Emergency Management Agency assists local areas with Hazard Mitigation Plans. In New Zealand, planners in district councils develop community floodplain management plans that include elevation requirements, tied to the building permit system, for buildings in specific local areas. These forces are attempting both to minimize impacts and to bring about local commitment to loss reduction.

Capital Requirements

It is difficult to imagine a hazard adjustment that does not involve some outlay of money, whether millions of dollars for dams and reservoirs or hundreds or thousands for developing and implementing an emergency management plan. The local share varies significantly, depending on project costs and availability of financial assistance.

In any evaluation of capital investments, cost–benefit analysis is appropriate (see Chapter 6). On the benefit side of the equation is avoidance of public and private loss, including property, life, and infrastructure. The benefits are enjoyed by everyone in a community, regardless of their support of the measure. In addition, the range of adjustments available to a community, at least theoretically, is far greater than that available to an individual because of economies of scale and being able to spread costs more widely.

The problem lies in the relationship between costs and benefits. While costs may be large or small, benefits are likely to be quite large if measured in losses avoided. However, benefits are perceived as having as low a probability of occurring as does an event (at least for an individual). The recurrence probability is quite different for the community, as pointed out earlier. Although collective response is usually the most efficacious and cost-effective, it is frequently more difficult to achieve because of difficulties in reaching consensus on both the need and suitable methods.

Maintenance of Community Attitudes and Adjustments

As we have seen, community attitudes usually are stirred by some influence (such as an event), and the adoption of community adjustments follows. As time passes, other priorities come to the fore and the salience of an event or hazard may decrease. Furthermore, collective action is much more complex than individual decision making. Given all these factors, it is not surprising that sustaining collective adjustments, particularly those that are institutional rather than structural, is a significant problem (Burton et al., 1993). Unless the need for the action is reinforced or long-term maintenance is required, adjustments may be dropped as they fall on the list of community priorities.

An example is provided by Kirkby (1974), who made a case study of Oaxaca, Mexico. Traditionally, the communal good took precedence over personal advancement, which was helpful to farmers during times of drought or flood. In a sense, the community provided "mutual insurance" to its members. However, communalism lost importance as individual advancement began to be valued and impacts of the event dwindled. Kirkby predicted that "adjustment will . . . rest more with the responsibility of the individual at one end of the spectrum and with the state at the other. The present intervening scale of adjustment, *at the community level*, is likely to disappear, only to reemerge when extreme and infrequent hazards strike" (Kirkby, 1974, p. 127).

Disasters also create difficulties for communities beyond the impact area. For example, evacuees may not want to return to their hazard-prone community, preferring conditions at their refuge. This has been a major problem around the world; at the end of the twentieth century, refugees from natural and human-induced disasters are at a record high. In 1974, Typhoon Tracy killed 65 people in Darwin, Australia; most of the town's population of approximately 47,000 was evacuated for over a year, some to communities more than 1000 kilometers (620 miles) away. The evacuation severely disrupted social structures as the men were allowed back first into the community to help rebuild, then later were followed by women and children; unfortunately, the separation lasted so long that many families had problems getting back together (Bryant, 1991). In addition, the fact that Darwin is the capital of and largest community in the Northern Territory had social and economic repercussions throughout the region. Similarly, evacuation puts pressure on those in new environments; a drought in Australia in 1982–1983 forced many farm families into cities where children were exposed to urban problems, including increased drug use and permissiveness (Bryant, 1991).

Summary

The variables that influence long-term community response, both negatively and positively, are numerous and varied. Some are internal to the community and represent widespread local interests; others are internal, but reflect different and frequently conflicting interests. Still others are external and serve either to constrain or facilitate community choice and response. What complicates decision making all the more is that these sets of variables—internal and external, conflicting and cohesive—are at work at virtually the same time. It is easy to understand why communities have such difficulty in determining their collectively preferred response to natural hazards.

MANDATED COMMUNITY ADJUSTMENT

There are instances where community adjustment is not a matter of determining a preferred response, but instead has been required or forced. In the United States, community adjustment for floods has been mandated through the NFIP. This program was enacted long after it was documented that the structural approach to flood control had not reduced losses, and that a mix of community adjustments with an emphasis on nonstructural mitigation measures might be more effective (White et al., 1958). Communities designated flood-prone by the Federal Emergency Management Agency must join the program or risk being refused disaster assistance after a flood. NFIP requires communities to adopt maps outlining flood-hazard areas and to enact ordinances that restrict development in the 100-year floodplain (unless flood-proofed). Although regulations vary from place to place depending on the risk, minimum requirements include

- permitting for all proposed new development;
- reviewing subdivision proposals to assure that they will minimize flood damage;
- anchoring and floodproofing structures to be built in known floodprone areas;
- safeguarding new water and sewage systems and utility lines from flooding; and
- enforcing risk zone, base flood elevation, and floodway requirements after the flood insurance map for the area becomes effective. (Federal Interagency Floodplain Management Task Force, 1992, p. 31)

Once these regulations are in place, flood insurance is available to residents of that community, with the rates determined by the risk levels depicted on Flood Insurance Rate Maps. If flood insurance is available in a community, homeowners in the designated 100-year floodplain must carry it and lenders must require flood insurance as a condition of mortgages on properties located there. Although there are a number of problems associated with NFIP (mostly in enforcement of requirements and land use regulations), it provides a means of modifying vulnerability as well as sharing loss among those at risk. Indeed, these regulations attempt to minimize reliance on flood-control structures, phase out uneconomic development of hazardous areas, and force floodplain residents to insure against their individual losses, thereby minimizing the need for relief.

There are other provisions of the NFIP. For instance, there may be federal buyouts of structures that have been significantly damaged. Flood-proofing is required for substantial improvements to existing structures. In addition, several changes to the program are designed to make it more effective in meeting its objectives. (The reader is referred to the National Flood Insurance Act of 1968 for specific requirements of the program.) What is particularly important is the emphasis on the community as well as the individual. Only after a community joins NFIP can an individual take out the flood insurance, which encourages collective action to minimize loss. As of mid-1995, there were more than 18,000 communities in NFIP with over 2.9 million flood insurance policies in effect, accounting for more than $290 billion in insurance coverage (Emergency Information Public Affairs, 1995a). Table 4.1 shows the number of flood insurance policies in force at the end of 1994.

Reaction to NFIP has been variable. Some communities have resisted joining the program until forced to do so by sanctions leveled by the Federal Emergency Management Agency (see Montz and Gruntfest, 1986). Others have entered the program willingly, and still others have joined but claim to be experiencing local problems as a result; many of these problems have been evaluated in numerous studies of NFIP, including its local impacts on communities as a whole and on various groups within them (Burby et al., 1988; Burby and French, 1985; Montz, 1983; Muckleston, Turner, and Brainerd, 1981; Sheaffer and Roland, Inc., 1981). In addition, NFIP may have generated some risky construction in coastal and hurricane-prone areas, which resulted in the Coastal Barrier Resources Act that eliminated flood insurance for coastal areas where development had not begun.

As might be expected, then, the results of NFIP have been quite varied and relate in large part to community context, including the availability of flood-free land for development, pressures for urban expansion, and communities' experience with flooding. To overcome some of these

TABLE 4.1. Number of National Flood Insurance Program Policies in Force (as of November, 1994)

State, territory, or other	Number of policies	State, territory, or other	Number of policies
Florida	1,138,266	Nebraska	10,324
Louisiana	263,953	Arkansas	10,081
Texas	222,597	Tennessee	9,940
California	207,487	Oregon	9,706
New Jersey	129,647	Colorado	9,500
New York	76,697	Delaware	9,410
South Carolina	72,019	Kansas	9,159
Pennsylvania	57,555	Iowa	8,718
North Carolina	53,583	Rhode Island	8,586
Virginia	49,107	Nevada	8,285
Georgia	37,708	Wisconsin	8,012
Maryland	34,635	New Mexico	7,225
Illinois	34,635	Maine	5,879
Mississippi	34,245	Minnesota	5,870
Massachusetts	28,195	North Dakota	4,560
Puerto Rico	25,973	New Hampshire	3,208
Alabama	23,619	Vermont	2,256
Ohio	22,793	Idaho	2,185
Connecticut	20,178	South Dakota	2,152
Hawaii	19,896	Virgin Islands of the	2,074
Indiana	19,715	United States	
Michigan	18,953	Alaska	1,991
Missouri	18,925	Montana	1,853
Arizona	17,651	Wyoming	1,265
Washington	15,426	Utah	1,094
Kentucky	15,110	District of Columbia	236
West Virginia	13,442	Guam	122
Oklahoma	12,722	American Samoa	18

Source: Association of State Floodplain Managers (1994). Copyright 1994 by ASFPM. Reprinted by permission.

problems, the NFIP now offers an incentive to communities to initiate flood alleviation measures, rewarding businesspersons and homeowners with a 5% discount on premiums when satisfactory remedial action is taken. As of 1995, 61 communities in 31 states were eligible (Emergency Information Public Affairs, 1995b).

The impetus for NFIP stems partly from the large losses from urban flooding that are experienced annually. However, its focus on the community as well as the individual builds on the need for collective action. As might be expected, some communities would drop out of the program if they were permitted, just as many individuals drop their flood insurance as soon as it is no longer a mandatory condition of a loan. Regardless whether it is appreciated, NFIP requires collective action and, in a sense, forces a collective attitude.

CONCLUSION AND INTEGRATION

Given the range of internal and external influences on communities, it is impossible to develop a single scheme to explain the development of community attitudes. Different community contexts, varying levels of hazardousness from one or more sources, variable levels of growth pressure, and contrasting experiences with different events further complicate a complex of individuals and interests working within an organized system. However, it is possible to relate the findings from studies of community attitudes and adjustment to develop some integrating themes that can guide our understanding of community dynamics in a natural hazards framework (Figure 4.3).

FIGURE 4.3. The community recovery process. At any given time, various issues within communities generate agreement and disagreement among different interests. After a disaster, however, agreement tends to become dominant as interest centers on a single issue or problem. Ultimately, the community returns to something close to its predisaster state, marked by differences of opinion on different issues (one of which may be hazard adjustment).

If the smaller boxes in Figure 4.3 represent various interests within a community (represented by the larger boxes), then the intersection of these interests depicts the area of least divergence of opinion. In the first diagram (Figure 4.3a), there is little agreement as various interests have their own priorities for the community; because a community is rarely concerned with only one issue at a time, their varied concerns are represented by the several boxes (A–D) in the pre-disaster community. The second large box (Figure 4.3b) represents the convergent values that have been reported in communities immediately following a natural disaster, where there is close agreement on priorities largely because all interests are focused on the problem at hand. As the community returns to "normal," we need to re-evaluate the first diagram (Figure 4.3c); we know that the return to normalcy occurs over different time-frames in different communities, depending on considerations such as the availability of local resources, amount and suitability of outside assistance, degree of outside interference, and level of local preparedness. Next, in the areas of agreement (whether of the whole or its parts), long-term adjustment to the hazard begins (as represented in Figure 4.3d). Combinations of adjustments are more or less agreed upon by varying interests, influenced by factors such as experience with the hazard, pressures for growth or development, level of communality, type and extent of available outside assistance, and requirements and regulations. The shaded area on the diagram represents the areas of greatest agreement at any given time; for our purposes, we are assuming that this is a set of collective adjustments to a hazard, no matter what the hazard. Thus, out of its set of priorities, we have a community that has developed some agreement on adjustment to a hazard, achieved through a convergence of community attitude on this topic of community concern.

However, the diagram presented here represents a static situation when we know that conditions change over time. As the community moves farther away from the event, attitudes shift, partly because the hazard loses its prominence as other concerns come to the fore. In addition, outside influences may cause a community to change its focus, either abruptly or more subtly, as in the case study of Oaxaca by Kirkby (1974). Finally, interest groups change over time with A, B, C, and D perhaps giving way to new organizational structures of W, X, Y, and Z. Therefore, in the complex, dynamic organization of a community, we can trace the development of community attitudes toward natural hazards as they are influenced by a variety of factors. The difficulty lies in predicting how that response will lead to adjustment, and how that will or will not change over time.

5 Public Policy and Natural Hazards

INTRODUCTION

In the previous two chapters, we examined first how individuals perceive and respond to natural hazards, then how communities react to the threat of hazardous events. The community and individual responses discussed in Chapters 3 and 4 are the result of complex decision making and as we saw, adjustments to hazards do not evolve simply from the threat of extreme geophysical events, but occur within a complex web of social, psychological, economic, and political factors. For individuals, perception plays an important role in determining whether action is taken. There also are facilitating conditions for action at the community or public level. First, the hazard must be identifiable and recognized as a threat to society; second, decision makers (whether collectives of residents and victims or designated decision makers) must be aware of potential mitigation strategies; and third, there must be a common will to act. If we can understand the various forces operating within society, perhaps we can explain the rationale behind different mitigation strategies.

In both Chapters 3 and 4, we focused on the psychological, sociological, and economic factors that interact in defining appropriate and acceptable responses. In this chapter, we change the scale again to look at response and reaction to natural hazards through public policy. Policy, which can be made at any level of government, sets the framework within which actions are taken. We are concerned with what that framework is, why it is constructed as it is, how it will be implemented, and what the impacts of implementation are or will be.

The policy process is dynamic and multifaceted, involving a number of steps. For our purposes, the most important are perceiving a problem, getting it on the political agenda, bringing together multiple interests, developing alternative solutions or strategies, determining a means of implementation, implementation, evaluation, and revision or termination (Kruschke and Jackson, 1987). For natural hazards, the first three steps are relatively simple on the heels of a disastrous event, but it is difficult to sustain perceptions and interests. Appropriate alternatives vary, depending on geophysical considerations. Implementation is undertaken in a number of ways, including but not limited to regulations, mandates, and financial incentives. Evaluation may or may not be done consciously; with natural hazards, once an event occurs we often get an unanticipated evaluation of hazards policy, reflected in losses or costs avoided (or not avoided). Obviously, the development and implementation of public policy may come from different levels of government, from local to national or even international.

Based on what we know about the policy process in general and about natural hazards policy in particular, it is appropriate to ask whether a model exists that predicts when and at what level public intervention will take place. Further, public policies include a range of measures, such as financial incentives to encourage participation in alleviation projects (the carrot approach) and compulsory actions (the stick approach). Why do some policies embrace cooperative approaches toward hazard mitigation while others favor a mandatory approach? Is there an ideal level of public intervention or best practice to ameliorate the impacts of hazards? Can we explain why some administrative units do nothing to mitigate hazards while others adopt sophisticated programs to reduce loss?

In this chapter, we cover several themes on public decision making. These include recognition of hazards as public problems, goals of policy options, alternative implementation strategies, constraints on public policy, roles of interested parties in decision making and policy formulation, and legislative involvement. Throughout, we combine theoretical models with practical applications to help explain public policy. We do not give a detailed account of public policy and decision making (well described by Petak and Atkisson [1982]), but rather paint a broad picture of policy issues related to natural hazards.

PUBLIC PROBLEM, PUBLIC POLICY

Natural hazards are ubiquitous, posing an ever-present threat to livelihoods around the world. However, regardless whether hazards are of the same or different types, we cannot expect similar or consistent policy responses to them. Responses vary from the do-nothing approach to comprehensive plans comprising structural and nonstructural measures. What creates such differences between societies, making one area or community adopt policies directed toward action while another ignores the problem altogether?

Historically, it has been argued that the physical dimensions of hazards precipitate public responses, fostering policies, programs, and planning for hazard mitigation and relief. In theory, the more frequent and catastrophic the event(s), the more likely it is that a community will enact policies to ameliorate their impacts. To some extent this is true; more frequent and catastrophic events are more likely to trigger public concern, and certainly no action will be taken if the hazard is not defined as a public problem. If frequent floods threaten human life and regularly cause property damage, some public response is probable; if the incidence of floods is low, the problems may be small, public response may not be forthcoming, and the impacts may be viewed as the responsibility of individuals.

Thus, scale is an important factor. In the United States, most significant flood-related policy and legislation has followed directly from devastating events. Federal flood-control policy was revisited and redirected after a number of catastrophic hurricanes and coastal storms in the early 1960s (Burby and French, 1985), which began a policy shift away from flood control and the technological fix toward flood insurance and floodplain management. With other hazards, policy has shifted from disaster relief to prevention and mitigation; earthquake policy at the federal level shows this pattern (Table 5.1). Not surprisingly, it is the large price tags associated with events that have increased recognition of hazards and placed them on the public agenda. In policymaking, the magnitude of a problem frequently is defined by the extent of loss caused by events. A problem becomes public because of the widespread nature of a hazard or significant losses and costs to society.

TABLE 5.1. Policy Responses to Events

Event	Response
1933 California earthquake	The Field Act, specifying building codes
1964 Alaska earthquake	National Academy of Sciences study of earthquake and broader assessments of earthquake research and research needs
1971 San Fernando earthquake	Bills directed at expanding support for prediction research; Federal Office of Emergency Preparedness develops set of earthquake preparedness guidelines
1975 Chinese earthquake predictions and information on Palmdale bulge	Earthquake Hazards Reduction Act of 1977

Sources: Mileti and Fitzpatrick (1993) and Palm (1995).

However, scale or magnitude of a hazardous event is not the only motivation for hazard recognition. For example, take expansive soils that shrink and swell under different moisture conditions. Each year, structural damage to the tune of $5.6 billion results from construction on expansive soils in the United States (Alexander, 1993), but there is no national program to address that threat (although some local building codes attempt to minimize damage). The first step in policy development is recognition that a hazard represents a public threat (which in itself means different things to different people since one person's problem is often another's solution [Petak and Atkisson, 1982]). In the case of expansive soils, the problems are spatially diffuse and do not generally threaten lives.

Hazards are only a subset of the problems with which policymakers must deal. Even assuming that hazards are identified as public problems, policymakers must establish priorities among them because of scarce resources. Simple criteria would be magnitude of the problems and the ease with which they can be addressed; significant problems should be considered before less significant ones and those with easy solutions before those requiring major initiatives. However, setting priorities can be an exercise in choices among values embedded in judgments, which is no simple task. In the United States, if annual exposure to risk is determined to be the prime consideration, then floods and tornadoes should be the focus of policy, but if preventing loss of life in single events is of greatest concern, earthquakes should take priority. Using the criteria of total economic loss or structural damage would suggest that floods and expansive soils should be the focus of policymaking. Finally, policy could be based on a moral standard of universal health and safety; for example, public policy could authorize intervention wherever lives and property are at risk through no fault of the individuals concerned, but decline protection to voluntary risktakers (which would necessitate an assessment of hazard vulnerability).

In addition, public policies cannot be introduced without due consideration of spatial and temporal attributes of a hazard. Because hazards are identified, at least initially, at the local level, public responsibility should rest with local government, but large-scale disasters raise the level of concern and suggest that national policies are more appropriate. Recognition of a problem might be related therefore to the spatial extent of a hazard. Public policy also must reflect the appropriate temporal scale of response; for instance, policies could be based on recognition of current problems or the magnitude of potential future problems. Petak and Atkisson (1982) operationalize the problem of setting priorities by posing four questions:

1. How many people or how much area is adversely affected?
2. How intensely are these people and areas affected?

3. Will the situation get better or worse if nothing is done?
4. Is the greatest gain to be derived from dealing with existing exposures or preventing future ones?

Geophysical processes neither offer a foundation for policy nor dictate the significance of hazards. Instead, public policy must focus on human factors and the societal risk related to natural hazards. Further, the vulnerability of communities and marginalization of social groups (discussed in Chapters 3 and 4) are part of the policy process. In essence, hazards can be incorporated in policymaking for all of society's needs. The policymaker may then make decisions based on who is affected, the extent of the impacts, the values affected, and the perceived outcome of either implementing a new policy or doing nothing (Petak and Atkisson, 1982).

GOALS OF POLICY

As mentioned earlier, hazards policy can be initiated and implemented at any level of government, and it frequently crosses those levels. The widespread occurrence of some hazards requires a centralized view while in other situations, hazards are best handled at the local level, with or without direction, guidance, or regulation from above. Yet, to view physical characteristics as the sole determinant of policy implementation is to ignore the historic divisions of responsibility and authority among levels of government. The goals of policy also serve to specify the appropriate level or levels of responsibility and the appropriate relationships among public authorities.

In general, the goal of any hazard-related policy is to reduce exposure and vulnerability, but how that goal is realized proves complicated. Policies can be directed toward any part of the hazards complex, from preventive policies that focus on planning and mitigation to relief policies aimed at lessening the economic impact on victims. Usually, there is not a single policy goal; much of the policymaking process involves defining the desired result. A broad goal of reducing vulnerability has a number of options embedded within it, each of which must be considered in relation to the nature, values, and operations of the political system as well as the needs and constraints associated with a particular hazard.

Thus, one of the challenges in hazard policy is to define specific goals, the sum total of which will allow us to reduce vulnerability. One difficulty follows from the number of sectors affected by natural hazards; housing, business, agriculture, transportation, and public health are all vulnerable to damage. Thus, policy goals must address both the "big picture" and sectoral needs and values. At the central government level, goal-setting must seek to avoid serious conflicts between sectoral and hazard policies. Similar conflicts exist between different levels of government. A central government policy goal of reducing property damage from natural events may result in decreased building in hazard-prone areas or a condemnation of land so that alleviation projects can be built; either result may reduce property taxes to the local community. Similarly, a central government goal of reducing vulnerability may conflict with a local goal of distributing loss.

It is difficult to evaluate goals without assessing strategies for achieving them because it is the implementation of a strategy that has the direct impact, not the goal itself. However, it is important to realize that different goals suggest different strategies. Further, different organizations, groups, and elements of society may have different goals while the policy process starts with defining what is desired socially, economically, morally, and politically. What is desired by one sector or level of government may not be desired by others. There are a number of ways goals can be achieved, and the preferred strategies probably will vary between sectors and between levels of government as well.

POLICY IMPLEMENTATION: ALTERNATIVE STRATEGIES

In varying degrees, the mix of options for implementing hazard policy incorporates regulatory mechanisms, programmatic initiatives, planning, and financial packages. Another layer in the process is the political structure and context in which strategies are evaluated, chosen, and implemented. In particular, policies may be implemented with different focuses and responsibilities at different levels of government. Much of this complexity stems from the fact that hazards are "everybody's business."

> Flood control has grown to be a big business. Whether one lives near a river or in a house on the top of a hill, every person in the United States has a stake in this enterprise. When all the installed and proposed big dams and levee systems on the major rivers are considered together with land management and the small dams upstream, flood control and its related programs probably constitute one of the largest single activities of the federal government other than national defense. It is, indeed, everybody's business. (Leopold and Maddock, 1954, p. 3)

Everyone, it seems, has a stake in hazards and mitigation. People in the hazard zone are involved either directly through personal experience or indirectly through missed employment opportunities or adverse economic impacts. Because hazards are not confined to the area of the geophysical event, some individuals and businesses gain from others' misfortunes (see Chapter 6). Finally, there are others who support or oppose the implementation of different policies for other reasons.

Categories of Policy Implementation

Natural hazard policies can be grouped into categories based on combinations of the following elements: levels of government, means of achieving goals, and timing with regard to events. Figure 5.1 illustrates how these elements fit together and represents the theoretical range of implementation strategies. In practice, some blocks are more common than others, but all are possible.

Three groups of government are represented (although there could be more). Both central and local governments implement regulations directed toward hazard management and some states, such as California, take an active role in policy development and implementation. Thus, states could be considered central government in the case of some earthquake regulations (such as state standards for essential service facilities). Shared

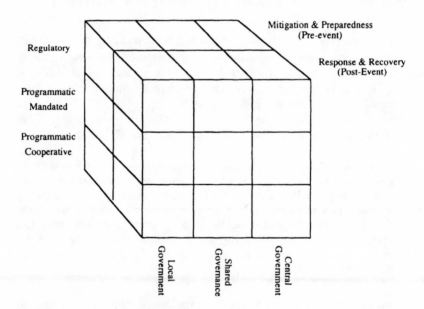

FIGURE 5.1. Forms of natural hazards policy implementation. The individual blocks represent the mix of factors that define how, when, and by whom policy is implemented. Empirical examples exist for each block, but analysis of the most effective mix still is needed.

governance is included as a separate category because there are policies where a true partnership exists.

The distinction between pre- and postevent groups is quite clear when strategies are implemented immediately following a disaster, such as allocating relief. However, the distinction becomes less clear with other types of postevent policy response aimed at reducing loss from the next occurrence. For example, as a result of ground shaking, there may be implementation or stricter enforcement of building codes to protect against earthquake damage.

Three groups suggest the means by which goals can be achieved. The regulatory approach defines what must be done within a "command and control" framework that requires compliance with standards or other measures. This is similar, but not always equivalent to mandates, which are orders or charges that may or may not be tied to specific regulations; indeed, mandates also can be programmatic. Program activities "can be viewed as tactics available for executing an implementation strategy" (May and Williams, 1986, p. 33). The figure shows that programmatic strategies can be mandated or cooperative; in addition, some may be a combination. Cooperative approaches exist where levels of government or even different interests within a single level work together to achieve the goals of hazard policies, unconstrained by a regulatory or mandated framework. Of course, cooperation also may be mandated or result from programmatic initiatives. Therefore, although the diagram implies that these are mutually exclusive groups, that is not an accurate picture in all instances. Still, Figure 5.1 presents a conceptual framework for discussion of policy implementation.

Using blocks in the figure as a point of departure, the discussion below presents examples of different policies. Though the examples may seem to suggest "success stories" in policy implementation, that is not necessarily the case; given the interests affected, there are both winners and losers. In addition, what works in one place may not work or be appropriate in another because of different political, social, and economic contexts. Finally, many other examples could have been used; the sample here serves to illustrate the range of options and issues involved.

Regulatory Approaches

Regulatory approaches to natural hazards generally restrict where and how building in hazard-prone areas can occur. Indeed, many natural hazards lend themselves to zoning legislation because the spatial extent of their occurrence is so well defined. Given hydrological and topographical information, flood hazards can be delimited relatively easily and, with little effort, areas prone to earthquakes, volcanic eruptions, tsunami, coastal erosion, avalanches, landslides, and hurricanes also can be determined accurately (as discussed shown in Chapter 2). Zoning ordinances are designed to restrict development of hazardous areas to those uses that will not suffer extensive losses and to encourage the siting of more intensive land use in relatively safe areas. Less spatially bounded hazards (particularly the meteorological events of snow, hail, drought, and tornadoes) are not generally conducive to zoning; instead, different regulations (usually in the form of building codes) make sense. Most of these regulations are implemented at the local level where most land use decisions are made. However, in some cases, the policy is initiated by the central government, to be implemented by local government under oversight; one example is the National Flood Insurance Program (NFIP).

Zoning

Zoning for hazards would seem more than justified by public needs. The unwary would be protected, the natural system would suffer fewer changes, and money would be saved on rescue and rehabilitation. Unfortunately, most zoning has been implemented for socioeconomic reasons rather than hazard mitigation. One of the major problems associated with zoning ordinances has been a question of legality, especially in those nations where the rights of individual landownership are prized. In the United States, zoning has been accepted slowly and reluctantly. Although it was first tried in 1916, it was not until 10 years later that *Euclid v. Ambler Realty* established its legality; the court compared zoning to ordinances upholding public health, safety, and welfare. However, 30 years later, questions of constitutionality still were being raised (Platt, 1976). Indeed, in 1992, the U.S. Supreme Court ruled against the State of South Carolina, which had attempted to restrict development in areas subject to hurricane damage (Platt, 1992). Still, zoning remains an important policy tool through which damage from natural hazards has been and can be managed.

Zoning ordinances have probably been used more extensively for flooding than any other natural hazard. In general, however, authorities have been rather slow to adopt such measures, as the development of floodplain zoning in the United States illustrates. Although *Euclid* established the legality of zoning, it did not guarantee its acceptance. For example, zoning was considered for New Albany, Louisville, and Cincinnati in the 1930s, but the proposals came to nothing (Anon., 1937). At the time, technology was seen as the solution to most hazards, and evidence was not hard to come by to support the perceived success of engineering (Bennett, 1937). (The danger of a purely structural response to the flood hazard was recognized fairly early by Segoe [1937] who, in an advanced comment for the time, described in detail the process now called the "levee-effect" and suggested how it could be avoided through land use planning.)

In the 1950s, various authors recognized the need for floodplain planning. Kollmorgen (1953) advocated the use of settlement control, which he regarded as more beneficial in the long run than flood control. Murphy (1958) adopted a similarly strong attitude by suggesting that floodplain development should be permitted only when its economic advantages outweighed the disadvantages and where there was no serious threat to health or life; as he noted, only limited zoning then existed to control flooding and only 7 states had legislated against encroachment on the stream channel itself. Several zoning proposals were developed, such as one for Lewisburg (Anon., 1937), and there were calls for local government to take greater control of urban development (Meistrell, 1957). However, zoning ordinances did not become more accepted generally until a decision by the General Assembly of States to control floodplain land use for the purpose of reducing loss (Dunham, 1959).

In the United States, the federal government has tried to encourage a more consistent policy toward flooding since the Federal Flood Control Act of 1936. In 1966, a unified national program for managing flood losses was established, whose task force concluded that traditional structural measures had helped to reduce loss, but that additional action on land use was needed to promote appropriate economic development on floodplains. The primary aim of the project was to make those who occupied floodplains responsible for their own actions.

During the 1970s, federal commitment to land use planning as a viable means of flood-related loss reduction strengthened (U.S. Water Resources Council, 1971, 1972, 1979). Since introduction of NFIP in 1968, more than 18,000 communities in all states and jurisdictions have joined it and adopted some zoning (Committee on Banking, Finance, and Urban Affairs, 1994). At times, the Federal Emergency Management Agency (FEMA) has taken a hard line by suspending communities for noncompliance; for example, in 1995, Ocean Shores, Washington, was notified of impending probation because of deficiencies in its floodplain management program. (It should be noted that out of an estimated 11 million households in flood zones, there are only 2.9 million insurance policies in effect [Committee on Banking, Finance, and Urban Affairs, 1994]; the cost of insurance premiums serves as a disincentive to many individuals. However, it is the land use component [the stick], not the insurance component [the carrot] that concerns us here.) Although many communities participate in NFIP, too many enforce minimum regulatory standards and do not take "an active role in steering new development away from flood-prone lands" (Riebau, 1990, p. 4), in large part because there has been no incentive to do so. Attempts have been made to alter this situation with the development of the Community Rating System; under this modification of NFIP, communities that successfully demonstrate to FEMA that their efforts to reduce flood losses or promote the purchase of flood insurance are well above minimum requirements may be granted reduced insurance rates.

Floodplain zoning divides a unit of land into specified areas for the purpose of regulating the intensity or type of land use. The zoning plan may incorporate a number of districts, but for flooding, a dual zone is perhaps most common. The top diagram in Figure 5.2 illustrates these two divisions: a floodway, which represents the stream channel and those portions of the adjoining floodplain necessary to provide reasonable passage of flood flows; and the flood fringe, which occupies the zone immediately outside the floodway. The size of the zones usually is based on analysis of hydrologic data (particularly historical flood records) and local topographical factors, from which the expected limits of floods of different magnitude can be determined. The United States normally uses the 100-year flood event as the limit for zoning, but standards can vary according to the acceptable level of risk for particular land uses (as the bottom diagram in Figure 5.2 shows). The Flood Insurance Rate Map (FIRM) for a section of West Des Moines, Iowa, illustrates how these concepts translate into usable documents for managing floodplains and delineating levels of risk necessary for setting insurance rates (Figure 5.3). In some locations, there may be very little or no land lying outside the flood zones, which poses problems for land use planners; in the Florida Keys, for example, there is no land available that is not subject to coastal flooding, so all development violates NFIP standards to some degree.

In other countries, detailed zoning and national strategies for floodplain planning are less common. In the United Kingdom, a different system of planning evolved that, like its U.S. counterpart, was relatively ineffective until recently. Until the twentieth century, little attention was given to flood problems and, particularly during the rapid urban expansion of the industrial revolution, cities continued to spread onto surrounding low-lying lands. Any official response was limited to small-scale structural measures. In 1933, the Medway Letter recommended that all new urban development be constructed 8 ft above known flood levels, but its recommendation usually was ignored. With passage of the Town and Country Planning Acts, planned urban development became more feasible, and government circulars encouraged planning and local drainage authorities to work together more closely. The Ministry of Housing and Local Government reiterated those recommendations in 1962 and 1969, but because of the voluntary nature of circulars, liaisons between responsible organizations did not really improve. Penning-Rowsell and Parker (1974) blame planners' poor appreciation of the hazard for the introduction of only a very few floodplain zoning programs in Britain. In Nottingham, a zone designated

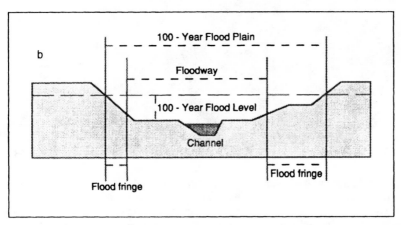

FIGURE 5.2. Floodplain structure. Floodplains consist of two parts (a). These natural features have been incorporated into floodplain management requirements. In the National Flood Insurance Program, the designated 100-year floodplain in any community includes both the floodway and the flood fringe (b). *Source:* Dunne and Leopold (1978). Copyright 1978 by W. H. Freeman and Company. Adapted by permission.

"prone to periodic flooding" has been excluded from further development and other communities have put local restrictions on development (as noted by Harding and Parker [1974] in Shrewsbury), but there is no unified national program for floodplain management.

In Canada, the Flood Damage Reduction Program (FDRP), initiated in 1978 (revised in 1982 and amended in 1985), was designed to protect residents of floodplains and to reduce the costs of damage to federal,

FIGURE 5.3. Flood insurance rate map (FIRM) for West Des Moines, Iowa. Zone A represents the 100-year floodplain, which is subdivided to show different levels of risk (such as Zone A5). Zone B is the 100- to 500-year floodplain, and Zone C is that area outside designated flood hazard areas. *Source:* U.S. Department of Housing and Urban Development (1979).

provincial, and local governments (Anderson, 1990). The core of the program seeks to discourage further development in flood-prone areas and is implemented through the local conservation authority. FDRP focuses on identifying and mapping flood areas (based on hydraulic and hydrologic analyses) and distributing the maps to the public along with other information about the program (Dufournaud, Millerd, and Schaefer, 1990). Mapping costs are shared (50% federal, 40% provincial, and 10% municipal) and maps are generally produced at a scale of 1:2000, using 1-m (3-ft) contours with half-meter lines added for increased detail.

Once floodplains are designated on these maps, the FDRP effectively prohibits further development on them. No new federal or provincial buildings can be constructed and mortgage sources that rely on federal or provincial support (such as the Canada Mortgage and Housing Corporation) can no longer lend money for properties built in these areas. New construction in the designated areas is not covered for flood damage (although existing dwellings continue to be covered). Because Canadians do not support federal responsibility for flood losses, the program calls on the provinces, territories, and municipalities to adopt stringent floodplain regulations. The federal government provides the means to attain these goals by allocating funds for mapping and by restricting mortgages as an incentive to reach enforcement goals.

The Ontario FDRP, which is the largest of the provincial programs, had spent $13.4 million (Canadian) determining flood areas by 1990, made 25 identifications of 55 waterways, and produced 79 public information maps (Anderson, 1990). Dufournaud et al. (1990) suggest that the Ontario FDRP has been economically beneficial, especially to local municipalities and particularly in comparison to the costs of structural measures.

Although there are variations on zoning for other hazards, they usually are not implemented as zoning ordinances. For example, any development within California's Special Studies Zones requires prior approval from the pertinent city or county agency, based on geologic investigations; located on historic fault lines, these zones are usually 0.25 mi (400 m) or less in width. In most other cases, land use zoning proposals serve as input to other planning and hazard management programs (and are discussed below under programmatic policies).

Locational Permits

Besides zoning, there are legislative and administrative tools that enhance planning. Restriction of land use can be accomplished by location and construction permits, issued by the local authority. These are used fairly extensively to curtail high population density and excessive investment in areas of significant risk as well as to motivate investment in "safe" areas. Each building application is reviewed on its own merit, within the context of a system that may incorporate hazard criteria. In the United Kingdom, urban development on floodplains often is prevented in this way and, although this method does not favor a unified program, it does emphasize the importance of local context.

A similar concept is the open space requirement that has been adopted widely in many urban development plans. Originally devised for the aesthetic enrichment of the urban environment, there is absolutely no reason an open space policy cannot be coordinated with hazard mitigation to prohibit or limit intense development of hazardous areas. Despite some legal problems attributed to the apparently selective nature of legislation,

White maintained that the benefits to society far outweighed the costs (see Marx, 1977).

Open space regulations can fulfill virtually the same role as zoning ordinances in hazard mitigation, and are particularly useful in preserving floodways. In Cedar Rapids, Iowa, parklands and recreational areas have been maintained for some time along the Cedar River (Gardner, 1969), which is typical of a number of communities (Byrne and Ueda [1975] describe the benefits of open space in Tempe, Arizona, along the Indian Bend Washway). In Dallas, Texas, a greenway was considered a viable measure to reduce the flood hazard in a fully developed neighborhood of low-cost housing (Novoa and Halff, 1977). In some areas, open space is considered so important to the health, safety, and welfare of the community that urban plans incorporated it with little difficulty. In Tokyo, an urban green-space policy was implemented to minimize the spread of fire (in the 1923 earthquake, many of the 123,000 deaths and much of the damage resulted from fires rather than seismic activity). Similarly, the particular problems of fire in Arctic environments, where water may not be readily available during emergencies, has inspired some open space requirements in Alaska and Canada (UNECE, 1980).

However, there have been some legal challenges to this regulatory device. For example, a requirement by Tigard, Oregon, that, under certain circumstances, portions of land be reserved for open space and landscaping was successfully challenged (Platt, 1994). Not all open space is automatically called into question, but it is not a perfect measure. Again, local factors dictate the characteristics of plans.

Design Standards

Having established which land uses should be allowed in particular hazard zones and the degree or intensity of use, further control of urban morphology can be maintained through measures that require developers to meet certain minimum standards. In this way, additional protection can be afforded structures exposed to risk. Design criteria may be applied to whole units of land (as in subdivision regulations) or may involve specific construction standards.

Subdivision

Subdivision regulations are frequently used to define the basic minimum standards for urban development. In the United States, power of enforcement lies with local governments and planning boards, which have been authorized by many state legislatures to adopt design standards. The developer then has to meet those requirements in order to qualify for a building permit. Invariably, these regulations apply to new development and the subdivision of "new" land into individual lots; they prescribe minimum acceptable standards for public facilities and ensure adequate water and sewage service. Although such measures were introduced primarily for contemporary town planning, there is no reason hazard regulations should not be included in them.

In California, a subdivision ordinance requires local communities to reject development proposals that do not meet certain conditions. First, proposals must be consistent with the general and specific plans for the community; second, the site must be physically suitable for the proposed land use and density of development; third, the proposed development should not substantially damage the environment; and fourth, it should not lead to serious public health problems. This legislation was enacted by the state for several reasons, among them hazard mitigation. For instance, the regulations clearly state that subdivision ordinances should ensure slope stability and prevent extensive interference with natural slopes by grading (Nilsen, Wright, Vlasic, Spangle, and Spangle, 1979). Subdivision ordinances also control development in earthquake and flood zones (Blair, Spangle, and Spangle, 1979; Waananen and Spangle, 1977).

A carefully administered subdivision policy could bring about more effective management of hazardous areas. New development could be curbed to eliminate piecemeal expansion by coordinating individual and private development and prohibiting plans for high population densities in areas at risk. However, such regulations have not always been applied adequately, and problems have arisen when land has been segmented with little regard for local topography (Larson and Nikkel, 1979). Too often, subdivision regulations have not guided development appropriately, especially when confronted with unique situations (UNDRO, 1977).

In Hahn's Peak Basin, Colorado, several thousand hectares were subdivided into numerous small units, but little development occurred because sewage and water capabilities were so limited (Utterback, 1977). Figure 5.4a illustrates a similar case in California where a developer, apparently infatuated with the neat block system, totally disregarded the underlying topography; the plan was not pursued because of the states strict subdivision regulations, and Figure 5.4b shows how the project eventually was implemented with advice from local officials. In subdivisions, the community retains control over spatial and temporal aspects of development, but does not assume responsibility for producing the detailed information required for each proposal; the private developer must provide those data.

Building Codes

The effectiveness of zoning and other legislative measures can be enhanced through the use of building codes, which establish minimum design

FIGURE 5.4. Development planning. Topography can make development difficult and hazardous (a). However, it is possible to design development that is in keeping with topographic constraints (b), thereby minimizing hazards. *Source:* California Resources Agency (1971). Copyright 1971 by California Resources Agency. Reprinted by permission.

standards for structures in hazardous areas. Codes can specify not only structural design but also construction methods and materials. In combination with a zoning ordinance, appropriate levels of control can correlate building codes with designated zones of high risk, enforcing more stringent design requirements where appropriate. In addition, such programs can encourage cities already located in hazard areas to adopt measures of protection.

Unlike zoning ordinances, which may be based primarily on geophysical criteria, design standards require more subtle assessments of the
interaction between buildings and events. There is a wealth of knowledge
on the behavior of buildings in hazardous environments, which should be
used as the basis for building codes. Evidence from different earthquake
events suggests that some buildings perform consistently well in earthquakes, particularly those constructed of steel and wood frames which tend
to be light and flexible and, with careful design, can withstand some degree
of both horizontal and vertical ground movements. By contrast, masonry
and brick constructions have a relatively poor record due to their inflexi-
bility and great weight. In Japan, the response of buildings to seismic
activity has received considerable attention and has led to the adoption of
more stringent building codes (Nichols, 1974); codes were tightened after
a large earthquake in northern Japan in the late 1960s, but the extent of
destruction in Kobe in 1995 put even the stricter standards in question.

In the 1971 earthquake in California's San Fernando Valley, buildings
of 20–50 stories performed surprisingly well. Newmark (1970) produced
similar findings on tall buildings, and Douglas (1978) suggested that 5
stories might in fact prove more hazardous during earthquakes than taller
heights. However, tall buildings may resonate with the frequency of ground
waves so that the force applied at the bottom is multiplied at the top
(Ambrose and Vergun, 1985; Smith, 1992), as was experienced in the
Mexico City earthquake of 1985. Besides building height, other factors to
consider include the height and geometry of connecting wings, the relationship of the building axis to ground motion, and local geology.

In the United States, building codes were introduced in the 1933 Field
Act and have been updated fairly regularly with improved knowledge of
building design and construction. Many nations have adopted similar
standards to those in the United States, and the USSR, Japan, and over 30
other nations employ some form of building code to mitigate the effects
of seismic activity (UNDRO, 1978). In 1971, California passed the Seismic
Safety Element Act, making it mandatory for all communities in the state
to include building codes in urban planning (Griggs and Gilchrist, 1977);
it gives priority to unreinforced masonry structures, buildings constructed
prior to the building code jurisdiction, and critical facilities such as
emergency buildings, schools, and high-occupancy structures. Of these
three, unreinforced masonry buildings probably present the largest problem because of political difficulties in enforcing regulations (Alesch and
Petak, 1986).

Building codes have not always been entirely successful, partly because
of problems with compliance and enforcement. While these are not
surprising problems in less developed countries where buildings are frequently constructed with little technical supervision (Smith, 1992) or with
whatever materials are available, examples can be found in developed

countries as well. For example, although the South Florida Building Code has been praised as one of the best in providing protection against hurricane winds, the tremendous destruction caused by Hurricane Andrew exposed its thoroughly inadequate enforcement. If dwellings had been built to code, the damage would have been reduced significantly. This revelation received a great deal of media attention because of dramatically different responses to the event by what were thought to be substantially similar houses. Some survived relatively intact with only minor damage while whole roofs were blown off others, and both residents and some local officials accused developers of cutting corners, thus evading standards. Similarly, in California in 1971, several supposedly earthquake-proof structures failed, including two hospitals.

Many structures need alterations to bring them into conformity with elementary safety standards. Bolt, Horn, MacDonald, and Scott (1977) estimated that in 1974, nearly 200,000 children in Californian were attending schools that failed to meet even 1933 standards. In several places (notably Greece, Turkey, the Middle East, and Latin America), problems have arisen from poor design and use of unsatisfactory materials. The Managua earthquake in 1973 showed some of the benefits of reinforced concrete structures over local adobe construction, and communities in Nicaragua, Guatemala, and Iran have experienced high death rates because similar structures have collapsed (Bolt et al., 1977; Kubo and Katayam, 1978). However, particularly in recent urbanization, there also are problems with construction materials. Indeed, Sachanski (1978) questioned the resistance of some of the newer, synthetic materials to seismic motion. The 1994 earthquake in Northridge, California, revealed weaknesses in many "modern" construction methods; some steel-framed buildings developed hidden cracks where the structural beams had been welded together, seriously diminishing their potential to withstand future earthquakes. Studies continue of safer materials and design for structures, from steel-framed to prefabricated concrete buildings.

Building codes designed for other hazards are less common, but in some areas minimum ground-floor levels have been imposed to restrict flood damage. In the United States, any new development in designated floodplains or modifications to existing structures must meet codes that include raising utilities and other damageable property above the 100-year flood level. These same regulations apply in coastal areas to protect against floods caused by storm surges. Of course, enforcement varies (Committee on Banking, Finance, and Urban Affairs, 1994).

Although buildings can be made resistant to quite strong winds, it is more difficult to design them to withstand sea surge and storm waves. However, Bretschneider (1972) calculated hurricane-related waves and sea surge as a basis for design standards. Even minimal standards to strengthen buildings should be enforced to provide some structural protection to their inhabitants. Few countries have anything approaching a national policy. In Townsville, Australia, building codes were introduced to protect against high winds, and those standards later were raised following a tropical cyclone in 1971 (Oliver, 1978). Severe winds also have led to the adoption of more stringent codes in Boulder, Colorado; following violent wind-storms in 1969 and 1972, legislation was passed that improved roof and fence construction and required mobile homes to be tied securely to the ground. In desert environments of the Middle East, building codes are designed to withstand high wind velocities and assure maximum shade (Cones, 1979). Criticism that such regulations make construction costs too expensive is not well founded; it has been estimated that costs rise by only 6% for protection against winds up to 240 km/h (150 mph) while the savings in damages would be approximately 60% (UNDRO, 1980).

Considerable care is needed in Arctic environments where the eco-logical balance can be destroyed so easily. Building codes there are de-signed to protect the permafrost from melting, which would cause struc-tures to collapse. A classic example of Arctic urban planning is Inuvik in northern Canada, which was a planned new town. In its code, this model settlement incorporated detailed studies of the physical characteristics of the area and the most advanced standards for construction methodology. Houses were built on stilt foundations above a gravel base while all services (water, heating, and sewage) were incorporated in well-insulated utilidors (Price, 1972). Although problems still arise, this community plan repre-sents a genuine attempt to balance extremes of the physical environment with societal goals.

Despite the obvious benefits of safe building design, structural stand-ards can be problematic, and some have been severely criticized as unnec-essarily restrictive. Building codes raise construction costs, if only margin-ally, but that can exclude low-income families from certain areas and thereby reinforce their vulnerability. Codes also have been viewed as inflexible and enforced with little or no regard for changing circumstances or knowledge (Hageman, 1983). Others have seen building codes as too permissive; Clawson and Hall (1973) recommended that they be used more extensively to reduce the impact of natural hazards. Another kind of problem arises with multiple hazards. In parts of California, for example, a wood-frame house with a cedar shake roof provides relatively good resistance to seismic shaking, but would prove prime tinder to the state's frequent wildfires. Alternatively, a masonry house with a heavy tile roof would provide protection during the fire season, but would be far more likely to collapse during an earthquake. Effective building codes must take into account local context and the hazardousness of a place, but when related to degree of risk, their advantages would seem to outweigh the disadvantages.

Summary

There are a number of regulatory mechanisms through which hazards policies can be implemented. In the United States, most regulations are implemented at the local level, although the enabling legislation may be at a different level (as with the NFIP). While all the regulatory schemes discussed here are designed for mitigation and preparedness, most were initiated in the aftermath of a disaster. It would seem obvious that this type of policy is the most cost-effective because assuming it is enforced, it can be virtually guaranteed. In many cases, there would be little incentive to take action without regulations.

Programmatic Options

In contrast to regulatory mechanisms, programmatic options comprise those activities that attempt to develop operating procedures for hazards management. These may be mandated programs (similar to NFIP) or they can be of a more cooperative nature. They may involve some element of zoning, but they are multifaceted (e.g., they might incorporate emergency management provisions as well). Table 5.2 illustrates some of the ways in which programs can be administered as well as differences based on implementation structures. It shows three examples of programs administered by FEMA that vary based on whether they are mandated or cooperative.

Land Use Plans and Programs: Zoning

In areas of potential earthquakes, any urban planning and zoning should be based on a sound physical evaluation of the event. This should include details of past seismic activity and an assessment of the probability of future

TABLE 5.2. Selected Federal Emergency Management Agency Programs: Mandated and Cooperative

	Floodplain regulation	Dam safety	Earthquake preparedness
Mode of operation	Regulatory	Mobilization	Collaboration
Authorization	Congress	Executive order	Congress
Implementation process	Regulatory partnership	Owner responsibility; state oversight	Partnership among levels of government

Source: Adapted from May and Williams (1986). Copyright 1986 by Plenum Press. Reprinted by permission.

activity (Karnik and Algermissen, 1978). Spatially, such analysis presents few difficulties at the macroscale, but at the local level many additional physical parameters must be studied; for instance, surface fault displacement is relatively uncommon, but other phenomena are more likely to occur (such as ground shaking, slope collapse, landslides, and liquefaction of surface materials). Detailed microgeological studies are essential to effective risk assessment, permitting areas of potential problems to be mapped and zoned accordingly.

Although a relatively rare practice, zoning of seismic areas can be a fairly effective means of mitigating the hazard. As with flood hazards, the goal would be to avoid settlement of high-risk areas by dense populations and complex urban structures; zoning ordinances would define a series of areas by degree of risk. High-risk zones would include areas of active faults, alluvial soils, unconsolidated landfill, and unstable slopes; their uses should be of extremely low intensity or risk (such as agriculture, recreational open space, or parking lots). A second zone might constitute a moderate-risk environment where low-density usage (such as single-story dwellings) could be permitted, consistent with a zoned area for ground shaking or potentially active faults. Outside these areas more intensive development could be permitted, built to structural design standards.

In many seismically active areas, zonation based on risk has been undertaken in order to understand the nature and spatial extent of the problem and to plan for emergency services. Risk zoning has not been widely adopted as a method of regulating land use, but the delineation of areas based on seismic risk fits very well into the planning process. Tokyo has been mapped for soil types, liquefaction potential, ground response characteristics, and demographics in order to determine relative vulnerability (Nakano and Matsuda, 1984). In Wellington, New Zealand, the regional council has undertaken a hazard assessment that includes maps of active faults and areas subject to ground shaking, liquefaction, and ground damage (Kingsbury, 1994).

The United States has carried out some detailed research in this field. Geological factors, surface deformations, differential soil settlement rates, liquefaction potential, and slope stability have been examined to establish zones of risk. For example, the United States Geological Survey (USGS) produced a series of such maps for San Mateo County, California. One of the biggest problems has been to assess the degree of acceptable risk based on the expected return periods of events of different magnitudes. For instance, questions arise over the potential activity of fault lines; in New Zealand, a fault is considered active for urban planning purposes if it has moved in the last 20,000 years. For high-risk facilities, that figure may prove too generous; when constructing thermonuclear facilities, the U.S. Atomic Energy Commission (now the Nuclear Regulatory Commission) considered a fault active if it had moved either once in the last 35,000 years or more than once in 50,000 years (Petrovski, 1978).

Some of the most comprehensive zoning programs for seismic risk may be found in the United States. Although there is no unified national program, the federal government has encouraged the adoption of zoning ordinances. Around San Francisco Bay, extensive studies have been made of earthquakes, and detailed maps have been drawn of various risk zones. These include mapping of geological structures that affect the speed of seismic waves, potential liquefaction, the relative stability of upland slopes, and flood potential from dam and other failures (Blair, Spangle, and Spangle, 1979). From these details, an overview and zonation of risk have been produced, which can be used in conjunction with land capability maps. These systematically evaluate economic, social, and political factors of an area and are helpful in land use planning and in developing zoning ordinances.

Other Zoning Programs

Although zoning ordinances have been considered for other natural hazards, few have been implemented. For instance, tropical cyclones frequently affect the same areas and the spatial extent of their direct and indirect effects can be established with some precision. Unfortunately, thousands of square kilometers of land are involved, making restricted use neither socially nor economically feasible. Immediately along the coastline, however, zoning may be a more reasonable proposition, particularly useful in mitigating the effects of the tidal surge that often accompanies tropical storms. In the United States, NFIP applies to hurricane-prone areas and delimits zones according to the probable height of storm surge. In Japan, the town of Nagoya designated five coastal zones; in the high-risk zone, little development is permitted (particularly of houses or hotels), and in the other zones, ordinances control building design, the height of ground level, and construction materials (Oya, 1970). Similar plans exist for tsunami. For example, Hilo, Hawaii, introduced zoning laws to control land use in tsunami-prone areas (Morgan, 1979) and, in parts of California threatened by tsunami, some mapping of the risk has been accomplished (Blair et al., 1979). Following damage sustained from a tsunami generated by the 1964 earthquake in Alaska, Crescent City, California, rezoned land to avoid further losses from tsunami (Alexander, 1993).

Management of coastal erosion, a constant threat to many communities, has been attempted through zoning. Although communities have often debated this issue (such as Bolinas, California [Rowntree, 1974]), discussion has not always led to appropriate action. Similarly, South Carolina has experienced legal problems in trying to restrict development of areas subject to coastal erosion. The federal government has tried to encourage greater control of such areas under the Coastal Zone Management Act of 1972, but its effectiveness is questionable.

In mountainous areas, zoning has been proposed to reduce the effects of such hazards as avalanches and slope instability, but implementation is contingent upon the initial identification of hazardous areas. Restrictions on urban development have been fairly common once the local population has been alerted to the problem, even if there has not been comprehensive planning for such disasters. Switzerland and Norway have treated the issue more seriously than many countries and have restricted further urban development in known avalanche tracks (Ramsli, 1974; Visvader and Burton, 1974). At the local level, there has been some advance in zoning programs. de Quervain and Jaccard (1980) proposed zoning in the Alps, but conceded that major problems would continue to exist in established communities.

In the United States, the federal government made disaster funds and grants from the former Department of Housing and Urban Development contingent upon detailed physical surveys (Nilsen, Wright, Vlasic, Spangle, and Spangle, 1979). Hazard zoning of mountains has been proposed or undertaken in both California and Colorado. Other policies have focused on slope stability; for example, standards on slope steepness and density of construction have been formulated for such communities as Portola Valley and Pacifica, California, to guide planners and developers (Table 5.3). Until 1973, Vail, Colorado, grew rapidly with little concern for the avalanche hazard, but since then the local government has prevented the siting of real estate (including high-density condominiums) in known avalanche tracks (Ives and Krebs, 1978).

Some degree of zoned planning has been applied to some new desert

TABLE 5.3. Slope Steepness and Density Provisions for Portola Valley and Pacifica, California

Average slope (%)	Portola Valley		Pacifica
	Gross area per dwelling unit, in ha (acres)	Required minimum parcel area, in ha (acres)	% of site to remain in natural state
1 and under	0.46 (1.13)	0.41 (1.02)	
10	0.55 (1.36)	0.49 (1.22)	32
15	0.62 (1.52)	0.55 (1.36)	36
20	0.70 (1.73)	0.63 (1.55)	45
25	0.81 (2.00)	0.72 (1.79)	57
30	0.96 (2.37)	0.86 (2.13)	72
35	1.18 (2.91)	1.06 (2.63)	90
40	1.52 (3.76)	1.38 (3.42)	100
45	2.15 (5.32)	1.99 (4.91)	
50 and over	3.68 (9.09)	3.52 (8.70)	

Source: Nilsen, Wright, Vlasic, Spangle, and Spangle (1979).

settlements. Towns in these areas face particular problems from high wind velocities, dust storms, and flash floods. The need for careful zoning is of paramount importance (Cones, 1979).

Zoning and Hazard Disclosure

Some areas have developed and implemented policies of designating areas or locations hazard-prone and requiring or suggesting public disclosure of the existence or level of hazardousness. Designation of the hazardousness of an area can be based on the results of models, the spatial extent of previous events, the existence and location of characteristic landforms, and other environmental indicators. Disclosure can take a number of forms, including the drafting and public distribution of hazard area maps, requiring realtors or mortgage lenders to inform potential buyers of the hazardousness of property, and tagging the deeds of properties with indicators of hazardousness; even signs suggesting previous events, such as flood levels, may generate awareness of a hazard. Fears that hazard designation and disclosure will have an adverse effect on property values have not generally been supported by research (Montz, 1993; Palm, 1981), although the dynamics of local real-estate markets and the availability or lack of alternative sites may affect those values (Montz, 1987; Muckleston, 1983).

There are several reasons why designation of hazardousness and disclosure of that information might be undertaken. Disclosure helps to avoid or control development in particularly hazardous areas. Thus, it would not apply to all property transactions, only those above a given threshold, which is the case in several communities in New Zealand (Montz, 1992). Another purpose of disclosure is to inform prospective buyers so that they will be aware of the risks they will assume if they purchase a property (Palm, 1981); in California, realtors must disclose designation as a Special Studies Zone. The goal of the first type of disclosure, as practiced in some New Zealand communities, is to minimize losses through control of development, while the goal of the second is to have informed buyers who can decide what measures, if any, are appropriate to lessen the risk.

Warning and Emergency Management

Zonation and land use planning are most useful where a hazardous area can be spatially delineated. In such cases (as well as with some spatially diffuse hazards), warning, evacuation, and emergency management are critical, with one or another being more salient for different hazards. While these activities may be seen simply as adjustments (thus, not part of the policy process), ideally they should be incorporated in planning and backed by the political will to invest in them. At the outset, a financial decision is required to invest time, money, and staff in a probabilistic venture; afterward, these activities can be separated from the normal mode of operation (and operators) or they can be integral components. Certainly the latter is preferable, but there are sound economic, political, and social reasons this does not occur or occurs only with great effort (many of which have been touched on in earlier chapters).

Several examples can serve to illustrate warning systems and their effects. Given the ability to identify and track tropical storms and based on knowledge of their patterns of movement, the National Hurricane Center in Miami, Florida, forecasts hurricane landfall and issues watches and warnings for reaches of coastline. Since 1983, the probability that a hurricane would pass within 65 miles of a given location during a certain time period has been added to the warning (Baker, 1995). Experience with Hurricane Hugo in 1989 suggested the need for improved tracking forecasts and refined skills for prediction of intensity; the 24-hour forecast timing of landfall was off by 6 hours, and the acceleration of the storm was not detected soon enough for inland communities to respond (Powell, 1994). Although such information is critical for local emergency management officials, there has been some concern that probabilities would be poorly understood by the public; however, research suggests that people respond to local officials' advice rather than probabilities (Baker, 1995). Indeed, in Hurricane Hugo, local officials relied on the forecasts rather than the probabilities in making decisions; the relative accuracy of the forecasts was critical in the success of evacuation and other emergency management efforts (Baker, 1994).

The ability to detect tornado outbreaks has improved significantly in recent years with the introduction of Doppler radar. In contrast to conventional radar based on reflectivity signatures, Doppler radar "involves combined interpretation of reflectivity as well as velocity signatures" (Burgess, Donaldson, and Desrochers, 1993, p. 214). Because the velocity of raindrops within a thunderstorm can be measured with Doppler instruments, the rotation of a potential tornado can be detected. While the system is not infallible, it increases the probability of detecting a damaging tornado to 94% (Alexander, 1993). However, having a radar system is not sufficient, as the 1987 tornado in Saragosa, Texas, demonstrated. Emergency tornado warnings issued by the National Weather Service in Midland, Texas, did not reach most residents of Saragosa for a number of reasons, including the way in which they were issued, the lack of an emergency warning system in the community, and the failure to issue bilingual warnings to the area's Spanish-speaking population (Aguirre, Anderson, Balandran, Peters, and White, 1991). Out of a community of 428, 30 people were killed and 121 were injured.

In contrast to warning systems based on federal forecasts, some systems require a local commitment to invest in technology and develop response plans. Flash flood warning systems are probably the best example.

These can range from systems where volunteers monitor stream and precipitation gauges to fully automated rain gauge and river-level systems that link National Weather Service forecasts with automated alarm capabilities. Of course, the system is only as effective as the response plan that is tied to it, particularly the warning dissemination and public awareness components.

Relief

The provision of relief involves programs that can be directed from one level of government to another or from public and private sources to disaster victims. Many relief programs are well-established and are put into operation immediately after an event, making food, medical supplies, building materials, clothing, and money quickly available. This form of assistance works best when programs are organized in advance of disasters, with Red Cross assistance. More ad hoc efforts, which in too many cases can hardly be termed "programs," tend to create problems such as providing inappropriate clothing, food that is unacceptable because of religious beliefs, outdated drugs, or creating situations where aid agencies are "tripping over" each other (Schmitz, 1987). Even well-planned relief programs can experience logistical and political problems in getting supplies to victims.

One issue, particularly with international aid, is that it is not always apparent exactly what is needed. Victims are often from another culture, they are frequently a "long way off," and the direct impacts and extent of the disaster may not be clear (Schmitz, 1987). Inevitably, mistakes are made: the wrong aid is sent and frustration can result in both donor and recipient nations. Disruption of political and economic organizational structures can exacerbate problems by stalling or inhibiting aid distribution. (Schmitz [1987] reviews a litany of mistakes that have occurred in international aid, including the problem of receiving too much.) It is easy to see how criticism of aid arises, but given the chaos of any disaster environment, we should praise such organizations as the United Nations Disaster Relief Office (UNDRO), the United Nations Department of Humanitarian Affairs (UNDHA), the Red Cross, and the Red Crescent, as well as countries that donate resources (Table 5.4).

Aid can flow from a variety of sources: directly from one country to another, from countries and filtered through the United Nations, from NGOs, and from voluntary organizations. For example, following the typhoons in Vietnam in 1995, which caused over $21.2 million in damage (Emergency Information Administrator, 1995b), the country received $50,000 from UNDHA; various sums from Canada, Denmark, Germany, Italy, Japan, Luxembourg, New Zealand, and the United Kingdom; and $20,000 from the Nippon Foundation, an NGO (Emergency Information

TABLE 5.4. International Disaster Aid Reported to the United Nations Disaster Relief Organization, 1985

Major donor governments	Major recipient governments
United States	Algeria
Italy	Angola
Federal Republic of Germany	Argentina
Saudi Arabia	Bangladesh
United Kingdom	Benin
Canada	Botswana
Netherlands	Chad
Australia	Chile
Switzerland	Colombia
Norway	Comoros
Sweden	Cuba
Denmark	Ecuador
Belgium	Ethiopia
Japan	Fiji
Bulgaria	Ghana
USSR	Guatemala
	Jamaica
	Mali
	Mauritania
	Mexico
	Mozambique
	Nepal
	Niger
	Philippines
	Somalia
	Sudan
	Togo
	Vanuatu
	Vietnam

Source: Schmitz (1987). Copyright 1987 by Council on Foreign Relations. Reprinted by permission.

Administrator, 1995c). Flooding in Benin in 1995 drew aid from UNDHA, the World Health Organization, Italy, Luxembourg, Norway, and the Nippon Foundation (Emergency Information Administrator, 1995a). Similarly, Taiwan, Japan, the United States, France, and Germany donated money to South Africa following severe flooding along the Kwapata and Slangspruit tributaries of the Umsunduzi River, which killed 154 people and left nearly 5,000 homeless (Emergency Information Administrator, 1996). A typical breakdown of relief, as reported to UNDHA following floods in the People's Republic of Korea in 1995, is shown in Table 5.5.

International aid policies also are affected by domestic political leanings. Nations tend to favor allies so that more aid generally flows to "friendly" nations (Schmitz, 1987). Furthermore, some aid comes with

TABLE 5.5. Contributions to Democratic People's Republic of Korea following Floods, 21 December 1995 (in U.S. $)

Organization	Contribution
United Nations organizations:	
Department of Humanitarian Affairs (DHA)	$50,000 emergency grant (provision of blankets, clothing, and kitchen sets); financing one DHA-UNDAC team leader
Food and Agriculture Organization (FAO)	Emergency assistance, under FAO Technical Cooperation Program, to support affected farmers' dry-season vegetable production (value = $201,000)
UN Development Programme (UNDP)	$50,000 (provision of blankets, clothing, and kitchen sets)
UN Educational, Scientific and Cultural Organization (UNESCO)	$40,000
UN Population Fund (UNFPA)	$150,000
UN Children's Fund (UNICEF)	$300,000
Universal Postal Union (UPU)	2,100 postal bags, 14 electronic scales (value = $26,087)
World Food Programme (WFP)	$2,168,000 (provision of rice) drawn from emergency food reserve, to be replenished by donor countries
World Health Organization (WHO)	$100,000 (provision of emergency health kits)
World Meteorological Organization (WMO)	$10,000
Intergovernmental organizations:	
European Community/ European Commission Host Organization (EC/ECHO)	$200,000 through UNICEF; $182,082 through MSF
International organizations:	
International Federation of Red Cross and Red Crescents Society (IFRC)	$50,000 disaster relief emergency fund
Governments:	
Australia	$75,188 (AUD 100,000) through DHA
Austria	$24,435 (AUS 250,000) through DHA
Belgium	$7,300
China	Relief goods worth $3,623,188 (CNY 30,000,000)
Denmark	$43,706 from DHA; $183,824 (DK 1,000,000) from WFP, financing one DHA-UNDAC expert
Finland	$23,295 (FIM 100,000) through DHA/WFP, financing one DHA-UNDAC expert
Germany	Baby food worth $68,027 (DM 100,000)
India	$100,000
Indonesia	$20,000
Iran	$100,000
Ireland	$50,000 through IFRC
Italy	Provision of relief goods from DHA Pisa warehouse, valued at $83,652
Japan	$120,000 through DHA; $120,000 through WHO; $260,000 through UNICEF
Luxembourg	Provision of relief goods from DHA Pisa warehouse, valued at $18,387

TABLE 5.5. *(cont.)*

Organization	Contribution
Malaysia	$25,000
Netherlands	$62,893 (NLG 100,000) through IFRC; $63,694 (NLG 100,000) through UNICEF
Norway	$220,000 through DHA
Pakistan	$160,000
Russia	Airlift of 27 tn of food, blankets, and medical supplies (value = $136,278)
Singapore	$20,000
Sweden	$113,798 (SK 800,000) through UNICEF, financing one DHA–UNDAC expert
Switzerland	$106,941 through DHA for 4-month mission of DHA relief coordination team; $147,540 (CHF 180,000) through CARITAS; 7,000 tn of rice, valued at $2,459,000 (CHF 3,000,000)
Syria	Grains worth $5,850,000
United Kingdom	$194,585 through DHA; $39,605 through IFRC
United States	$225,000 through UNICEF
NGOs:	
ADRA International	$700,000
American AID	$240,000
American Friends Service Committee	$10,000 through DHA
American National Council of Churches	$100,000
Association of Medical Doctors of Asia	Clothing worth $70,000; emergency health kits
CARITAS Austria	$30,000
CARITAS Hong Kong	1,400 tn of rice (value = $500,000)
Catholic Relief Service	$50,000
Church of Jesus Christ of the Latter Day Saints	80,000 lb winter clothing, 35,000 lb powdered milk, and medicine (value = $166,000)
Direct Relief International	$200,000
Food for the Hungry	Medical supplies (value = $925,000)
International:	
Medecins sans Frontieres (MSF)	Medical equipment with assessment team (value = $1,000,000)
National Red Cross/Red Crescent Societies	$1,011,580 through IFRC
New Apostolic Church, Germany	Medical supplies (value = $214,286)
Nippon Foundation (formerly Sasakawa Foundation)	$20,000 through DHA; $1,306,535 through IFRC
Rotary International Zurich	$5,000 through DHA
Samcheng Hai Association	$200,000
World Council of Churches	$10,000
World Vision International	In-kind contribution (value = $726,000), including 1,000 tn of wheat flour ($280,000), 4 noodle machines ($30,000), and medical supplies ($416,000)

*Red Cross/Red Crescent Societies included here are from Austria, Canada, China, Denmark, Finland, Germany, Iceland, Japan, Republic of Korea, Netherlands, Norway, Poland, Slovakia, Sweden, and the United Kingdom. *Source:* Emergency Information Administrator (1995d).

strings attached. The donor's transportation system (be it airplanes or ships) must be used, or aid may be linked to other activities, or it may create obligations on the part of recipients. The needs of disaster victims are sometimes secondary to the political baggage of donor and recipient nations. Having said this, most aid does go to less wealthy nations (as shown in Table 5.4).

Many hazards researchers have criticized relief programs for causing dependency (Williams, 1986), dampening interest in other forms of hazard mitigation that might be more effective (White, 1975), and creating as many problems as they solve. If inappropriately distributed, food aid during a famine can have a significant negative impact on local producers who have food for sale, resulting in economic disincentives for food production that, in the long run, only worsen the food shortage. In theory, aid should not induce dependency, but in practice, that is sometimes inevitable and given the difficulties, perhaps it is not surprising that disaster aid fluctuates between too little and too much. Nonetheless, better understanding of victims' needs and better planning could reduce waste and improve response significantly, and a relief program that is set up in advance of disasters can be much more effective in meeting victims' immediate needs than can less planned and organized efforts. A change in policy also is required. Greater coordination through an organization such as UNDHA would minimize the political machinations that compromise some relief efforts; if donor nations provided aid to a centralized, nonpolitical relief agency, half the problem would be solved. (Distribution within disaster areas is a more difficult problem to overcome.)

Cooperative and Mandated Programs

Programs can be cooperative, based on agreements and sharing of responsibilities; mandated, where an entity is coerced or required to implement a given program; or combine cooperative and mandatory elements. These categories are analytically useful, but cannot be easily separated in practice.

Fiscal Inducements

Financial policies can be instrumental in hazards management. They include positive and negative incentives through subsidies, loans, and differential tax systems, and can be formulated to encourage appropriate levels of development in some areas and discourage unnecessary investment in hazardous areas. Some policies encourage or require insurance against hazards or some other means of covering losses when they do occur, and such fiscal policies also alter the benefits and costs of location.

Taxation has been proposed as a basis for urban planning for a long time (Ascher, 1942) despite questions of legality. These are "discriminating" schemes, and in the United States, a fundamental principle of legislation is that it must be applied equally and uniformly. Historically, though, tax incentives have been used to preserve open space (U.S. Army Corps of Engineers, 1976), and certainly they could be applied more liberally to manage hazardous environments. However, there are dangers with such policies. It may seem reasonable to set a more onerous tax rate for hazardous development (such as floodplain encroachment), if only to make those in the area more fiscally responsible for their own actions. However, punitive taxation may merely encourage high-density investment and use of the area, which could lead to even greater hazard-related losses. To counteract this possibility, other measures should be used to attract development elsewhere or restrict the subdivision of land units and intensification of land use. Similarly, open land in and around cities often is assessed at high tax rates because of the great demand; owners may seek to relieve their tax burden by selling the land for further development, thus exacerbating hazard problems (Waananen and Spangle, 1977). A lower tax assessment on undeveloped hazard-prone property might reduce the economic pressure to sell, though further measures could be required to restrict development of this relatively cheap land.

For the most part, there has not been widespread use of financial measures to control hazardous development. Subsidies and low mortgage rates have been used to influence urban development, but usually the purpose has been to relieve economic hardship and unemployment in depressed areas or to promote new development. If such measures are to be effective in reducing loss, they will require a new community focus on mitigation of natural hazards.

Insurance

Several programs make hazard insurance available, handled by the public sector separately from or (more likely) in cooperation with private insurance companies. These programs can be cooperative, coercive, or both. In the United States, the NFIP began as a cooperative program where the federal government subsidized flood insurance for homeowners in communities that adopted floodplain management regulations. Following Tropical Storm Agnes in 1972, it became clear that few communities were joining the program and, as a result, there was a change from a cooperative to a mandated approach through the Federal Disaster Protection Act of 1973. As discussed in Chapter 4, communities identified as flood-prone must implement and enforce floodplain management regulations, and homeowners in designated 100-year floodplains must carry flood insurance. Noncompliance results in sanctions, including refusing disaster aid to the uninsured. In some instances, however, such as after the 1993 midwestern floods, the sanctions have not been enforced, which weakens the program by promoting the belief that the government will always bail out disaster victims.

In the United States, earthquake insurance is available through the private market, usually as an endorsement to a homeowner's policy. These policies usually have relatively high deductibles amounting to as much as 10% of the value of the insurance. To date, purchase has not been widespread, with reported rates of adoption varying from 5% of homeowners in some areas (Kunreuther et al., 1978) to 40% in others (Palm, 1995).

There are a number of differences between availability and purchase of earthquake and flood insurance in the United States, the most important for our purposes being that earthquake insurance is not included in a public program. NFIP represents a policy of distributing losses among potential flood victims rather than among all taxpayers, tied to the further goal of reducing vulnerability through land use management. Though recently NFIP has been self-supporting, Hurricanes Hugo and Andrew and the events of 1993 have clearly put pressure on its resources. In contrast, private insurance companies in the United States have agitated recently for mandating earthquake insurance throughout the country. Earthquake insurance has not been successful because only those in high-risk areas have taken out policies, which, given the high deductible, is really not surprising (Palm, 1995). Insurance companies wish to spread the financial burden over the country, irrespective of local social and geological conditions.

Public Loss Recovery Programs

In many cases, it is beyond the ability of local government to repair damages to public infrastructure resulting from an event. The United States has a procedure that is implemented in such situations; when an event occurs, the magnitude and range of loss are assessed and application is made to the state and federal governments for a disaster declaration. In the 1993 midwestern floods, 487 counties in 9 states were included in a presidential declaration of disaster, and the declaration for the East Coast blizzard of 1996 covered 8 states. After a disaster is declared, federal resources and supplies become available to the affected area. However, there may be strings attached to the money, such as requiring a mitigation plan to reduce losses in the future. Similar policies have been adopted in New Zealand. During restructuring in the late 1980s and early 1990s, the central government put in place the Local Authority Protection Programme (LAPP). Under LAPP, the central government pays only 60% of restoration costs for essential services. Local authorities must not only cover the other 40%, but must demonstrate that they can meet that obligation before the central government's assistance will be made available. LAPP is an excellent example of a cooperative venture with mandated requirements.

Relocation

One of the general aims of legislation is to separate, so far as possible, the human-use system from the vagaries of nature. Taken to the extreme, this would entail the relocation of settlements to hazard-free sites. Clearly, that would not be practical for most urban centers, and it would be simplistic to suggest that all hazard losses could be eliminated in this way. Apart from the socioeconomic factors that would preclude such movement, there may be sound locational benefits to the existing site. Given the additional problems of finding a truly hazard-free site, relocation usually cannot be justified on a large scale, but despite the problems, some communities have moved to safer locations. Typical characteristics of such communities include a relatively small size, a well-defined natural hazard, available nearby land, and good economic support.

Flooding has been the stimulus for several relocations. In the United States two early examples were Shawneetown, Illinois, and Leavenworth, Indiana, but for the most part, the moves were failures because they lacked the full support of the total population (Murphy, 1958). More recently, Big Store Gap, Virginia, undertook a voluntary relocation supported by 80% of the population, and Rapid City, South Dakota, and Prairie du Chien, Wisconsin, have moved some properties off the floodplain (Natural Hazards Research and Applications Information Center, 1976). One of the most significant moves to date has been the relocation of the central business district of Soldiers Grove, Wisconsin. The local community undertook most of the planning and organization and, at the time, was very supportive of the move; however, subsequent studies have shown that not all local residents are enamored of the new town layout (Tobin, 1992; Tobin and Peacock, 1982). Following the 1993 midwestern floods, portions of several communities relocated rather than mitigate recurring floods (Tobin and Montz, 1994b); in Valmeyer, Illinois, 265 properties have been acquired by FEMA as the community moves out of the flood hazard area.

There are other examples of relocation. In the Indian state of Gujarat, 30 villages have been relocated to higher ground above the Narmadi and Tupti Rivers; a prime consideration in these moves was not to separate the local population from the fertile floodplains. In the more extensive lowlands of Uttar Pradesh, about 45,000 villagers have either moved or raised ground levels (UNDRO, 1977). In Alaska, the town of Valdez retreated from low-lying areas that were unstable in earthquakes and prone to subsidence and tsunami. In Italy, 9,000 residents of Pozzuoli were relocated in 1970 due to volcanic problems (Legget, 1973). In Hawaii, Hilo has been partially relocated following extensive tsunami damage in 1960 that killed 61 people and caused $22 million in damage. In all these cases, relocation reduced loss.

Direct Public Acquisition

Direct public acquisition is one way to facilitate hazard mitigation policies. Authorities at any level can purchase land to control future development. While this may be costly in the short term, it can lead to considerable savings in disaster relief over the long run. Acquisition can entail buying established urban lots as they become available or obtaining new land undergoing urbanization, for which authorities may introduce "preemption rights" legislation giving public agencies first option on the purchase of any land. A more extreme measure adopted by some communities is the compulsory purchase order, which permits government to control or enforce land use change and urban development. Switzerland has exercised preemption rights extensively to protect agricultural land from urbanization and, in the United Kingdom, the compulsory purchase order has been used fairly widely for urban renewal.

It would be relatively easy to apply such measures to hazard mitigation. Having purchased the land, policymakers could more freely choose strategies consistent with an urban plan. Local authorities might take primary responsibility for development or, alternatively, strictly control the designs of subcontracting private developers. Once acquired, development of anything from small-scale facilities (to encourage acceptable private investment) to large-scale residential projects or even entirely new settlements can be required.

Public acquisition need not be negative or coercive. Certainly, land can be turned over to less intensive uses such as recreational parks or parking facilities, which many communities have done. In the United States, FEMA has begun to sponsor local authorities' acquisition of real properties on floodplains, which was done in Arnold, Missouri, and other communities after the 1993 floods.

Short of acquisition, authorities may turn to land registration. Since the land market in many countries is so complex (involving public and private activities), it is often difficult for public officials to maintain an accurate inventory of the number, value, and types of land transactions. Such records are essential for effective land control. Registration of land would fulfill this need and help officials to introduce measures such as zoning ordinances, building codes, and financial policies. In New Zealand, one local authority tags the deeds of properties located in designated hazard zones to ensure that subsequent owners are aware of hazards

(Montz, 1993). The use of Geographic Information Systems (GIS) to register and map changing land use can greatly facilitate this measure.

CONSTRAINTS ON PUBLIC POLICY

There are many alternative policies that can be adopted by any or all levels of government, and there are numerous ways in which policies can be implemented. However, development and implementation of hazard mitigation policies are not simply a matter of coming up with the right idea. Mitigation strategies often are constrained by societal needs and limited resources. Day-to-day concerns are usually at the forefront of community decision making, and elected officials must address such issues as traffic, schools, and commercial enterprise before advancing apparently exotic items. Unfortunately, this often relegates hazard mitigation discussions to postdisaster periods. Thus, we see flood alleviation projects implemented immediately after an event, houses raised on stilts after a hurricane, or building codes strengthened after a particularly devastating earthquake. In other words, the hazard is not identified as a public problem or the risk fully appreciated until after a disaster has occurred.

It is worth contemplating, therefore, what constitutes a public problem and which factors might lead policymakers to define it as such. It is possible, for instance, that hazards are viewed as public problems whenever there is a significant threat to lives or property. In that case, flooding of rural areas and parklands would not evoke public concern, but the inundation of large residential or commercial and industrial areas would almost certainly lead to demands for public action. Other concerns might be frequency of occurrence, magnitude, or spatial extent of a geophysical event; technical parameters; financial and economic concerns; legal, administrative, and ownership restrictions; social and psychological factors; and political interests. Each of these places constraints on public decision making and can be used to assess proposed projects. (For discussions of these considerations, see James and Lee [1971]; May and Williams [1986]; O'Riordan and More [1969]; Rossi, Wright, and Weber-Burdin [1982]; and Smith and Tobin [1979]).

Physical and Technical Constraints

These constraints apply most directly to issues of engineering feasibility. In a sense, all alleviation projects are technically feasible because, given sufficient resources, we could implement virtually any project we wanted. If we wished to redirect the Yangtze River to eliminate flooding or to build a dome over Tulsa, Oklahoma, to protect inhabitants from tornadoes, we probably could do so. Obviously, these are ill-conceived ideas, not least because money could be spent more effectively on other projects. The point is that most projects are technically feasible. However, the physical environment restricts the range of policies that can be implemented. For example, it is possible to install a flood warning system on most rivers, but in a flashy drainage basin, the rate of onset might not allow time to issue warnings and mobilize action.

A detailed assessment of the physical environment should be made before project implementation. Failure to undertake adequate study has created many problems in urban areas and exposed many people and considerable property to risk. The urbanization of earthquake-prone areas and the encroachment of residential and commercial property onto floodplains exemplify the increasing exposure of populations to risk (see Chapter 7 for a discussion of risk). Many public policies have actually increased vulnerability and exacerbated impacts. For example, new town construction on steep slopes in parts of Japan has raised the potential for landslides (Nakano, 1974), and flood control structures have induced development in "safe" areas (Montz and Gruntfest, 1986; Tobin, 1995; and White et al., 1958). Of course, a physical assessment does not guarantee successful policies. A policy of encouraging pastoralism in the Sahel seemed to be supported by research on water resources, but it was soon learned that the regional ecology could not support so many permanent settlements; the ensuing droughts killed thousands of people (Haldeman, 1972; Schmitz, 1987). On other occasions, research findings have been ignored. During a major earthquake in 1964, buildings in the Turnagain Heights neighborhood of Anchorage, Alaska, suffered extensive damage when soils began to liquefy; this possibility had been predicted by the USGS several years earlier.

Nonetheless, detailed studies should identify which extreme geophysical events might occur in a particular area, determine their spatial extent, and ascertain their intensity, duration, and speed of onset. Much of this information is already available but should be made site-specific to form the basis of policy. Plans should be contingent upon detailed analysis of the local environment because such information can affect the efficacy of different measures. For instance, zoning may be ideal when hazards are spatially well defined (such as floods), but unsatisfactory for a meteorological hazard (like tornadoes). The relative patterns and intensities of different hazards should help to determine the appropriate strategies. Flash floods can cause intensive local problems resulting in severe structural damage, but more extensive downstream flooding can cause problems of long-term duration (such as sedimentation); a policy emphasizing structural design would need to assess the floods' intensity and other characteristics. Similarly, policies aimed at mitigating earthquake damage are confounded by the multifaceted effects of seismic activity (see Chapter 2), which can have a profound influence on earthquake safety, especially in urban areas. For example, several large buildings in Japan and Alaska have collapsed due to the liquefaction of underlying clays. Very detailed local studies are required to identify all potential problems associated with earthquakes. Furthermore, no policy should isolate a single hazard as there are delicate links between events (Hewitt and Burton, 1971; Montz, 1994); for instance, earthquakes can precipitate avalanches, slope failures, fires, and floods. Ciborowski (1978) advocated consideration of the "chain of events" in hazards research, noting that both Tokyo and San Francisco experienced extensive fires following catastrophic earthquakes. Similarly, in 1995, there were extensive fires in Kobe, Japan, following an earthquake.

Assessment of technical factors also is critical to the success of public policies, especially when large-scale structural measures are being considered. Without due attention to the limitations of building design and construction standards, there is every possibility that structures will fail during low-probability events. The huge monetary investment in the structure will be lost, and many people and much property may be placed at risk. There are many examples where this has occurred. In 1926, when the St. Francis Dam in California failed, 426 people were killed because inadequate attention had been given to the local geology (Walters, 1971). The Teton Dam in Idaho failed in 1976 because of failure to consider springs and ground-water hydrology, and the Van Norman Dam in California rotated after an earthquake. In Britain, the Woodhead (1850), Holmfirth (1852), and Skelmorlie Dams (1925) failed because of unforeseen geological weaknesses (Amey, 1974), and the Dale Dyke (1864) and Dolgarrog Dams failed because of poor design or construction (Amey, 1974; Smith, 1971).

Other failures have occurred because the geophysical event exceeded the design capacity of the structure. In the summer floods of 1993, 70% of the 1,576 levees in the upper midwestern United States failed, many because the magnitude of the event exceeded design standards (Tobin and Montz, 1994b). (It should be noted that only 20% of the federally approved levees failed whereas over 75% of the private levees were breached.) However, it is not practicable to design mitigation projects to withstand the lowest probability event; doing so might even be undesirable economically and aesthetically. It is possible to estimate the maximum probable flood, the potentially highest intensity earthquake, or the most severe hurricane for a given location, but the reduction in risk does not warrant such drastic action. It is therefore up to policymakers to determine an acceptable level of risk.

One confounding problem is the lack of detailed historical records.

Though it is possible to estimate low-probability events (as discussed in Chapter 2), the lack of accurate data increases the potential for error. For example, after the eruption of Mount St. Helens in 1980, flood flows on the Toutle River exceeded the estimated maximum probable flood. It seems that calculations had failed to consider the possibility that all the ice and snow would melt almost instantaneously. Seldom do we have sufficiently lengthy records to determine accurately the return periods for events of different size.

Financial and Economic Constraints

Policymakers are perhaps most constrained by limited financial resources and competing demands for them. If money can be found to pay for a project, then financial feasibility is satisfied, which means only that resources are available for the project, not that it is economically viable. While financial feasibility would seem contingent on technical feasibility, some projects have been implemented that do not work as expected. For instance, some dams on tributaries of the upper Mississippi River control downstream flows much less than was projected. On the other hand, some projects are technically and economically sound, but the policies are not implemented because financial backers cannot be found. Frequently, the funding needed for such ventures is invested in projects or programs for which results are more readily seen.

Economic feasibility is satisfied if the total benefits outweigh the total costs of implementation (discussed in detail in Chapter 6). If carried out in sufficient detail, a cost–benefit analysis of different projects can help to determine the most efficient strategy. Unfortunately, there are many examples of cost–benefit analysis failing to incorporate all the costs and benefits, therefore providing inadequate results. Further, a small change in the discount rate can alter the feasibility of particular measures significantly. For example, La Farge Dam on the Kickapoo River in Wisconsin was authorized after a favorable cost–benefit analysis, but after $52 million had been spent and the dam was almost complete, evaluation showed that it was no longer economically viable; the dam now stands as a monument to inadequate attention to economic constraints (Tobin and Peacock, 1982). In other cases, social, environmental, and aesthetic costs have been overlooked, which also compromises projects. Finally, we must consider more carefully who benefits from and who pays for particular policies and their implementation.

In spite of these criticisms, a careful cost–benefit analysis can provide invaluable guidance. It is useful for individual projects, and also helps to discriminate among policies, even when they have different goals. For instance, it is possible to compare the value of a hazard mitigation project with that of a new transportation link. Used correctly, the technique helps to distribute resources between competing demands.

Legal, Administrative, and Ownership Constraints

Other constraints on public policy include questions of legality and administration. Many nations maintain a firm belief in individual property rights that can seriously impair the effectiveness of land use planning, and the power of the state or local authority to acquire property or regulate land use may be severely limited. In Costa Rica, for example, the constitution states, "Property is inviolable; no one may be deprived of what is his except for legally proven public interest after compensation in accordance with the law." This constitutional right is common to many countries, including Paraguay, Kuwait, the Philippines, Turkey, and Peru (UNDRO, 1980). Thus, even elementary zoning legislation to restrict land use may be challenged for proof that it enhances the general public welfare. More active measures, such as public acquisition of land, may be curtailed even more stringently.

There has been hostile reaction to regulation of private property in the United States, as witnessed in a number of court cases (Platt, 1994). However, in view of the impact hazardous development can have on citizens, there are strong arguments to support public action. In a study of coastal erosion in Denmark, Bruun (1972) suggested that hazard planning should be based on criteria similar to those guiding water supply and public access; in other words, where development on the coastline was likely to be detrimental to others, it should be prevented.

Some hazard-related policies have been constrained by limited enforcement. Well-intentioned legislation has failed in the past because funds were not available for implementation and administration. Simply put, the policies could not be enforced (see above). Nelson (1980) found that because of limited finances, there were too few inspectors in California to enforce building codes for earthquake safety. Also in California, the development and grading of slopes have increased their instability, largely because there has been little attempt to enforce local ordinances (Griggs and Gilchrist, 1977). In some poorer nations, enforcement problems are serious; the administrative structure at the local level may be unable to comply with sophisticated planning policy. Beirut experienced rapid, unplanned expansion along with a growing squatter community because local legislation was not enforced (UNDRO, 1980), and some of the structural damage in the 1985 earthquake in Mexico City has been attributed to nonadherence to construction specifications (Havlick, 1986).

Other Constraints

Many other constraints affect adoption and implementation of policy. Certainly, social and psychological factors influence which policies are implemented and how (related closely to hazard perception and community response). Similarly, all policies are constrained politically. That is, the political climate in which they are considered largely determines what they address, their cooperative or mandated character, and who is made responsible for implementation and enforcement. A discussion of political constraints is beyond the scope of this text as it would involve analysis of the policy process in many contexts, but it is important to recognize that what is possible and effective in one place may not be politically feasible elsewhere. In the end, hazard policies cannot be understood fully without attending to local context.

LEVELS OF GOVERNMENT INVOLVEMENT

Given that natural hazards represent public problems that can be addressed through a variety of policy instruments, we might ask why certain levels of government get involved and others do not. In the United States, for instance, flood insurance comes under the auspices of the federal government, which also has spent billions of dollars on large-scale flood mitigation. In Canada, the national government has made it policy not to get involved in flood alleviation. Two similar societies have chosen opposite approaches to the flood hazard. How can this be explained?

The implementation of carefully designed programs to mitigate loss would appear to be both socially and economically beneficial to any urban center, but as indicated above, few communities have adopted comprehensive strategies for hazard reduction. Many influences on urban planning drastically affect the effectiveness of such strategies; these range from political awareness and commitment at the local or national level to legislative and fiscal constraints. Should planning be administered at the national level to encourage local authorities to adopt a positive attitude toward urban hazard mitigation through land use management, or should regional and local administrations accept such responsibility? Within the poorer nations, the tremendous losses suffered would suggest that some form of national directive is required, especially since a substantial portion of gross national product may be going toward hazard relief. In addition, hazards constitute extensive problems that transcend administrative boundaries, generating more than local interest, but enforcement of various measures may be local.

In some instances, local or regional responsibility for hazard management has failed to reduce loss. In the United States, land use planning is the prerogative of state and local authorities, but they have failed to prevent rising losses. As a result, a closer liaison has developed between national and local governments. The federal government has adopted a firmer line by insisting that local communities implement local zoning ordinances as a condition of participating in NFIP and qualifying for its assistance. In the United Kingdom, government circulars have tried (often with little success) to encourage cooperation between local authorities and water organizations to control floodplain development (Smith and Tobin, 1979). Other studies suggest that locally oriented projects are more effective, not least because they are more site-specific in outlook. Several cases of floodplain planning have been highly successful because of sustained effort at the local level (National Science Foundation, 1980), and the Community Rating System within NFIP builds on these successes.

Effective policy implementation for hazard mitigation probably can be achieved only through a combination of national directives and detailed local studies. In a decentralized system, the influence of the central body is limited so that cooperation among different administrative levels is necessary for a consistent hazard policy to be pursued. In a centralized system, general planning policies may be well defined, but the important local circumstances may be overlooked. Thus, while there is no one solution for all hazards, a national commitment would appear to be necessary, but no policy should be implemented without a detailed survey of local requirements.

The commitment and attitude of local authorities also can be major determinants of the effectiveness of policies and planning. On occasion, plans have changed due to external pressures; for example, Xenia, Ohio, tried to implement local zoning ordinances following severe tornado damage in 1974, but was pressured by certain commercial interests into waiving its plans (Spangle, Mader, and Blair, 1980). In Managua, Nicaragua, following the major earthquake of 1972, a program to reduce risk by deconcentrating the downtown area was adopted; although it probably has had long-term benefits, at the time it caused considerable hardship to low-income families, having been proposed by a Mexican team that probably gave inadequate consideration to their immediate needs.

Local attitudes and support also can influence implementation of land use policy. For instance, Nelson (1980) implied that local apathy, combined with only limited funds, brought earthquake disaster planning in California to a virtual halt. In many communities, concern for individual freedom predominates to the extent that people are allowed to locate anywhere they choose. Similarly, firm individualists often believe that those who live in hazardous areas should be responsible for themselves, which is not always satisfactory; many people in a community may have been affected by unplanned development that increased their vulnerability. For instance, urbanization of floodplains is known to increase the incidence of flooding (Hollis, 1974; National Science Foundation, 1980). The costs of rescue missions, relief, rehabilitation aid, and general repairs to roads and services usually are borne by the community at large rather than just those individuals affected directly. In Brisbane, Australia, over 22% of the cost of damage was absorbed indirectly by the national government (Butler and Doessel, 1980). On a larger scale, the next major earthquake in San Francisco probably will affect the whole of the United States through taxation and disaster donations (Bowden and Kates, 1974). A laissez-faire attitude toward urban development is not conducive to hazard mitigation.

Given all the constraints on policy development and implementation (and particularly on urban land use planning), perhaps it is not surprising that few comprehensive plans have been implemented. In poorer nations, certainly, there are questions of priorities. Can these nations afford to restrict development, and is their administrative structure capable of supporting and enforcing such a program? They face numerous problems, many more pressing than the control of future or potential hazards, but a long-term strategy eventually would aid rather than hinder economic development (if only in reducing significant loss). Squatter communities add to problems of public safety and health, and some way must be found to enhance the livelihoods and personal environments of marginalized people. In wealthy nations, failure to overcome planning constraints and recognize the damaging potential of unmitigated hazards increases risk in nearly every urban community. A more equitable and intelligent approach is needed.

THEORY AND PRACTICE

Some element of risk exists in all settlements, and no measure of planning can reduce it to zero. The extended droughts in the United Kingdom in 1976 and 1995 demonstrated that even humid environments can experience severe water shortages, while many other communities around the world have suffered from rare but catastrophic events. Planning for low-probability events may not be possible because a community usually is faced with many more pressing social problems, but for more frequent and identifiable hazards, some form of community response is advisable if damage is to be ameliorated. One of the fundamental questions in policy formulation is, at what level of risk is planning necessary? The level of concern demonstrated by a community over one particular hazard is important for effective hazard management, and, once need has been established, strategies based on various criteria can be formulated.

The Need for Broader Planning

Several studies (cited in Chapter 1) indicate that losses from hazards are rising despite increased expenditures on hazards. This apparent anomaly would suggest that something is fundamentally wrong with society's approach unless the incidence of extreme geophysical events also has risen, which does not appear to be the case (although global climate change may significantly change this picture). Two hypotheses might be tested to account for these anomalous trends. First, mitigation measures have failed to achieve even a modicum of success and, second, continued urban development in hazardous areas has aggravated the problems.

Although many adjustments have significantly reduced loss, alleviation projects have not always had their desired effect. A single-purpose, large-scale structural measure may reduce losses and save lives, but its subsequent effects on community development may cause other problems. Such measures (for example, dams and embankments) often stimulate investment in hazardous areas. The adjustment instills a false sense of security that the hazard has been eliminated, thereby opening the way to further economic development, but when its design capacity is exceeded, catastrophic losses may result. Single-purpose measures that attempt to control nature are insufficient; other approaches (based perhaps on some form of land use management and loss reduction and distribution) are needed for full effectiveness. The problem has been recognized for some time (Segoe, 1937), but often is overlooked by communities seeking to minimize losses.

Urban development in new areas also has added to rising losses. Even when initiated by local authorities, construction frequently is undertaken without regard for environmental constraints. Even when individuals or communities consider exogenous factors, decision making too often is clouded by limited experience and a failure to appreciate the full costs of particular decisions. Before any development decision is finalized, its relative advantages and disadvantages should be evaluated so that all the risks will be balanced. Clearly, there are advantages to many hazardous locations (such as fertile soils around a volcano or the aesthetic value of some river locations), but these should be carefully weighed against their hazards. Better urban planning is needed to prevent further encroachment into hazardous areas and to minimize future losses of life and property. In addition, we can assume that losses eventually will occur so that some means of recovery must be in place.

Risk Assessment

The risk of occurrence of any situation must be evaluated, requiring risk assessment and analysis of risk acceptability (discussed in detail in Chapter 7). It is the magnitude of risk that makes natural hazards the subject of public policy.

As pointed out earlier in this chapter, it is not difficult to map hazard-prone areas for events with distinct locational characteristics. Various physical criteria can be mapped to show areas of relative safety and degrees of risk. Hazard maps can then form the basis of sound urban planning for hazard mitigation, resting on detailed studies of the physical environment and their careful application to the human-use system. Details on the frequency and expected intensity of an event would add to the risk assessment of any area.

There is nothing new in this. A similar methodology was proposed by McHarg (1969), who suggested that planning policies could be based on a series of maps, each showing different parameters so that several factors could be considered simultaneously. Arora (1976) describes the application of this methodology to environmental planning in Carlisle, Massachusetts, but only fairly recently has attention been given to its usefulness in hazard planning. Land capability studies have been made to assess the potential risk of different areas (Laird et al., 1969) and comprehensive vulnerability maps, such as one for Greater Manila (UNDRO, 1979), have been constructed to indicate the relative risks from various hazards in different parts of an area. Perhaps the most comprehensive hazard maps have been drawn for California, which has been subject to exceptionally detailed studies. Over time, several series of maps depicting areas susceptible to damage from the many processes associated with seismic activity have been developed (Blair et al., 1979; Association of Bay Area Governments, 1995), and in some cases, composite maps of risk have been produced (Blair et al., 1979). GIS enhances our analytical capabilities far beyond those imagined by McHarg (French and Isaacson, 1984; Montz, 1994), and the Internet has made these maps more widely available (as the map in Figure 5.5 shows).

Thus, it is not methodology that hinders analysis. Given that communities normally wish to maximize the use of available land and minimize the detrimental effects of development, it would seem reasonable to expect consideration of potential hazards in any decision making. However, rational policies are possible only if decision makers are aware of local hazards. A preliminary step should be the identification of potential areas of risk.

A further problem lies in issues of risk acceptability, including which level of government makes that determination. The risk of flooding in a particular community can be calculated, and we frequently talk about the

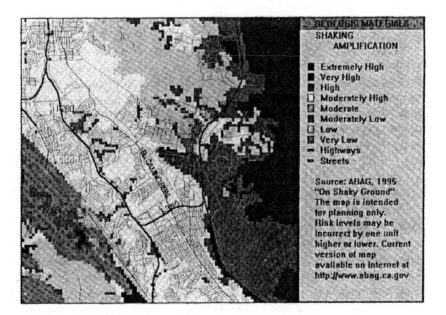

FIGURE 5.5. Risk areas in south San Francisco, California. This map (based on shaking amplification from seismic activity) is a black and white version of a color map available on the Worldwide Web. Other risk maps for this and other areas are available on the Web as well. *Source:* Association of Bay Area Governments (1995). Copyright 1995 by ABAG. Reprinted by permission.

100-year or 1% flood. However, because there are many flood-prone communities within a state or region, the risk of losses for a state is greater than that for individual communities. At the federal level, there is much greater certainty that damaging floods (or any other event, for that matter) will occur over a given period of time than there is for a given community. Each level of government has a different level of risk (with three levels of government, there are three levels of risk), and it is that risk to which policy is a response.

Two criteria are of fundamental importance in assessing risk acceptability. First, should an economic value be placed on human life? Any decision (be it for a dam, a levee, or a building ordinance) is based on its perceived effectiveness, which includes an estimate of the number of lives it will save. Nevertheless, an element of risk remains, and indirectly a valuation is placed on those who will die in the next event. It is quite probable that with stricter zoning or greater financial investment in mitigation, fewer deaths might result. What value to place on human life presents many problems, not all of which have been resolved satisfactorily (see Foster, 1980).

The second and perhaps main criterion on which decisions are made is economic. Planning for hazard alleviation has clear benefits in terms of reduced loss, but must be balanced against the costs of implementing and maintaining a policy. If land is to be left "open," then costs may include those benefits forgone by not developing the land. If buildings are to be hazard-proofed to withstand all foreseeable events, costs increase with the level of protection. This ratio of benefits to costs would seem reasonable if decision makers were seeking to maximize benefits, but the full impact of hazards is rarely considered. Even when individuals accept monetary responsibility for occupying a hazardous location, they rarely pay fully for any losses, as shown in an Australian study (Butler and Doessel, 1980). Other evidence of the failure of this approach, whether by design or due to external factors (see below), is the continuing rise in losses. If basic economic principles were effective, high-risk areas should be avoided and low-risk ones developed. Since communities have continued to expand into hazardous areas and then to call on government for assistance, clearly policies are not working and further controls are necessary.

Apart from purely economic constraints, the social acceptability of any proposal would seem important to its success (unless it is to be enforced by strict and rigid police powers). For example, zoning has been described as a major planning device, but its success is contingent upon acceptability of the concept. To separate human activities from the extremes of nature is a sound concept, and in the long run can reduce losses from natural hazards significantly. However, every zoning decision depends on an appraisal of the risk involved. It is not feasible, nor would it be desirable to exclude all urban development from hazardous areas and, in permitting some development, an element of risk is left. It is up to the policymaker to evaluate the level of acceptable risk so that the advantages of the location will exceed its disadvantages.

Political factors also affect planning. In some countries, the acceptability of different measures may be in doubt depending on differing constitutional beliefs. Planning may involve zoning, open space, subdivision requirements, or even building codes. Alternatively or in conjunction with those measures, public officials may wield financial inducements, perhaps to coerce developers into a more rational pattern of land use. Finally, governments may take direct action to prohibit unwise development, for example, by acquiring land or strictly controlling investment. While most of these measures can be implemented through national plans or at the local level, they all have the same objective: the rational and safe development of urban areas. However, there are inherent problems in political structures. In fact, there are two, distinct political worlds: "normal" politics where disaster policies have little political salience and are relegated to low priority, and the politically active period immediately after an event (May and Williams, 1986). The politically most popular actions are to augment relief efforts and promote reconstruction while pushing preparedness and mitigation into the background.

Comprehensive Planning and Policy Development

Communities are not usually prepared to leave the future to chance, and many now adopt comprehensive plans to guide future activity. Even Houston, Texas, which for a long time was a strong proponent of unregulated development, has established some zoning ordinances. However, planning should not be static, but must be continual in order to encompass a wide range of evolving factors. Since physical processes are constantly changing, planning must be vigilant to changing problems, and any policy must be sufficiently dynamic and flexible to deal with new problems. For example, urban development can alter the frequency and location of some hazards, and floodplain encroachment can aggravate the flood hazard by blocking the floodway and raising water levels. The introduction of a structural measure in one part of a drainage basin may have serious repercussions on flooding in another part. Development can affect coastal erosion elsewhere, as studied in the Netherlands (ten Hoopen and Bakker, 1974), just as slope failures in California have been caused by urban development and improper grading (Griggs and Gilchrist, 1977). According to Gupta and Rastogi (1976), the construction of Koyna Dam in India apparently caused an earthquake that killed 200 people and injured over 1,500, and they document nearly 20 other cases around the world where seismic activity was associated with dam and reservoir construction. Policies must be sufficiently flexible to cope with these changing and sometimes unexpected circumstances.

Planning also should be comprehensive in design, ideally weighing all societal goals against the constraints of nature. However, comprehensive planning is a fairly recent consideration despite various proposals for it in the past. Babroski and Goswami (1979), Sheaffer (1969), and Weathers (1965) describe more comprehensive approaches to flood problems that are part of a trend in flood alleviation away from single-solution engineering structures to comprehensive management incorporating a variety of measures and planning (for example, see Dzurik, 1979). Davis (1978) suggested a more substantial methodology for floodplain management to take account of the whole drainage basin rather than focusing on individual communities. Following the 1993 floods in the Midwest, the issue of comprehensive, coordinated management of floodplains resurfaced and efforts to achieve it were initiated (Interagency Floodplain Management Review Committee, 1994). Recent work on planning for hazards reduction calls for considering the suite of hazards that occur at a location rather than concentrating on individual hazards (Burby et al., 1991; Montz, 1994).

Planning is probably the most comprehensive means of implementing hazard mitigation policies. However, it is not the only means, nor does it work in all cases. Thus, it is not a comprehensive plan, but rather a comprehensive policy that will be most successful, incorporating elements of land use management, construction regulations, financial inducements or sanctions, and cooperation among all actors. However, this is complex, because it involves hierarchical levels of government and the private sector, contrasting goals, and different implementation possibilities. In addition, there are constraints within or around which any hazard mitigation policy has to work. Nonetheless, hazards are public problems involving potentially very large losses, which we can expect to increase with the rapid urbanization that is occurring in many countries, creating major future catastrophes. Concerted policy responses are required, and we must recognize why they are so difficult to develop and implement.

Many studies have been made of hazards, and there is considerable information available on hazard policies and planning. This chapter has tried to pull the two together more closely, for it seems that planning failures and irrational development have occurred primarily because of neglect of contextual matters rather than lack of information. Nevertheless, some developers and planners remain ignorant of the risks associated with hazards and others, motivated by greed, choose to ignore this information. Various authors have advocated greater consideration of the physical environment, which is admirable, but it should not go so far as to ignore completely the human-use system. There may be a question of priorities among the many other problems which society has to face, and this may be particularly true for developing nations currently undergoing rapid urbanization. Thus, natural hazards should not be taken out of the context of all challenges facing societies.

Nevertheless, well-conceived policies that include elements of planning can reduce hazard losses in the long term. There is a moral obligation to encourage the adoption of measures that can reduce human tragedy, and a focus on hazard victims—especially those who are socially, economically, and politically marginalized—may provide a new starting point for effective hazard mitigation policies.

6 The Economic Impacts of Hazards and Disasters

INTRODUCTION

So far, we have addressed natural hazards as actual and potential events to which humans respond based on a variety of factors, ranging from personality characteristics to the policies of governments at all levels. This chapter takes a somewhat different approach by evaluating impacts in a specific arena, that of economics. Evaluation of the economic impacts of extreme natural events provides a measure of their costs to society. That is, the magnitude of disasters can be assessed physically, socially, or economically; costs quantify the monetary value of property loss and the monies associated with adjustments or mitigation. Combined, of course, with human loss, economic loss shapes attitudes toward reducing personal or group vulnerability to future events. We can be more or less motivated to seek adjustments, based in large part on experienced and anticipated loss relative to the costs of adjustment.

The losses attributed to natural hazards were discussed in Chapter 1, but it is worth emphasizing that the larger events account for a considerable percentage of total deaths and damages. In the United States, Hurricane Andrew which devastated southern Florida in 1992, the Mississippi River floods of 1993, and the California earthquake of 1994 exceeded all previous events economically. In Bangladesh, a tropical cyclone in 1970 killed at least 300,000 people and 280,000 head of cattle, and destroyed crops worth $63 million (Frank and Husain, 1971). The story, of course, is never ending and future earthquakes, hurricanes, and floods will surpass even those records.

While it is clear that economic impacts are important considerations, there are some serious problems with an overreliance on them in decision making or in understanding the range of impacts associated with natural hazards. As discussed below, total dollars lost in property damage or spent in cleanup, even if they could be measured accurately, do not delineate fully the range of losses experienced in an economy. In most places, it is difficult to obtain reliable figures for economic loss because of difficulties in precisely determining the costs of emergency action and repairs; as a result, we often represent the severity of an event by the number of lives lost. Some economic losses are not easily measured, such as increased personal transportation costs due to locating temporary housing at a distance from the permanent home or work, or impacts on the market value of subsistence agriculture. In addition, estimates of how much an individual life is worth are open to considerable debate. While understanding the range of economic impacts associated with natural events is essential to understanding their effects and subsequent responses, we also recognize that they do not tell the whole story. However, economic impacts have direct relevance to policy and action, and therefore must not be overlooked.

THE RANGE OF CONSIDERATIONS

Natural hazards have economic impacts at several scales, the most obvious being the losses incurred by a locality from damaged infrastructure and property. Those losses also may be felt over a wider geographic area (depending, for example, upon the magnitude of the disaster, the importance of the affected area to the region, and the functional relationship between the locality and other levels of government), and adjustments have both costs and benefits that may not be shared equally by those in the hazardous area and elsewhere. Intuitively, one would expect events to have negative economic impacts (costs) and hazard alleviation programs to have positive effects (benefits), but that pattern does not always hold.

This chapter does not provide a detailed financial accounting of losses from natural hazards. Such data suggest the severity of events, but remain snapshots of particular points in time rather than documentaries of the long-term economic impacts of natural hazards on society, providing insufficient detail to evaluate the total benefits or costs over the long term. While we need to know the cost of individual events, we also must consider the record of costs over space and time and integrate it into our analysis. In addition, hazardous events produce what are termed "direct" and "indirect" costs and benefits to society, but many are not taken into account when the economics of natural events are summed. Consequently, this chapter goes beyond simple accounting to bring together economically based research and what we have learned from case studies in order to develop an integrated approach to evaluating economic impacts.

As noted above, the economics of natural hazards must be examined with respect to impacts at different spatial and temporal scales. Natural disasters cause losses to individuals, neighborhoods, and communities, but the impacts go far beyond the disaster location. While individuals often experience damage to housing and personal property, communities experience similar problems with infrastructure, such as transportation networks and utility lines. Such losses can have regional or even national impacts, depending on the spatial extent of the event, economic linkages between the affected area and other areas, and the amount of relief funds allocated.

Clearly, the scale of a physical event can be important as larger events usually have a broader impact. Further, the probability of occurrence increases as the scale changes from the local to the national and even international levels (Figure 6.1). As probability of occurrence increases, so, too, does the probability of economic loss. While larger events have impacts beyond the directly affected locale or region, the cumulative effects of increased occurrences at higher levels also can be significant. However, if one considers the proportion of economic assets affected by a disastrous event, the local level suffers more. Though this chapter focuses on economic impacts in the directly affected area, the various spatial scales at

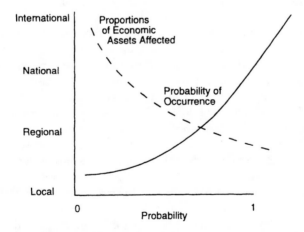

FIGURE 6.1. Relationship between disaster-related loss and spatial extent. Economic impacts can extend far beyond a disaster area. The probability of experiencing economic loss from natural events increases with increasing spatial area. However, while the impact of loss increases in absolute terms with increasing spatial area, it decreases in relative terms. Thus, the proportion of an area's assets impacted by a disaster is greatest at the local level.

which economic loss may be felt must be appreciated to understand its full range.

Temporal considerations also are important in identifying immediate loss and long-term impacts. Economic impacts do not end once an event is over and structures have been restored. Long-term impacts, including the costs and benefits associated with adjustments, also must be evaluated because they relate directly to the perceived seriousness of a hazard and, therefore, to willingness to pay (or not to pay) to avoid future loss. However, some costs and benefits are not easily measured, such as reduced stress or number of lives saved. Others are related indirectly to the hazard itself, such as changes in the tax base because of land use restrictions. Thus, spatial and temporal dimensions are considered throughout this chapter.

ECONOMIC IMPACTS OF EVENTS

Immediate Losses from an Event

Traditionally, losses from natural hazards have been categorized by type and measurability. That is, costs are associated with particular functions such as domestic, industrial and commercial uses, then disaggregated into components such as damage to houses, destroyed possessions, and lost production. Although these would seem amenable to measurement, the ability to assess total losses accurately is severely curtailed by inadequate and nonstandardized data collection.

Indeed, methods of projecting damage from natural hazards are notoriously deficient. The term "inexact science" has been used to refer to the estimation of earthquake losses, based on the recognition that estimated figures are dependent upon who did the accounting and which criteria were included (Ayre et al., 1975). For example, data from the Bakersfield, California, earthquake of 1952 showed that the figure for actual losses represented only 36% of replacement loss (that is, it would cost considerably more for people to replace damaged possessions than the assessment showed). Which of these—actual loss or replacement costs— should be used for hazard losses? Similarly, Ayre, Mileti, and Trainer (1975) found that initial estimates of losses were invariably higher than later ones; in Alaska, immediate losses from the 1964 earthquake were placed at $620 million, but by September had fallen to $335 million. Clearly, the wide range in estimates for just one disaster makes comparisons between events extremely difficult. Similarly, statistics on relief do not provide accurate data on losses. For instance, between 1989 and 1994, the Federal Emergency Management Agency (FEMA) distributed nearly $7.5 billion from its relief program and even more aid from special programs (Table 6.1).

While these figures indicate very large losses, they do not necessarily reflect the full spatial and temporal extent of economic impact from these events.

To understand the economic costs of natural hazards, we must examine in detail the full range of loss. For instance, the occurrence of an event invariably brings about an immediate crisis, often resulting in loss of life or property. Following these "direct" losses and depending on the type of event, there is usually a cleanup period that constitutes an "indirect" cost, and sometimes benefit, of the disaster. In addition, there may be "intangible costs" such as loss of historical sites and buildings, stress, or inconvenience. Consequently, we can classify losses in two categories based on measurability (that is, tangible and intangible losses) and type (that is, direct and indirect losses), remembering that the significance of each subcategory varies spatially and temporally along with the characteristics of the natural hazard.

Direct loss results from physical contact or association with the physical aspects of a hazard. It includes damage to buildings and their contents, transportation routes, and farmland; for instance, the destruction of dwellings by a tornado would constitute a direct loss. Indirect loss results from the breakdown of physical or economic linkages in the economy and might include lost production, loss of income and business, and delays in transportation (Ward, 1978). The disruption of services and normal economic activity can be wide-ranging and have impacts well beyond the

TABLE 6.1. Relief Allocations by the Federal Emergency Management Agency for the 10 Worst Disasters, 1984–1994

Event	Location	Year	Relief allocations ($)
Northridge earthquake	California	1994	2.9 billion
Hurricane Andrew	Florida, Louisiana	1992	1.8 billion
Hurricane Hugo	North and South Carolina, Puerto Rico, Virgin Islands of the United States	1989	1.3 billion
Midwestern floods	9 states	1993	1.0 billion
Loma Prieta earthquake	California	1989	764 million
Hurricane Iniki	Hawaii	1992	255 million
Tropical Storm Alberto	Alabama, Florida, Georgia	1994	221 million
Winter storms, mudslides, flooding	California	1992	197 million
Civil unrest and fires	California	1992	160 million
Northeast coastal storm	Connecticut, Delaware, New Jersey, New York	1992	137 million

Source: FEMA (1995).

directly impacted locality; for example, the consequences of tornadoes for small communities can be severe if the economic base is destroyed (Green, Parker, Thompson, and Penning-Rowsell, 1983; Handmer, 1988b; Smith, 1989). In larger disasters, the economic ramifications can resonate much farther. The 1995 earthquake in Kobe, Japan, disrupted shipping throughout Asia because of the critical role the city's port played as a transshipment point; consequently, several industries (including some computer components manufacturers) were affected around the world and shortages of specialty steels were experienced before the city could begin to rebuild. We might contemplate the potential consequences of disasters at other economic centers; a major flood in London, England, or a devastating typhoon in Tokyo, Japan, would have repercussions throughout world financial markets.

Tangible losses, which can be measured relatively easily, may include such direct and indirect effects as damage to buildings and lost possessions. Intangible losses, on the other hand, are more difficult to assess and include impacts such as stress, fear, anxiety, and inconvenience. Efforts to quantify tangible versus intangible losses are fraught with difficulties. We already have seen some of the problems associated with identifying immediate, or tangible losses, and intangible losses raise such problems to a new level. Because of the difficulty in putting an accurate monetary value on stress or inconvenience, intangibles often are based on some percentage of direct tangible losses, but there is no agreement on what this proportion should be. A wide range of estimates can be found in the literature, from a low of 15% to a high of 75% (Handmer, 1988a); Chambers (1975) even suggested that intangibles might prove to be twice the tangible figure. However, there is no reason to expect that a standard relationship exists between direct and indirect or tangible and intangible losses for all disasters, and for all places.

Combining these categories provides a topology of loss from extreme natural events that describes both the nature of losses and our ability to estimate them accurately (Table 6.2). Direct tangible losses are the most straightforward, and therefore the most commonly reported in loss accounting. Indirect and intangible losses are more difficult to measure

TABLE 6.2. Topology of Losses from Natural Hazards

Form	Measurability	
	Tangible	Intangible
Direct	Damage to buildings	Loss of archaeological site
Indirect	Loss of production	Emotional stress

Source: Adapted from Green, Tunstall, and Fordham (1983), Handmer (1988), and Ward (1978).

because of the type of losses and the extended time frames over which they may occur. Indirect losses (such as lost industrial production) are measurable and estimates frequently are made, but an accurate accounting of all indirect loss is quite difficult. For example, a loss of industrial production may not be a loss at all, but rather a redistribution to another industry (Handmer, 1988b); that is, industries not damaged by a flood or tornado may experience increased business as buyers look for alternative suppliers. Similarly, local businesses may benefit from reconstruction or, alternatively, the benefits may shift to another community so that though the region does not suffer, the locality does. In contrast, intangible losses usually are not measured and, if they are, there is some level of subjective evaluation involved (Green et al., 1983). Yet, these losses may be more important to victims because they cannot be easily recovered. They also may be more long term in nature and may constitute a large proportion of total damage (Green et al., 1983).

Clearly, we can estimate the economic loss due to an extreme natural event, but that in no way describes its full impact. Assessing loss in terms of disruption would go a long way toward overcoming some of the difficulties. Aspects of disruption that might prove useful include the time required for the homeless to be resettled, the number of refugees created by the event (and associated impacts), and the length of time victims remained without adequate housing, food, or power. Such impacts are measurable as are related economic indicators, such as productive time lost because buildings or equipment or both were damaged or because workers were unable to get to work due to disrupted or damaged transportation systems (Handmer, 1988a). Other indicators might include the local property market and mortgage loans. For example, in a study of flooding in Jackson, Mississippi, Anderson and Weinrobe (1980) found little impact over the short term on the local residential housing market; only a few residents defaulted on mortgage payments, and a small number of houses were sold in a damaged condition. Business-related indicators might include shipping delays related to lost access and production, higher prices and costs, and credit problems (Alesch, Taylor, Ghanty, and Nagy, 1993). For example, during the 1993 floods in the upper Midwest of the United States, between 1,500 and 2,000 barges were tied up on the Mississippi and Missouri Rivers, resulting in an estimated loss in revenues of $3–4 million per day, according to the Secretary of Transportation (Flood Response Meeting, 1993). (Illustrating the problem of assessing loss accurately, the National Weather Service [1993] projected losses at only $2 million per day.) At the same time, other transportation systems benefited considerably from the flooding because grain was being moved by rail and road; thus, losses to one sector translated into benefits for another.

However, indicators of economic impact should be used cautiously.

First, some indicators may not reveal the wider impacts of an event; as the mortgage example shows, most flood victims managed to meet their loan commitments and properties still were bought and sold, but because money was spent on recovery, it was not available for other purposes (including long-term investments). Second, indicators are not always readily comparable to more traditional economic measures (such as costs of cleanup, solid-waste disposal, or restoration of a water filtration plant to working order). Third, although indicators may give a more comprehensive picture of the overall economic impact of an event to a region, nation, and victims, they do not lend themselves to traditional accounting methods, which makes measurement and comparison difficult.

So far, this discussion has looked only at losses incurred, omitting those avoided. However, a range of individual and collective adjustments can be undertaken in anticipation of an event, and sometimes as is occurring, to reduce economic loss. Any discussion of immediate economic impacts must incorporate those adjustments; more detailed discussion follows in this chapter, but here their place in the loss topology is merely introduced.

Just as there are tangible and intangible losses from an event, there are tangible and intangible costs and benefits associated with various adjustments. The equation in Figure 6.2 reflects the totality of losses as they are influenced by an event and the adjustments adopted. The costs and benefits associated with the adjustments (see the section on cost–benefit analysis below) fit well into the topology in Table 6.2 since they, too, must be considered in terms of measurability. A complete accounting, then, of the immediate losses incurred in an event must incorporate all costs, including intangible losses and the costs of adjustments. However, this rarely is done.

Thus, it is quite difficult to tally economic losses for natural events. Although the severity of an event usually is presented by the total economic losses and lives lost, reports are typically inaccurate and may well be misleading. Dollar losses are not easily summed, particularly in the days and weeks of confusion immediately following an event. Yet, estimates provided even months or years later also are inaccurate or misleading because they tend to use traditional accounting methods that ignore or mask indirect and intangible losses. The infusion of relief funds and supplies further complicates analysis of economic impacts.

Relief following an Event

Disaster relief is an important component of recovery from natural events and the literature on it is vast, ranging from reports at the local level to

Total loss = L_{DI} + L_{In} + CA_{DI} + CA_{In} − BA_{DI} − BA_{In}

where

L_{DI} = Direct and indirect losses
L_{In} = Intangible losses
CA_{DI} = Direct and indirect costs of adjustment
CA_{In} = Intangible costs of adjustment
BA_{DI} = Direct and indirect benifits of adjustment
BA_{In} = Intangible benefits of adjustment

For example

Direct losses = property damage
Indirect losses = lost industrial production
Intangible losses = lost or destroyed keepsakes
Direct costs of adjusment = amount paid to hazard-proof a building
Indirect costs of adjustment = false sense of security brought about by adjustment
Direct benefits of adjustment = avoided economic losses
Indirect benefits of adjustment = continued industrial production
Intangible benefits of adjustment = reduced stress

FIGURE 6.2. Calculating total loss. It is possible to estimate total losses from natural disasters by including not only direct and losses, but also direct, indirect, and intangible costs and benefits of adjustments. Omitting the costs and benefits of adjustments would distort the scope and nature of losses. *Source:* Adapted from Mukerjee (1971).

international evaluations of efforts, successes, and problems. (On local relief, see Berke, Kartez, and Wenger [1993] and Pereau [1990], and on international relief, see Cuny [1983] and de Goyet [1993].) Because relief generally is tied directly to the amount of loss sustained or believed to have been sustained, it provides an indication of economic impacts.

Consisting of financial aid, food, medical supplies, and other necessities, disaster relief often is provided immediately after a natural disaster. Because this chapter concerns economic impacts, we focus on financial assistance, but other forms of relief can be equally important (as shown in Chapter 5). Financial aid is particularly beneficial to victims who otherwise could not afford to replace or repair damaged property. In addition, the money often is expended in the community, where it can help to trigger economic recovery.

However, relief funds rarely provide enough money to compensate all losses unless government agencies step in with added contributions. In flood relief, Relph (1968) found the average recoupment from a public fund to be as low as 4% of actual losses, reaching a maximum of only 26%. In a more recent analysis of public sector losses, Burby et al. (1991) showed that this level of recoupment had increased, at least for losses from large events that are readily documented. In the Whittier Narrows, California, earthquake of 1987, they estimated that federal disaster relief covered approximately 37% of total losses for which city and county governments filed claims, suggesting that the state and federal governments have been more generous than Relph's data suggest. However, their data covered only public losses, not those experienced by individuals. The latter data are difficult to collect, but bearing in mind indirect and intangible losses, perhaps Relph's figures are not inaccurate, particularly given the fact that at least in the United States, relief generally is not available unless a disaster declaration has been issued. There are hundreds of smaller events annually that cause damage requiring repair, but for which assistance is available only through insurance. Further research is needed to assess the significance of relief funds relative to losses.

To ensure that some assistance is available when needed, some nations have set up central relief funds or sequestered additional funding to help during emergencies. In the United States, for example, once disaster has been declared by the president, federal funds become available and are distributed roughly in proportion to the amount of economic loss sustained. Since the late 1980s, there has been an increase in the number of federally declared disasters and a disproportionate rise in costs (see Figure 6.3). However, initial estimates of economic loss have been shown to be highly unreliable and, in fact, it is in the interest of stricken communities

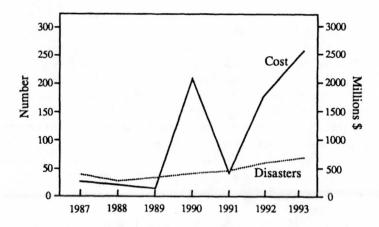

FIGURE 6.3. Federally declared disasters, 1987–1993. The number of disaster declarations has risen steadily. Although the costs associated with these declarations vary from year to year, recent trends are upward. *Source:* Abernathy and Weiner (1995).

to inflate estimated loss as much as possible. Disaster declarations also lead to private relief organizations' disbursement of money and mobilization of volunteers. Similar declarations also are made at some state levels. However, it is rare for these funds to match losses dollar for dollar.

Financial aid can be given in the form of loans (often at low interest) or grants. In New Zealand, disaster relief for natural events other than floods is distributed through the Earthquake and War Damage Commission, which receives a percentage of insurance money collected for property covered for fire (Ericksen, 1986). Most countries, however, do not have an organized disaster relief fund and rely on humanitarian contributions from individuals and governments. Some of the poorer nations, for example, have little option but to wait for humanitarian aid from wealthier countries, some of which is coordinated through the United Nations Disaster Relief Organization (UNDRO), established in 1972 to coordinate the disaster relief efforts of a number of organizations (Table 6.3). Between March 1972 and December 1977, UNDRO provided material assistance in 102 disaster situations (United Nations, 1979). Private relief organizations, such as the Red Cross and Red Crescent Societies, provide other forms of assistance (see Table 5.4).

Unfortunately, not all disasters generate immediate response though they may cause comparable deaths and damage (Smith, 1992). For instance, disaster assistance is mobilized easily following events like earthquakes and tropical cyclones which cause a large number of deaths; the media are immediately on the scene to record the devastation. In contrast, droughts and some floods seldom draw such a response even though there may be equally as many victims (i.e., survivors) who could benefit from aid. The difference is due in part to the visual quality of events; a drought can take a long time to develop and, at least initially, does not have the visual impact of earthquakes and hurricanes. However, political factors play an important role that is perhaps even more difficult to document.

It has been suggested that providing aid to disaster victims only perpetuates hazard problems. It is argued, for instance, that individuals come to expect some form of compensation following disasters, hence may not take the necessary remedial action to minimize losses (White 1966). While this may be true of some events and for some individuals, clearly it is not possible to remove risks completely. On a purely humanitarian level, therefore, some financial aid is appropriate, though recognition of and linkage to long-term development and social and cultural conditions is recommended (Berke et al., 1993; Schmitz, 1987; Smith, 1992). Another criticism of aid, directed particularly at poorer nations, is that it rarely reaches the people who need it. In the 1990s, for example, we have seen cases in Ethiopia and the Sudan where food aid was not distributed to drought-ravaged areas because of political and military turmoil. Discrimi-

TABLE 6.3. Examples of the United Nations Disaster Relief Organization's Response to Natural Disasters, 1979

Month	Country	Event
January	Bolivia	Flood
	Mozambique	Hurricane
February	Indonesia	Floods
	Indonesia	Landslides
	Indonesia	Volcanic eruption
March	Fiji	Hurricane
	Paraguay	Floods
	Tunisia	Floods
April	St. Vincent	Volcanic eruption
	Yugoslavia	Earthquake
May	Argentina	Floods
	Indonesia	Earthquake
	Malawi	Floods
June	Jamaica	Floods
July	Indonesia	Tsunami
	Nepal	Floods
August	Ethiopia	Floods
September	Dominica	Hurricane
	Dominican Republic	Hurricane
October	Colombia	Floods
	Egypt	Floods
November	Colombia	Earthquake
	Honduras	Floods
	Iran	Earthquake
	Yugoslavia	Floods
	Colombia	Earthquake
December	Mauritius	Hurricane
	Nicaragua	Floods

Source: UNDRO (1980).

nating between needs of hazard victims is essential (although perhaps very difficult) since the poor are usually disproportionately affected. There also are problems in defining the spatial extent of affected areas because of difficulties in determining the magnitude of impacts in the hours and days immediately after an event. Dudasik (1980) suggested considering disbursement of relief to those affected but outside the area of immediate impact. Organizational and jurisdictional failings have exacerbated the vulnerability of local populations and perpetuated hazard problems.

There are reported problems everywhere with the timing and distribution of disaster relief, but what different areas and different organizational structures share is a tendency to target immediate needs. In concentrating on immediate economic losses, relief favors direct tangible losses. This is a useful practice if indirect loss can be addressed and recovered, and disruption minimized, after direct loss has been handled, but problems with the amount, distribution, and timing of relief sometimes compromise its effectiveness with respect to indirect and long-term loss. In some instances, relief succeeds in providing much-needed supplies, but initiates other serious problems (though not always more problems than it has solved). In Fiji, a history of government-provided relief following hurricanes served to dismantle traditional responses to disaster (Campbell, 1984). In 1886, the colonial government supplied both food and its encouragement to plant quick-maturing crops such as cassava; unfortunately, because cassava is one of the most vulnerable crops during a hurricane, that advice made many Fijians more vulnerable to the hurricane hazard. More recently, food rationing programs in Fiji have centered on rice and flour, rather than the more nutritious staple root crops which are likely to be available in areas of the country not directly affected by a particular hurricane. Similarly, it has been argued that no matter what the location, the availability of provisions and the systems established to distribute them during postdisaster relief efforts essentially create an economic system that competes with the local, pre-disaster system that, ironically, the relief system is attempting to restore (Cuny, 1983).

In sum, disaster relief is a direct response to the losses incurred in an area from a natural event. Despite its problems, the provisions distributed by relief programs usually are sorely needed (Table 6.4). As Cuny (1983) pointed out, "The question is not basically one of need, but rather the

TABLE 6.4. Official Development Assistance (ODA) in Recipient and Donor Countries, 1989–1991

	Recipients			Donors	
Country	Average annual ODA (U.S.$ million)	1991 ODA per capita	Country	Average annual ODA (U.S.$ million)	1991 ODA per capita
Cuba	50	4	Canada	2,465	96
Guatemala	217	20	United States	10,111	45
Mexico	147	2	Austria	408	71
Argentina	217	8	Netherlands	2,383	167
Brazil	189	1	Norway	1,100	276
Burkina Faso	327	41	Sweden	1,974	246
Chad	278	47	United	2,824	56
Egypt	3,890	86	Kingdom		
Mauritania	222	100	Australia	1,008	61
Sudan	805	32	Japan	9,662	88
Bangladesh	2,013	18	Kuwait	354	187
China	2,099	2	Saudi Arabia	2,134	108
India	1,678	2	United Arab	109	346
Malaysia	356	25	Emirates		
Pakistan	1,153	10			

Source: Adapted from World Resources Institute (1994).

manner in which the need is met" (p. 99). Relief deals most effectively with direct, tangible losses and represents an important contribution to disaster recovery in the short term, but its long-term impacts and relationship to indirect losses are less clear.

Long-Term Hazard-Related Impacts

Our knowledge of the long-term economic impacts of natural hazards is quite limited in comparison with what we know about the immediate impacts. The reasons for this are twofold. First, there are very few longitudinal studies of the impacts of an event or a series of events in an area; instead, case studies undertaken at one point in time tend to dominate the hazards literature. Second, as the time elapsed after an event increases, so too does the influence of other factors that may mask or even render moot any disaster-related impacts. However, this is not to say that long-term, hazard- or disaster-related impacts are not important. Indeed, in combination with disaster-independent changes, they work to define the economic situation of an area and, eventually, to define the economic and social conditions in which the next event will take place. In Banks Islands, Vanuatu, Campbell (1990) traced changes in population size, crop diversity and resilience, food storage, and surplus versus deficit production. Although these changes were caused largely by agents external to the local tropical cyclone hazard, they defined "disaster preconditions" which, in turn, influenced communities' ability to cope with disaster. Numerous examples around the world illustrate the dynamic relationship between hazards and vulnerability; changes in population numbers and characteristics and in economic trends make areas more or less vulnerable to natural events. In the dynamic relationship between natural hazards and socioeconomic systems, the recurrence of an event at a given location will have different impacts than did previous events (for better or for worse, depending on the nature of the changes). This is of particular concern in developing countries because of increasing populations and widespread poverty that result in increased physical and economic vulnerability to hazards (Kreimer and Munasinghe, 1991).

Not surprisingly, findings on the long-term impacts of disasters tend to conflict. Some argue that communities may actually benefit from a disaster because capital is brought into the community for reconstruction and rehabilitation (Dacy and Kunreuther, 1969). Following Tropical Storm Agnes in 1972, Wilkes-Barre, Pennsylvania, received a huge influx of aid that may have assured the town's revitalization. Similarly, several communities in Alaska benefited from the considerable federal aid that followed the 1964 earthquake (Ayre et al., 1975).

Other long-term studies contend that whether positive or negative, pre-disaster trends continue after a disaster (Haas, Kates, and Bowden, 1977). In essence, the socioeconomic conditions prevailing at the time of the disaster override its economic ramifications. For example, Buffalo Creek, West Virginia, was economically depressed before a severe flood in 1972, and the natural event only aggravated the existing situation. On the other hand, a disaster may be the spur to community revitalization, as occurred in Soldiers Grove, Wisconsin (Tobin, 1992). There, the community initiated a major program that incorporated flood alleviation and economic development; it included relocating the business district and some residents off the floodplain and closer to the main road instead of constructing a larger levee system. Although initial cost estimates were similar, the benefits of relocation outweighed those of the levee system (David and Mayer, 1984).

The important point is that it is extremely difficult to generalize about how a natural disaster will affect a community economically over the long term. The reasons for the differences lie in a variety of arenas, including characteristics of the event (particularly severity of impact), local pre-disaster economic conditions, and ability of the affected area to respond efficaciously. Each of these is influenced in turn by a number of factors, some totally independent of the hazard and some over which the victims have little control. Isolating disaster-related impacts from other influences becomes more difficult as time passes. Developing an integrated sense of long-term impacts is hindered by the local and regional contexts in which disasters occur, and the changes that occur within those contexts over time.

Predicting Loss

In theory, given what we know about the nature of losses caused by natural events and the susceptibility of areas with different densities and types of development to a range of natural hazards, it is possible to predict losses (at least, direct and indirect tangible losses) from an event. Information needed to predict loss includes characteristics of structures (relative to physical characteristics of the natural agent), inventories of the contents of all buildings (no matter what the use), documentation of susceptible infrastructure. For example, stage–damage curves can be used to estimate losses from floods for buildings in a community (Penning-Rowsell and Chatterton, 1977; White, 1964). (Figure 6.4 shows a stage–damage curve for different residences.) While stage–damage curves can be constructed following a flood, they also can be defined synthetically to predict losses from future flooding (Smith, 1981; Taylor, Greenaway, and Smith, 1983; Ward, 1978). The curves therefore denote potential losses and can be

FIGURE 6.4. Stage–damage curves for representative establishments. The magnitude of loss is affected not only by depth of floodwater but also by type, use, and value of structures. Stage–damage curves, which are based on historical records, make it possible to predict losses before an event or to estimate them afterward. *Source:* White (1964). Copyright 1964 by University of Chicago Press. Reprinted by permission.

calculated for various structures and building uses, with the caveat that they deal only with tangible direct losses. Indirect losses can be calculated, but again, only with difficulty because of problems in differentiating between outright losses and redistributions. Typically, indirect loss is treated as a percentage of direct loss in the estimations (Smith, 1988).

One way to predict losses is to develop a network depicting the range of damage-causing factors for an event, then to trace their pathways to evaluate impacts. Figure 6.5 does this for hurricanes in Fiji (though that was not its original purpose). This "mapping" of the hurricane hazard gives a view of direct and indirect impacts and incorporates both short- and long-term assessments. Certainly, if desired, dollar values could be assigned to each of the areas of impact to yield a prediction of economic loss, though actual loss would vary with the severity of the event. Different iterations, incorporating varied physical parameters of events, also could be undertaken to provide a range of predictions. Note that intangible losses are not included (which is not a criticism, but a recognition of the difficulty in delineating all areas of impact); further problems arise in assigning dollar values to production losses outside the commercial sector, such as those associated with subsistence crops.

Most predictions of loss do not concentrate solely on economic impacts, but rather pursue different scenarios in order to evaluate the overall impacts of a given event in a community under various conditions of adjustment, or lack of adjustment (Ericksen, 1975). However, economic impacts—primarily immediate impacts—are an important concern, and they are quite complex. As Cochrane (1974) pointed out with regard to indirect effects of an earthquake, "we want to know how the disaster disturbs productive capacity; how this disturbance filters through other industries and affects their output; how these combined effects influence employment and profitability and hence, demand; how these in turn affect taxes collected; and how each of these effects then influences production" (p. 7). Indeed, a multiplier effect may extend from primary losses (that is, damage to or destruction of property) to secondary and tertiary levels (Ward, 1978). For example, indirect impacts associated with the failure of lifeline systems have been shown to outweigh the direct costs of system repair (Eguchi and Seligson, 1993). Such indirect impacts include the interruption of business, environmental damage, and "collateral" damage, perhaps resulting from fires that cannot be extinguished efficiently. These impacts are difficult to measure, in part because an inventory of all

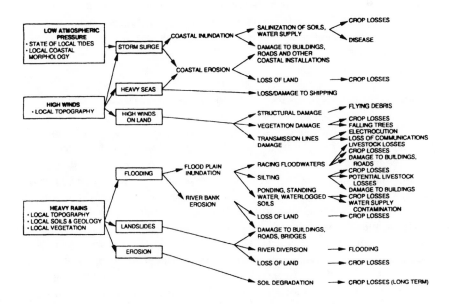

FIGURE 6.5. Predicting economic loss. Prediction is enhanced by modeling networks that trace economic linkages through a community. Scenarios of events of different magnitudes can be traced through the network. Unfortunately, not all elements of the network are equally measurable, nor are the forms of loss similar. *Source:* Campbell (1984). Reprinted by permission of author.

potentially affected businesses and the extent to which they may be affected is required. In addition, some indirect impacts, such as environmental damage, are difficult to quantify monetarily. No wonder estimates of this nature are not often undertaken.

Predictions are most often developed in order to evaluate policy options, particularly concerning adjustments and in cost–benefit analysis of adjustment options (discussed below). Prediction also helps to analyze the economic impacts of various mitigation options, such as floodplain designations (Burby et al., 1988; Montz, 1987; Muckleston, 1983) and policies for earthquake hazard mitigation policies (Alesch and Petak, 1986; Milliman and Roberts, 1985). These analyses tend to focus on indirect impacts of events and adjustments rather than predictions of direct economic losses.

Predictions are difficult because of the number of assumptions that must be made about an event and the affected area, exacerbated when the latter cannot be easily defined. For example, we know that floods are generally restricted to floodplains and that earthquakes usually occur near fault zones, but even those assumptions become less certain with increasing magnitude of an event. The difficulties are only increased when the location of an event cannot be predicted, either because it can occur anywhere (as in the case of tornadoes) or because it can affect a large area with boundaries that are spatially and temporally hard to define (as occurs with droughts).

Long-Term Impacts on the Housing Market: A Special Case

While it may be difficult to obtain a view of the overall long-term economic impact of a particular natural hazard, specific areas of the local economy have been isolated and investigated in some detail. Much work has been done on housing markets and how they respond, over the long term, to an event or to the presence of a hazard threat. Most of this research has been limited to flood and earthquake hazards, primarily because of their location-specific nature.

The underlying premise of these studies is that the existence of a hazard or the occurrence of an event has a negative effect on housing values. A theoretical examination of the earthquake hazard suggested movement of housing locations toward areas of lower damage, housing values falling in high-risk areas and rising in low-risk areas (Scawthorn, Iemura, and Yamada, 1982). That argument is supported empirically by a study of Los Angeles and San Francisco, where houses in Special Studies Zones (i.e., areas designated as particularly prone to earthquakes) sold for less than did those outside the zones (Brookshire, Thayer, Tschirhart, and

Schulze, 1985). Similar results have been obtained for housing on flood-plains, although there have been contradictory results in different communities; in general, houses located in flood hazard areas have been found to sell for less than houses outside them (MacDonald, Murdoch, and White, 1987; Shilling, Benjamin, and Sirmans, 1985).

Building on these findings, it might be assumed that land values decline to the extent that a natural event reduces the utility of the land, as evidenced in the market price for that property. This decline is an instance of the capitalization of an environmental externality (the natural event). For example, a flood damages structures on flood-prone land, which represents a reduction in the utility of the property and is reflected in lower property values (Figure 6.6). How far values are reduced depends on a number of factors, especially the severity of the event and local economic conditions, including the vitality of the real-estate market.

However, the graph also depicts a recovery process, represented by lines A, B, and C. The extent to which property values recover and the time period over which recovery occurs depend on physical characteristics of the flood (particularly frequency and severity) as well as local market conditions. For example, line A might be characteristic of areas with repeated flooding so that recovery is never completed before the next event hits; hence, land values in flood-prone areas remain low relative to non-flood areas. Line C represents the expected recovery pattern for a rare event where land values decline initially due to the flood, but recover rather

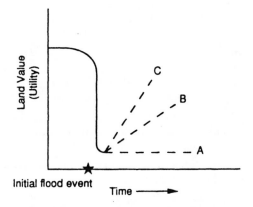

FIGURE 6.6. Theorized impact of flooding on property values. A represents an area of frequent flooding, and C an area with a "once-in-a-lifetime" catastrophic event. B represents a flood with a magnitude and frequency somewhere in-between A and C. *Source:* Tobin and Montz (1994a). Copyright 1994 by American Water Resources Association. Reprinted by permission.

quickly to levels near pre-flood values; the market, then, reflects the expectation (or rather, lack of expectation) of a future event and only temporarily incorporates the externality of flooding. Line B represents flood frequency somewhere in between the rare and the frequent.

Our example of a flood brings frequency into consideration, something that is not included in some studies. The severity of an event also must be recognized in evaluating long-term impacts. We expect that the more severe the flood (in terms of greater depths, longer durations, or faster velocities), the more apparent the capitalization process because of the greater damage. In other words, the flood event reduces the utility of the land, which translates into lower house prices; thus, the flood hazard is capitalized in the house value. The recovery from the hazard, then, is related in part to the degree of damage.

These ideas have been tested empirically in several flood settings (Montz and Tobin, 1988, Tobin and Montz, 1988, 1994a). In cases of rare flooding, such as in Linda and Olivehurst, California (which flooded when a levee was breached), selling prices for houses eventually returned to and even exceeded, pre-flood levels, though the recovery period varied significantly. The initial declines in values and the length of the recovery periods were greatest for those properties flooded to greater depths. In Wilkes-Barre, Pennsylvania, flooded properties eventually surpassed pre-flood values, but they did not experience the same proportional increase as did nonflooded properties or those that experienced lower depths of flood-water. Thus, even seven years after the event, the flood hazard was capitalized into property values. This is in contrast to Linda and Olivehurst, California, where for the most part, speed and extent of recovery were inversely related to the depth of flooding; that is, houses flooded to greater depths took longer to return to near pre-flood values. In Des Plaines, Illinois, which experiences frequent flooding, values of flooded properties reflected the number of times they had been flooded (never, once, or twice), but the entire market exceeded pre-flood levels rather quickly. However, those flooded more frequently took longer to attain the new market equilibrium because of interruptions caused by the floods. Figure 6.7 depicts the changes traced in the three communities.

Studies outside the United States, particularly in Australia and New Zealand, indicate difficulties in separating out flood-related market impacts from other socioeconomic and environmental variables that influence housing values (Lambley and Cordery, 1991; Montz, 1992). Nonetheless, evidence from a number of locations suggests that depreciating effects of events have an immediate impact and, often, continued influence over the long term.

This theoretical framework has been tested only with flooding, but it

House Value Trends in Des Plaines

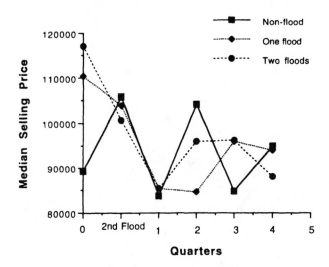

House Value Trends for Linda and Olivehurst

continued

FIGURE 6.7. House value trends after flooding. Empirical testing of the model in Figure 6.6 illustrates its validity for three communities. All experienced a fall in house values after flooding. *Source:* Tobin and Montz (1994a). Copyright 1994 by American Water Resources Association. Reprinted by permission.

House Value Trends in Wilkes-Barre

FIGURE 6.7. *(cont.)*

should prove applicable to other natural hazards where location in or outside a hazardous area (such as a seismically active area or coastal region) is salient. The framework allows for examination of the long-term impacts of natural events on one aspect of a community—the housing market—and therefore begins to build our knowledge of temporal impacts of natural events. It also directly addresses economic impacts that are not usually considered in tallies of losses.

THE ECONOMIC IMPACTS OF POLICY OPTIONS AND ADJUSTMENTS

Many economic impacts of events are obvious: damaged and destroyed buildings, sewage treatment plants put out of operation, industry and business temporarily suspended. Different mitigation measures and adjustments can be undertaken to avoid or minimize these and other losses, but such choices bear costs as well. The costs may be quite diverse, such as engineering expenses, increased building costs, decreased tax revenues, or additional allocations to planning and warning systems.

If money is going to be expended to implement an adjustment (such as to build a dam or retrofit a building), decision makers want to be assured that the benefits will be worth the costs. However, the balance is not as straightforward as it may appear at the outset because adjustments have indirect and intangible benefits and costs (see Figure 6.1). In addition, temporal and spatial considerations must be assessed because adjustments may have different long- and short-term impacts or may benefit one area while causing greater problems for another. To use another flood example, dredging a river may decrease flood potential where the dredging is undertaken, but may well increase it further downstream; also, the efficacy of dredging will be reduced over time by natural processes.

There are, then, numerous issues surrounding the economic implications of adjustments. Nevertheless, protective works have saved millions of dollars in property damage. For instance, the levee systems along the Mississippi and Missouri Rivers prevented many losses from flooding in 1993. The Interagency Floodplain Management Review Committee (1994) reported that without such protection, low-lying areas in Rock Island and Moline, Illinois, as well as in Kansas City, Missouri, would have been flooded. As a result of protection by a large flood wall, St. Louis was spared from flooding although the floodwaters crested almost 20 feet above flood stage.

However, there has been no overall assessment of the efficacy of hazard alleviation measures. Many projects are implemented based on what are perceived to be favorable benefit–cost ratios, but there is no follow-up to validate the original estimates. Some broad analyses have been attempted. For example, Hertler (1961) estimated that benefits from flood alleviation projects in the United States had exceeded costs by a ratio of 1.33 since 1918. The Tennessee Valley Authority claimed an even greater benefit–cost ratio (1.98) and suggested that savings amounted to $517 million with costs at $261 million (Tennessee Valley Authority, 1967). However, structural adjustments are more readily evaluated economically than nonstructural adjustments because there are direct, tangible costs for building a structure and benefits are expected to result from it (usually in the form of losses avoided). The implementation and enforcement costs of nonstructural adjustments are not always easy to estimate, though such measures also are expected to generate direct benefits. Both types of adjustment have indirect costs and benefits that must be considered.

Cost–Benefit Analysis

Cost–benefit analysis was developed in its present form by the Navigation Boards in the United States, which used the approach to evaluate harbor policies in the 1920s. In 1936, in light of the ever-increasing costs of flood alleviation, cost–benefit analysis was incorporated in the Flood Control Act, which specified that "the benefits, to whomsoever they accrue, must justify the costs" (Sewell, Davis, Scott, and Ross, 1962). Dixon (1964) demonstrated the value of cost–benefit analysis graphically, showing that size of a project and availability of funding might influence project selection (as shown in Figure 6.8).

Traditionally, structural adjustments that are implemented by governments rather than individuals have been subject to cost–benefit analysis. The stream of costs associated with building, maintaining, and operating a structure (such as a dam) is compared to the stream of benefits to be gained (with a dam, usually in flood losses avoided). That is, benefits consist of losses that will not be sustained. An alternative way of looking at this is through the "with and without" principle, or what the hazard situation would have been with or without the project. If the benefits exceed the costs, the adjustment is considered economically viable. Usually, however, analysis is not centered exclusively on whether a project is viable, economically or otherwise. It also is based on the level of protection considered reasonable, which is specified in the design standards (such as protection against a 100-year storm or an earthquake of magnitude 6.5). Throughout

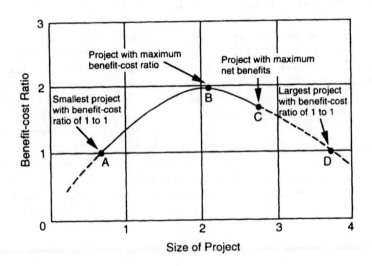

FIGURE 6.8. Benefit–cost ratios. Different project goals have different benefit–cost ratios. For example, maximizing the benefit–cost ratio (B) yields a different result than maximizing net benefits (C). Numerous other options that would fit along this curve are not represented here. *Source:* Dixon (1964). Copyright 1964 by McGraw-Hill Book Company. Reprinted by permission.

the planning, design, and implementation stages, there is recognition that the measure or structure will *not* protect against all events and that, over the long term, losses will occur.

Rarely have indirect costs been included in cost–benefit analysis; the same is true of indirect benefits, but perhaps less so. As early as 1958, a group of geographers from the University of Chicago pointed out the problems of ignoring indirect costs by documenting the increased building in flood-prone areas that had followed erection of protective structures (White et al., 1958). This danger from structural responses to the flood hazard had been recognized earlier by Segoe (1937), who described in detail the processes we now call the "levee effect." That is, hazard protection leads to perceived safety, hence to increased development in the protected area. If the results of cost–benefit analysis were true, flood losses should have decreased after investment in flood control structures, but losses increased even as expenditures increased. For example, in the St. Louis area, the Monarch–Chesterfield levee originally was constructed as an agricultural levee; it was upgraded in the 1980s to meet National Flood Insurance Program (NFIP) standards and development, including an industrial park, gradually occurred behind it. The levee failed in 1993, and the 67 flood insurance claims from property owners behind it totaled $13.2 million, or almost 5% of all claims for the nine states flooded that summer (Interagency Floodplain Management Review Committee, 1994).

The same pattern has been found elsewhere. Figure 6.9 depicts the trends in flood losses and protection outlays in New Zealand. The lines on the graph indicate that construction of flood-control works usually follows high levels of loss. As the loss line increases, so, too, does the line showing protection costs (though the increase is not as smooth because of particularly large projects in specific years). The relationship between these two lines illustrates the paradox that losses increase even when more money is invested in structures to lessen loss. Several factors may be contributing to this situation, including increased urbanization of watersheds, floods exceeding design levels of the structures, and development induced in the newly protected areas. Cost–benefit analyses may not incorporate all these effects, some of which are indirect and therefore difficult to estimate, let alone to measure directly.

Such problems have been well documented in the literature as have others, including the distribution of benefits and the appropriateness or efficacy of adjustments. However, it may be that these are problems of the past, particularly with regard to structural measures but also with respect to applying cost–benefit analysis. First, major new structures are being built only rarely, partly because of their high costs and the changing cost-sharing ratios in some places (notably the United States) where local governments have become responsible for a higher proportion of costs. Second, prob-

FIGURE 6.9. Flood losses and protection costs in New Zealand, 1950–1985. It is an interesting but explainable paradox that flood losses continue to increase even as we invest more into flood control. The data here are for New Zealand, but we would see similar patterns anywhere. *Source:* Ericksen (1986). Copyright 1986 by New Zealand Crown. Reprinted by permission.

lems in determining the full range of costs and benefits have called into question the use of such analyses. Finally, the utility of a discount rate to discount future expenses to their present value has been questioned. The assumption that values necessarily will be lower in the future ignores the benefits to be gained by saving resources for the future, which clearly conflicts with the emphasis on sustainability in current decision making on resource use. It should be noted that projects can be made more or less viable economically by changing the discount rate; the hypothetical example in Table 6.5 shows that a change of 2% in the discount rate changes the outcome from a net benefit of $3.5 million to a loss of $3.5 million.

Assessing the actual costs and benefits accruing from particular projects is not easy, and attention needs to be given to issues of double counting (Chambers and Rogers, 1973). For example, if damages not accrued are on the benefits side of the equation, then appreciation in land values should not be included. The argument is that development that helped the land appreciate in value (for example, construction of new housing) would have occurred anyway, but possibly elsewhere. Thus, one area may have lost and another gained because the mitigation project was implemented, but the overall effect on the region or community was zero. Consequently, it may not always be valid to include increased development as a benefit of a hazard adjustment. If problems associated with developing hazard zones are to be included in analyzing adjustments, more attention must be given to the distribution of benefits and costs over time, across the hazard area, and within the population.

Many other issues arise in cost–benefit analysis that further complicate the analytical procedures. For instance, social costs are important and the values of particular societies must be taken into account. It may be socially acceptable to implement some projects even if the cost–benefit ratio fails the economic tests. For example, projects that alleviate problems in squatter communities but have low economic returns (such as the La Paz Municipal Development Project) would seem socially, if not economically acceptable (Kreimer and Preece, 1991). Similarly, all too often planners have viewed disaster prevention efforts as "unaffordable extras," values that are very difficult to measure, or both, but when incorporated in development planning such measures may become affordable and understood as decreasing the likelihood of damage (Anderson, 1991).

Another difficulty associated with cost–benefit analysis stems from the fact that implementing a project means that investment is forgone elsewhere, known as the "opportunity costs" of a project. For example, when a decision is made to construct coastal flood walls to protect against the tidal surge of hurricanes, money may no longer be available to alleviate

TABLE 6.5. Results for Project Feasibility of Using Different Discount Rates

Proposed project:	Flood-control dam		
Estimated costs:	$28,000,000		
Estimated benefits:	$2,000,000/yr for 50 yr		

Discounting procedure	Discount rate	Present value of benefits	Results
A	6%	$31,500,000	Benefits exceed costs by $3,500,000 over life of project
B	8%	$24,5000,000	Costs exceed benefits by $3,500,000 over life of project

flood problems inland. Similarly, hazard adjustments may "take" money away from other projects such as economic development.

A further concern with cost–benefit analysis lies in the distribution of costs and benefits. Are the costs and benefits distributed fairly? Who benefits from implementation of a flood alleviation measure or a coastal retention wall? Is it the taxpayers who have contributed to the project or is it a few homeowners in particular locations? Similarly, who loses because the money is spent on one project and, as a result, money is not available for projects elsewhere or to meet some other public good? This question is related to opportunity costs, but goes beyond economics to address issues of equity. For example, Foster (1976) found that some flood alleviation measures were highly beneficial to small groups, but that for the most part, they were socially suboptimal. As he pointed out, some measures (such as engineering works) placed a greater financial burden on the taxpayer than on those at risk, and he suggested that government-sponsored programs of insurance, zoning, and flood-proofing could return an element of equity to flood alleviation strategies. Thus, a project may have a high cost–benefit ratio favorable to construction, but may not be warranted because of the relationship between the winners and losers.

In cost–benefit analysis, one also must consider the value of human life. In particular, structural measures are implemented to specified design standards, which do not attain 100% safety from natural hazards. Consequently, any engineering adjustment to a natural hazard puts a value on human life, at least indirectly, because some risk remains that the structure will fail or its capacity be exceeded. As Layard (1962) commented, "Perhaps the most difficult item of all to value is human life. Yet it is quite clear that countless policy decisions affect the incident of death and none of them aim to minimize its incidence regardless of cost. So each decision implies some valuation of human life" (p. 26).

However, putting a value on life is not an easy task. Prest and Turvey (1965) suggested that it would be better to save a productive than an unproductive person. Though they hastened to add that that would be socially and ethically unacceptable, their argument is sound from an economic standpoint and has been pursued by others. Mushkin (1962) proposed the median wage as a baseline for life, and others have countered that the average wage would be a better measure.

More recently Cropper and Portney (1992) suggested that lives should be discounted much like costs, deciding, for instance, whether a life saved today should be valued equal to one 25 or even 100 years from now. In their study, 38% of respondents chose to save 100 people today rather than 4,000 people in 25 years, and 47% preferred to save 100 now rather than 7,000 in 100 years. In effect, Cropper and Portney found that individuals chose to discount lives much like money. Table 6.6 shows the calculated

TABLE 6.6. Calculated Discount Rate for Human Lives

Time horizon (yr)	Sample size	Lives saved at T equivalent to 1 life saved today	Implicit discount rate
T = 5[a]	475	2	0.168
T = 10[a]	480	3	0.112
T = 25[b]	462	6	0.074
T = 50[c]	528	11	0.048
T = 100[b]	442	44	0.038

[a]Data from U.S. national poll.
[b]Data from Maryland poll.
[c]Data from Washington poll.
Source: Cropper and Portney (1992). Copyright 1992 by Resources for the Future. Reprinted by permission.

discount rate for human lives based on several studies. In statements made to researchers, many people indicated that they thought the pressing problems of today far more important than those in the future. Although such a consideration may appear to be far afield of cost–benefit analysis for hazard mitigation, it is not. If structures provide protection, then they save lives, minimize injuries, *and* protect property. Interestingly, only the last is considered in most of these analyses.

Cost–benefit analysis usually is not performed for nonstructural adjustments, though there have been notable attempts to do so (for example see Milliman and Roberts, 1985). There are several reasons for this. First, it is sometimes difficult to attribute specific costs to measures such as zoning or setting building standards, which serve to redistribute land use or shift costs to individuals rather than the public. The actual extent of the costs depends on future decisions about building or land uses, which will not occur at one point in time. Consequently, not only must the costs be estimated, but the estimates must incorporate a series of decisions that will extend over an indeterminate length of time. Second, land use and building requirements are implemented when it is believed that the public good will be served. Such policies inevitably raise questions of individual freedom versus the public good. For example, should a state or community be free to insist on rigorous building codes to prevent future earthquake or hurricane damage? In California, fairly stringent building codes are enforced to ensure public safety, even in private buildings. The questions become even more acute when land is taken out of private ownership for the public good. The construction of a dam, building of a sea wall, or removal of buildings from a floodplain may involve the taking of private property. The "takings" issue is of major concern in the United States; in *Lucas v. South Carolina Coastal Council,* the U.S. Supreme Court reversed a South Carolina Supreme Court ruling "that upheld the constitutionality of the South Carolina Beachfront Management Act against a 'takings' challenge" (Platt, 1992, p. 8).

Several authors have examined the economics of different projects. Foster (1976) demonstrated the efficiency of nonstructural measures in flood alleviation projects; many structural adjustments were not efficient when social factors were included, but nonstructural measures (such as zoning, flood-proofing, and insurance) usually proved more socially optimal. Krzysztofowicz and Davis (1986), who considered the efficiency of a flood forecasting and warning system from a systems approach, recommended that investment be directed more toward behavior than technology. Others have stressed the need to look at all adjustments as systems, particularly those involving behavioral changes and individual responses (Mileti and Krane, 1973). Rettger and Boisvert (1979), who compared flood disaster loans with flood insurance, found that despite variations in premiums and interest rate subsidies, both provided equal coverage at equivalent costs; however, they distributed public and private costs differently. They also concluded that flood insurance rates should be tailored to the local hydrological regime, although the authors cautioned that theirs was a single case study and further research was needed to establish the validity of their findings. However, insurance companies are reluctant to underwrite flood or hazard insurance because of the financial risk. In a major catastrophe (such as a Category 5 hurricane striking Miami, Florida, or an earthquake of Richter magnitude 7.5 in St. Louis, Missouri), many insurance companies would be severely stretched, if not bankrupted. At existing coverage, projected insurance industry losses from a major earthquake along the northern San Andreas and Newport–Inglewood faults in California have been estimated at $31 billion and $52.4 billion, respectively (Lecomte, 1989). The ramifications of such losses would be felt throughout the insurance and reinsurance markets and could have national, perhaps global consequences.

Although it is difficult to define the costs and benefits of protecting the public good, for hazards we could look at public losses avoided. Again, however, we are faced with the problem of incremental decision making over time, where the stream of costs and benefits associated with a policy decision is difficult to determine over the long term. In addition, there are significant "technical and informational problems associated with conducting benefit–cost analyses on complex problems involving many probabilities and requiring many assumptions" (Alesch and Petak, 1986, p. 221). Simply put, we do not always have the methods or the knowledge needed to evaluate the complex relationships involved. However, there still is reason to attempt cost–benefit analyses. As Alesch and Petak argued, it can shed light on the expected outcomes of adjustment alternatives and therefore can further our understanding by facilitating analysis of costs and benefits under different probability scenarios.

Some problems with cost–benefit analysis are well illustrated in Long Beach, Los Angeles, and Santa Ana, California which considered ordinances requiring unreinforced masonry buildings to be brought to standards associated with the level of seismic hazard. These proposals met with fierce opposition on both political and economic grounds (Alesch and Petak, 1986). The economic arguments centered on how various policymakers viewed the risks of low-probability and high-consequence events relative to the costs of different alternatives; their decision making was complicated by the absence of solid evidence on expected costs of mitigation and expected losses without mitigation. Indeed, advocates tended to focus on the benefits of hazard reduction while opponents focused on the adverse consequences, particularly costs. In addition, there were groups who would be affected positively and others who would be affected negatively, with consequences extending from the short to the long term. These groups included building owners who would be required to reinforce unreinforced buildings, occupants of the buildings whose rent would likely increase as a result of the improvements, and policymakers who would need to consider, either explicitly or implicitly, the balance between the social costs and social benefits of implementing or not implementing the ordinances. Bringing all these interests together is difficult enough where technical information is available, but the difficulties increase with the level of uncertainty associated with hazard planning. In addition, it is a relatively easy task to estimate the costs of reinforcement, but it is far more difficult to estimate benefits over time, let alone to allocate the benefits among the affected parties.

Requirements for nonstructural adjustments circumvent the cost–benefit issue to a large degree. In New Zealand, for example, building in designated hazard zones was regulated formerly under the Local Government Act and the Town and Country Planning Act, both of which were superseded by the Resource Management Act in mid-1991. Under these acts, regional and district councils have been required to exclude development from areas subject to hazards and to issue building permits, subject to conditions, where the land is prone to hazards such as erosion, subsidence, flooding, and landslides. Conditions include elevation of structures above expected flood levels and foundation design by geotechnical engineers. Here, cost–benefit analysis does not enter into the picture at all as the public good—that is, protecting development from damage, protecting lives, and minimizing the need for relief funds following an event—is protected legislatively. Similar types of legislation exist in many other nations.

Where adoption of adjustments is not legislated, individuals undertake informal cost–benefit analyses in their personal decision making. Generally, homeowners flood-proof, retrofit, or tie down their houses only if the benefits are perceived to be greater than the costs. As discussed in Chapter 3, the potential for individual action is related directly to clear perception of a hazard. Certainly there are indirect costs and benefits associated with individual decisions, but they do not usually extend beyond the homeowner and may or may not enter into the decision-making process directly. That process is changed when homeowners are forced by local regulations to build or retrofit a building to bring it up to specified standards. In such cases, costs are incurred by those in the hazardous area rather than by the community as a whole; in fact, some short-term benefits may accrue to those outside the hazard zone when individuals purchase materials and hire others to do the necessary renovations. However, legislation has been criticized as an ineffective means of changing human behavior; indeed, Natsios (1991) suggests that human behavior toward hazards only will be changed through positive economic incentives, not through negative or restrictive legislation. That is, the marketplace can be used to influence individuals to adopt sound remedial actions.

This discussion has centered on economic analyses associated with the adoption of adjustments, that is, with the estimated costs and benefits of undertaking a hazard mitigation option to avoid future loss. However, economic impacts do not end with implementation. There also may be operating, maintenance, or enforcement costs associated with adjustments. These can be estimated and in fact, should be part of the initial debate. However, there also may be long-term costs and benefits of an indirect nature, which may not be considered by decision makers.

The Indirect Economic Impacts of Adjustment Choices

Because most adjustments influence how or where land is used, impacts on the utility of that land are to be expected. If the occurrence of a natural event is seen to have a negative impact on the value of property at risk, then hazard alleviation programs should have a positive impact. However, it also has been suggested that disclosing the hazardousness of a location (for instance, through floodplain designations, identification of seismically active areas [such as Special Studies Zones], or mapping susceptibility to land slippage) depreciates its value. However, this has not been supported empirically; it appears that in contrast to events, disclosure has very little impact on land values.

Structural adjustments to flooding have been shown to encourage a false sense of security, based on the belief that the protected area has been completely flood-proofed (White et al., 1958). Assuming that this perception carries over into the market, we can assume that protected land is more valuable economically than unprotected land. While it is difficult to separate flood protection from other locational and hydrological factors that influence property values, differences have been found in the value of protected and unprotected property with the former experiencing higher values (Damianos, 1975).

However, for the most part, emphasis has shifted from structural flood protection to nonstructural measures that regulate land use and building. In the United States, the NFIP regulates development on areas designated as 100-year floodplains. Because the land market exists in the private sector and the NFIP is implemented in the public sector, the impacts of the latter on the former are clearly indirect (Burby et al., 1988). Some have asserted that designation of land as hazardous would devalue it. However, the findings are variable, with some case studies indicating a depressing effect and others indicating no effect whatsoever (Montz, 1987; Muckleston, 1983). These differences are not really surprising if one considers the effect that zoning of any kind has on property values (Dowall, 1979; Ohls, Weisberg, and White, 1974). In all cases, the dynamics of the local real-estate market, the availability of land outside the regulated area, and any amenities offered by the "marked" land can offset any devaluations that might otherwise be evident. Understanding local context, therefore, is extremely important.

In Los Angeles, a real estate submarket appeared to be in operation in the mid-1980s, centered on old brick buildings that typically did not meet building standards (Alesch and Petak, 1986). Because of this, buyers had difficulty obtaining conventional mortgages, hence resorted to other methods such as land contracts to transfer the properties. Although there is little empirical evidence on this submarket, what is known suggests that these properties sold for higher prices than would be expected given their noncompliance. Low interest rates also stimulated the market. In the end, both buyers and sellers gained from the arrangement. This example illustrates both indirect economic impacts of building regulations and the compromising of mitigation strategies that is inherent to human behavior—if one agrees with Natsios (1991), who observes, "People find ways to take financial advantage of programs and to evade regulations that frustrate the most brilliant policy and program analyst. Human behavior is not as simple or controllable as some policy makers think" (p. 112). On the other hand, perhaps it was the probabilistic nature of the risk rather than a concerted decision to take advantage that led to the situation with brick, unreinforced buildings.

Although most studies of floodplain designations concern already developed land, a notable exception is the work of Burby and his colleagues, who have worked extensively on the impacts of floodplain designation on the value of vacant land (Burby et al., 1988; Holway and Burby, 1990). They argue, and quite logically so, that the regulations associated with floodplain designation increase the costs of development, thereby reducing land values; more specifically, they argue that "zoning floodplains for lower density development, implementing building regulations more stringent than the minimum required by the NFIP, and providing clear local leadership of programs each contribute to lowering floodplain land values" (Holway and Burby, 1990, p. 259). Their findings support this contention, even in cities with very different economic, demographic, and political situations.

In California, similar studies followed implementation of legislation requiring disclosure of a property's location in a Special Studies Zone. Some of the results we reported above, but interestingly, there have been conflicting findings. Some studies report lower house prices (Brookshire, Thayer, Tschirbart, and Schulze, 1985) while others indicate that there has been essentially no impact from disclosure (Palm, 1981). Some of these differences may be attributable to the areas studied; that is, different communities and neighborhoods probably have very different real-estate markets. No matter what the reason for the differences, they illustrate the difficulties inherent in documenting indirect impacts empirically as well as the need to control for variables that influence property values independently of a hazard (a difficult task in itself). However, if we are to understand the full range of impacts associated with adjustments so that we can evaluate alternatives fully, we must have this kind of information.

INTEGRATION

In this chapter, we have shown that any analysis of the economic impacts associated with natural hazards involves considerably more than measuring the immediate and direct losses resulting from a geophysical event or assessing the tangible losses avoided through implementation of some mitigation measures. In particular, we have focused on intangible loss (such as stress) and indirect loss that might occur in areas spatially removed from the disaster site or later in time. However, the picture becomes cloudy when we combine concern for loss with tangible and intangible and direct and indirect benefits. In essence, it is extremely difficult to account for all the impacts of natural hazards as the stream of benefits and costs cascades through the socioeconomic system. At the same time, the scale of impact varies, depending not only on the severity of the event but also on the interconnectedness of the location directly affected. Similarly, the temporal scale of the impact varies depending on many factors, not least being the vitality of the economy.

Yezer and Rubin (1987) suggest that such impacts can be assessed at the local scale through expectation theory. They have attempted to develop an economic model of indirect effects based on the connection between past events and future expectations. "The most important result of the application of an expectation model to indirect economic effects of disasters is the conclusion that *the unanticipated component of disaster experience leads to economic change, while the anticipated component of disaster experience does not result in indirect economic effects*" (Yezer and Rubin, 1987, p. 62). The argument follows that if a community is located in an area where the probability of flooding is once every 3 years but no flood occurs, the local economy will be stimulated because actual disasters fall below expectations. Conversely, two floods in the same 3-year period would have a negative impact because actual events would exceed expectations. Of course, if there were just one flood, there would be no change in the local economy because reality and expectation would match. These ideas are applied to hypothetical, flood prone communities, but there is no reason other hazardous areas should not be included. As discussed in Chapter 3, however, expectations or perceptions of risk are based on a variety of factors, and we cannot expect all floodplain occupants to have an equal assessment of the hazard. Nevertheless, expectation theory provides a framework around which we might examine how extreme geophysical events affect small communities.

Beyond the local scale, other approaches must be taken. Our discussion in this chapter has illustrated the relationship between development and disasters, one that is too frequently ignored in development planning and decision making. However, it has become clear around the world that short-term natural events can have long-lasting socioeconomic impacts that too frequently thwart efforts to improve conditions. Indeed, in some cases in the United States and elsewhere, communities find themselves worse off after an event, in spite of an infusion of funds and supplies. These cases illustrate a tendency to separate the natural event from larger considerations of development even though the two are integrally related. Too often, this is recognized only after the fact. Thus, a development perspective can illuminate the economic impacts, repercussions, and alterations over both the short and long term.

SUMMARY

Economics enters into any analysis of natural hazards. Certainly we are initially concerned with the economic value of losses caused by natural events. Indeed, our responses to loss usually are framed in relation to structural and nonstructural measures that might mitigate immediate impacts and prevent future loss. These measures are evaluated largely in terms of economic criteria, whether they consist of immediate financial relief or long-term risk reduction. Traditionally, if the stream of benefits exceeds that of costs, the mitigation measure is implemented, but such an approach tends to focus on direct, short-term economic variables. In addition, little consideration has been given historically to the entire range of impacts associated with implementation.

As illustrated in this chapter, economic impacts can and should be evaluated from a number of perspectives, including the long-term impacts of events, the long-term impacts of alleviation and prevention projects and programs, and the distribution of costs and benefits. We grossly underestimate the costs of events by focusing on immediate losses (as reported in dollar values of property damage, for instance). Similarly, we grossly overestimate the benefits of alleviation options by ignoring the distribution of those benefits among all groups, over the long term. We have begun to frame our economic analyses differently, but we need to continue to develop methods to ensure that such considerations become integral to the accounting process.

7 Risk Assessment

WHAT IS RISK?

Thus far, we have discussed hazards as they exist in nature and as they affect and are affected by human perceptions, actions, and institutions. However, there is another element that is crucial to our understanding of natural hazards and of the ways we seek to manage them: that of risk. The terms risk, risk analysis, and risk assessment are pertinent to many fields besides natural hazards research and encompass many processes, including scientific calculations of probabilities of occurrence, expert evaluations of possible consequences, laypersons' understandings of risk, and various types of risk management. As a result, the literature on risk is extensive and diverse, with a sizable component devoted entirely to risk analysis (Covello and Mumpower, 1985). We cannot include all the facets of risk in one chapter, nor do we want to. Instead, our focus is on definitions, issues, and management of risk from a natural hazards perspective.

The nature of the risk associated with natural hazards can shape both individual and collective perception and action. Indeed, as compared to other risks, the level of disaster risk is an important determinant of whether communities or individuals take action to reduce it. Thus, it is imperative to understand risk not only technically, but also as it is perceived by those in hazardous areas and as it is managed (or not) through risk mitigation, sharing, and avoidance strategies. In order to do this, some background is necessary on how risk is determined and on what we know about its natural and technological sources.

Hazard Risk

Sometimes, risk is equated erroneously with hazard, and perceived risk with hazard perception. In fact, risk is part of hazard, but the two terms are not synonymous. Risk is an important component of hazard analysis and risk analysis forms an important subdivision of the study of natural hazards.

To put the two in perspective, we might consider the elements of risk analysis. Frequently, risk is seen as the product of some probability of occurrence and expected loss. In Chapter 2, the probability of recurrence of particular geophysical events was assessed through historical trends; for example, from lengthy historical records it is possible to determine the approximate size of a 100-year flood and to estimate the probability of certain-sized events occurring in any given year. While this information is useful in evaluating technical risk, it does not indicate the numbers of people exposed to a hazard or the losses expected from a specific event. To get a better assessment of hazard risk, details on vulnerability must be incorporated in the analysis. Statistically, this relationship can be expressed as

$$\text{risk} = \text{probability of occurrence} \times \text{vulnerability}$$

This relationship was used by Van Dissen and McVerry (1994) to evaluate earthquake risk in New Zealand; they defined probability as the likelihood of an earthquake occurring (based on results of a seismicity model) and vulnerability as the damage potential for property, using a damage ratio.

While this formula represents a useful attempt to include additional factors affecting risk, it fails to incorporate geographic differences in population size and density (or what might be termed exposure) as well as communal adjustments undertaken to minimize loss. Mitchell (1990) conceptualizes hazards as a multiplicative function of risk, exposure, vulnerability, and response:

$$\text{hazard} = \text{f}(\text{risk} \times \text{exposure} \times \text{vulnerability} \times \text{response})$$

where

risk = the probability of an adverse effect
exposure = the size and characteristics of the at-risk population
vulnerability = the potential for loss
response = the extent to which mitigation measures are in place

In combination, these elements serve to explain differences in hazardousness from place to place and from time to time. If we adopt Lowrance's definition of risk as "a measure of the probability and severity of harm" (Lowrance, 1976, p. 8) and combine it with Mitchell's conceptualization, it is not difficult to imagine a small risk (a very low probability of occurrence in a given period of time) but a severe hazard. This might be the case for an unprepared, densely settled population. Even with relatively constant probabilities of occurrence (such as a seismic probability), different measures of vulnerability can significantly affect the estimated magnitude of a hazard (as we saw with the 1995 earthquake in Kobe, Japan). Alternatively, at a given risk, a hazard may be lessened if the vulnerable population is protected by mitigation measures or has the financial or other resources to recover from loss. Thus, risk is only part of hazard, but we must understand risk in order to grasp the complexities of hazards.

Just as risk is only one component of hazards, risk also is complex. It comprises two elements that must be considered separately and together. These are (1) a choice of action and (2) an outcome, which includes a probability of occurrence and a consequence (or magnitude).

Choice of Action

Human life requires choices among actions that are linked inextricably to risk (Dinman, 1980). Every decision and resultant action, whether voluntary or involuntary, is an intricate part of human existence that ultimately exposes people to risk. In reality, no one is ever completely safe no matter what decisions are made, although clearly some individuals are safer than others. For example, inhabitants of squatter settlements on the unstable slopes of Lima, Rio de Janeiro, and Hong Kong, or on the floodplains of Santiago, Karachi, Caracas, Delhi, Manila, and Mexico City (Torry, 1980) are obviously more vulnerable than people residing in substantial dwellings in less hazardous environments. In contrast, choosing to purchase property on a steep, unstable cliff along the California coastline represents a decision to take advantage of its scenic vistas as weighed against the potential costs of landslides. The decision to locate in a particular area is not necessarily made freely, but is usually determined by socioeconomic forces that often are beyond the control of individuals. Thus, risk involves choices, but those at risk are not always the ones who make the choices. As discussed by Slovic (1987), voluntary risk (as in the example of Californians seeking scenic views) is much more acceptable than involuntary risk (illustrated by the floodplain dwellers in Santiago).

Of course, many decisions place people at risk. People face risks every day, many apparently voluntary such as choosing to drive (rather than walk) to work, or to smoke, or to live along the coast. However, decision makers may find their choices restricted by economic, cultural, social, political, and even religious constraints. Further, even some voluntary risks may not be deemed fair in the eyes of society. Sagoff (1992) contrasts the risk associated with purchasing a lottery ticket with that of owning a home adjacent to a chemical factory. Although both situations are risky, the lottery ticket holder cannot expect society to reimburse any losses while it is generally agreed that the homeowner should not be held accountable for his or her loss if an accident occurs at the factory.

Knowledge of the available choices is an important factor in choosing a course of action. Although the relationship is much more complex than suggested here, it is important to recognize that risk involves choice. Who makes that choice, based on what information, for whom, and with what results are critical to answering questions about why people continue to occupy hazardous areas. Obviously, these are complex questions.

Outcomes: Probability and Consequence

Besides choice, some attention must be given to outcomes of decision making. For instance, those who elect to live in hazardous areas may not fully appreciate the risk. While their initial choice may have been voluntary, their knowledge of potential outcomes may have been incomplete. For example, the retiree, who becomes a homeowner in Florida may be unaware of the seriousness of threats from hurricane-force winds, tidal surges, or rising sea level. Such a decision may be perceived as rational by the individuals concerned, who may expect to live there for a limited time (perhaps only 5 or 10 years) thus reducing the probability of being there during a major storm.

Predicting outcomes is not easy for either individuals or society. Outcomes have a number of characteristics that can vary significantly depending, among other factors, on geographic location and time of occurrence. For example, take the physical aspects of a hazard; the magnitude, timing, and extent of a geophysical event influence the decision making of individuals involved. Although risks associated with natural hazards are not usually viewed as positive, if the timing is right, the flooding of agricultural lands can increase fertility without damaging crops. However, a flash flood calls for immediately vacating the hazard zone, and failure to respond appropriately can result in death. Over 140 people died, in the Big Thompson Canyon in 1976, many because they made the wrong decision to try to outrun the flood in their automobiles (Gruntfest, 1977). Actions can have negative or positive outcomes, which affects the element of risk; those individuals who attempted to drive down the canyon increased their risk while those who climbed to safety reduced theirs. Societal decisions also can have positive and negative outcomes; building a sea wall to prevent sea surge from hurricanes, as was done in Galveston, Texas, increased development in the "safe" zone, thereby placing more people and property at risk.

Thus, there is uncertainty associated with outcomes, an uncertainty founded not only in the physical dimensions of hazards but also in human decision making. For instance, we cannot guarantee that a disaster of a particular magnitude will occur at a specific time or place; for most hazards, it is not possible to forecast events accurately. Similarly, although we can identify some areas as prone to particular geophysical events, we cannot be certain that any area is free from a specific hazard (as discussed in Chapter 2). Human behavior further complicates the range of possible outcomes, such that the extent of property damage or number of lives lost cannot always be determined accurately prior to an event.

Nonetheless, there exist some projections for deaths from events. In California, scenarios for earthquakes of different magnitudes at different times of day have been used to estimate deaths (Federal Emergency Management Agency, 1980; U.S. Department of Commerce, 1973). Similar models do not generally exist for other hazards at other places. To counter this omission, we must think in terms of probabilities of outcomes or consequences. For example, it is possible to calculate the likelihood of dying from a particular action or event (Table 7.1). However, while these numbers are useful in comparing risks, they are based on the same outcome (death) and do not include a temporal component; without it, such figures can be grossly misleading because exposure is not considered. For example, the risk of being killed in a quarry is 1 in 3,100 per year at any given time, but risk increases to 1 in 80 over a 40-year work period (Dinman, 1980). In general, death rates are mean-

TABLE 7.1. Probability of an Individual Dying in Any One Year

Hazard	Probability of death
Smoking 10 cigarettes/day	1 in 200
All natural causes, age 40	1 in 850
Any kind of violence or poisoning	1 in 3,300
Influenza	1 in 5,000
Accident on the road if driving in Europe	1 in 8,000
Leukemia	1 in 12,500
Earthquake, if living in Iran	**1 in 23,000**
Playing field sports	1 in 25,000
Accident at home	1 in 26,000
Accident at work	1 in 43,500
Floods, if living in Bangladesh	**1 in 50,000**
Radiation, if working in radiation industry	1 in 57,000
Homicide, if living in Europe	1 in 100,000
Floods, if living in Northern China	**1 in 100,000**
Accident on railway, if traveling in Europe	1 in 500,000
Earthquake, if living in California	**1 in 2,000,000**
Hit by lightning	**1 in 10,000,000**
Wind storm, in northern Europe	**1 in 10,000,000**

Note: Geophysical hazards are shown in boldface. Source: Coburn, Spence, and Pomonis (1991). Copyright 1991 by United Nations Development Programme. Reprinted by permission.

ingless without qualifying information on factors that might affect exposure, such as occupation, gender, and age.

Overall, risk should be viewed as existing on a two-dimensional plane for any specific location, the extremes of which are high probability–low consequence and low probability–high consequence risks (Figure 7.1). An example of the former might be a thunderstorm in Florida; a level 5 hurricane making landfall near Miami Beach would illustrate the latter. Certainly, any number of natural and technological hazards could be depicted at different points on the plane. The examples chosen serve to illustrate extremes at a given location.

In its simplest form, risk can be defined as the product of probability of occurrence and magnitude of an event. This might be considered "technical risk" because it combines the two elements of risk in a logical, mathematically sound manner, thus yielding a means of comparison. The difficulty with this concept, however, lies in the fact that identical values may represent either high probability–low consequence or low probability–high consequence risks. For example, an earthquake of magnitude 7.5 on the Richter scale may have a return period of approximately 100 years at a particular location, i.e., .01 probability of occurring in any given year.

FIGURE 7.1. Relationship between an event's probability of occurrence and the extent of probable consequences. Different hazards would occupy different points on this diagram; the two depicted here illustrate potential extremes.

According to the formula, the risk could be described as 0.075. Let us assume also that a 3.75 magnitude earthquake has a .02 probability (that is, a 50-year event), which translates into a risk factor of 0.075. Similarly, an earthquake with a .05 probability and a magnitude of 1.5 also would have a risk of 0.075.

Despite identical risk, these events would have significantly different outcomes. If we are unconcerned about attitudes toward these differences, technical risk is an appropriate measure. However, here we are not unconcerned with society's views and perceptions of risk. In fact, these are our focus because they influence attitudes, actions, and ultimately vulnerability. Hence, Whyte (1982) suggested altering the risk formula from

$$\text{risk} = \text{probability of occurrence} \times \text{magnitude}$$

to

$$\text{risk} = \text{probability of occurrence} \times \text{magnitude}^n$$

where
n = social values.

Whyte's modification allows for inclusion of social concepts of risk and consequently, recognizes variation in perception in different contexts. The difficulty, of course, lies in trying to put a value on n. Nonetheless, it expands our view beyond a narrow, technical measure of risk to one that recognizes the importance of different social and cultural interpretations and outcomes, and, with the formula presented earlier, illustrates the numerous ways risk can be conceptualized. As we have emphasized, outcomes can be and frequently are changed, most obviously through mitigation; such actions as retrofitting buildings to withstand shaking, refining emergency response systems, and planning recovery procedures change the nature of the earthquake risk because they reduce adverse consequences and change perception. It has been argued that when n is sufficiently high, mitigation measures are sought in an attempt to lower the ultimate value of n. For instance, if the magnitude of the 7.5 earthquake event were squared, then risk would rise to 0.56.

Uncertainty

One difficulty that pervades all consideration, evaluation, and analysis of risk is the level of uncertainty with which we must contend. Indeed, uncertainty is inherent in all aspects of risk. It lies in the probabilistic nature of occurrences, in outcomes, and in the efficacy of various choices. Uncertainty is problematic for several reasons. First, because it is found in all elements of risk, reducing it is not a simple matter; different approaches are required for different elements. Second, the level of uncertainty is not the same for each element or all hazards. Third, people differ in their stamina for uncertainty, and their differences are seen in the elements of risk. For instance, some individuals may be more comfortable with uncertainty of occurrence than with uncertainty of outcome while for others the reverse may hold; in both cases, their decisions would reflect their differences. Finally, uncertainty may be increased by combined risk (discussed later in this chapter). As the risks become more complex, uncertainty increases (as do all the problems noted above). Some researchers have addressed such uncertainty directly. For example, Wilson and Crouch (1987) showed that the risk of getting cancer from drinking water at the U.S. Environmental Protection Agency's chloroform standard is 6×10^{-7}; however, there is an uncertainty level by a factor of 10. Similarly, they reported that risks of cancer from cigarette smoking, background radiation, and eating large quantities of peanut butter also had uncertainty levels by a factor of 3.

Uncertainty plays an important role in our estimations of risk, definitions of which risks we are willing to face, and ability to understand what risk means. Individuals often reduce their personal vulnerability by misinterpreting probabilities. For example, an event of low probability may be completely negated or perceived as a cyclical event that will not occur again in a certain number of years; thus, many individuals feel safe after a 100-year flood. For those charged with managing risk, uncertainty is a complication that must be recognized and addressed.

Risk Communication

In the discussion above, knowledge of choices and outcomes surfaced as an important consideration. Indeed, the extent to which people understand the nature of a given risk and the choices available to them can affect the level of risk to which they are exposed. Experts and the media are the major sources from which the public derives its knowledge of risk. However, experts tend either to over- or underestimate the public's ability to evaluate risk and choice; in either case, the result is the same: appropriate information in an understandable form is not communicated to the public. As Morgan (1993) comments, "If anyone should be faulted for the poor quality of responses to risk, it is probably not the public but rather risk managers in government and industry" (p. 4). Psychological studies show that people process information through existing knowledge; experts should exploit that insight in risk communication. Morgan suggests that the only way to convey risk appropriately is first to find out what people already know, then to develop warning messages based on that information, and finally to test them for successful communication.

One example of the need for experts to plan how risk is communicated arises from what has been termed "unconventional predictions" (Showalter, 1994). A classic case arose when Iben Browning, who was not a seismologist, predicted an earthquake for December 3, 1990, near New Madrid, Missouri. Because this had been the site of one of the largest earthquakes in the contiguous United States, Browning's prediction aroused some fears; in fact, a large number of people reacted to his warning in spite of its lack of scientific underpinnings. Officials should prepare for unconventional predictions by planning how to communicate accurate risk information based on scientific probabilities and potential for occurrence.

For its part, the media also creates problems. Journalists may have a limited understanding of risk and probability, which makes their reporting prone to error. The focus on sensational occurrences (which actually present very low risks to the general public) also causes inaccurate knowledge of probabilities and outcomes. Indeed, the attention devoted to events by the media bears little relationship to their severity (Adams, 1986; Combs and Slovic, 1979).

Undoubtedly, a large part of the problem is due to the uncertainty that surrounds both expert and lay knowledge of risk. Scientific knowledge of probabilities of occurrence, available choices, and possible outcomes is lacking. Although there is a demonstrable need for clear communication of risk, there are many difficulties in achieving it.

Risk Analysis: Differing Views/Different Decisions

Having a working definition of risk is a start, but we cannot stop there. Even more difficult than establishing a logical or functional relationship among variables is analysis of the results. Once a numeric or nonnumeric value has been determined, the complex task remains of analyzing what it means to those affected or to other decision makers. One of the key issues in understanding risk and accomplishing risk assessment is the differing views people hold on the importance of various risks, which will be discussed in some detail later in this chapter when different types of risk are considered. Here, it suffices to say that regardless of individuals' experience and training, it is not the scientific definition of risk on which they base decisions about which actions to take or to which hazards they will knowingly expose themselves. Thus far, the standard risk analysis paradigm—in which one chooses from among alternatives that have different potential outcomes, the probability of which can be measured—has not proved entirely satisfactory in explaining behavior.

For example, Green, Tunstall, and Fordham (1991) have explored the differing views of risk held by engineers, emergency planners, and the public. While engineers tend to view risk as a measure of probability of occurrence of an event, emergency planners appear to be more concerned about the risks associated with different public responses to official actions such as warnings. In contrast, views of risk held by the public are much more difficult to categorize because they vary based on experience, among other factors. (These differences affect the success of risk communication as well.) Each of us bases our decisions on our own assessment of risk. While perhaps starting from the same point (in this case, the calculated risk value) we conclude or make decisions at different points. Knowledge is important, but only part of the process; if risk involves choice among actions and outcomes, each of these has characteristics that also influence views of risk. It may be that differing views of risk relate to different levels of importance accorded one or other of its components. For instance, some experts may evaluate risk by focusing on choices, with values attached to those choices. Others, perhaps laypersons, may focus on outcomes, particularly the manageability of outcomes (Smith, 1990). Thus, it is not merely differences in types and levels of risk that explain differences in attitudes and decisions. One also must look to the importance given the components of risk by decision makers.

For example, take the influence of religion. Many individuals are bound by cultural and religious beliefs, and respond to risks accordingly. We are all familiar with risktaking by Christian Scientists, who reject medical support intervention in favor of prayer. Religious individuals may view natural events as divine acts and believe that little can be done to prevent them. In many traditions, deaths from disasters are attributed to wrongdoing, such as immoral lifestyles or irreligious behavior. It is not unusual for disasters to be interpreted by some faiths as manifesting divine wrath; the Lisbon earthquake of 1755 was followed by just such claims as were the 1906 San Francisco earthquake, the 1993 floods in the upper midwestern United States, and the Californian earthquakes of 1994. The relationship between death and religion is complex (Spilka, 1985); the extent to which it affects risktaking and decision making in hazardous situations is intriguing and has not been fully explored.

MEASURES OF RISK

The first step in interpreting the extent and significance of differences in risk requires greater precision in terminology. When people talk about risk, they often are referring to different things. There also are several measures of risk (Starr, Rudman, and Whipple, 1976), which frequently are used interchangeably. The different measures include real, statistical, predicted, and perceived risk. Real risk is perhaps the most difficult to determine because of the role played by time, that is, real risk is determined by future circumstances as well as past history of actual occurrences. Statistical and predicted risk are "objective" estimates based on observed frequencies and theoretical probabilities, respectively (as discussed in Chapter 2). The former (which has been embraced by the insurance industry) is grounded in scientific method and uses common statistical techniques. In contrast, where there is a lack of experience with or knowledge of frequencies and outcomes, predicted risk relies on simulation models. It is used most often for events that have an extremely low probability of occurrence.

Our primary concern here lies with perceived risk, or the subjective value to which people react and respond. Research on risk perception shows that quantitative measures are less important than the qualitative attributes of a risk. People tend to evaluate risks in a multidimensional but subjective manner and as a result, some risks become "socially amplified" while others are "socially attenuated" (Kasperson et al., 1988). In other words, some risks may be perceived as greater than scientific measures would suggest (that is, they are overestimated) because their effects are judged to be unacceptable (whether socially, economically, psychologically, or otherwise). For example, after the nuclear power plant accident at Three Mile Island, similar plants around the world were shut down, checked for safety, and restarted more frequently than prior to the accident, despite the fact that this procedure is the most risky of the operational stages. Other risks may be underestimated or attenuated, and thus receive less public concern and attention than they merit; examples would include raised speed limits and, until the late twentieth century, cigarette smoking.

A number of factors contribute to risk perception, including exposure, familiarity, preventability, and dread (Coburn, Spence, and Pomonis, 1991). Again, the differential weights associated with these factors illustrate the divergence between lay and expert estimates of risk. These differences are frequently evident in estimates of the risks associated with land uses proposed for particular places, especially noxious or unwanted facilities such as landfills, solid waste incinerators, and jails. Expert estimates frequently ignore the social and cultural context within which risks are evaluated by the public, and misconstrue the role of individual and group perception.

As discussed in Chapter 3, perception is a complex concept that provides a basis for understanding responses to risks and hazards. While difficult to measure precisely (which frustrates many experts in risk analysis), there are social, cultural, and psychological components to laypersons' estimations that go a long way toward explaining why some risks are judged to be acceptable and others unacceptable or why some risks are socially amplified and others are attenuated. Acceptable risk is not different from perceived risk, although acceptable risk depends on perception of both risks and benefits; that is, risk, benefits, and costs must be perceived before risk can be judged either acceptable or unacceptable. Further, risk perception is dynamic; new choices or information about outcomes lead to new perceptions.

Perceived Risk

Within the large body of research on risk, much attention has been given to determining and weighing the factors that affect risk perceptions. This is an enormous challenge because of the range of social, psychological, physical, technological, and cultural factors involved and the interactions among them. For example, Slovic (1987) showed that the perceived risk of 30 different activities and technologies varied significantly among social groups; college students and members of the League of Women Voters ranked nuclear power the most risky, whereas experts ranked motor vehicles the highest risk and nuclear power only twentieth. Risk means different things to different people; the importance of understanding risk perception and the factors that have an impact on it, cannot be overstated. If we do not understand risk perception, we can neither comprehend nor anticipate responses to risk, which complicates risk reduction.

Determining which social, psychological, and environmental factors influence risk perception is not easy. A number of techniques have been used to discriminate between factors, including social and attitudinal surveys (usually based on individual questionnaires) and various scales (based on psychometric analysis and multidimensional scaling). (Applications of scaling methods are discussed in Ajzen and Fishbein [1980], Slovic, Fischhoff, and Lichtenstein [1985], and Tversky and Kahneman [1982].) Each of these techniques has been criticized, either methodologically or in specific applications. Although we should interpret the results with caution, the data are enlightening and, on the whole, allow for evaluation of the complexity of factors involved.

Table 7.2 categorizes factors affecting risk perception, based on the nature of the risk, the nature of its consequences, and individual characteristics. The three categories show that characteristics of a risky activity and its consequences are critical to perception. Individual characteristics might be seen as an overriding layer against which factors associated with the risk and consequences interact. This typology is not entirely neat as some factors fit into more than one category, but groupings differentiate factors in a way that is meaningful for analysis and discussion. However, the reader should not infer that these are discrete factors that are individually relevant in every situation; they are not independent, but rather interact with a number of other factors to shape perception. Different combinations

TABLE 7.2. Some Factors Affecting Risk Perception

Nature of risk	Nature of consequences	Individual/social characteristics
Voluntary or involuntary	Immediate or delayed	Familiarity with risk
Known to science	Chronic, cumulative, or	New risk or
Measure of control over risk	catastrophic effects	experienced risk
(controllability)	Common or feared	Degree of personal
Changing character of	consequences	exposure
risk	Severity of consequences	Perceived ease of
Availability of alternatives	Size of group exposed	reducing risk
Necessity of exposure	Distribution (equity) of	Occupational hazard
Possibility for misuse	exposure	
	Effect on future	
	generations	
	Global catastrophic nature	
	Average number of	
	people affected	
	Reversibility of consequences	

Sources: Adapted from Slovic, Fischhoff, and Lichtenstein (1979), Covello, Flamm, Rodricks, and Tardiff (1981), and Griffiths (1981).

of the same factors can result in different, yet perfectly rational decision making. In addition, not all factors apply in all circumstances (Fischhoff, Slovic, Lichtenstein, Reed, and Coombs, 1978; Starr, 1969). Finally, other socioeconomic and situational traits must be considered, including age, income, education, and gender, which have been found to influence perception almost irrespective of the 23 factors on the list. As discussed in Chapter 3, these additional variables influence perceptions of hazards, but it is important to consider the related, yet not identical perceptions of risk. In particular, we are concerned with perceptions of risk probabilities and outcomes, as they stem from characteristics of the intrusion, while views of personal vulnerability come into play later.

The nature of the risk has been found to have an important influence on perception, though there is some discrepancy in the results of various studies. For instance, Starr (1969) indicated that perception and acceptability of risk are influenced by whether the activity that includes risk is voluntary. While Slovic, Fischhoff, and Lichtenstein (1979) could not correlate perceived with voluntary risk for more than 30 hazards, there is some evidence that the extent to which people face risks voluntarily is salient to perception (Slovic, Fischhoff, and Lichtenstein, 1980). This relationship is complicated because some risks may appear to be voluntary (such as locating in a flood-prone area) when, in fact, the individuals concerned may have few options; economic constraints may determine their available choices. Therefore, consideration also might be given to questions of fair and unfair risks, although there are difficulties in drawing such general distinctions because of the context in which risktaking occurs (Sagoff, 1992). Could we consider the risk of tornadoes to be fair because they can occur virtually anywhere, while living in squatter settlements on unstable slopes would be considered unfair? The categorical distinction blurs when the effect of housing structure on tornado damage is added.

These issues illustrate the complexity of risk perception as well as the differential influence that various factors can have. In light of the many possible combinations of variables given in Table 7.2, the search for order in perceived risk is indeed an onerous task. At the same time, risk is not static. It changes as we learn more about it, as alternative activities or mitigation strategies are developed, and as the physical environment changes, and risk perception changes with it.

However, we cannot focus solely on the nature of the risk. Risk perception also is influenced by the perceived consequences of a particular activity or location. The risks differ in severity and probability of consequences. Perceived outcomes such as immediacy of a threat, dread, numbers affected, and potential for catastrophe all greatly influence risk perception. Consequently, individual perception of risk rarely corresponds with technical risk. These "misperceptions" may be related to fear of consequences, rather than probabilities of occurrence. For instance, many people express great fear of flying, but are relatively unconcerned about driving to and from an airport although statistically, the latter is far riskier than the former; similar patterns emerge with natural hazards. In addition, familiarity and denial play an important role in risk perception; there is reassurance in the perception that others are worse off because they face greater hazards. For example, Californians express surprise that Midwesterners can live with tornadoes cutting swaths of devastation through communities, while Midwesterners cannot comprehend how Californians can live where structures are leveled by earthquakes. Obviously, many more variables influence perception than are discussed here.

To some extent, the popular media also affect perception by highlighting risks and fatal incidents. The resulting image of risk encourages the perception that every element of life (especially in "other peoples' environments") is extremely hazardous. Again, these relationships do not work independently or consistently; their effects vary from risk to risk, from person to person, from place to place, and from time to time. For example, Lave and Lave (1991) showed that public perception of the flood risk fails in some aspects because of inadequate communication, suggesting that current government publications are not likely to be understood by those at risk; they found that people who worked and had higher levels of education knew more about flooding and were more likely to have flood insurance.

Studies have identified several systematic biases in lay risk perception. These include the memorability or imaginability of a hazard (related to an availability heuristic, discussed below), overconfidence in risk judgment, and tendencies toward underestimating uncertainty (Tversky and Kahnemann, 1982). Combined with other factors, these biases contribute to the acceptance of certain risks while others are deemed unacceptable. Thus, one question to which risk analysts have returned repeatedly is, how safe is safe enough?

Accepted and Acceptable Risk

A distinction must be made between accepted and acceptable risks. Some risks are viewed as the consequence of living in a particular location, but though they are an accepted part of a lifestyle, it does not necessarily follow that they are acceptable. For example, Bangladeshis may accept the hazardousness of their existence without feeling that the risks are acceptable. Even those who choose to live in hazardous locations (such as on unstable slopes or eroding cliffs) may come to accept the risk, but consider it unacceptable. In fact, we face risks every day, and we accept their outcomes, but this does not necessarily make them acceptable. When choice is limited, the balance of perceived risks and benefits is severely constrained, but it still can be argued that the risk is accepted.

By contrast, acceptable risk is determined by the decision-making process. Benefits (perhaps of increased safety) are balanced against the costs of reducing risk or restricting hazardous activity (Fischhoff et al., 1978; Fischhoff, Lichtenstein, Slovic, Derby, and Keeney, 1981). Thus, acceptable risk is based on perceived risk and benefits, particularly perceived benefits. However, there are problems with this definition of acceptable risk. For instance, once the term has been applied, it may wrongly imply that a risk is acceptable to everyone. In fact, the distribution of risks and benefits can be quite inequitable (see Chapter 6). For example, charges of environmental racism are based on the inequitable distribution of risks. We must therefore consider to whom a risk is acceptable and evaluate the social, political, and economic contexts in which risk is decided to be acceptable. It may be that a risk only appears to be acceptable when it is merely accepted, necessary, tolerable, or unknown.

Necessary risks exemplify accepted risks that, in a different context, may not be acceptable. Necessary risks are those we face unavoidably (for instance, as a result of our occupation, income, or ages), and their outcomes are not changed easily by our own volition. Some necessary risks exist because political decision makers have determined that they are necessary to attain socially desirable objectives. An example of a necessary risk might be flood-proofing public buildings rather than moving them off the floodplain; although flood-proofing allows them to remain centrally located, the workers and visitors are vulnerable. The workers take a necessary risk, which is accepted for employment but is not necessarily acceptable.

Tolerable risk represents temporarily acceptable risk. An individual may be prepared to tolerate a risk because it is confined to a brief time period or associated with a short-term activity. For example, Midwesterners tolerate extreme wind chill, knowing that it is seasonal. Individuals who remain outside videotaping tornadoes could be described as finding it a tolerable risk.

Finally, there are risks about which we currently know very little or nothing. As we come to learn more about magnitudes and probabilities, it may appear that these risks were accepted, if not acceptable, when in fact they were unknown. This possibility illustrates the need for dynamic risk analysis; at different times, risks may be accepted, acceptable, necessary, or tolerable. It has been argued that risks fall along a spectrum of accepted through acceptable (Dinman 1980).

Some attempts to measure accepted and acceptable risks have focused on economics, especially willingness to pay and revealed preferences. It has been argued that societies arrive at "optimum solutions" to hazard risks (Slovic, 1987) through a process of evolved tolerance (Alexander, 1993). In other words, the existing conditions for a particular community may reflect the currently accepted level of risk. If society is willing to pay for reduced risk through the construction of a mitigation project, that provides evidence of acceptance. The actual, accepted risk level may be reached through trial and error as conditions and attitudes change and as different projects are implemented; the strengthening of building codes in earthquake areas is an example of this trial-and-error search for an acceptable level of risk.

While the actions of society may reveal much about communal attitudes towards risk, they do not necessarily reveal an optimum solution. Sagoff (1992) argues against the concept of revealed preference (discussed in Chapter 3), contending that past actions are simply those that have occurred; for example, in eighteenth-century mills in England, intolerable working conditions were accepted by workers who had little choice. Those conditions did not result from societal decision making but rather a few mill owners dominated the decision-making process. Similarly, many people living in hazardous environments (such as the squatter residents of urban agglomerations) may not willingly accept the risk, but they have little alternative but to tolerate their situation.

There also appear to be different acceptance levels for voluntary and involuntary risks. Generally, people are more willing to accept higher risks from voluntary actions than for involuntary ones. Starr (1969) reported that given similar types of benefits, the accepted levels of risk were a thousand times higher for voluntary activities. While there are criticisms of his study, its findings echo a common theme: people tolerate risk for others. For example, it is estimated that in the United States, approximately 200,000 excess deaths occur each year because of smoking, yet this was socially acceptable until late in the twentieth century (Dinman, 1980). Similarly, in 1996, many states raised speed limits on certain roads, demonstrating a willingness to tolerate more traffic deaths. In contrast, action has been taken to minimize deaths from traffic accidents by passing laws requiring seat belts, which could save up to 20,000 lives per year.

The question is where does society draw the line? Although most laypersons find most risks unacceptably high, individuals continue to participate in hazardous activities (Slovic, 1987). However, attitudes are always changing (Morgan, 1993). We have seen increased seat belt usage, reduced smoking, and improved diets that have reduced many risks. On the other hand, these are relatively inexpensive to implement and may not be comparable to large-scale mitigation projects. The willingness of society to pay for reduced risk remains a useful measure of accepted risk.

Availability of Information and the Role of the Media

An important factor that greatly influences perceptions of risk is the accessibility of information, termed the "availability heuristic" (Tversky and Kahnemann, 1982). Although it is not listed separately in Table 7.2, it is an important component of several factors such as familiarity with a risk and average number of people affected. This heuristic suggests that events are judged to be likely or frequent if they are easy to recall; in other words, the more available information is about the occurrence of an event, the more likely it is that people will expect the event to recur. Personal experience with disasters (especially with more than one event) can translate into more accurate perceptions of risk, but because many natural hazards have a low probability of occurrence (with larger magnitude events occurring with much less frequency), most people do not have this reinforcement from personal experience. Consequently, their information is obtained from the media, which influence risk perception through the reporting, or nonreporting, of events.

Studies of media reporting have found an emphasis on life-threatening events, which are not closely related to statistical frequencies (Combs and Slovic, 1979); catastrophic events like tornadoes, homicides, and fires tended to be reported disproportionally often. Thus, the information that is easily available and accessible to laypersons often gives an inaccurate and distorted view of risk. It is little wonder that perceived risk can differ so much from real or statistical risk. The media are only one influence on perception, albeit a very important one. Other studies have shown that people are reasonably accurate when asked to order a number of common or well-known hazards based on injuries and deaths, but when asked to rank them in terms of risk, the results are much less accurate (Morgan, 1993).

Summary

Risk is more than probability of an event, though that often is how it is measured. Any analysis of risk must include vulnerability, including absolute and relative measures of the population and property at risk. Even when we include these variables, however, there is difficulty in managing risk. One might assume that once risk is defined and measured, appropriate steps to minimize it can and will be taken, but that depends on the nature of the risk and decisions to be made. Of course, decision makers (whether public or private, individual or collective, expert or lay) approach the problem with different experiences, fewer or greater constraints on choice, and perhaps even different views of what is meant by risk. Further, risk is dynamic. All these factors influence the ability to manage a given risk, assuming we even have the opportunity to do so. As a critical component of any comprehensive analysis, risk confounds, but does not entirely diminish our capacity to address natural hazards.

THE CHANGING NATURE OF RISK

Risk from Technology and Risk from Nature

The risk associated with natural and technological hazards often is discussed interchangeably. Indeed, when considering risk, it matters little which type of hazard is of concern. Table 7.1 illustrates different levels of risk and the probability of death associated with various technological and natural hazards; these data show that the risk of dying from various natural hazards is not particularly different from risks associated with other hazards or diseases. However, much of the research on risk and hazards has addressed technological rather than natural hazards, and natural hazards research often has borrowed or applied those results.

To some extent, that is reasonable. Table 7.3 illustrates the similar grounds on which risks from both types of hazard can be measured; although the terminology is slightly different, the variables are essentially the same. For example, the greater the quantity of a pollutant or chemical released, the greater the risk in most cases; the physical magnitude of

TABLE 7.3. Variables in Risk Assessment

Technological hazards	Natural hazards
Probability of release of a harmful substance	Probability of occurrence
Quantity of harmful substance released	Magnitude
Dispersion of a harmful substance and resulting concentrations in the environment	Spatial extent
Population exposed to release of a harmful substance	Population in the risk area
Uptake of harmful substances by humans and other organisms	Occurrence of geophysical event
Relationship between dose of harmful substance and adverse toxicological effects	Vulnerability or damage potential
Measurement error	Measurement error

Source: Adapted from Talcott (1992).

natural events also affects risk. This is not to say that the nature of the risk is not important. For instance, the level of uncertainty associated with each common variable differs among hazard types, and physical characteristics of the risk source can have a distinct influence on perception, acceptance, and management of risk. Because natural and technological hazards differ in many respects, we can expect differences in the nature of the risk and in risk perception. The focus here will remain on natural hazards, making reference to technological hazards where appropriate.

Figure 7.2 compares some natural and technological hazards, based on the relationship between frequency of occurrence and number of casualties. Although these data suggest that natural disasters cause much greater loss of life, for the most part natural hazards are not seen to be as great a risk as many technological hazards. Other comparisons lead to very different conclusions; for instance, many technological accidents surpass natural hazards. The 1984 catastrophe in Bhopal, India (when a chemical release killed more than 2,500 people in the first few days and injured or disabled thousands more) far exceeds most natural hazards in its severity. Similarly, the failure of the nuclear power plant at Chernobyl has had long-term consequences on local populations and the landscape. Another way of comparing these two categories of hazards is by cumulative deaths; many more people die in automobile accidents in the United States than in all natural hazards combined, but the fact that the deaths do not occur all at once alters our perceptions of the risk. Indeed, it is instructive to think about the differences among hazards that lead to such varied perceptions.

Natural and technological hazards often influence each other. Indeed, "natural and technological hazards coexist and can combine synergistically" (Showalter and Myers, 1994, p. 181). Technological hazards have caused natural events, including earthquakes in Saskatchewan brought on

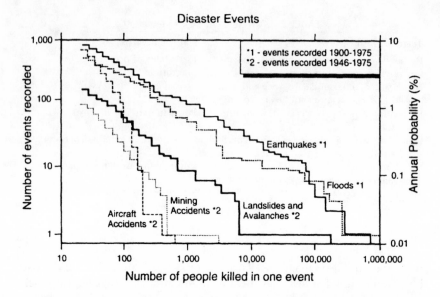

FIGURE 7.2. Comparison of natural and technological hazards. When considered on a similar scale, the former greatly exceed the latter in death tolls. *Source:* Coburn, Spence, and Pomonis (1991). Copyright 1991 by United Nations Development Programme. Reprinted by permission.

by underground mining (Stevens, 1988) and earthquakes caused by filling reservoirs (El-Sabh and Murty, 1988). Natural hazards also can cause technological failures, such as when an earthquake causes a dam to break or floodwaters put fuel tanks at risk (Gruntfest and Pollack, 1994). The floods in Texas in October 1994 led to fuel leaks that ignited and spread fire along extensive reaches of the San Jacinto River.

It is the latter situation, where extreme geophysical events create technological risks, that is of increasing concern to researchers, regulators, and emergency managers. These emergencies, termed "na-tech" events, present some real difficulties in planning and management, largely because of the complexities of combined risk. The threat of na-tech events has been growing, as shown in Figure 7.3). Through 1989, earthquakes were responsible for the majority of na-tech events though floods accounted for the most disasters (see Chapter 2). Nevertheless, the 1993 floods in the Mississippi River system certainly increased the visibility of na-tech events with numerous accounts of threatened fuel tanks in the flood hazard areas and attempts by industry, notably Phillips Petroleum, to mitigate the na-tech hazard (Gruntfest and Pollack, 1994). Na-tech hazards are particularly relevant to the legal context, as litigation surrounding the Phillips Petroleum propane tanks in St. Louis illustrates.

We are dealing, then, with what might be considered combined risks. It is logical to consider the risk of each independently, but it also makes sense to consider the risk resulting from their interaction, which becomes increasingly complex as the number of variables increases. This is difficult enough when considering the risk from multiple natural hazards at a location. Incorporating risk from technological hazards increases the

FIGURE 7.3. Na-tech incidents in 20 states, 1980–1989. Natural events can cause technological incidents, including releases of hazardous materials. Earthquakes cause the most na-tech incidents (a). The temporal trend is somewhat disturbing, suggesting an apparent, though not steady increase (b). However, this could be a function of improved reporting. *Source:* Showalter and Myers (1994). Copyright 1994 by Society for Risk Analysis. Reprinted by permission.

complexity, perhaps disproportionately (see the data on earthquake-generated incidents in Figure 7.3). With technological hazards, one must consider the risk of disaster due to technological failure or human error as well as physical parameters such as seismic fragility, or the potential for and effects of failure of structures and equipment from different intensities of ground motion (Reiter, 1990). The risk is changed in terms of both probability of occurrence and consequences. During the Mississippi River floods of 1993, a propane tank in Jefferson City, Missouri, broke loose from its moorings, cracked open, and was carried along with the floodwaters; had the fuel ignited, the consequences of flooding would have paled in comparison to those from the resulting explosion. Of course, some of this risk can be mitigated, but that requires sufficient recognition of the combined risk to undertake pre-event vulnerability analysis (Showalter and Myers, 1994).

Despite similarities, characteristics of the risk associated with these hazards differ. Generally, a technological hazard is "controllable" in that safeguards are incorporated in the technology; their failure is required for an event to occur. By contrast, natural hazards are not generally controllable, although loss from them may be. This difference affects risk both statistically and, more important, perceptually. Mitigation measures for natural hazards are designed to lessen risk and vulnerability, perhaps by changing physical characteristics from uncontrollable to controllable; such measures also change the risk from natural to technological, so that it is the failure of the technology rather than the natural system that causes an event. The sea wall designed to prevent sea surges in the event of a hurricane offers protection from an event of particular size, but when a larger event occurs the structure may fail, presenting a technological rather than natural problem.

Changing Risks by Mitigation

The implementation of hazard mitigation projects affects all aspects of risk: actual, scientific, technical, and perceived. Implementation often generates a false sense of security (see Chapter 3), lowering the perceived risk. For example, take the changing nature of risk from earthquakes. The seismic hazard is identified with the magnitude and intensity of earth movements and the resulting ground shaking, deformation, soil liquefaction, and landslides, with effects including property damage, injury, and death. The probability of negative consequences depends on the magnitude of the geophysical event and the preventative measures taken within the affected region (Reiter, 1990). Because an earthquake is uncontrollable, society must depend on adjustments to reduce the consequences, and those adjustments change both perceived and actual risks even though the seismic processes remain constant. Thus, risk is reduced because vulnerability is altered, but probability of occurrence of an event remains the same. The adoption of mitigation measures effectively lowers the risk, at least for earthquakes of a given magnitude, but if an earthquake exceeds the design standards, the measures may fail.

Floods provide another example. In the United States and elsewhere, engineered structures have been used historically as mitigation measures, providing a means of controlling the flow of rivers and therefore controlling the hazard. This both lowers and changes the nature of the risk. Because of flood-control measures like dams, levees, and flood walls, the actual risk of flooding from small floods is decreased at a given place. However, the development of dams and levees also has created a technological hazard. The risk of flooding has been redirected to the risk of dam or levee failure. The problem is exacerbated by the false sense of security provided by flood-control measures.

It may be argued that the failure of a dam or levee is a natural disaster because the result is the same: destruction, property damage, injury, and perhaps loss of life. However, the nature of the flooding is different. Except with flash floods, natural floods are more likely to develop gradually, allowing time for warning and evacuation. With technological failure, however, flood heights and the velocity of floodwaters often are increased, leaving less time for warning and evacuation and leading to greater losses. Risks may well be underestimated if the probability and consequences of technological failure are not considered.

Risks from Multiple Natural Hazards

Most places are subject to more than one natural hazard. Many parts of California are subject to earthquakes, floods, drought, landslides, and wildfire. Bangladeshis experience hurricanes, riverine flooding, storm damage, and occasional tornadoes. Many countries in Western Europe are exposed to blizzards, flooding, drought, and earthquakes. Unfortunately, the multiple hazard perspective is rarely adopted in assessing hazardousness or riskiness. Instead, the focus has been on the risk posed by single hazards, which does not provide a sufficiently comprehensive understanding of the overall risk that exists at a given place; it can lead to gross underestimates of risk and hazardousness and may result in inadequate risk management.

The classic work on multiple hazards at a place was undertaken by Hewitt and Burton (1971) in a study of London, Ontario. Although considered by the authors at the time to be an exploratory effort, this research remains relevant today. Hewitt and Burton considered the joint-risk magnitude for all hazards, which they suggested would provide the integrated view of vulnerability needed for planning purposes. Further, they considered the cumulative effects of smaller hazards, which frequently are thought to pose only small risks. The authors provided a valuable analytical framework for the interaction of various components of the physical and human systems, which inspired further studies. For instance, in Rotorua, New Zealand, their framework has been adapted for a planning context, as shown in Table 7.4 (Montz, 1994); a number of factors contribute to the multiple hazard risk there, some related to the nature of the geophysical events (such as probabilities of occurrence) and others related to probable magnitude of the consequences (such as the spatial extent of the impact).

These studies illustrate some of the difficulties associated with evaluating risk from multiple hazards at a place. First, as pointed out by Hewitt and Burton, the data on hazard characteristics are often grossly inadequate and present problems in comparing different geophysical events. Second, there is a distinct paucity of satisfactory analytical models, which compro-

TABLE 7.4. Comparison of Hazard Characteristics in Rotorua, New Zealand

	Probability		Spatial dimensions				
Hazard	Known or calculable	Independent/ dependent	Known	Variable within city	Widespread/ limited[a]	Onset intensity[b]	Forecast/ warning abilities
Earthquakes	Modified Mercalli recurrence intervals	I, D	Y?	N/Y[c]	W	Very high	None
Volcanic activity	Y	I	Y	N[d]	W	High–moderate	Moderate–none
Geothermal activity	Return periods	I, D	Y	Y	L	Very high–high	None–moderate
Subsidence	N	I, D	N[e]	Y	L	High	None–moderate
Hydrogen sulfide gas	N	I	Y	Y	L	Moderate–high	Poor–moderate
Seiches	N[f]	D	Y	Y	L	Moderate–high	None
Stream flooding	Y	I	Y	Y	L	Moderate	Poor–moderate

[a]Extent or size of the area likely to be affected by an event.
[b]After Hewitt and Burton (1971).
[c]Depends of location of movement and magnitude of event.
[d]Except perhaps the depth of tephra.
[e]Except relating to geothermal activity.
[f]Models for determining amplitude under various conditions exist, but have not been widely applied.

mises risk assessment. Although progress has been made in both these areas since 1971, contemporary research on multiple hazards remains somewhat constrained. Third, the differences between natural hazards are not always fully appreciated (Hewitt, 1969), which is particularly evident when probabilities of occurrence of different geophysical events are included. In some instances, natural hazards lead to discrete events with independent probabilities (as might be the case with hurricanes and floods); other events are linked (such as thunderstorms that bring hail, lightning, and tornadoes). The difficulties in multiple risk analysis are exacerbated when the probabilities are not independent (as Table 7.4 illustrates), for example, a seiche or tsunami resulting from an earthquake. Not all earthquakes (even oceanic ones) cause tsunami, but the probability of a tsunami is dependent on that of an earthquake while the reverse is not true.

Some models have been developed to address multiple probabilities. Jacobs and Vesilind (1992) analyzed the many risks of environmental damage due to chemical spills, and their model was adapted for the analysis of independent multiple hazards in New Zealand (Montz, 1994). Figure 7.4 shows the first step in evaluating multiple risks, that is, focusing on the probability of occurrence of multiple independent events for a neighborhood in Rotorua. In order to refine the model, consideration of the uncertainty associated with the probability of dependent events is required, which is illustrated in Figure 7.5. As shown, hydrothermal eruptions can result from natural hydrogeological changes or they may be caused by earthquakes or volcanic activity that force the hydrogeological changes. The uncertainties associated with dependent probabilities certainly complicate attempts to develop multiple risk measures, but without them, we can neither understand nor manage risk in an area.

Summary

It is tempting to lump all hazards together in risk analysis, but that often does not work. There is no doubt that technological and natural hazards are integrally related in that one type can cause, aggravate, or mitigate the other. Further, multiple hazards at a site increase the analytical complexities; whether dealing with relationships between types of risk or with multiple risks at a location, we can expect a synergism that increases difficulties disproportionately to the number of risks under analysis. Having presented the linkages between natural and technological hazards and linkages between different natural hazards at a place, the remainder of this chapter focuses on risk management for natural hazards. However, it is worthwhile to ask how the discussion might differ if the focus were on multiple hazards at a place.

FIGURE 7.4. Multiple hazard probabilities. Communities are subject to multiple hazards each with a probability of occurrence. Analysis of the combined probabilities of several independent events revealed this spatial pattern of risk for a neighborhood of Rotorua, New Zealand. *Source:* Montz (1994).

RISK MANAGEMENT

No matter how risk is defined, risk management seeks to minimize, distribute, or share the potentially adverse consequences. For individuals, risk management requires knowing characteristics and magnitudes of the risks. This knowledge usually is gained from experience, and "the essence of risk assessment is the application of this knowledge of past mistakes (and deliberate actions) in an attempt to prevent new mistakes in new situations"

(Wilson and Crouch, 1987, p. 267). Just as individuals want to reduce exposure to loss, so do private and public entities, including government at all levels because of the numbers of people and amount of property at risk. As Morgan (1993) so aptly put it, "risk management . . . tends to force a society to consider what it cares about and who should bear the burden of living with or mitigating a problem once it has been identified" (p. 35). The term "risk management" describes a decision-making process that involves defining need, recognizing the options that are available and acceptable, and choosing an appropriate alleviation strategy. For the sake of discussion here, we focus on public risk management, but we also could consider private industry or individuals.

It is important to recognize that uncertainty plays an important role in risk management. First, there is considerable uncertainty associated with the scientific data gathered from extreme geophysical events (see Chapter 2); the relative size and return periods for particular events are not always easy to calculate, and other dimensions of geophysical events may be even more difficult to comprehend. Second, there is added uncertainty associ-

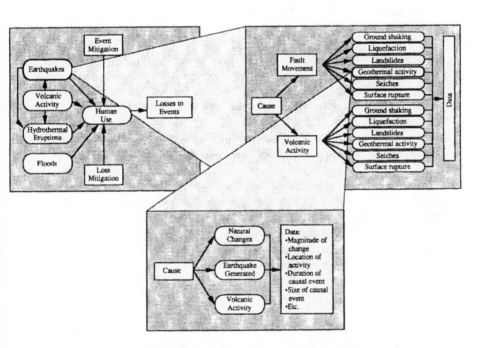

FIGURE 7.5. Dependent hazard probabilities. Not all natural hazards have independent probabilities of occurrence. The layers in this diagram illustrate some linkages in Rotorua, New Zealand. Numerous layers of data are required to develop the basis for analyzing scenarios of events. *Source:* Montz (1994).

ated with information collected within the human-use system. Whether public or private, risk management must take into account legal, technological, economic, social, and ethical considerations if truly effective policies are to be implemented. Gathering the necessary data adds considerably to the uncertainty and further contributes to management difficulties. When the physical and human sources of uncertainty are combined, the problem is amplified.

Uncertainty also becomes an integral part of risk management when mitigation projects are implemented because the magnitude and significance of uncertainty are woven into policy. However, the analytical techniques essential to risk definition and the policymaking process become secondary; the potential errors emanating from uncertainty are subsumed within policy and, by default, are no longer considered important. Thus, there is an integral relationship between risk management and public policy and, since a crucial aspect of risk management is adopting mitigative policies, they cannot be discussed separately. As a result, a number of policies are included in this chapter that were discussed in Chapter 5.

Elements of the Risk Management Process

The process of risk management is diagrammed in Figure 7.6. It begins with identifying the nature of the risk, a critical step though it may appear to be self-evident; there can be no attempt at risk analysis or management without having first identified the need for them. In the public sector, there also must be the political will to follow through with the rest of the process. If the commitment to risk management is weak or nonexistent or if there is strong opposition to using public resources to deal with the risk, then there will be no movement beyond this point. In the private sector, trade-offs are estimated, balancing investment in risk management against alternative investments; if risk management is found to have fewer benefits, the process will come to an end. Of course, in both the public and private sectors, need can be revisited at any time, as frequently happens when more information becomes available. Ultimately, better information changes perceptions of risk management.

Identifying Exposure

Exposure can be defined in two primary contexts, the physical and financial. Physical exposure includes spatial assessments of those areas at risk from geophysical events of given magnitudes. Risks at given locations differ according to the size and recurrence intervals of particular hazards; because larger events usually affect larger areas, it is useful

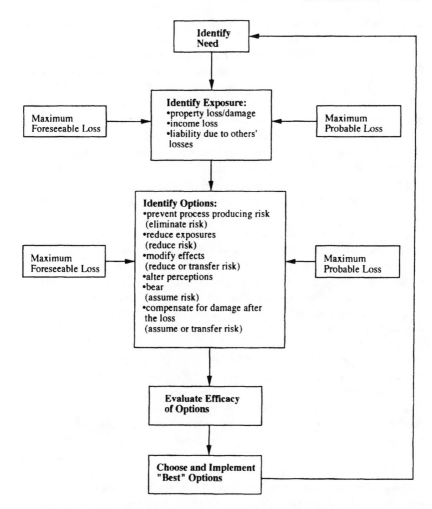

FIGURE 7.6. The risk management process. Each step requires decisions by policymakers. Many times this process is undertaken formally, as depicted here, but often it is an informal process in which the different stages are not easily discernible. *Sources:* After Burby et al. (1991) and Morgan (1993).

to know which types of events pose risks to an area and at what level of severity. (See the discussion on physical characteristics of natural hazards in Chapter 2, and that on multiple hazards above.) In this instance, exposure is defined in terms of frequency and severity of occurrence. Because professional risk managers (especially insurers) deal with uncertainty by increasing the price, the greater the uncertainty, the higher the price; thus, the more accurately the nature of the risk can be described, the better (Estall, 1994).

Financial exposure entails a risk management strategy to reduce financial loss. Thus, it is important to know the details of potential losses, such as the value and type of property at risk, the expected extent and type of damage, lost income, damage to infrastructure, cost of disrupted social and economic activities, and the expense of emergency response. These losses must be assessed for public and private concerns and for various geographic areas. There also may be liability concerns if responsibility for the disaster can be traced to a specific cause. For example, if the failure of a levee leads to extensive damage, there may be legal action if it can be proved that the system failed because of inadequate maintenance.

Analyses of exposure take a commitment of human and financial resources. In many cases, resources are diverted to other activities for any number of reasons such as limited national wealth, which explains many differences among countries. Within countries and other political entities, there are issues of the political will to act, the socioeconomic status of those at greatest risk, and risk perception. The discussion that follows is based on the best-case scenario where, if resources are needed for analysis, they can be found.

For geographic exposure, hazardous areas can be mapped to show probability of occurrence or some other risk measures. This is done for the flood hazard in the United States through the use of Flood Insurance Rate Maps (FIRMs) (see Figure 5.3), which are based on hydrologic modeling. Using rainfall records of storm events to generate flood flows, the flow is routed downstream to determine the areas and depths of flooding for the 100- and 500-year events. The maps show technical risk based on the expected depth and probability of occurrence of the 100- and 500-year floods. These maps are then used to set insurance rates, based in turn on depth of flooding (risk) and type of property. Similar maps are constructed in coastal areas where hurricanes present a hazard.

Our ability to map spatial differences in risk is probably best refined for floods and hurricanes, and to some extent volcanoes and earthquakes (as shown in Figure 5.5). Other hazards (such as tornadoes, snowstorms, and drought) present more problems that add to uncertainty, in part because they are less spatially defined. In addition, the secondary hazards associated with some geophysical events (such as earthquakes and thunderstorms) make the mapping of risk levels problematic. Still, some maps have been produced that detail the potential severity of various earthquake hazards (French and Isaacson, 1984); areas at risk from volcanic mudflows and landslides are well mapped in some areas as are seismic shaking and liquefaction in San Francisco, Salt Lake City, and Portland, Oregon. Less detailed, but still useful, are the Special Studies Zone maps in California (Cross, 1988).

Indices also have been developed to represent different levels of risk. In New Zealand, a standardized coastal sensitivity index (CSI) has been developed to rank sections of coast for susceptibility to coastal hazards (Gibb, 1994). It is derived from

CSI = elevation + storm wave runup + gradient + tsunami runup + lithology + landform + long-term trend + short-term fluctuation.

Many of these studies do not address the probability issue directly, but rather focus on what likely will happen, where it will occur, and when an event might take place. For instance, if it is estimated that a geophysical event of magnitude x will affect an area of size y, the human impact will be assessed based on an inventory of the number of structures and people in the area. A refined analysis would include a determination of structure types, construction standards, and the ability of the population to undertake remedial action. It is also important to determine the probability of occurrence of an event of magnitude x in order to have a measure of exposure against which the costs and benefits of various risk management options can be compared.

An inventory of property at risk is necessary in order to determine financial exposure. Typically, property is categorized by its estimated vulnerability to damage, involving such considerations as type of construction material, age of the structure, and number of stories. Public and private buildings are differentiated as are various uses. Because of the potentially large financial exposure, public infrastructure is particularly important but problematic. It comprises systems with varied components exhibiting different vulnerability characteristics (Burby et al., 1991), and, because these systems usually consist of networks, the probability of failure of one node or link affects many other nodes and links. In 1994 and 1995, the destruction of bridges during floods and earthquakes in California greatly exacerbated problems in the affected communities. Damage to the public infrastructure also can increase damage to other types of property.

Having determined spatial variations in risk and having completed an inventory of property at risk, the two can be combined to analyze potential losses. The precision of results is dependent on the quality of the data and, clearly, a Geographic Information System (GIS) or some other data management system would facilitate processing the volume and complexity of data. Using systems such as this, several scenarios could be modeled to evaluate exposures under different conditions. Two are shown in Figure 7.6, estimating maximum foreseeable losses, or what might be termed the "worst-case analysis," and maximum probable losses, or "the worst loss to be expected under 'average' conditions" (Burby et al., 1991, p. 65); many other scenarios could be evaluated as well. The goal of analysis, however, is a measure of the magnitude of risk to which an area is exposed. If the risk is not high or is not perceived to be high, then the risk management process will stop. However, if decision makers believe that the risk is unacceptable, risk management options will be identified and evaluated.

Identifying and Evaluating Options

There are several ways in which risk can be managed. General categories include elimination, reduction, transfer, and assumption of risk (Burby et al., 1991). Within these categories there are further options, as shown in Figure 7.6; as with identifying exposure, the results of these options vary depending on the risk threshold under consideration, identified here by the maximum foreseeable and maximum probable loss. In addition, not all options are necessarily available at a given place at a given time. Lack of available resources, limited technological capabilities, lack of interest by outside investors, and a population that has (or believes it has) no power to effect change are among the reasons options may be unavailable.

With most hazards, it is not possible to eliminate the risk completely by modifying the physical processes. With floods, a river can be controlled to the extent that the probability of high-frequency events is greatly reduced, but the ability to control low-probability floods is questionable. The 500-year flood is always a threat, and this event would certainly exceed most structural designs. Ability to control the physical environment is even more constrained with other geophysical events, such as earthquakes and tornadoes. Nevertheless, risk can be reduced by adopting safety measures or by avoiding or eliminating risky activities, but it might involve large investments that are not justified in light of the technical risk. It is only for high frequency–high consequence events that major investment would be considered; even then, the benefits would have to be significant. More often, efforts focus on reducing risk to an acceptable level rather than eliminating it completely.

Exposure can be reduced by limiting activities in hazardous areas. For instance, building on floodplains, in seismically active areas, and along coasts in unprotected zones can be avoided, which reduces levels of exposure. Of course, the level of acceptable risk must be defined first. In the United States, development within a designated 100-year floodplain is restricted; the same is true in designated hurricane hazard areas and in tsunami runup zones. For hazards with less delimited zones, building codes can reduce exposure; tie-down regulations in tornado-prone areas and landscaping requirements in areas subject to chaparral fires are two examples. This sort of action both reduces exposure and modifies effects (or losses), thus, altered location, materials, or style of building should be considered in estimating probable losses from events.

Another way to reduce exposure—especially financial exposure—is by transferring the risk, which also can modify the adverse financial effects of an event. The most common means of transferring risk is through insurance. For instance, local governments frequently carry commercial insurance for damage to real and personal property (Burby et al., 1991). Insurance provides some certainty that funds will be available to cover losses incurred from an event; for a fixed annual cost, those at risk can meet unbudgeted costs of a catastrophic nature. Of course, the greater the risk and the greater its uncertainty, the greater the annual cost; indeed, "some risk exposures are apparently so unattractive to insurers, that coverage is not readily available or, otherwise, only at an outrageous price" (Estall, 1994, p. 30). The insured are also taking a risk, namely, that the funds will be available when needed. If the insurers are not solvent or if claims exceed reserves and reinsurance is unavailable, then the insured lose. There are many instances of insurance companies failing because of disasters. In the United States in the 1920s, a group of insurance companies was forced out of business because of heavy losses (Rommel, 1960; Vaughn, 1971). That pattern has continued, especially as insured losses for natural disasters have increased in recent years (Table 7.5), although the losses associated with Hurricane Andrew skew the 1992 figures considerably. The effect of Hurricane Andrew on the insurance industry is illustrated by reports that 8 insurance companies with significant coverage in Florida collapsed within months of the hurricane (Flavin, 1994).

There are many problems with relying on insurance as a risk management option. First, insurance may not be available for some hazards. In the United States, flood insurance was not generally available until passage of the National Flood Insurance Act of 1968. A person or community was

TABLE 7.5. Insured Losses from Natural Disasters in the United States, 1984–1993

Year	Total Insured Losses ($)
1984	1,500,000,000
1985	2,900,000,000
1986	900,000,000
1987	900,000,000
1988	1,400,000,000
1989	7,600,000,000
1990	2,800,000,000
1991	4,700,000,000
1992	23,000,000,000
1993	5,700,000,000

Source: Property Claims Services, as cited by Flavin (1994).

unlikely to purchase flood insurance unless the risk was perceived as substantial; thus, individuals or communities who chose to purchase flood insurance generally resided in high-risk areas. The private insurance industry was unwilling to accept such a great risk (with such homogeneity of risk, a disaster produces a large number of claims). Consequently, when the National Flood Insurance Program (NFIP) was implemented, flood insurance was made available by the federal government at subsidized rates, provided that communities met certain floodplain management requirements. Since its inception, flood insurance has moved back into the private sector, but companies are reinsured by the government, which protects them against catastrophic losses. Still, the perception of risk was such that residents sometimes procrastinated on taking out insurance until flooding was imminent, and once the flood had passed, often dropped their insurance. (This may represent rational behavior for downstream residents who may have several days or even weeks of warning, but it does not work for those in flash flood environments.) To close this loophole, the Federal Emergency Management Agency amended NFIP rules to require a 30-day waiting period for flood insurance to take effect (Federal Register, 1995). In other countries, flood insurance is sometimes available through the private market; for instance, in Australia, limited coverage is available in Queensland and the Northern Territory, but is not available elsewhere (Smith and Handmer, 1989).

Earthquake insurance is another story. It is available in the United States only on the private market and, as would be expected, where the risk is high it is very expensive. People living in earthquake-prone areas must pay the full cost of the insurance and, as a result, coverage in California is not extensive. Although wide ranges are reported among counties, it is estimated that adoption rates for the state are well below 50% of homeowners (Palm, 1995). Following the earthquakes in California in 1994, the insurance industry asked that all homeowners' policies across the country include earthquake insurance, which would reduce premiums for those in high-risk areas and distribute the costs across a much broader population, but to date this has not occurred. In New Zealand, flood and earthquake insurance for homeowners is available through "normal" homeowners' policies. Crop insurance in the United States is becoming increasingly difficult for farmers to afford, largely because given the rising costs of disasters, reinsurers are becoming more reluctant to offer it. "The government, which has acted as a reinsurer of last resort, has lost about $3 billion on such policies since 1980" (Anon., 1995, p. 68).

Risk also can be transferred from one level of government to another by making another organization or agency responsible for emergency action. In the United States, declaration of a disaster area effectively transfers responsibility. At the national level, a presidential declaration means that financial assistance in the form of grants and loans becomes available from the federal government; for example, in 1993, the Federal Flood Relief Act provided over $6 billion for victims of the midwestern flooding. The NFIP also represents an intergovernmental transfer of risk because of the way in which the federal government underwrites it. At the state level, governors can declare disaster areas, thereby permitting state aid and mobilization of other resources. Of course, such funds do not cover all losses; other means, beyond intergovernmental transfers of risk, must be found as well.

In industrialized countries, relief frequently is applied to restoration of infrastructure, but it is also available to homeowners and other victims. International relief is another form of risk transference, from one government or agency to another. As discussed in other chapters, there are many problems associated with international aid, including the question whether it actually transfers risk or allows it to continue to increase, at least for part of the population.

There are other ways to transfer risk. One example is a mutual pool into which there are a number of contributors; its purpose is to build up sufficient reserves to meet a claim by any contributor, and claims cannot be made until an adequate pool exists. Usually, contributors self-insure for high frequency–low consequence events. Low frequency–high consequence events may wipe out a pool entirely. It is the moderate frequency–moderate consequence events for which this option is best suited (Burby et al., 1991). Pools work well because costs can be kept relatively low due to risk-sharing among participants, but they can backfire if one participant is at greater risk than the others and depletes the reserves; that is, if risk among participants is diverse, some participants may end up subsidizing the high-risk participant. Research in the United States has shown a reluctance by local officials to invest in such a pool (Burby et al., 1991).

New Zealand is interesting because of the mix of methods through which exposure is reduced and how it has evolved over time. Physical exposure is reduced through land use restrictions and building codes, notably under the Resource Management Act of 1991 and the Building Act of 1991. The Resource Management Act concerns land use and the placement of buildings, and the Building Act covers construction. There are some difficulties in how these two acts work together in practice, but both take into consideration natural hazards, particularly flooding.

More important is the way in which means to reduce financial exposure have changed in light of increased losses and decreased willingness of central government to continue to underwrite them. In 1944, New Zealand passed the Earthquake and War Damage Act. All homeowners with fire insurance paid 5 cents per $NZ100 coverage into a common fund, on the premise that because earthquake damage was widespread and unpredictable, the risk could reasonably be spread among all homeowners (O'Riordan, 1971). Following public pressure (especially from those in areas of lower risk) and some damaging events, the Act was amended in 1949 to cover other natural events. Thus, New Zealand acquired a national insurance program for natural hazards.

More recently, New Zealand passed the Earthquake Commission Act of 1993, which is consistent with the central government's interest in reducing postdisaster exposure and in having local and regional governments take more responsibility for the consequences of land use decisions. Among other changes, the Act restructured the existing Earthquake Commission and put a cap on payments from its fund, ending the national insurance program for natural hazards. Instead, under the new Earthquake Commission and the National Disaster Recovery Plan of 1992, local and regional governments must look to other ways of reducing exposure through land use decisions, private insurance, or other means. For example, central government is encouraging communities to prepare themselves for losses through the Local Authority Protection Program (LAPP), which involves insuring the community through the Local Government Association Insurance Corporation Ltd., a kind of mutual pool (Montz, 1994). Of course, it is not known whether this pool will have sufficient funds when needed by a community or communities or whether central government will provide additional relief in the event of a catastrophic disaster.

A final option is to assume the risk. On a global scale, this is probably the most common option because of a lack of access to alternatives. Indeed, bearing the cost is the only choice that many have until or unless some form of relief becomes available. Even in places where there are choices, assuming the loss may be the only practical alternative because of the high cost of eliminating or transferring risk or the lack of technical and economic feasibility. In some cases, risk may be eliminated or transferred up to some level, beyond which it is assumed. For individuals, assuming risk involves drawing on personal resources to cover losses when an event occurs. Of course, some people choose this option in spite of the availability of other choices because of how they perceive the risk and their own vulnerability; compared to the inputs required for other options, they are comfortable living with the risk to which they see themselves exposed.

Governments face a different situation with regard to assuming risk. Most obvious is the fact that their losses can be very large due, among other things, to damaged infrastructure. In addition, since repairs to the public infrastructure can affect the recovery process, it is critical that governments be able to overcome their losses as quickly as possible. Having said that, at a given level of risk, the costs of the other options may be too high for a government to commit funds to them. Governments also may choose to eliminate or transfer part of the risk and to assume the risk that remains; since it does not make economic sense to insure against absolutely all risk (and because we may not completely understand the nature or magnitude of risk at a place), assuming risk makes sense.

There are several ways in which governments assume risk. One is simply by default, resulting from lack of knowledge of a risk. Officials also may decide to assume all or part of the risk to which an area is exposed, which can be accomplished through a self-insurance program in which reserve funds exist to cover losses. Such funds can be created through various devices, such as accumulating contingency funds (based on a percentage of expenditures or some other formula) or financing through capital markets (Burby et al., 1991). Obviously, the ability of government to dedicate funds to such a reserve depends on a community's financial situation. The success of this option depends on the ability of a community to determine its needs with reasonable accuracy, based on the risk to which it is exposed.

Comparing and Choosing Options

The options detailed above are not mutually exclusive. Assuming that choices are available, individuals, governments, and the private sector usually favor some mix of options. The selection of an appropriate mix is based on several factors, including exposure, the frequency and severity of events, and cost-effectiveness. Further, we must recognize that "which strategy is best depends in large part on the attributes of the particular risk" (Morgan, 1993, p. 38). When it is difficult to eliminate risk because of cost or technical feasibility or both, another strategy or set of strategies is needed. This is the case with earthquakes, where we cannot eliminate the risk by modifying the geophysical process, and with floods, where we can control some but not all hydrological processes.

The choice of options also is bounded by the results of formal or informal cost–benefit analysis. It is risky to invest money, personnel, or other resources in an activity to protect against future losses from an event, the occurrence of which is probabilistic (see Chapter 6), a problem that is exacerbated by multiple hazards. Thus, measures of cost-effectiveness and risk come into play.

Yet, we know that most local governments do not take adequate measures to protect against financial loss. The reasons include "inadequate appreciation of the potential for catastrophic losses, inadequately developed risk management capacity, and insufficient resources for loss prevention" (Burby et al., 1991, p. 118). The risk management process is a complex one involving economic, technical, and perceptual factors. Though the options are well developed and, in theory, well understood, adoption is not. In a study of the degree to which risk management had been success-fully implemented after the 1964 Alaska earthquake, two obstacles to success were identified: organizational and political (Natural Hazards Research and Applications Information Center, 1985). Organizational obstacles were associated with scientific uncertainty, a decision-making model based on the rational actor, and ambiguous policy directives. Political obstacles were more difficult to overcome; they included leaders lacking knowledge, a lack of commitment to implementation on the part of political leaders, and inadequate definition of what level of government was or should be responsible for it. Nevertheless, there are those who advocate using economics to clarify choices about risk (Freeman and Portney, 1989).

Although risk management appears to be a straightforward process (see Figure 7.6), it is anything but that in practice. A person, community, or industry can go through the process in a systematic, logical sequence, but the ultimate decision on which strategy to adopt brings in other factors. With governments, these are political. In addition, it must be recognized that a decision to do nothing is in fact a decision to assume the risk or perhaps to transfer it (through relief expected from other levels of govern-ment). However, this is not risk management, but avoidance, which too many times may be seen as the most cost-effective option (based, of course, on a restricted view of cost-effectiveness). To higher levels of government, it may not be cost-effective and, if these higher levels of government decide to implement only what they view as cost-effective, they may not accept risk transfers in the future (witness New Zealand).

All risk choices are compromised in some way, presenting difficulties for disaster preparedness. May (1989) summarized such problems in cultural, systemic, and individual models. Inherent in the cultural model are conflicts that surround decision making and stem from the need to compromise; given varied interests and perceptions, not everyone will be completely satisfied and the risk will not necessarily be minimized. Simi-larly, societies' demand for goods and services adds to community risks while individuals generally miscalculate risk. Managing risk is intrinsically difficult.

THE RISKY BUSINESS OF MANAGING HAZARDS

Risk analysis and management transcend hazard types. While specific components of analysis differ among hazards (related largely to the ability to identify physical exposures), the process of identifying, choosing, and implementing risk management strategies does not really differ. However, there are many important differences in organizational and political structures that serve either to facilitate or complicate the process. Added difficulties lie in identifying the extent to which an area is at risk, within acceptable levels of accuracy. It is little wonder that if they are managed at all, risks tend to be managed on a hazard-by-hazard basis. Certainly, some actions and strategies transcend particular hazards, such as emergency preparedness plans. However, other forms of risk management—such as transferring, eliminating, or assuming risk—usually are undertaken with a particular hazard in mind. It is the perceived risk of different hazards (as opposed to technical measures of risk) that motivates managers to attempt risk reduction.

Risk ends up in the policy arena because it is through policies—especially public policies—that risks most frequently are managed. Various groups are exposed directly and indirectly to a public risk, and losses can far exceed the abilities of affected areas to underwrite. Further, different levels of government and different groups have different definitions of risk; though they may all use the same term, they are discussing different types of risk. As a result, responsibility frequently falls to the public sector. This does not make it any easier to manage risk; indeed, it may be a complicating factor. However, it is there that considerations of technical, social, economic, political, and geophysical factors can come together.

8 An Integrated Approach to Natural Hazards

INTRODUCTION

In this text, we have taken an integrated approach to natural hazards, incorporating elements from both the natural and social sciences to examine how different societies have responded to extreme geophysical events. Throughout, we have tried to emphasize the underlying theoretical and conceptual frameworks through which the understanding of hazards has been enhanced. Our concepts have come from a variety of disciplines, reflecting the interdisciplinary nature of natural hazards research. We have run the gamut from geological, hydrological, meteorological, and other earth sciences to perspectives provided by geography, anthropology, sociology, economics, politics, psychology, and other social sciences. We have drawn from the medical literature and included references to the arts and humanities in order to broaden our understanding of natural hazards, their occurrence, their impacts, how they are viewed, and how societies cope with them. If the scope has been far-reaching, it has reflected the interrelated nature of the elements associated with natural hazards.

It is our contention that we can learn much from an integrated perspective. Experiences with individual types of hazards and events provide lessons for other hazards and events. Many of these lessons have widespread and seemingly timeless applications; others are more place-, time-, or hazard-specific. Still, it is through integrating what we know and interpreting salient elements of that knowledge that we can develop a framework for analysis of hazards and our relationships to them. In the end, of course, there is no panacea for natural hazards, but our understanding of the human and physical components comprising natural hazards is increasing and, in some places, losses are decreasing. However, in many locations, losses have been increasing over time, so much remains to be learned and done. The link between theory and application is a critical one.

There have been numerous and continuing calls for greater consideration and development of theory in natural hazards research. Some have criticized researchers for failing to develop a comprehensive and explicit theory. For example, White and Haas (1975, p. 5) wrote, "Natural hazards research in our nation is spotty, largely uncoordinated, and concentrated in physical and technological fields." From their perspective, the natural hazards research agenda had failed to incorporate sound social scientific principles. Too much attention, they argued, had been devoted to case studies unrelated to general theories. While this criticism might be viewed as somewhat harsh since we have learned much from case studies, the reference to "uncoordinated social perspectives" still has credibility more than twenty years later. And they (1975, p. 5) continued, "Unfortunately, much of the social research has been sporadic and limited to an investigator's interest in local problems and narrow theory. No broad body of knowledge has been created, nor have earlier findings been updated in terms of underlying social and economic changes in the United States." Again, these comments are still pertinent today. Indeed, fifteen years after White and Haas made their plea, Palm (1990, p. 18) noted that "what has been lacking is a perspective that integrates individual and societal response."

There appears to be agreement that social scientific theory in natural hazards research has been and continues to be in need of improvement. Thus, we should consider the following questions as we look to integrate and interpret what we know and what it means. To what extent have general theories applicable to hazardous situations been developed? Have the social sciences made a significant contribution to "solving" or at least understanding the intricacies of natural hazards? What are the burning questions that remain unanswered?

We have made some progress. Hazards research has evolved over the years to take on an interdisciplinary approach, soundly based on natural and social scientific principles. The seemingly disparate findings of case studies have grown more focused as researchers have sought commonalities among hazards and between places. Indeed, the series of case studies compiled by White (1974), although strongly criticized by some (Waddell, 1977), provided significant baseline data from various cultural contexts. Unfortunately, subsequent hazards research failed to develop either predictive or explanatory models pertinent to a range of cultural traits.

Of particular importance has been the significance of context in understanding natural hazards. As Mitchell (1989a) argues, there are many contexts in which hazards occur, of which two stand out: hazards as real-world problems, and hazards as complicated by issues of scale. As real-world problems, they are not easily bounded since they tend to overlap with other societal issues; these were described by Mitchell as "interlocking crises," such as environmental mismanagement, depletion of resources, hunger, geopolitical tensions, and widening gulfs between wealthy and poor nations. In many senses, therefore, hazards cannot be viewed in isolation. At the same time, other contextual aspects of hazards either exacerbate or attenuate community vulnerability; for instance, the impacts of a severe windstorm in England were quickly overshadowed by pressing economic and political crises so that necessary changes in the emergency-response system were not made. As discussed in Chapter 7, the time of occurrence, social factors, political processes, and precise location also affect disaster outcomes in a variety of ways (Mitchell, Devine, and Jagger, 1989). Thus, hazard problems are not easily defined because they interact with all aspects of society, making mitigation difficult. In consequence, some scientists have argued that natural hazards constitute a normal state of affairs (see Chapter 1).

In addition, hazard mitigation is complicated by scale and the spatial organization of society. The different ways in which place is organized leads to competing groups divided by religion, language, class, ethnicity, and historical context; to these can be added distribution of wealth, conflicting cultural norms, and political structures. In essence, behavior and response to natural hazards are contingent upon the social, economic, and political realities of place. Therefore, hazards must be examined through contextual filters that account for spatial differences and broader social needs.

Further, it can be argued that hazards scientists have a moral obligation to pursue *applied* research, seeking out real solutions to societies' problems. Rising populations, finite resources, spatial inequalities in resource distribution, the marginalization of specific groups, and greater global interdependence all contribute to a hazardous world beset with continuing risk and increasing vulnerability. There is an ongoing conflict between the extremes of the geophysical world, which does not recognize political and administrative boundaries, and the human-use system. As so aptly described by Wells (1924, p. 1100), "Human history becomes more and more a race between education and catastrophe." Although Wells was referring to warfare, the image fits natural hazards and may indeed constitute the real "war of the worlds." Education and research are fundamental to understanding natural hazards and to coping with them, no matter what the context. Context, however, defines how they are to be undertaken and applied. Indeed, the burgeoning world population (stretching resources to their limits) ongoing ethnic warfare and conflicts, and possibly changing climatic vulnerability provide the context in which society must come to grips with hazards. Continuing study of natural hazards is crucial; the human-use system not only increases our vulnerability to catastrophe, but potentially changes the physical dimensions of many hazards as well.

In this final chapter, we discuss these issues, pulling together some of the concepts discussed earlier and unifying different aspects of natural hazards. At the same time, consideration is given to the appropriate contextual constraints.

THEORETICAL AND CONTEXTUAL FRAMEWORKS

Within the boundaries of theory to date, we can pull together those elements that have been found to have salience and develop a conceptual framework against which theoretical constructs and contextual analyses might be tested. The diagram in Figure 8.1 includes elements, variables, and relationships that have been presented in earlier chapters in an attempt to provide a relational schema for integration of what we know and what it means. Although the figure may appear to suggest that all elements are of equal concern in all situations, that is not the case; indeed, not all variables may be relevant in a given situation, and some are certainly more significant than others. For instance, political and economic factors, over which the individual has little control, dominate in some hazardous situations or in post-event activities and priorities; in other instances, it is the relationship between the physical environment and social characteristics that defines what happens, either pre- or post-event. In addition, context surrounds all these hazard dimensions; it is not a filter through which they are connected to loss reduction. Furthermore, this diagram is not all-inclusive; it suggests some elements that come into play and proposes relationships among them. Still, we believe it pulls together many of the points made in earlier chapters and provides an instrument for further analysis. We discuss the various components of the diagram in the sections below.

Physical Characteristics of Natural Hazards

Natural hazards research has gleaned much from examining the physical forces that comprise disasters and it has leaned heavily on the natural sciences. There are many common features shared by natural hazards, even those as apparently divergent as floods and droughts or tornadoes and earthquakes. For example, it is possible to classify natural hazards according to measures of frequency, intensity, duration, and speed of onset, each

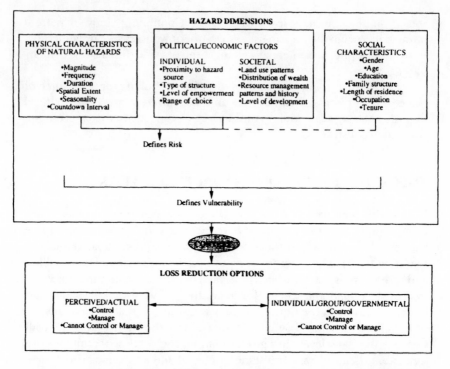

FIGURE 8.1. Natural hazards and human response in context. Many variables and relationships define the hazard environment, and the common elements among places can be documented, analyzed, and compared. Context is critical to defining how relationships play out at a place. This diagram incorporates some of the operative variables and depicts the relationships among them that help us explain human response. Though context is shown as a filter here, in reality it affects all hazard dimensions.

of which can have a direct impact on response patterns (as described in Chapter 2). All other factors being equal or held constant, we might well assume that the greater the frequency of the hazard, the more likely some form of response will be undertaken since the hazard will be recognized as a prevailing problem; therefore, mitigation would be a feature of communities subject to frequent disasters. Similarly, the more intense or the longer the duration of a disaster, the greater the likelihood of action being taken, assuming there are not other factors (political–economic or social) working against such relationships. It is generally the case that appropriate responses are contingent upon a full appreciation of physical processes. For example, hazard warning systems are ineffective if the speed of onset is more rapid than anticipated and planned for, though in other circumstances they save lives and property.

Unfortunately, it is not possible to develop a scientifically predictive model of response based solely on physical or environmental criteria. Some descriptive models exist, such as that proposed by Kates (1962), but they are not particularly useful beyond the areas studied. Indeed, basing any hazard response model purely on physical factors has little chance of success since many other variables influence individuals' and communities' decisions to make adjustments to the hazard (as Figure 8.1 illustrates). In other words, the social, economic, and political contexts affect decision making. Having said this, however, the significance of physical variables should not be overlooked nor should the search for predictive models stop. The greater the frequency and the more intense the hazard experience, the louder the calls for action; it is these calls that capture the attention of the political machinery and ultimately can stimulate action.

Furthermore, to some extent, physical criteria often form the basis for mitigation. In the United States, for example, the 100-year floodplain has become the standard for flood alleviation. The National Flood Insurance Program utilizes the 100-year floodplain with respect to zoning ordinances (although a different and more conservative measure is used by the U.S. Army Corps of Engineers as a design standard for structural flood control). While the 100-year flood sets a uniform national standard, it does not take into account local hydrologic, economic, or demographic characteristics that may serve to define the flood hazard for a given community (Lord, 1994). Nonetheless, physical criteria (specifically a measure of magnitude and frequency) are utilized for floodplain management and structural flood-control measures are tied to engineering and hydrological principles.

Similarly, alleviation measures for other hazards are based on physical criteria. Building codes are often associated with projected earthquake or tornado magnitudes, and sea walls are designed with given heights of tidal surges in mind. In fact, design standards must take into account the physical forces associated with natural hazards at a place. However, determining the design level for safety is part of the human realm, and managers ultimately must make decisions about the value of life and property. Although physical criteria are important, they do not provide the complete answer.

Political–Economic Dimensions

Whether during postdisaster periods or long-term responses, hazard–zone activities are constrained by the social, economic, and political realities of a given place. The most basic example is the money needed to undertake mitigation projects and provide relief, which may not be forthcoming or available in poorer communities. Yet, it is not just the absolute availability of money for development, planning, hazard mitigation, or disaster response that is a problem; more often, it is the distribution of resources—especially financial resources—that defines vulnerability. For instance, following the 1987 flood in Bangladesh, housing of the landless poor was more easily destroyed than that of landowners, because of either how or where the structures were built in relation to the river. As a result, "shelter is an immediate need and the poor or near landless go on borrowing from money lenders by mortgaging whatever they have, including any land" (Chowdury, 1988, p. 299), even when they have lost up to 50% of their annual income because of the flood. There are similar, though perhaps not always as devastating examples everywhere. Following the 1989 Loma Prieta earthquake in California, a preexisting shortage of low-income housing in the community of Watsonville was exacerbated by the damage and destruction of almost 1,000 units (Bolin and Stanford, 1990). These problems were not caused by the disaster, but were brought to the fore by it.

We can also point to development that occurs when inadequate attention is given to a natural hazard, or where the expense to protect against it cannot be justified or frequently is not even considered. As a result, vulnerability is high, stemming from the fact that the hazard represents only one of a series of crises facing society, in fact all demands must be assessed according to need. Overpopulation, traditional land use patterns, availability of materials and technology, and the need for industrial development all represent issues that are important and that demand planning and the commitment of financial, human, and natural resources. Embedded at a deeper level, these social, economic, and political factors interact to produce philosophical constructs that govern how society functions and, finally, determine levels of vulnerability as well as the marginalization of many groups. These constraints affect every level of society and become obvious after an event has occurred and the distribution of impacts can be analyzed.

We might consider these constraints socioeconomic "traps" that perpetuate risk and foster increased vulnerability. We have seen that not all individuals are equally vulnerable; the risk to all inhabitants of hazard-prone areas is high (as indicated by the frequency of disasters), but some groups and individuals are more vulnerable than others. At the global scale, floods, droughts, earthquakes, and other hazards recur with great frequency, bringing repeated incidents of death and destruction. The logical solution would seem to be to "get out of harm's way," but unfortunately, this option is not open to all individuals and, indeed, often is inappropriate from a societal perspective. For instance, rural land distribution systems and the decreasing ability of the rural poor to provide for themselves and their families, combined with increased investment in urban areas as well as increased opportunities for employment, lead to the development of urban squatter settlements which are frequently located in the most hazardous sites. Yet, as Havlick (1986, p. 8) points out, "Precise figures for most squatter settlement populations are almost impossible to find—because many shantytown residents are considered to be noncitizens. . . . " The locus of social, economic, and political power is such that many cannot effect change to better their own situation or to promote structural change, which has important ramifications for natural hazards. The traditional response to hazards has been to return communities—along with prevailing socioeconomic structures—to those conditions existing prior to the hazardous event, continuing the cycle of disaster–damage–repair–disaster and doing nothing to reduce vulnerability. The risk of small events may be reduced by mitigation measures, but the stress remains and the same individuals suffer the consequences. Blaming such individuals for their predicament is not constructive. Instead, we should examine the underlying reasons such hazard-prone areas continue to be inhabited.

Individual and Societal Locational Choices

Let us consider several of the elements in Figure 8.1 that relate to the individual's range of choice and to land and resource management by "society." Regulations to prohibit activities in hazardous locations serve to limit future disaster losses, but they do not tackle the underlying causes that foster development of hazardous areas. Inhabitants of hazardous areas frequently have only a narrow range of locational choices. In some instances, individuals are unable to afford the higher costs of moving to safer areas; a few may gather sufficient resources to move after a disaster, but most will not. This assumes, of course, that there also are safe areas to which people can move, in fact, there may be no "safe" areas available at sufficiently low cost. Beyond economic considerations, there may be cultural and social reasons, over which individuals may have no control, that also explain why people continue to occupy extremely hazardous sites.

Arguments that people should not continue to live in hazardous areas are moot when they have little or no choice. It seems obvious that planning for hazards and even disaster relief programs must include sufficient resources for all victims so that adequate new locations can be found. Too often, planning for hazards is nonexistent and disaster relief is concentrated on rebuilding quickly and in place. Consequently, marginalized individuals are usually unable to initiate changes in their living conditions. Control rests with those having political and economic power who may "fix" the symptoms of the problem (such as a flood or earthquake) by repairing structures and returning the community to the pre-event status quo, but who do nothing to remedy the causes. The hazard is perpetuated, victims remain part of the disaster–damage–repair–disaster cycle, and the scene is set for recurring problems.

Building on the theme of land value and risk, there may be perceived economic advantages and other benefits to remaining in the hazard-prone area. Studies have shown that property values may be lower than in "safe" areas because the risk has been capitalized into the value of the land. These negative impacts of hazards may last for several years, as witnessed by the studies of floodplains and land values (see Chapter 6). The consequences of this capitalization may mean that hazard-prone areas develop different socioeconomic traits compared to other areas; for example, many flood-plains are characterized by low-income households, the elderly, young families, people on assistance, and mobile homes (Interagency Floodplain Management Review Committee, 1994). Therefore, even if choices are available, socioeconomic conditions may encourage individuals to seek short-term gains from less expensive property in the hazardous areas, temporarily sacrificing the long-term community goals of a safer environment (Costanza, 1987). Not surprisingly, economic factors influence locational decision making, with the alternatives considered being defined by the context in which the decision is made. A hazard threat may or may not enter into this decision making.

In some instances, with a shortage of safe land and a rapidly increasing urban population (as is the case in much of the world), priorities are on meeting day-to-day needs. When an event occurs and a disaster results, concern turns to hazardousness to the extent possible. Still, there remain many "compelling disincentives" to instituting a comprehensive program of hazard mitigation, including the need to replace structures quickly and at the lowest cost, the lack of a good database for establishing a baseline of hazardousness, and the continual growth of urban populations (Havlick, 1986).

In other instances, response and reaction may reflect the gambler's fallacy (discussed in Chapter 3). A disaster has occurred, and the generally held view is that another one is not expected for some time. This is particularly common with respect to floods. It is not unusual, for instance, to hear flood victims state that they have experienced the 100-year flood, and consequently will not experience a similar flood in their lifetimes. As many victims of the 1993 floods of the upper Mississippi River found out, this is a fallacious argument; flooding occurred again just two years later (in 1995) in some communities. During the postdisaster period in 1993, some of these people took out flood insurance only to let it lapse one year later. As the period between disasters increases, this problem gets even worse because the hazard is forgotten, replaced by more immediate concerns. Thus, socioeconomic forces work against effective mitigation, and the failure to enforce flood insurance programs contributes to recurring disasters.

The levee effect also contributes to increased hazard losses. The construction of structural measures to mitigate hazards or even the implementation of nonstructural adjustments (such as zoning ordinances) can lead to increased development on the "safe" side of the project. For example, constructing sea walls to withstand tidal surges of hurricanes, or zoning land to accommodate ground shaking and subsidence, may actively encourage development in "safe" areas. Of course, all projects have limitations as defined by the design standards, which will fail when compromised. Consequently, losses can be catastrophic and, over the long term, possibly greater than if the project had never been implemented because of the increased development; there are many examples of increased losses, such as when levee systems fail. This is not a new phenomenon; Segoe (1937, pp. 55–56) reported "a danger that by reason of increased occupancy and values in the areas subject to floods after construction of the system of headwater reservoirs, the damages over a long period of years might be greater rather than less, in spite of the 300–, 400–, or 500–million dollar expenditure that these works might cost." In spite of these warnings from 60 years ago, little has been done to curb development in protected areas. It appears that all mitigation projects can stimulate this effect, leading to changes in the hazard milieu.

All of us find ourselves in an environment defined by economic conditions and projections, social customs and norms, and political ideologies and realities. Some of these offer opportunities while others define constraints for our decisions and futures. Thus, when analyzing hazardousness as it varies from place to place or decisions that are made or not made to mitigate hazardousness, we cannot ignore the political, economic, and social factors that put boundaries around individual and societal choices. We are not trying to suggest that the individual is always an unwilling and unknowing victim of societal decisions. Rather, we hope to illustrate the relationship between the individual, society, and the physical environment, as well as the context for that relationship. Certainly, the individual is important, as we discuss in the next section.

Social Dimensions

The key elements in this category relate to individuals' situational contexts (discussed in Chapter 3). Specifically, these are the variables that, separately and in combination, influence vulnerability and our ability to modify it. Although it may seem that these variables are not much different from those political and economic factors relating to the individual discussed earlier, they relate more to individual and household attributes than to the political economy. (Needless to say, they are closely related.) Although we discussed many of the variables in Chapter 3, one example will serve to illustrate the relationships among the categories.

The Trinity River in Liberty County, Texas, floods frequently, including four times between 1989 and 1992. Approximately 10 subdivisions were affected each time and, in the 1992 flood, more than 270 residences were inundated, causing great disruption as water still had not receded three weeks after initial flooding. The affected houses differ in value, with some subdivisions consisting of more affluent homes and others of mainly poor quality structures. Many of the houses have been elevated on stilts or fill. Those elevated on stilts (usually with federal assistance) suffered little or no damage, while the few houses remaining at ground level or those raised only minimally suffered to a great extent. Nonetheless, disruption to the community was considerable as many residents were housed in shelters, motels, and with friends for the duration of the flood.

Many of the less affluent residents of these subdivisions could not relocate without substantial financial assistance. At the same time, many do not want to leave as they have lived there for a number of years and enjoy the river environment in spite of its drawbacks. Thus, situational factors work to keep people in a very hazardous situation; they like it there, they want to stay, and even if they wanted to relocate they would have great difficulty, even with financial assistance from the Federal Emergency Management Agency (FEMA). Based on economic and social factors, there is a local distinction between flood-prone and nonflood-prone properties in the county (Montz and Tobin, 1992), and any attempt to relocate floodplain dwellers would likely be met with hostility from both sides. Many of the residents want to stay, and many in the county want to keep them there; one real-estate agent noted that relocation would benefit the floodplain residents greatly because FEMA would have to give them housing well above the current value of their homes, as that is all that is available in the county. In the end, social characteristics, combined with political and economic factors, serve to keep residents of the Trinity River floodplain in Liberty County at risk, and they also keep disaster relief funds coming to these people.

We also know that these social factors affect perceptions of a hazard, which affect the choices that ultimately are made. Indeed, it is through the perceived world that individual decision making occurs. This perceptual context is dynamic and variable over time and space, and is perhaps most difficult to model from a scientific perspective. Questionnaire surveys of disaster victims have attempted to elicit perceptual worldviews, but these have been merely snapshots of a changing and evolving scenario. What people say they will do under certain circumstances does not always coincide with what actually happens later (Deutscher, 1973) and, in many cases, the influence of broader political, economic, cultural, and social factors on perception has sometimes been missed. Consequently, further work is needed to put perception and situation in context. As shown in Chapter 3, people tend to behave according to perceived rather than objective reality; if a tornado hazard is considered a low risk or hurricanes are perceived as survivable, the individual concerned may take little or no action to mitigate impending events. However, if much of one's perceived reality is constrained from the outside, then we have dealt with only one part of a complex situation and studies of perception do not get at the entire picture.

The context in which decisions are made is critical to our understanding of risk and vulnerability (Figure 8.1 shows the elements that comprise context). Researchers often have frequently compartmentalized the three crucial components of the hazard complex: physical characteristics, political and economic factors, and social or situational characteristics. This has been necessary in order to address manageable research questions and build a base of knowledge within each component. In addition, hazards' researchers come from a number of disciplines and tend to focus on topics within their disciplinary interests. However, losses to natural disasters are increasing. Even as we learn more about each of the components, it is necessary to put them together in an attempt to develop our understanding of the relationships between components as they define risk and vulnerability.

Risk and Vulnerability

Because these topics are considered in detail throughout the book, discussion here is brief. However, they are critical to our conceptualization as it is risk and vulnerability that we are seeking to alter. As shown in Figure 8.1, it is the combination of physical characteristics and political factors that serves to define risk. Recalling our definition of risk in Chapter 7, it is from the physical characteristics that we can determine probability of occurrence while the political and economic factors determine severity of harm or outcome. (Social characteristics also come into play, but only as they are related to the other boxes; hence the dashed line). By contrast, vulnerability is determined by all the elements in various combinations; this suggests that if we alter one of the elements, we have altered vulnerability—and that is true. Measuring the extent to which vulnerability has been altered is a difficult undertaking, but certainly one that should be incorporated into any development planning as well as in hazard mitigation planning and disaster relief programs.

We have now defined what is meant by context. Risk and vulnerability are a part of context, and they are changed when any one element in any of the three categories is changed. Thus, we are working with a moving target because a change in any one factor often affects other factors. Given that hazards do not usually rank high in priority among problems with which society must deal but recognizing the great costs they impose on society, it is imperative that we work to develop our understanding of both the conceptualization in Figure 8.1 as a whole and its varied, component relationships. Our goal is to bring about loss reduction, usually through hazard mitigation, but we cannot do that if we do not understand both the players and the rules.

Hazard Mitigation in Context

Those evaluating hazard mitigation strategies must consider not only the broader impacts of implementing any project, be it structural or nonstructural, but also the context in which strategies will be implemented. Development plans must coordinate with hazard planning to prevent transferring frequent local problems into large-scale catastrophic losses. One approach would be to require protection of property located in "safe" areas, but at risk should the structure or project fail. In the United States, for example, the National Flood Insurance Program (NFIP) anticipates covering properties on the "safe" side of the levee that lie below the 100-year flood level (Interagency Floodplain Management Review Committee, 1994). Similar plans could be made for other projects and different hazards; the mitigation measure does not eliminate the risk completely so that people remain vulnerable to disaster. Providing the hazard-prone population can afford the insurance premiums, this policy would reduce vulnerability. Other policies should focus on land use regulations as a means of restricting unnecessary development. Of course, political will to develop such policies is not the only requirement for their successful implementation; human, financial, and technological resources are needed to ensure that appropriate measures are taken and that they are implemented properly. Not only does this involve the sharing of resources (including technology transfer), but it also requires their redistribution; for example, resources could be moved from safe areas to more hazardous areas, from one program to another, or from one set of priorities to another. The difficulties should be obvious, but they need to be faced.

Mitigation policies that restrict the use of land and property raise other issues. Land use regulations (such as zoning) limit what can and cannot be done with land. In this case, the political context can constrain activities. In some nations, land can be taken with little or no legal repercussions, but in others, the ramifications can be quite significant. In the United States, land can be assumed under public ownership through eminent domain if it benefits the larger population and just compensation is paid. However, claims by the state are inevitably faced with counterclaims. The denial of land use rights to property owners raises the takings issue; some legal cases have demonstrated that the state has authority to do this under hazard mitigation strategies, although some reassessments of individual projects may be necessary (Singer, 1990), while other court decisions have asserted the rights of the landowner (Platt, 1994). The rights of individuals in this instance must be evaluated against the common good and the need to underwrite residents and businesspersons in hazardous areas. By comparison, centralized political systems may have fewer problems removing people from hazardous environments or keeping them from moving there in the first place; for instance, the Chinese government has relocated thousands of people to construct dams and other water resource projects. In other instances, strong disincentives exist to keep people from moving to urban areas (Havlick, 1986). While not all of these were undertaken with natural hazards in mind, we can see that the success of hazard mitigation policies can differ, depending on political context.

Blaikie, Cannon, Davis, and Wisner (1994) suggest that mitigation be directed toward specific sectors of society so that the maximum number of people can be protected, and further propose that mitigation measures become more "active," at least in less developed countries, in order to encourage popular participation in hazard reduction practices. Passive actions would include planning controls and structural measures whereas active measures would include offering resource transfers as incentives to undertaking remedial action. They argue that in wealthier nations, passive mitigation can be effective, but grass-roots actions may be most effective in less wealthy nations. For example, in Bangladesh, the control of flooding has been improved in some locations by compartmentalization projects that allow controlled flooding of fields so that benefits to agriculture can be realized. These projects are regulated locally. However, the same can be argued of the wealthier nations, which have moved away from large-scale structural controls to small, nonstructural measures. For instance, in the United States, the Association of State Floodplain Managers (1992) strongly urges state and local controls for flood alleviation plans rather than federal involvement; local governments are considered more aware of local issues as well as having a vested interest in effective planning. In this case, local government parallels federal efforts by administering and enforcing floodplain regulations such as those pursuant to the NFIP; the states coordinate activities between local and federal governments, providing the basis for

comprehensive floodplain management strategies. Thus, the political context is mixed, and it may be that coordination and implementation of
hazard mitigation projects rely more on interested actors who are willing
to participate in the procedure than on the system itself. In addition,
mitigation under these circumstances can work only when all levels of
government agree that the problem is sufficiently large to merit such
attention. In many places, this recognition occurs only after an event, and
emphasis then is on recovery rather than mitigation. The extent to which
recovery activities become mitigation measures is extremely variable.

In the United States, one factor that may be fueling intergovernmen-
tal cooperation is the possibility of the state facing liability if mitigation
is not undertaken. The possibility of hazard victims suing for losses has
risen as hazards have become better understood scientifically; we know
where most hazards will occur, we can map zones of high and low risk,
and we can predict the frequency of events of different size with a degree
of accuracy never before possible. Many hazard-prone areas have been
mapped, such as the floodplains of the United States, seismically active
areas in California, and other risk-prone areas determined from historic
records (Foster, 1980). Although the physical basis of natural hazards is
understood more fully, the weakness in our mitigation strategies continues to lie in identifying degrees of vulnerability. However, failure to take
preventive action to maintain public health and safety may result in
liability suits.

For the most part, the examples above focus on developed countries
and nonmarginalized groups. We also need to ask other questions. Who
are the most vulnerable people? How will they be helped by a hazard
mitigation strategy? Will they have access to resources, money, and aid to
respond appropriately to the mitigation strategy? Groups within society
should be identified clearly, without stigma, so that services can reach those
for whom they were intended (Blaikie et al., 1994). This applies as much
or more to post-disaster programs as to hazard mitigation efforts. Of
course, if groups are identified for one program, they can be more easily
identified for the other, but this does not always happen. For example,
post-disaster studies show that aid in the United States does not always
reach the most needy. FEMA sets up programs to distribute money, provide
grants and loans, and other counseling, but individuals often are required
to fill out long bureaucratic forms and wait in line for action, which can be
a degrading experience for hazard victims. First, individuals are anxious
to deal with the disaster wreckage, not to spend an inordinate amount of
time waiting in line. The psychological benefits of physical activity are
probably more rewarding than filling out forms. Such problems were seen
after the 1993 floods in the United States as victims tried to get some form
of relief.

I have a great deal of grief, bitterness, and loss of faith as an indirect side effect of the flood, but this is not because the waters came gushing into my home and vandalized my belongings. It is not because the mud and oil made saving anything virtually impossible. It is not because "God struck us down." I am simply at a loss that there are still people in my neighborhood that are living out of cardboard boxes with their belongings stacked in a corner of their yet-to-be finished homes. I am angry that all the money that has poured in from fellow victims and sympathetic Americans over these great United States is being holed up somewhere or is being furnished to businesses, etc. I am furious that when a person really needs help, whether it is financial or other, they must wait 3 to 4 hours in line with hundreds of others in a searing 95-plus–degree day in the sun to enter some relief establishment only to be told that they forgot some ridiculous form of identification or that they must come back the next day and those that *do* get to go in have another 1- to 2-hour wait for an interview and are then allowed to be given whatever they came for. (Personal communication with a flood victim, Iowa, 1993).

Second, the information required by FEMA often is regarded as personal, and the government is viewed as invading privacy. Consequently, some individuals undoubtedly shy away from requesting help. Following floods in California in 1986, many immigrants did not apply for aid, reportedly because of fear of the federal government. The situation may be compounded by language difficulties, although FEMA has made strong efforts recently to communicate appropriately; for example, along the Trinity River in Texas, flyers advertising the locations and hours of FEMA offices were printed in Spanish and English. According to Bolin and Stanford (1990, p. 102), "it is those groups with marginal pre-disaster resources (ethnic minorities, elderly on fixed incomes, and other lower income households) who are most likely to use mass care facilities, and who have the greatest difficulty in finding suitable temporary and permanent housing." Thus, their need is greatest and so are their difficulties. These problems are multiplied at the international level, where there are not only language problems but social and cultural differences. For example, there is little use in sending wheat surpluses from the United States as food aid if those in need have no way of turning it into bread (Alexander, 1993). In addition, local economies must be recognized before sending disaster aid; in too many instances, the availability of food and other supplies following disasters serves to undermine local markets for those same goods, making it more difficult for the area to get back on its feet.

Thus, vulnerable groups should be identified in any mitigation planning. Where are the elderly, the young families, the poor, and others needing special physical and mental attention located? In addition, the first priority should be saving lives, not property and, as Blaikie et al. (1994) point out, the property of the poor is more critical than that of the wealthy.

Its loss is critical if one is left with nothing. Local governments are best placed to answer these questions and set priorities within their planning activities, but this assumes that they conduct development planning and hazard mitigation. Too frequently, that is not the case; political realities affect not only planning (or the lack of it) but response behavior. After the Neftegorsk earthquake in 1995, Russia refused foreign aid, preferring to rely on domestic resources rather than admit its inability to meet the problem. In other instances foreign aid has been refused, both by donor and recipient nations, purely for political reasons. To the victim in need of support, such political machinations must seem foolish.

DISASTER–DAMAGE–REPAIR–DISASTER CYCLE

The litany of disasters in Chapter 1 attests to colossal losses of life and property and widespread suffering throughout the world. Glickman et al. (1992) showed that deaths from natural hazards were averaging 56,000 per year, while the annual cost to the global economy was in excess of $50 billion. Temporal trends indicate that such losses are increasing, thereby exerting tremendous pressure on nations' resources that have repercussions far beyond those of the disasters themselves. Development projects are compromised and social plans disrupted as money is devoted to post-hazard repairs and rehabilitation.

Losses also are skewed spatially, with more deaths occurring in less developed countries than elsewhere. While economic losses are highest in the wealthy nations, this does not account for the significance of losses to the particular countries concerned. The impact can be especially devastating to less developed nations, some of which have experienced a 5% reduction in their gross national product during disaster years (Burton et al., 1993).

At the personal level, poorer individuals suffer more than the wealthy as they stand to lose a greater proportion of their possessions and assets. In the end, the repairs that are undertaken only perpetuate problems, fixing symptoms of the hazard rather than tackling deep-rooted causes. Inevitably, vulnerable groups that are exposed frequently to a hazard remain in the unenviable position of victims who are constantly threatened by it. Nothing is done, it seems, to alleviate the lot of the most vulnerable groups. This is not to say that relief aid is not valuable, and certainly many federal and voluntary agencies provide much-needed support during times of stress. Nevertheless, it contributes to and certainly aggravates future hazard problems because the pre-disaster situation is essentially recreated. Houses are rebuilt, roads and rail links repaired, and businesses reopened in exactly the same places as before and, even if adjustments are made to mitigate the hazard, some risks remain. This cycle is repeated time and again, frequently with the best of intentions, but it needs to be broken.

While often conducive to helping immediate community needs, the immediate postdisaster period does not encourage change in structural constraints that sustain social "traps" nor does it break the cycle. Financial aid and other support is spent on repairs, not on transforming the status quo. Recipients of this support are in a position neither to refuse nor to insist on different forms of aid; the aid relieves the ravages of the disaster, but pressure for change must come from others. At the same time, efforts to change the status quo must be sensitive to the needs of the victims. For example, relocation projects will not be successful or even welcome if those asked to relocate are not provided with sufficient funds to move. Similar problems concern wealthy residents affected by disasters. For instance, residents living on the eroding cliffs of California and Oregon, or in select suburbs which flood, may not want to move if emergency aid is always available; some encouragement must be made to get them to move to safer locations or to implement effective adjustments. Consequently, those in power must take the initiative to restructure emergency relief and hazard planning so that problems are not exacerbated. The root causes of hazard problems must be tackled rather than the symptoms.

This is not an easy task. As we have shown, the problems have been aggravated in recent years by a rapidly growing global population, increasingly concentrated in urban agglomerations that place more people at risk and raise disaster losses. Hazards will not go away and, no matter what mitigation projects are undertaken, disasters will always occur. We cannot control every aspect of nature so that all risk is permanently removed from people's lives. This is not to say that remedial action and adjustments to hazards are not profitable, or that losses of lives and property cannot be reduced significantly. However, measures must be instituted that minimize the impacts of disasters by considering not only the physical environment but the human-use system. Responsibility for disasters lies with social, economic, political, and legal factors, and is not confined to the physical world. Only with this recognition will we break the cycle.

SURVIVAL AMID CATASTROPHE

A recurring theme of this book has been the human ability to survive in the face of catastrophe. Despite large losses, individuals have survived and communities have rebuilt to fight the disaster battle once again. There is an optimism among many victims that contradicts the sense of doom and gloom in many hazards texts. Although hazards are an ever-present threat, the seemingly indomitable spirit of human society emphasizes survival regardless of the context. People are usually willing to act to combat disaster; it is finding the appropriate action that raises difficulties. This is not to say that some individuals do not experience extreme difficulties in hazardous environments, as witnessed by studies of stress (Tobin and Ollenburger, 1996); there also are instances of looting, hoarding, fights over water, and other antisocial acts. Yet, valiant efforts are made by those at risk to build flood walls out of sandbags, to extinguish the fires caused by earthquakes, and to rescue survivors.

However, individual action often is not most appropriate for reducing losses and mitigating hazards over the long term. Instead, this should be the role of hazard managers: the local or central government planners, emergency managers, political leaders, foreign development experts, and aid organizations. Traditionally, the goal of the hazard manager has been to reduce losses that accrue from extreme geophysical events (Figure 8.2), including both economic losses from damaged property and lost business and loss of life. This can be achieved, of course, by reducing the risk from natural hazards, minimizing exposure, and diminishing the vulnerability of hazard-prone populations. It has been accomplished by constructing structural measures such as sea walls to prevent tidal surge in hurricane-prone areas, erecting levees to mitigate riverine flooding, and enforcing building codes to strengthen buildings in earthquake areas. More recently, there has been greater emphasis on nonstructural measures to reduce losses, focusing on insurance policies, land use regulations, and even relocation policies.

Unfortunately, losses have continued to rise in spite of investment in alleviation adjustments. A new focus is required to prevent reconstruction and development in known hazardous areas. Figure 8.2, which illustrates the traditional view, should be altered to take into account a more comprehensive view of hazard mitigation as shown in Figure 8.3. This also requires a change in the accepted norms of society. Postdisaster relief and rehabilitation cannot be allowed to perpetuate the problems of the past by merely rebuilding what has been damaged. Ultimately, a change in perception and attitude is required at all administrative levels and by all parts of society. Politicians and hazard managers must address the needs and interests of marginalized groups and seek to minimize the vulnerability of those in hazard-prone areas. At the same time, individuals must be made aware of the hazardousness of particular locations and become willing to change behavior and, if necessary, location to reduce hazard losses. In effect, misperceptions of reality must be changed.

Clearly, this is no easy task since it requires more than simply implementing a structural or nonstructural alleviation project. Instead, mitigation goals, plans, and projects must encompass sociological and economic realities (that is, the context of a given place) and integrate new concepts

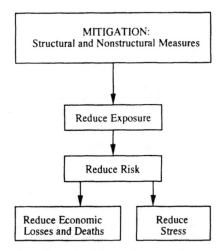

FIGURE 8.2. The classic model of hazard management. Only through mitigation can we begin to reduce losses from natural hazards, but the classic model has not been entirely successful thus far.

of loss reduction. Figure 8.3 broadens our definitions of hazard mitigation, as suggested by others. For example, Blaikie et al. (1994) advocate focusing attention on reducing vulnerability as a way to combat hazards, ameliorating not only aspects of the geophysical event but also modifying the human-use system. They put forward twelve principles to guide mitigation projects and reduce vulnerability (Table 8.1).

Coordinated planning at all levels of government will be required to reduce hazard losses. This is certainly not a new concept; many others have promoted comprehensive planning for natural hazards. The schema in Figure 8.3 and the principles suggested by Blaikie et al. (1994) focus on both reducing vulnerability and instituting comprehensive planning that involves all parties in order to avoid repetition of past mistakes. In a wider sense, then, hazard reduction plans must be reoriented to address long-term societal needs rather than concentrating on short-term fixes.

The major opportunity for implementation of hazard reduction measures occurs immediately after a disaster, when interest is running high. Disasters inevitably trigger calls for action. For example, earthquakes in the United States have gradually encouraged more rigorous building codes since the events of 1933. Similarly, the history of flood alleviation is replete with examples of postdisaster action; dams and levees have been constructed after flooding, not prior to events, and even after the 1993 floods on the Mississippi and Missouri Rivers, calls were made to reexamine federal approaches to flood mitigation. Reassessment should be under-

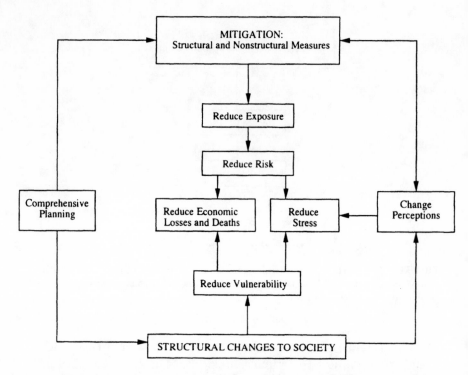

FIGURE 8.3. A revised model of hazard management. Loss reduction can occur only through structural change to society so that people are not forced into hazardous areas and the short-term benefits of a hazardous location do not turn into significant losses in the long run. The classic model of hazards management must be revised to incorporate planning aimed at identifying and bringing about structural changes, one result of which should be changed perceptions of hazards and mitigation.

taken regularly, but it took the catastrophe of 1993 to shake up political action. The report of the task force was exceptional, pointing out the problems inherent in the present system and making several recommendations for improvements (Interagency Floodplain Management Review Committee, 1994). Nevertheless, there comes a time when the disasters are forgotten, relegated to become benchmarks in people's lives as the horror and stress fade away. Gradually, recollections of the devastation are pushed aside in favor of other memories of the positive aspects of the disaster. Of course, the more serious the disaster, especially if it has involved deaths of relatives or friends, the longer the memory will prevail. Unfortunately, memories of the 1993 floods are already fading and, consequently, no major changes in flood mitigation policies in the United States have occurred. Some minor adjustments have been made with increased efforts toward relocation of certain communities, increased acquisition of flood-plain lands for wildlife reserves, changes in the waiting period for purchasing flood insurance, and the addition of incentives for adoption of mitigation strategies through reduced flood insurance premiums.

Thus, it is important to take action when interest is high. There is a danger, however, in using postdisaster response and interest to promote hazard mitigation. There tends to be a lack of comprehensiveness in such responses; proposals for dams and levees and rebuilding houses and roads may well keep many at risk when other actions might have a more widespread mitigative result. In addition, even when reassessments and reevaluations are initiated after disasters, actions often are taken at the same time that feed into the disaster–damage–repair–disaster cycle. Indeed, in 1993, levees were being repaired and rebuilt in the Midwest even as discussion of federal approaches to flood mitigation was beginning; thus, actions were taken that perpetuated past approaches while discussion about changing those approaches was in progress. There is great motivation to act because we want to be protected from the next event, but our actions also can worsen the next event. Thus, careful planning is required and it should be in place prior to the next event.

In the best scenario, mitigation strategies would be assimilated into standard planning practices and other development projects. In this way, hazards would become part of the normal functioning of society and would be treated in context. However, it would require a change in the political will and a redirection of goals to redefine hazards as a "normal state of affairs." Furthermore, mitigation strategies should be aimed at multiple hazards because so many disasters are the consequence of multiple events— the flood that initiates a gasoline fire, the earthquake that ignites large

TABLE 8.1. Guiding Principles for Managing a Reduction in Vulnerability to Natural Hazards

1. Vigorously manage mitigation.
2. Integrate elements of mitigation into development planning.
3. Capitalize on a disaster to initiate or develop mitigation.
4. Monitor and modify to suit new conditions.
5. Focus attention on protection of the most vulnerable.
6. Focus on the protection of lives and livelihoods of the vulnerable.
7. Focus on active rather than passive approaches.
8. Focus on protecting priority sectors.
9. Emphasize measures that are sustainable over time.
10. Assimilate mitigation into normal practices.
11. Incorporate mitigation into specific development projects.
12. Maintain political commitment.

Source: Adapted from Blaikie, Cannon, Davis, and Wisner (1994).

urban fires, the tornado that brings large hail damage (Alexander, 1993). Indeed, planning for multiple hazards at a place is an important principle to follow and is compatible with the arguments on vulnerability made by Blaikie et al. (1994). In fact, unless multiple hazards are incorporated into planning and hazard mitigation activities, efforts to reduce vulnerability may well be wasted.

We have suggested that local government may be the right level for implementing many mitigation strategies. However, it may take incentives and directives from the central government to instigate change as local governments, even with the political will, frequently lack sufficient resources to bring about the kinds of changes that are needed. Even international input must be considered as we search for ways to reduce exposure, whether to an occurrence of a natural event or to losses from it. This would include development planning as well as the coordinated distribution of relief.

THE INTERNATIONAL DECADE FOR NATURAL DISASTER REDUCTION

In recognition of the increasing vulnerability of world population to the effects of natural hazards, the United Nations General Assembly declared the 1990s the International Decade for Natural Disaster Reduction (IDNDR). The objective of IDNDR was to reduce loss of life, property damage, and social and economic disruption caused by natural disasters, especially in developing countries. (The five goals of the IDNDR are shown in Table 8.2.) The UN called on all countries to participate and many established their own National Decade for Natural Disaster Reduction,

TABLE 8.2. Goals of the International Decade for Natural Disaster Reduction

1. Improve the capacity of each country to mitigate the effects of natural disasters expeditiously and effectively, paying special attention to assisting developing countries in the establishment of early warning systems when needed.
2. Devise appropriate guidelines and strategies for applying existing knowledge, taking into account the cultural and economic diversity of nations.
3. Foster scientific and engineering endeavors aimed at closing critical gaps in knowledge in order to reduce loss of life and property.
4. Disseminate existing and new information related to measures for assessment, prediction, prevention, and mitigation of natural disasters.
5. Develop measures for the assessment, prediction, prevention, and mitigation of natural disasters (tailored to specific hazards and locations) through programs of technical assistance and technology transfer, demonstration projects, and education and training, and evaluate the effectiveness of those programs.

Source: National Research Council (1989, pp. 2–3).

though it was clear that this was seen by the UN to be a cooperative effort aimed at reducing the global toll of natural disasters.

In early development of IDNDR activities, emphasis was placed on technology and the dissemination of scientific knowledge, based on "the promise scientific and technical progress holds for understanding these hazards and mitigating their effects" (National Research Council, 1989, p. 2). Indeed, the goals listed in Table 8.2 focus on the application of new and existing knowledge though some recognition, albeit cursory, was extended to cultural and economic diversity. The early emphasis of IDNDR activities missed the mark in four areas, discussed below.

Technological and Structural Fixes

In large part, IDNDR was based on the belief that disaster reduction was possible through science, technology, and engineering. Missing was the application of social science. We know from our studies of perception and adjustments that technological "remedies" for natural events influence perceptions of the problem and tend to distort views of the potential severity of events, those of potential victims and decision makers alike, who may see the problem as solved. Further, such solutions allow us to ignore the broader social context, which goes a long way toward defining vulnerability. Indeed, it is in the social realm that the effects of technological fixes must be questioned, especially if "the resignation that there is no other place to relocate, the misunderstanding that the area is safe, the attitude that the hazard can be humanly conquered, and the reliance on insurance and institutional relief on the remote chance that something doesn't go as designed, are allowed to prevail and proliferate without being questioned" (Nelsen, 1994, p. 6).

Data Deficiencies

Application of technology and science to solving hazard-related problems is based on definitions of severity of the problem and need. These definitions, in turn, are dependent on adequate baseline data. However, such data are usually incomplete at best, if records exist at all. In addition, data that do exist frequently invoke conflicting findings (Mitchell, 1989b), partly due to faulty or inconsistent methodologies used to collect data or derive information. It is extremely difficult to determine the extent to which losses have been reduced if we do not know what losses have been historically. There also is a tendency for losses to be overestimated, particularly in the immediate aftermath of an event. The greater the losses, the greater the assistance that can be expected, but while there are rational reasons for misrepresenting losses, they complicate our ability to set priorities and evaluate alternative strategies. Emphasis should be placed on developing methodologies for collecting data that are accurate, consistent, and able to be carried out relatively simply; otherwise, we will continue to make poor decisions based on inadequate data. Sapir and Mission (1992) advocate a systematic data collection strategy (as shown in Figure 8.4) to improve preparedness and response programs at the global level. While their strategy is admirable and undoubtedly would enhance emergency relief distribution, it would be subject to the same data collection difficulties described above.

Isolated Management of Disaster Situations

As we have pointed out a number of times in this book, disaster response tends to be ad hoc and piecemeal. Comprehensive, unified approaches to hazard mitigation and disaster reduction are lacking. Again, there are many good reasons for this, the most significant perhaps being a humanitarian wish to help those in need as quickly as possible. However, this only fuels the disaster–damage–repair–disaster cycle. There has been difficulty in the United States in developing a unified approach to floodplain management, a need that became apparent once again following the 1993 floods on the Missouri and upper Mississippi Rivers. Developing a coherent, comprehensive scheme for other parts of the world—many with less time to spare after an event—is seemingly impossible, but necessary if the goals of IDNDR are to be met. Furthermore, focusing on the technological and engineering options only serves to avoid the comprehensive view that is required.

Resistant Political Forces

Context is important. In some political contexts, the emphasis is on short-term benefits, such as the next election. Long-term objectives, such as those associated with a unified program of hazards management, are lost to short-term projects that provide protection. For the same reasons, local interests (that is, those of the local taxpayer and voter) win out over the welfare of the general public. Therefore, we find many political systems precipitating a focus on short-term, local goals to the detriment of long-term, public welfare interests. An example of this occurred at Folly Island, South Carolina, where legislation passed after Hurricane Hugo—and aimed

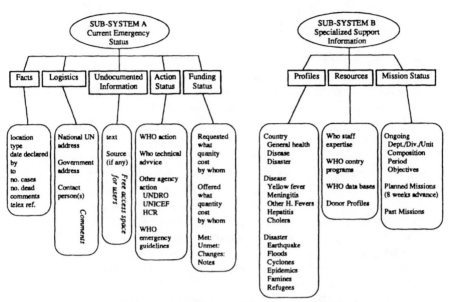

Emergency Management Information System: outline and contents

Emergency Management Information System: layout for databases on sub-system B

FIGURE 8.4. An emergency management information system. Disaster response, particularly the provision of aid, requires systematic and organized management. The capabilities of a management information system (MIS) can be directed toward this end, as the Emergency MIS here illustrates. *Source:* Sapir and Mission (1992). Copyright 1992 by Blackwell Publishers. Reprinted by permission.

at regulating future construction in coastal areas—contained loopholes that allowed homeowners to bypass some of the restrictions. This was aptly labeled a "political hazard" (Platt, Beatley, and Miller, 1991). Certainly, similar examples can be found worldwide.

In most places, short-term economic pressures place the priorities on issues other than the possible impacts of an event with a low probability of occurrence. In addition, development projects, funded if only in part by international agencies, bring in financial and technological resources to countries in an attempt to develop sectors of the economy. This assistance may or may not include hazard mitigation. Of course, there is great political currency associated with such projects because of the promise they hold for economic development. At times, the fact that they may increase vulnerability, or otherwise alter relationships between people and the environment, takes a back seat to the development goals. For example, in Bangladesh, a government official who advocated the right of citizens to be fully informed about the costs as well as the benefits of a large flood control project lost his government position in yet another example of a "political hazard" (Boyce, 1990).

While IDNDR certainly raised the level of awareness of hazards, the program itself was biased toward quick-onset events. Less attention was given to pervasive problems such as famine and drought, and initially there was little concern for global climate change (see Mitchell, 1988, for details). This represents another example of political factors influencing disaster management because pervasive hazards do not usually have the same visual impact as, for example, an area devastated by an earthquake in seconds. This bias may also reflect the emphasis on technological and structural fixes, which may be more relevant to quick-onset events.

Progress in Unifying Concerns

Since 1989, much progress has been made in IDNDR, though it is not possible to measure success at this time. There was a great deal of inertia to overcome, particularly involving political intransigence. Five years into the Decade, one disaster manager noted:

> When the *opportunity* for mitigating potential disasters is raised, there is an unease, an uncertainty, perhaps a bit of embarrassment about how to proceed. Whatever agency is being addressed, there is frequently a tendency to believe that mitigation of disasters is either "a specialist's" responsibility or "someone else's" concern. Even considering the existence of IDNDR as an institution to promote the greater awareness of hazards and the application of means to mitigate their worst effects, there is a sense of inertia, or

perhaps a hesitancy, to proceed with the application of resources against the politically uncomfortable thought of risk. (Jeggle, 1994, p. 1)

IDNDR has not made great strides in either disaster reduction or, as Jeggle suggests, the political will to act. However, positive outcomes can be attributed directly to actions that were initiated as a result of IDNDR and indirectly to the deficiencies seen in its stated goals. In fact, a 1994 World Conference on Natural Disaster Reduction suggested that development planners, politicians, economists, environmentalists, and relief and aid officials were putting greater emphasis on helping developing countries to build their own capacities to deal with disasters and on promoting disaster reduction (Elo, 1994). This is an important change from the original concern with technology transfer, engineering projects, and technical assistance (all of which have their place, but only as part of a larger effort that incorporates development issues and political and economic need and differences). As has been pointed out by development experts, outsiders do not develop societies, but they can help (Anderson and Woodrow, 1989). It appears that those associated with IDNDR have begun to recognize that.

Whether or not there are any measurable changes in disaster reduction that can be attributed to IDNDR, the Decade has had an impact. Not only did it bring together some 130 countries to consider the need for and means of disaster reduction, but it also placed disasters in a development context. Thus, a first step was made toward adopting a schema like that one in Figure 8.3, where vulnerability is reduced through structural changes in society and where comprehensive planning, rather than response to disasters, guides hazard mitigation.

Lest we be too optimistic, losses from natural disasters continue to be significant and to increase. A complex web of factors defines hazardousness and we cannot focus on one part of the web, or indeed on just one cooperative effort (like the IDNDR) to sort out the problems and reduce vulnerability. In addition, there are many actors with differing interests, differing resources, and differing abilities to bring about change. There are forces, external to the hazards themselves, that serve to increase vulnerability. The geophysical system is dynamic as well, which may affect vulnerability. In attempting to reduce hazardousness and vulnerability, we must focus on both the internal and external forces that fuel it. Progress will likely be slow because we must face huge problems like an ever-expanding world population that too often compels people to move to unsafe lands or to live in unsafe structures, and governments find themselves unable to cope with the rate of growth. Still, we are beginning to put the pieces of the puzzle together through integration of the physical, economic, social, and political elements that contribute to hazardousness or, alternatively, that can be directed toward decreasing vulnerability.

THE FUTURE OF HAZARD REDUCTION

Throughout the chapters of this book, we have examined different components of natural hazards—the physical environment from which extreme natural events emanate, individual and community perception and adjustment and the factors that influence them, the public policy arena, and economic factors that guide or determine decision making. It has been our purpose to illustrate the linkages among these components in order to emphasize the need for and difficulties associated with hazard reduction. Hazard reduction is required now more than ever before. It seems appropriate, then, that we conclude the book by identifying several axioms on which comprehensive hazard reduction policies might be based. Within these axioms, which provide the foundations for action, are found the various components of natural hazards. Thus, we have come full circle: from presenting the components, in all their complexity, to integrating them into a base on which hazard reduction policy recommendations can be made. Of course, many things will have to be in place for all of this to occur, not least of which are the political ability to recognize the axioms and the economic resources to develop the recommendations.

1. Mitigation practices may lead to reduced risk for some hazard-prone inhabitants, which in turn will reduce losses and stress from living in a hazardous location. The implementation of a mitigation project will also affect perception that, in turn, will reduce stress (Figure 8.3). However, the perception may be based on a false sense of security in the level of protection offered; indeed, losses may eventually rise because of this. The levee effect, where the protection offered by projects generates increased development and places more property at risk, is a perfect example. Only through comprehensive planning—incorporating not only the physical and technical aspects of the mitigation measures, but also the projected impacts on social, economic, political, legal, administrative, and environmental concerns— will hazard mitigation finally be effective. Further, hazard planning should be integrated with other planning practices at all levels of government; plans should be comprehensive and unified, incorporating structural and nonstructural adjustments while addressing appropriate changes in social structure to reduce vulnerability. That comprehensiveness must extend to multiple hazards, whether natural events that trigger other natural events or natural events that trigger technological events.

2. Enforcement of policies is absolutely essential. Not only must the political will be maintained throughout implementation, but the state must have the power to enforce policies. Groups and individuals must not be allowed to by-pass regulations to accommodate vested short-term interests. For instance, in the United States, compromises on zoning regulations have resulted in some communities being placed on probation or even excluded from the National Flood Insurance Program; without such enforcement, it would not be successful. It has been argued that communities that do not maintain rigorous floodplain zoning regulations should be excluded from emergency relief following disasters; to date, this has not been done for humanitarian reasons, but repeated events may force the issue. The same applies to other events. Reconstruction following earthquakes or tornadoes should meet strict standards though, of course, economic criteria and risk analyses should be considered as well; that is, there should be design standards that represent the highest level of protection that is cost-effective for a given level of risk. Avoiding or ignoring such standards in a rush to rebuild is not productive.

3. *Any hazard reduction program must be accompanied by an education program to broaden awareness of hazards and hazard mitigation.* Education programs need to be directed to public officials and politicians who often act as gatekeepers as well as to the lay public. Ultimately, this will have the effect of improving perceptions and changing hazard behavior. In part, it is hoped that increased awareness can take the place of experience and have the same effects on perception and action. It is important that all parties recognize the importance of safety and of minimizing losses from disasters. Part of this education must focus on the redistribution of resources needed for hazard reduction; financial and technical support that reduce the vulnerability of those at risk is good for society as a whole. Any education program must be ongoing to counteract motivational decay; as time passes, memory becomes distorted and recall of events grows less clear, but public education programs can keep the problem in the public eye and alert them to the potential for further disasters.

4. *Research into natural hazards provides a sound base for the development of comprehensive policies.* There is now a substantial literature on the physical and technical aspects of hazards, which needs to be continued. Similarly, it is only through further understanding of the social, economic, and political aspects of natural hazards that any real progress is going to be made in hazard mitigation. Social forces have compromised the effectiveness of mitigation measures through, among other things, the levee effect and other social traps that promote development of hazardous locations. Thus, planning must take into account the locus of power and how marginalized groups have little choice in some behavior patterns. We have shown throughout this book that we can integrate the research findings on a number of natural hazards to assist us in understanding the components, linkages, and development issues associated with hazards management.

While applied hazards research is required to solve many problems, we also should develop our conceptual base of analysis so that future research will facilitate broader understanding of the dynamic nature of hazard systems. Such research findings will go a long way toward promoting sound mitigation practices, but will require that hazards researchers transfer their information to hazard managers at all levels. Nevertheless, hazards research has already led to many important findings, and the axioms, outlined above, are a direct result of this significant body of work.

Natural disasters continue to take their toll. Our understanding of the human-use system as it affects and is affected by the natural system is changing every day. Still, we have a long way to go. The effects are severe and not unevenly distributed, and things seem to be getting worse in some parts of the world. Throughout time, there have been innumerable events that could be classified as natural disasters. Many die; others suffer economic losses. Although people invariably survive amid recurring catastrophes, for many catastrophe is a normal part of life.

References

Abernathy, A. M., & Weiner, L. (1995). Evolving federal role for emergency management. *Forum: For Applied Research and Public Policy, 10*(1), 45–48.

Adams, W. C. (1986). Whose lives count? T.V. coverage of natural disasters. *Journal of Communication, 36*(2), 113–122.

Aguirre, B. E., Anderson, W. A., Balandran, S., Peters, B. E., & White, H. M. (1991). *Saragosa, Texas tornado, May 22, 1987: An evaluation of the warning system* (Natural Disaster Studies, Vol. 3). Washington, D.C.: National Academy Press.

Ahrens, C. D. (1994). *Meteorology today: An introduction to weather, climate, and the environment* (5th ed.). Minneapolis, Minn.: West Publishing.

Ajzen, I., & Fishbein, M. (1981). *Foundations of information integration theory.* New York: Academic Press.

Alesch, D. J., & Petak, W. J. (1986). *The politics and economics of earthquake hazard mitigation: Unreinforced masonry buildings in Southern California* (Monograph 43). Program on Environment and Behavior. Boulder, Colo.: Institute of Behavioral Science, University of Colorado.

Alesch, D. J., Taylor, C., Ghanty, A. S., & Nagy, R. A. (1993). Earthquake risk reduction and small business. In Committee on Socioeconomic Impacts (Ed.), *Socioeconomic impacts* (Monograph 5, pp. 133–160). Memphis, Tenn.: Central United States Earthquake Consortium.

Alexander, D. (1993). *Natural hazards.* New York: Chapman and Hall.

Ambrose, J., & Vergun, D. (1985). *Seismic design of buildings.* New York: Wiley.

American Geological Institute. (1984). *Glossary of geology.* Falls Church, Va.: American Geological Institute.

Amey, G. (1974). *The collapse of Dale Dyke Dam, 1864.* London: Cassel.

Anderson, D. G. (1990). Mapping for Canada–Ontario flood damage reduction program. In *Proceedings of the Flood Plain Management Conference, Toronto* (Chapter 49). Toronto: Ontario Ministry of Natural Resources, Environment Canada, and the Association of Conservation Authorities of Canada.

Anderson, D. R., & Weinrobe, M. (1980). *Effects of natural disaster on local mortgage markets: The Pearl River flood in Jackson, Mississippi–April, 1979* (Natural Hazard Research Working Paper 39). Boulder, Colo.: Institute of Behavioral Science, University of Colorado.

Anderson, M. B. (1991). Which costs more: Prevention or recovery? In A. Kreimer and M. Munasinghe (Eds.), *Managing natural disasters and the environment* (pp. 17–27). Washington, D.C.: World Bank.

Anderson, M. B., & Woodrow, P. J. (1989). *Rising from the ashes: Development strategies in times of disaster.* Boulder, Colo.: Westview.

Anon. (1937). River cities zoning against the next flood. *American City, 52*(May), 109.

Anon. (1995). Old MacDonald had an option. *The Economist,* 11 February, 68.

Arora, C. R. (1976). Land-use maps for town planners. In P. Laconte (Ed.), *The environment of human settlements* (pp. 197–227). Oxford: Pergamon.

Ascher, C. (1942). *Better cities.* Washington, D.C.: National Resources Planning Board, U.S. Government Printing Office.

Association of Bay Area Governments. (1995). Geologic materials: Shaking amplification. *ABAG Online.* Oakland, Calif.: Association of Bay Area Governments. [Internet site: http://www.abag.ca.gov.]

Association of State Floodplain Managers. (1992). *Floodplain management, 1992, State and local programs.* Madison, WI: Association of State Floodplain Managers.

Association of State Floodplain Managers. (1995). Number of NFIP policies in force by state. *News & Views,* April, 5.

Atkisson, A. A., Petak, W. J., & Alesch, D. J. (1984). Natural hazard exposures, losses and mitigation costs in the United States, 1970–2000. *Wisconsin Academy of Sciences, Arts, and Letters, 72,* 106–112.

Ayre, R. S., with Mileti, D. S., & Trainer, P. B. (1975). *Earthquake and tsunami hazards in the United States: A research assessment* (NSF Monograph NSF-RA-E-75-005). Boulder, Colo.: Institute of Behavioral Science, University of Colorado.

Babroski, G. J., & Goswami, S. R. (1979). Urban development and flood control design. *Public Works, 110,* 78–81.

Bailey, W. E. (1995). The insurance industry and Hurricane Andrew. *Forum: For Applied Research and Public Policy, 10*(1), 22–25.

Baisden, B. (1979). Crisis intervention in smaller communities. In E. J. Miller & R. P. Wolensky (Eds.), *The small city and regional community: Proceedings of the 1979 Conference* (Vol. 2, pp. 325–332). Stevens Point, Wisc.: University of Wisconsin.

Baker, E. J. (1976). *Toward an evaluation of policy alternatives governing hazard-zone land-uses* (Natural Hazard Research Working Paper 28). Boulder, Colo.: Institute of Behavioral Science, University of Colorado.

Baker, E. J. (1994). Warning and response. In Committee on Natural Disasters (Ed.), *Hurricane Hugo: Puerto Rico, the U.S. Virgin Islands, and South Carolina, September 17–22, 1989* (Natural Disaster Studies, Vol. 6, pp. 202–210). Washington, D.C.: National Research Council.

Baker, E. J. (1995). Public response to hurricane probability forecasts. *Professional Geographer, 47*(2), 137–147.

Bardo, J. W. (1978). Organizational response to disaster: A topology of adaptation and change. *Mass Emergencies, 3*(2/3), 87–104.

Barry, R. G., & Chorley, R. J. (1992). *Atmosphere, weather, and climate* (6th ed.). London: Methuen.

Barton, A. H. (1969). *Communities in disaster: A sociological analysis of collective stress situations.* Garden City, N.Y.: Anchor.

Baumann, D. D., & Sims, J. H. (1974). Human response to the hurricane. In G. F. White (Ed.), *Natural hazards: Local, national, global* (pp. 25–29). New York: Oxford University Press.

Baumann, D. D., & Sims, J. H. (1978). Flood insurance: Some determinants of adoption. *Economic Geography, 54*(3), 189–196.

Baxter, P. J., Bernstein, R. S., Falk, H., French, J., & Ing, R. (1982). Medical aspects of volcanic disasters: An outline of the hazards and emergency response measures. *Disasters, 6*(4), 268–276.

Beinin, L. (1985). *Medical consequences of natural disasters*. New York: Springer-Verlag.

Bennett, C. S. (1937). Does flood protection pay? *American City, 52*(March), 57–59.

Berke, P. R., Kartez, J., & Wenger, D. (1993). Recovery after disasters: Achieving sustainable development, mitigation, and equity. *Disasters, 17*(2), 93–109.

Bernknopf, R. L., Brookshire, D., & Thayer, M. A. (1990). Earthquake and volcano hazard notices: An economic evaluation of changes in risk perceptions. *Journal of Environmental Economics and Management, 18*(1), 35–49.

Berz, G. (1988). List of major natural disasters, 1960–1987. *Natural Hazards, 1,* 97–99.

Blaikie, P., Cannon, T., Davis, I., & Wisner, B. (1994). *At risk: Natural hazards, people's vulnerability, and disasters*. New York: Routledge.

Blair, M. L., Spangle, W. E., & Spangle, W. (1979). *Seismic safety and land-use planning*. (U.S. Geological Survey Report 941-B.) Washington, D.C.: U.S. Government Printing Office.

Blake, P. A. (1989). Communicable disease control. In M. B. Gregg (Ed.), *The public health consequences of disasters, 1989* (pp. 7–12). Atlanta: Centers for Disease Control, U.S. Department of Health and Human Services.

Bluestein, H. B., & Golden, J. H. (1993). A review of tornado observations. In C. Church, D. Burgess, C. Doswell, & R. Davies-Jones (Eds.), *The tornado: Its structure, dynamics, prediction, and hazards* (Geophysical Monograph 79, pp. 319–352). Washington, D.C.: American Geophysical Union.

Bolin, R. (1994). Postdisaster sheltering and housing: Social processes in response and recovery. In R. Dynes & K. J. Tierney (Eds.), *Disasters, collective behavior, and social organization* (pp. 115–127). Newark, Del.: University of Delaware Press.

Bolin, R., & Stanford, L. (1990). Shelter and housing issues in Santa Cruz County. In R. Bolin (Ed.), *The Loma Prieta earthquake: Studies of short-term impacts* (Program on Environment and Behavior Monograph 50.) Boulder, Colo.: Institute of Behavioral Science, University of Colorado.

Bolin, R., & Stanford, L. (1991). Shelter, housing, and recovery: A comparison of U.S. disasters. *Disasters, 15*(1), 24–34.

Bolin, R. C., & Klenow, D. J. (1982–83). Response of the elderly to disaster: An age stratified analysis. *International Journal of Aging and Human Development, 16*(4), 283–296.

Bolin, R. C., & Klenow, D. J. (1988). Older people in disaster: A comparison of black and white victims. *International Journal of Aging and Human Development, 26*(1), 29–43.

Bolt, B. A. (1978). *Earthquakes: A primer*. San Francisco: Freeman.

Bolt, B. A. (1982). *Inside the earth: Evidence from earthquakes*. San Francisco: Freeman.

Bolt, B. A., Horn, W. L., MacDonald, G. A., & Scott, R. F. (1977). *Geological hazards*. New York: Springer-Verlag.

Bowden, M. J., & Kates, R. W. (1974). *The coming San Francisco earthquake: After the disaster* (Natural Hazard Research Working Paper 25.) Boulder, Colo.: Institute of Behavioral Science, University of Colorado.

Boyce, J. K. (1990). Birth of a megaproject: Political economy of flood control in Bangladesh. *Environmental Management, 14*(4), 419–428.

Brammer, H. (1990a). Floods in Bangladesh I: Geographical background to the 1987 and 1988 floods. *The Geographical Journal, 156*(1), 12–22.

Brammer, H. (1990b). Floods in Bangladesh II: Flood mitigation and environmental aspects. *The Geographical Journal, 156* (2), 158–165.

Bretschneider, C. L. (1972). Revisions to hurricane design wave practices. In American Society of Civil Engineers (Ed.), *Coastal Engineering Research Council, Thirteenth Coastal Engineering Conference* (pp. 167–195). New York: American Society of Civil Engineers.

Breznitz, S. (1984). *Cry wolf: The psychology of false alarms*. Hillsdale, N.J.: Erlbaum.

Brookshire, D. S., Thayer, M. A., Tschirhart, J., & Schulze, W. D. (1985). A test of the expected utility model: Evidence from earthquake risks. *Journal of Political Economy, 93*(2), 369–389.

Bryant, E. A. (1991). *Natural hazards*. Cambridge: Cambridge University Press.

Bruun, P. (1972). The history and philosophy of coastal protection. In American Society of Civil Engineers (Ed.), *Coastal Engineering Research Council, Thirteenth Coastal Engineering Conference* (pp. 33–74). New York: American Society of Civil Engineers.

Bullard, F. M. (1979). Volcanoes and their activity. In P. D. Sheets & D. K. Grayson (Eds.), *Volcanic activity and human ecology* (pp. 9–48). New York: Academic Press.

Bullard, F. M. (1984). *Volcanoes of the earth* (2d ed.). Austin: University of Texas Press.

Burby, R. J., & French, S. P. (1985). *Flood plain land use management: A national assessment* (Studies in Water Policy and Management 5). Boulder, Colo.: Westview.

Burby, R. J., Bollens, S. A., Holway, J. M., Kaiser, E. J., Mullan, D., & Sheaffer, J. R. (1988). *Cities under water: A comparative evaluation of ten cities' efforts to manage floodplain land use* (Program on Environment and Behavior Monograph 47). Boulder, Colo.: Institute of Behavioral Science, University of Colorado.

Burby, R. J., with Cigler, B. A., French, S. P., Kaiser, E. J., Kartez, J., Roenigk, D., Weist, D., & Whittington, D. (1991). *Sharing environmental risks: How to control governments' losses in natural disasters*. Boulder, Colo.: Westview.

Burgess, D., Donaldson, R. J., Jr., & Desrochers, P. R. (1993). Tornado detection and warning by radar. In C. Church, D. Burgess, C. Doswell, & R. Davies-Jones (Eds.), *The tornado: Its structure, dynamics, prediction, and hazards* (Geophysical Monograph 79, pp. 203–221). Washington, D.C.: American Geophysical Union.

Burton, I. (1961). Invasion and escape on the Little Calumet. In G. F. White (Ed.),

Papers on flood problems (Department of Geography Research Paper 70, pp. 84-92). Chicago: University of Chicago Press.

Burton, I. (1962). *Types of agricultural occupance of flood plains in the United States* (Department of Geography Research Paper 75). Chicago: University of Chicago Press.

Burton, I., & Kates, R. W. (1964). The perception of natural hazards in resource management. *Natural Resources Journal, 3,* 412-414.

Burton, I., Kates, R. W., & White, G. F. (1993). *The environment as hazard* (2d ed.). New York: Guilford.

Butler, J. R. G., & Doessel, D. P. (1980). Who bears the costs of natural disasters? An Australian case study. *Disasters, 4,* 187-204.

Byrne, M. G., & Ueda, J. Y. (1975). On rampage through a suburb. *Landscape Architecture, 65,* 304-311.

California Resources Agency. (1971). *Environmental impact of urbanization of the foothill and mountainous lands of California.* Sacramento: California Resources Agency.

Campbell, J. R. (1984). *Dealing with disaster: Hurricane response in Fiji.* Honolulu: Pacific Islands Development Program, East-West Center.

Campbell, J. R. (1990). Disasters and development in historical context: Tropical cyclone responses in the Banks Islands, Northern Vanuatu. *International Journal of Mass Emergencies and Disasters, 8*(3), 401-424.

Canino, G., Bravo, M., Rubio-Stipec, M., & Woodbury, M. (1990). The impact of disaster on mental health: Prospect and retrospect analysis. *Journal of Mental Health, 19*(1), 51-69.

Chambers, D. N. (1975). Procedures for determining the design flood in engineering works. *Proceedings of the Institute of Civil Engineers, 58,* 723-726.

Chambers, D. N., & K. G. Rogers. (1973). *The economics of flood alleviation* (Local Government Operational Research Unit Report C. 155). London: Government Publications.

Changnon, S. A., & J. M. Changnon. (1992). Temporal fluctuations in weather disasters: 1950-1989. *Climate Change, 22,* 191-208.

Choudhury, A. K. M. K. (1987). Land use in Bangladesh. In M. Ali, G. E. Radosevich, & A. A. Khan (Eds.), *Water resources policy for Asia* (pp. 203-215). Rotterdam: Balkema.

Chow, V. T. (Ed.). (1964). *Handbook of applied hydrology.* New York: McGraw-Hill.

Chowdury, A. M. R. (1988). The 1987 flood in Bangladesh: An estimate of damage in twelve villages. *Disasters, The Journal of Disaster Studies and Management, 12*(4), 294-300.

Ciborowski, A. (1978). Some aspects of physical development planning for human settlements in earthquake-prone regions. In UNESCO, *The assessment and mitigation of earthquake risk* (pp. 274-284). Paris: UNESCO.

Clawson, M., & Hall, P. (1973). *Planning and urban growth: An Anglo-American comparison.* Resources for the Future. Baltimore: Johns Hopkins University Press.

Clifford, R. A. (1956). *The Rio Grande flood: A comparative study of border communities in disaster* (Disaster Study 7, National Research Council Publication 458). Washington, D.C.: National Academy of Sciences.

Coburn, A. W., Spence, R. J. S., & Pomonis, A. (1991). *Vulnerability and risk assessment* (prepared for United Nations Development Programme Disaster Management Training Programme). Cambridge: Cambridge Architectural Research Limited.

Cochrane, H. C. (1974). Predicting the impact of earthquakes. In H. C. Cochrane, J. E. Haas, M. J. Bowden, & R. W. Kates (Eds.), *Social science perspectives on the coming San Francisco earthquake: Economic impact, prediction, and reconstruction* (Natural Hazard Research Working Paper 25). Boulder, Colo.: Institute of Behavioral Science, University of Colorado.

Cohen, M. (1991). Urban growth and natural hazards. In A. Kreimer & M. Munasinghe (Ed.), *Managing natural disasters and the environment* (p. 93). Washington, D.C.: World Bank.

Combs, B., & Slovic, P. (1979). Newspaper coverage of causes of death. *Journalism Quarterly, 56*, 837–843, 849.

Committee on Banking, Finance, and Urban Affairs, House of Representatives. (1994). *The status of the National Flood Insurance Program, Hearing before the Subcommittee on Consumer Credit Insurance* (103d Congress). Washington, D.C.: Government Printing Office.

Cones, M. (1979). Planning settlements for upland arid regions: An overview of environmental and building considerations. In D. Soen (Ed.), *New trends in urban planning* (pp. 283–288). New York: Pergamon.

Coombs, D. S., & Norris, R. J. (1981). The East Abbotsford, Dunedin, New Zealand landslide of August 8, 1979. *Bulletin de Liaison des Laboratories des Ponts et Cheusses*, Spécial *10* (January), 27–34.

Costa, J. E., & Baker, V. R. (1981). *Surficial geology: Building with the earth.* New York: Wiley.

Costanza, R. (1987). Social traps and environmental policy? Why do problems persist when there are technical solutions available? *Bioscience, 37*, 407–412.

Covello, V. T., Flamm, W. G., Rodricks, J. V., & Tardiff, R. G. (1981). *The analysis of actual versus perceived risk.* New York: Plenum.

Covello, V. T., & Mumpower, J. (1985). Risk analysis and risk management: An historical perspective. *Risk Analysis, 5*(2), 103–118.

Coyne, J. C., & DeLongis, A. (1986). Going beyond social support: The role of social relationships in adaptation. *Journal of Consulting and Clinical Psychology, 54*(4), 454–460.

Crandell, D. R., Mullineaux, D. R., & Miller, C. D. (1979). Volcanic-hazard studies in the Cascade Range of the western United States. In P. D. Sheets & D. K. Grayson (Eds.), *Volcanic activity and human ecology* (pp. 195–219). New York: Academic Press.

Craven, P., & Wellman, B. (1973). The network city. *Sociological Inquiry, 43*(3–4), 57–88.

Cropper, M. L., & Portney, P. R. (1992). Discounting human lives. *Resources,* Summer, 1–4.

Cross, J. A. (1985). Residents' awareness of the coastal flood hazard: Lower Florida Keys case study. *Flood hazard management in government and the private sector* (Special Publication 12, pp. 73–78). Boulder, Colo.: Institute of Behavioral Science, University of Colorado.

Cross, J. A. (1988). Hazard maps in the classroom. *Journal of Geography, 87,* 202-211.

Cross, J. A. (1990). Longitudinal changes in hurricane hazard perception. *International Journal of Mass Emergencies and Disasters, 8,* 31-47.

Cuny, F. C. (1983). *Disasters and development.* New York: Oxford University Press.

Cutrona, C., Russell, D., & Rose, J. (1986). Social support and adaptation to stress by the elderly. *Journal of Psychology and Aging, 1*(1), 47-54.

Dacy, D. C., & Kunreuther, H. (1969). *The economics of natural disasters: Implications for federal policy.* New York: The Free Press.

Damianos, D. (1975). The influence of flood hazards upon residential property values. Ph.D. diss., Virginia Polytechnical Institute.

David, E., & Mayer, J. (1984). Comparing costs of alternative flood hazard mitigation plans: The case of Soldiers Grove, Wisconsin. *Journal of the American Planning Association, 50*(1), 22-35.

Davis, D. W. (1978). Comprehensive floodplain studies using spatial data management techniques. *Water Resources Bulletin, 14,* 587-604.

de Boer, J. (1990). Definition and classification of disasters: Introduction of a disaster severity scale. *The Journal of Emergency Medicine, 8,* 591-595.

de Goyet, C. (1993). Post-disaster relief: The supply-management challenge. *Disasters, 17*(2), 169-176.

DeGraff, J. V. (1994). The geomorphology of some debris flows in the southern Sierra Nevada, California. *Geomorphology, 10,* 231-252.

de Man, A., & Simpson-Housley, P. (1988). Correlates of responses to two potential hazards. *The Journal of Social Psychology, 128*(3), 385-391.

de Quervain, M., & Jaccard, C. (1980). Les catastrophes dues aux avalanches et leur prévention. *UNDRO News,* November, 4-8.

Deutscher, I. (1973). *What we say/what we do: Sentiments and acts.* Glenview, Ill.: Scott, Foresman.

Dinman, B. D. (1980). The reality and acceptance of risk. *Journal of the American Medical Association, 244,* 1226-1228.

Dixon, J. W. (1964). Water resources: Part I, Planning and development. In V. T. Chow (Ed.), *Handbook of applied hydrology* (pp. 1-29). New York: McGraw-Hill.

Don Nanjira, D. D. C. (1991). Disasters and development in East Africa. In A. Kreimer & M. Munasinghe (Eds.), *Managing natural disasters and the environment* (pp. 82-89). Washington, D.C.: World Bank.

Douglas, J. H. (1978). Waiting for the Great Tokai Quake. *Science News, 113,* 282-286.

Dowall, D. E. (1979). The effect of land use and environmental regulations on housing costs. *Policy Studies Journal, 8,* 277-288.

Drabek, T. E. (1983). Alternative patterns of decision-making in emergent disaster response networks. *International Journal of Mass Emergencies and Disasters, 1*(2), 277-305.

Drabek, T. E. (1986). *Human system responses to disaster: An inventory of sociological findings.* New York: Springer-Verlag.

D'Souza, F. (1982). Recovery following the Southern Italian earthquake, November, 1980: Two contrasting examples. *Disasters, 6*(2), 101-109.

Dudasik, S. W. (1980). Victimization in natural disaster. *Disasters, 4*(3), 329-338.

Dufournaud, C., Millerd, F., & Schaefer, K. (1990). A benefit cost analysis of selected projects under the Canada–Ontario flood damage reduction program. In *Proceedings of the Flood Plain Management Conference, Toronto* (Chapter 23). Toronto: Ontario Ministry of Natural Resources, Environment Canada, and the Association of Conservation Authorities of Canada.

Dunham, A. (1959). Flood control via the police power. *University of Pennsylvania Law Review, 107,* 1098–1132.

Dunne, T., & Leopold, L. B. (1978). *Water in environmental planning.* San Francisco: Freeman.

Dworkin, J. (1974). *Global trends in natural disasters, 1947–1973* (Natural Hazard Research Working Paper 26). Boulder, Colo.: Institute of Behavioral Science, University of Colorado.

Dynes, R. R. (1974). *Organized behavior in disaster* (Disaster Research Center Monograph Series 3). Columbus: Disaster Research Center, Ohio State University.

Dynes, R. R. (1975). The comparative study of disaster: A social organizational approach. *Mass Emergencies, 1*(1), 21–31.

Dynes, R. R., Haas, J. E., & Quarantelli, E. L. (1964). *Some preliminary observations in organizational responses in the emergency period after the Niigata, Japan earthquake of June 16, 1964* (Research Report 11). Columbus: Disaster Research Center, Ohio State University.

Dynes, R. R., De Marchi, B., & Pelanda, C. (1987). *Sociology of disaster.* Milan: Franco Agneli Libri.

Dynes, R. R., & Quarantelli, E. L. (1972). *A perspective on disaster planning.* Washington, D.C.: Defense Civil Preparedness Agency.

Dzurik, A. A. (1979). Floodplain management: Some observations on the Corps of Engineers attitudes and approaches. *Water Resources Bulletin, 15,* 420–425.

Eagleman, J. R. (1983). *Severe and unusual weather.* New York: Van Nostrand Reinhold.

Earney, F. C., & Knowles, B. A. (1974). Urban snow hazard: Marquette, Michigan. In G. F. White (Ed.), *Natural hazards: Local, national, global* (pp. 167–174). New York: Oxford University Press.

Eguchi, R. T., & Seligson, H. A. (1993). Lifeline damage and resulting impacts. In Committee on Socioeconomic Impacts (Ed.), *Socioeconomic impacts* (Monograph 5, pp. 69–106). Memphis, Tenn.: Central United States Earthquake Consortium.

Elo, O. (1994). Making disaster reduction a priority in public policy. *Natural Hazards Observer, 17*(6), 1–2.

El-Sabh, M. I., & Murty, T. S. (1988). *Natural and man-made hazards.* Boston: Reidel.

Emergency Information Administrator. (1995a). *Benin–floods* (DHA Situation Report no. 2, 6 November). Volunteers in Technical Assistance, Disaster Information Center. [Received from incident@lan.vit.org.]

Emergency Information Administrator. (1995b). *Vietnam–Typhoon Zack* (DHA-Geneva Information Report no. 1, 6 November). Volunteers in Technical Assistance, Disaster Information Center. [Received from incident@lan.vita.org.]

Emergency Information Administrator. (1995c). *Vietnam–Floods and typhoon* (DHA–Geneva Information Report no. 3, 1 December). Volunteers in Technical Assistance, Disaster Information Center. [Received from incident@lan.vita.org.]

Emergency Information Administrator. (1995d). *Democratic People's Republic of Korea–Floods* (DHA–Geneva Situation Report no. 12, 21 December). Volunteers in Technical Assistance, Disaster Information Center. [Received from incident@lan.vita.org.]

Emergency Information Administrator. (1996). *South Africa floods* (DHA–Geneva Information Report no. 2, 17 January). Volunteers in Technical Assistance, Disaster Information Center. [Received from incident@lan.vita.org.]

Emergency Information Public Affairs. (1995a). *Flood insurance award recognition* 26 June. [Received from eipa@fema.gov.] Federal Emergency Management Agency.

Emergency Information Public Affairs (1995b). *Flood insurance community list* 2 November. [Received from eipa@fema.gov.] Federal Emergency Management Agency.

Endler, N. S., & Magnusson, D. (1976). Toward an interactional psychology of personality. *Psychological Bulletin, 83,* 956–979.

Ericksen, N. J. (1974). Flood information, expectation, and protection on the Opotiki floodplain, New Zealand. In G. F. White (Ed.), *Natural hazards: Local, national, global* (pp. 60–70). New York: Oxford University Press.

Ericksen, N. J. (1975). A tale of two cities: Flood history and the prophetic past of Rapid City, South Dakota. *Economic Geography, 54*(4), 305–320.

Ericksen, N. J. (1986). *Creating flood disasters? New Zealand's need for a new approach to urban flood hazard* (Water and Soil Miscellaneous Publication 77). Wellington, New Zealand: National Water and Soil Conservation Authority.

Ericksen, N. J. (no date). Natural hazards: Basic concepts (Natural Hazards Resource Guide). Hamilton, New Zealand: Department of Geography, University of Waikato.

Erickson, K. (1976). *Everything in its path: Destruction of community in the Buffalo Creek flood.* New York: Simon and Schuster.

Estall, R. (1994). Natural hazard risk management strategies and available insurance protection. In A. G. Hull & R. Coory (Eds.), *Proceedings of the Natural Hazards Management Workshop, Wellington, 8–9 November, 1994* (Institute of Geological and Nuclear Sciences Information Series 31, pp. 27–32). Lower Hutt, New Zealand: Institute of Geological and Nuclear Sciences.

Federal Emergency Management Agency. (1980). *An assessment of the consequences and preparations for a catastrophic California earthquake: Findings and actions taken.* Washington, D.C.: U.S. Government Printing Office.

Federal Interagency Floodplain Management Task Force. (1992). *Floodplain management in the United States: An assessment report* (FIA-17). Washington, D.C.: Federal Emergency Management Agency.

Federal Register. (1995). *Federal Register, 60,* 19 (January), 5583–5586.

Festinger, L. (1957). *A theory of cognitive dissonance.* Stanford: Stanford University Press.

Fischhoff, B., Lichtenstein, S., Slovic, P., Derby, S. L., & Keeney, R. L. (1981). *Acceptable risk.* New York: Cambridge University Press.

Fischhoff, B., Slovic, P., Lichtenstein, S., Reed, S., & Coombs, B. (1978). How safe is safe enough? A psychometric study of attitudes towards technological risks and benefits. *Policy Sciences, 9,* 127–152.

Flavin, C. (1994). Storm warnings: Climate change hits the insurance industry. *World Watch,* 7(November/December), 10–20.

Flood Response Meeting. (1993). Flood relief and recovery mobilization meeting (transcript). Held at Arnold, Missouri, 17 July. The White House: 75300. 3115@compuserve.com.

Flynn, B. W. (1994). *Psychological interventions.* Paper presented at the After Everyone Leaves: Preparing for, Managing and Monitoring Mid- and Long-Term Effects of Large-Scale Disasters conference, Minneapolis, Minnesota.

Foster, J. H. (1976). Flood management: Who benefits and who pays? *Water Resources Bulletin, 12*(5), 1029–1039.

Foster, P. D. (1980). *Disaster planning.* New York: Springer-Verlag.

Frank, N. L., & Husain, S. H. (1971). The deadliest cyclone in history. *Bulletin of the American Meteorological Society, 52*(6), 438–444.

Frazier, J. W., Harvey, M. E., & Montz, B. E. (1983). A causal analysis of urban flood hazard behaviors. *The Environmental Professional, 5*(1), 57–71.

Frazier, J. W., Harvey, M. E., & Montz, B. E. (1986). Flood policy preferences: A systemic analysis of some determinants. *The Environmental Professional, 8*(2), 70–87.

Freeman, A. M., III, & Portney, P. R. (1989). Economics clarifies choices about managing risk. *Resources,* Spring, 1–4.

French, J. G. (1989). Hurricanes. In M. B. Gregg (Ed.), *The public health consequences of disasters, 1989* (pp. 33–37). Atlanta: Centers for Disease Control, U.S. Department of Health and Human Services.

French, J. G., & Holt, K. W. (1989). Floods. In M. B. Gregg (Ed.), *The public health consequences of disasters, 1989* (pp. 69–78). Atlanta: Centers for Disease Control, US Department of Health and Human Services.

French, J., Ing, R., Von Allmen, S., & Wood, R. (1983). Mortality from flash floods: A review of National Weather Service reports, 1969–81. *Public Health Reports, 98*(6), 585–588.

French, S., & Isaacson, M. (1984). Applying earthquake risk analysis techniques to land use planning. *Journal of the American Institute of Planners, 50,* 509–522.

Fuson, K. (1993). The Great Flood of '93: Hell and high water. *Des Moines Register,* 22 August, 2–5.

Gabler, R. E., Sager, R. J., Brazier, S. M., & Wise, D. L. (1987). *Essentials of physical geography* (3d ed.). Philadelphia: Saunders College Publishing.

Gardner, D. K. (1969). The role of open spaces in floodplain management. In M. D. Dougal (Ed.), *Floodplain management: Iowa's experience* (pp. 137–146). Ames: Iowa State University Press.

Garling, T., & Golledge, R. G. (1993). Understanding behavior and environment: A joint challenge to psychology and geography. In T. Garling & R. G. Golledge (Eds.), *Behavior and environment: Psychological and geographical approaches* (pp. 1–15). Amsterdam: Elsevier.

Geipel, R. (1982). *Disaster and reconstruction: The Friuli (Italy) earthquakes of 1976.* London: Allen and Unwin.

Gibb, J. G. (1994). Defining coastal hazard areas and zones to control subdivision, use, and development. In A. G. Hull & R. Coory (Eds.), *Proceedings of the Natural Hazards Management Workshop, Wellington, 8–9 November, 1994* (Institute of Geological and Nuclear Sciences Information Series 31, pp. 43–46). Lower Hutt, New Zealand: Institute of Geological and Nuclear Sciences.

Gillespie, D. F., & Perry, R. W. (1976). An integrated systems and emergent norms approach to mass emergencies. *Mass Emergencies, 1*(4), 303–312.

Giorgis, D. W. (1987). Drought crisis management: The case of Ethiopia. In D. A. Wilhite & W. E. Easterling, with D. A. Wood (Eds.), *Planning for drought: Toward a reduction of societal vulnerability* (pp. 519–523). Boulder, Colo.: Westview.

Glickman, T. S., Golding, D., & Silverman, E. D. (1992). *Acts of God and acts of man: Recent trends in natural disasters and major industrial accidents* (Center for Risk Management, Discussion Paper 92-02). Washington, D.C.: Resources for the Future.

Goenjian, A. K., Najarian, L. M., Pynoos, R. S., Steinberg, A. M., Manoukian, G., Tavosian, A., & Fairbanks, L. A. (1994). Posttraumatic stress disorder in elderly and younger adults after the 1988 earthquake in Armenia. *American Journal of Psychiatry, 151,* 895–910.

Graves, P. E., & Bresnock, A. E. (1985). Are natural hazards temporally random? *Applied Geography, 5,* 5–12.

Gray, W. M. (1975). *Tropical cyclone genesis* (Atmospheric Science Paper 234). Fort Collins: Colorado State University.

Grazulis, T. P. (1993). *Significant tornadoes, 1680–1991: A chronology and analysis of events.* St. Johnsbury, Vt.: The Tornado Project of Environmental Films.

Green, C. H., Parker, D. J., Thompson, P., & Penning-Rowsell, E. C. (1983). *Indirect losses from urban flooding: An analytic framework* (Geography and Planning Paper 6). London: Flood Hazard Research Center, Middlesex Polytechnic.

Green, C. H., Tunstall, S. M., & Fordham, M. H. (1991). The risks from flooding: Which risks and whose perception? *Disasters, 15*(3), 227–236.

Greer, G. (1984). *Sex and destiny: The politics of human fertility.* London: Secker & Warburg.

Grether, D. M., & Mieszkowski, P. (1974). Determinants of real estate values. *Journal of Urban Economics, 1,* 127–146.

Griffiths, R. F. (Ed.). (1981). *Dealing with risk: The planning, management, and acceptability of technological risk.* Manchester: Manchester University Press.

Griggs, G. B., & Gilchrist, J. A. (1977). *The earth and land use planning.* Duxbury, Mass.: North Scituate.

Gruntfest, E. (1977). *What people did during the Big Thompson flood* (Natural Hazard Research Working Paper 32). Boulder, Colo.: Institute of Behavioral Science, University of Colorado.

Gruntfest, E. (1987). Warning dissemination and response with short lead times. In J. W. Handmer (Ed.), *Flood hazard management: British and international perspectives* (pp. 191–202). Norwich, England: Geo Books.

Gruntfest, E., & D. Pollack. (1994). Warnings, mitigation, and litigation: Lessons for research from the 1993 floods. *Water Resources Update, 95,* 40–44.

Gubrium, J. F. (1973). *The myth of the golden years: A socio-environmental theory of aging.* Springfield, Ill.: Thomas.

Gupta, H. K., & Rastogi, B. K. (1976). *Dams and earthquakes.* Amsterdam: Elsevier.

Haas, J. E., Kates, R. W., & Bowden, M. J. (Eds.). (1977). *Reconstruction following disaster.* Cambridge, Mass.: MIT Press.

Hageman, R. K. (1983). An assessment of the value of natural hazard damage in dwellings due to building codes: Two case studies. *Natural Resources Journal, 23,* 531–547.

Haggerty, L. J. (1982). Differential social contact in urban neighborhoods: Environmental vs. sociological explanations. *The Sociological Quarterly, 23,* 359–372.

Hagman, G. (1984). *Prevention better than cure: Report on human and environmental disasters in the Third World.* Stockholm: Swedish Red Cross.

Haldeman, J. M. (1972). *An analysis of continued semi-nomadism on the Kaputiei Maasai group ranches: Sociological and ecological factors* (Working Paper 28). Nairobi: University of Nairobi.

Hall, M. L. (1992). The 1985 Nevado del Ruiz eruption: Scientific, social and governmental response and interaction before the event. In G. J. H. McCall, D. J. C. Laming, & S. C. Scott (Eds.), *Geohazards: Natural and man-made* (pp. 43–52). London: Chapman Hall.

Halstead, M. (1900). *Galveston: The horrors of a stricken city.* Chicago: American Publishers Association.

Handmer, J. (1988a). Approaches to flood damage assessment. In N. J. Ericksen, J. Handmer, & D. I. Smith (Eds.), *Anuflood: Evaluation of a computerized urban flood loss assessment policy for New Zealand* (Water and Soil Conservation Authority Miscellaneous Publication no. 115, pp. 30–45). Wellington, New Zealand: National Water and Soil Conservation Authority.

Handmer, J. (1988b). Measuring flood losses: General issues. In N. J. Ericksen, J. Handmer, & D. I. Smith (Eds.), *Anuflood: Evaluation of a computerized urban flood loss assessment policy for New Zealand* (Water and Soil Conservation Authority Miscellaneous Publication no. 115, pp. 20–29). Wellington, New Zealand: National Water and Soil Conservation Authority.

Harding, D. M., & Parker, D. J. (1974). Flood hazard at Shrewsbury, United Kingdom. In G. F. White (Ed.), *Natural hazards: Local, national, global* (pp. 43–52). New York: Oxford University Press.

Hastenrath, S. (1987). The droughts of northeast Brazil and their prediction. In D. A. Wilhite & W. E. Easterling, with D. A. Wood (Eds.), *Planning for drought: Toward a reduction of societal vulnerability* (pp. 45–60). Boulder, Colo.: Westview.

Havlick, S. W. (1986). Building for calamity: Third World cities at risk. *Environment, 28*(9), 6–10, 41–45.

Heathcote, R. L. (1983). *The arid lands: Their use and abuse.* London: Longman Group.

Hertler, R. A. (1961). Corps of Engineers' experience in relation to flood plain regulation. In G. F. White (Ed.), *Papers on flood problems* (Department of Geography Research Paper 70, pp. 181–202). Chicago: University of Chicago Press.

Hershfield, D. M. (1961). *Rainfall frequency atlas of the United States for durations from 30 minutes to 24 hours and return periods from 1 to 100 years* (U.S. Weather Bureau Hydrologic Services Division, Technical Paper 40). Washington, D.C.: Government Printing Office.

Hewitt, K. (1969). *Probabilistic approaches to discrete natural events: A review and theoretical discussion* (Natural Hazard Research Working Paper No. 8). Toronto: University of Toronto.

Hewitt, K. (1983). The idea of calamity in a technocratic age. In K. Hewitt (Ed.), *Interpretations of calamity from the viewpoint of human ecology* (pp. 3–32). Boston: Allen and Unwin.

Hewitt, K., & Burton, I. (1971). *The hazardousness of a place: A regional ecology of damaging events*. Toronto: Department of Geography, University of Toronto.

Hodge, D., Sharp, V., & Marts, M. (1979). Contemporary responses to volcanism: Case studies from the Cascades and Hawaii. In P. D. Sheets & D. K. Grayson (Eds.), *Volcanic activity and human ecology* (pp. 221–248). New York: Academic Press.

Holahan, C. J. (1982). *Environmental psychology*. New York: McGraw-Hill.

Holen, A. (1991). A longitudinal study of the occurrence and persistence of post-traumatic health problems in disaster survivors. *Stress Medicine, 7,* 11–17.

Holland, G. J. (1983). Tropical cyclone motion: Environmental interaction plus a beta effect. *Journal of Atmospheric Science, 40,* 328–342.

Hollis, G. E. (1974). The effect of urbanization on floods in Canon's Brook Harlow, Essex. In K. J. Gregory & D. E. Walling (Eds.), *Fluvial processes in instrumented watersheds* (Institute of British Geographers Special Publication 6). London: Institute of British Geographers.

Holway, J. M., & Burby, R. J. (1990). The effects of floodplain development controls on residential land values. *Land Economics, 66*(3), 259–271.

Horton, J. (1994). *Disaster mitigation: Relocation in Illinois.* Paper presented at the After Everyone Leaves: Preparing for, Managing and Monitoring Mid- and Long-Term Effects of Large-Scale Disasters conference, Minneapolis, Minnesota.

Hossain, M. (1987). Agro-socio-economic conditions in Bangladesh. In M. Ali, G. E. Radosevich, & A. A. Khan (Eds.), *Water resources policy for Asia* (pp. 183–201). Rotterdam: Balkema.

Houghton, B. F. (1982). *Geyserland: A guide to the volcanoes and geothermal areas of Rotorua* (Geological Society of New Zealand Guidebook no. 4). Petone, New Zealand: Deslandes.

Huerta, F., & Horton, R. (1978). Coping behavior of elderly flood victims. *The Gerontologist, 18*(6), 541–546.

Huntington, E. (1919). *The evolution of the earth and its inhabitants*. New Haven, Conn.: Yale University Press.

Ingwerson, M. (1986). Quake readiness. In R. H. Maybury (Ed.), *Violent forces of nature* (pp. 9–18). Mt. Airy, Md.: Lomond.

Interagency Floodplain Management Review Committee. (1994). *Sharing the challenge: Floodplain management into the 21st century.* Washington, D.C.: Federal Emergency Management Agency.

International Federation of Red Cross and Red Crescent Societies. (1993). *World disaster report, 1993.* Dordrecht: Martinus Nijhoff.

Ives, J. D., & Krebs, P. V. (1978). Natural hazards research and land use planning responses in mountainous terrain: The town of Vail, Colorado. *Arctic and Alpine Research, 10,* 213–222.

Jackson, R. H. (1974). Frost hazard to tree crops in the Wasatch front: Perception and adjustment. In G. F. White (Ed.), *Natural hazards: Local, national, global* (pp. 146–150). New York: Oxford University Press.

Jacobs, T. L., & Vesilind, P. A. (1992). Probabilistic environmental risk of hazardous materials. *Journal of Environmental Engineering, 118*(6), 878–889.

James, L. D., & Lee, R. R. (1971). *Economics of water resource planning.* New York: McGraw-Hill.

Jeggle, T. (1994). Is the IDNDR a slow-onset disaster in the making? *Natural Hazards Observer, 17*(4), 1–2.

Johnson, W. F. (1889). *History of the Johnstown flood.* Philadelphia: Edgewood.

Johnston, D. M., & Nairn, I. A. (1993). *Volcanic impacts report: The impact of two eruption scenarios from the Okataina Volcanic Centre on the population and infrastructure of the Bay of Plenty, New Zealand* (Bay of Plenty Regional Council Resource Planning Publication 93/6). Whakatane, New Zealand: Bay of Plenty Regional Council.

Johnston, R. J. (1991). *Geography and geographers: Anglo-American human geography since 1945.* London: Edward Arnold.

Karnik, V., & Algermissen, S. T. (1978). Seismic zoning. In UNESCO, *The assessment and mitigation of earthquake risk* (pp. 11–47). Paris: UNESCO.

Kartez, J. D. (1984). Crisis response planning: Toward a contingent analysis. *Journal of the American Planning Association, 50*(1), 9–21.

Kasperson, R. E., Renn, O., Slovic, P., Brown, H. S., Emel, J., Goble, R., Kasperson, J. X., & Ratick, S. (1988). The social amplification of risk: A conceptual framework. *Risk Analysis, 8*(2), 177–187.

Kates, R. W. (1962). *Hazard choice and perception in floodplain management* (Department of Geography Research Paper 78). Chicago: University of Chicago Press.

Kates, R. W. (1971). Natural hazard in human ecological perspective: Hypotheses and models. *Economic Geography, 47,* 438–451.

Keller, E. A. (1988). *Environmental geology* (5th ed.). Columbus, Ohio: Charles Merrill.

Khan, M. I. (1991). The impact of local elites on disaster preparedness planning: The location of flood shelters in Northern Bangladesh. *Disasters, 15*(4), 340–354.

Kilbourne, E. M. (1989a). Heat waves. In M. B. Gregg (Ed.), *The public health consequences of disasters, 1989* (pp. 51–61). Atlanta: Centers for Disease Control, U.S. Department of Health and Human Services.

Kilbourne, E. M. (1989b). Cold environments. In M. B. Gregg (Ed.), *The public health consequences of disasters, 1989* (pp. 63–68). Atlanta: Centers for Disease Control, U.S. Department of Health and Human Services.

Kingsbury, P. A. (1994). Earthquake and geological hazard strategy. In A. G. Hull & R. Coory (Eds.), *Proceedings of the Natural Hazards Management Workshop, Wellington, 8–9 November 1994* (Institute of Geological and Nuclear Sciences Information Series 31, pp. 51–54). Lower Hutt, New Zealand: Institute of Geological and Nuclear Sciences.

Kirkby, A. V. (1974). Individual and community responses to rainfall variability in Oaxaca, Mexico. In G. F. White (Ed.), *Natural hazards: Local, national, global* (pp. 119–127). New York: Oxford University Press.

Kitizawa, K. (1986). Earthquake prediction and public response. In R. H. Maybury (Ed.), *Violent forces of nature* (pp. 31–43). Mt. Airy, Md.: Lomond.

Kling, G. W., Clark, M. A., Compton, H. R., Devine, J. D., Evans, W. C., Humphrey, A. M., Koenigsberg, E. J., Lockwood, J. P., Tuttle, M. L., & Wagner, G. N. (1987). The 1986 Lake Nyos gas disaster in Cameroon, West Africa. *Science, 236,* 169–176.

Koenig, R. L. (1993). Administration's plan on levees is misguided, Sen. Bond says. *St. Louis Post-Dispatch,* Friday, August 27, 8A.

Kollmorgen, W. M. (1953). Settlement control beats flood control. *Economic Geography, 29,* 208–215.

Korgen, B. J. (1995). Seiches. *American Scientist, 83* (July–August), 330–341.

Kotsch, W. J. (1977). *Weather for the mariner* (2d ed.). Annapolis, Md.: Naval Institute Press.

Krause, N. (1987). Exploring the impact of a natural disaster on the health and psychological well being of older adults. *Journal of Human Stress,* Summer, 61–69.

Kreimer, A. (1980). Low income housing under "normal" and post-disaster situations: Some basic continuities. *Habitat International, 4*(3), 273–283.

Kreimer, A., & Munasinghe, M. (1991). Managing environmental degradation and natural disasters: An overview. In A. Kreimer & M. Munasinghe (Eds.), *Managing natural disasters and the environment* (pp. 3–6). Washington, D.C.: World Bank.

Kreimer, A., & Preece, M. (1991). Case study: La Paz municipal development project. In A. Kreimer & M. Munasinghe (Eds.), *Managing natural disasters and the environment* (pp. 32–35). Washington, D.C.: World Bank.

Kruschke, E. R., & Jackson, B. M. (1987). *The public policy dictionary.* Santa Barbara, Calif.: ABC-Clio.

Krzysztofowicz, R., & Davis, D. (1986). Towards improving flood forecasting–response systems. In R. M. Maybury (Ed.), *Violent forces of nature* (pp. 169–189). Mt. Airy, Md.: Lomond.

Kubo, K., & Katayam, T. (1978). Earthquake resistant properties and design of public utilities. In UNESCO, *The assessment of mitigation of earthquake risk* (pp. 171–184). Paris: UNESCO.

Kunreuther, H. (1976). Limited knowledge and insurance protection. *Public Policy, 24,* 227–261.

Kunreuther, H., Ginsberg, R., Miller, L., Sagi, P., Slovic, P., Borkan, B., & Katz, N. (1978). *Disaster insurance protection: Public policy lessons.* New York: Wiley.

Lacayo, R. (1989). Is California worth the risk? *Time,* November 6, *134*(19), 18–24.

Laird, R. T., Perkins, J. B., Bainbridge, D. A., Baker, J. P., Boyd, R. T., Huntsman,

D., Straub, P. E., & Zuker, M. P. (1969). *Quantitative land capability analysis* (U.S. Geological Survey Report 945). Washington, D.C.: U.S. Government Printing Office.

Lambley, D. B., & Cordery, I. (1991). Effects of floods on the housing market in Sydney. In *Proceedings of international hydrology and water resources symposium, 1991* (pp. 863–866). Perth, Australia.

Larrain, P., & Simpson-Housley, P. (1990). Geophysical variables and behavior: Lonquimay and Alhue, Chile: Tension from volcanic and earthquake hazards. *Perception and Motor Skills, 70*(1), 296–298.

Larson, C. J., & Nikkel, S. R. (1979). *Urban problems.* Boston: Allyn and Bacon.

Laska, S. B. (1985). At-risk residents' knowledge and beliefs about structural and non-structural flood mitigation actions. In *Flood hazard management in government and the private sector* (Special Publication 12, pp. 52–59). Boulder, Colo.: Institute of Behavioral Science, University of Colorado.

Laska, S. B. (1990). Homeowner adaptation to flooding: An application of the general hazards coping theory. *Environment and Behavior, 22*(3), 320–357.

Lave, T. R., & Lave, L. B. (1991). Public perception of the risks of floods: Implications for communication. *Risk Analysis, 11*(2), 255–267.

Layard, R. (Ed.). (1962). *Cost-benefit analysis: Selected readings.* Harmondsworth, England: Penguin.

Lecomte, E. (1989). Earthquakes and the insurance industry. *Natural Hazards Observer, 14*(2), 1–2.

Legget, R. F. (1973). *Cities and geology.* New York: McGraw-Hill.

Leopold, L. B., & Maddock, T. (1954). *The flood control controversy: Big dams, little dams, and land management.* New York: Ronald Press.

Lipman, P. W., & Mullineaux, D. R. (Eds.). (1982). *The 1980 Eruptions of Mt. St. Helens, Washington* (U.S. Geological Survey Professional Paper 1250). Washington, D.C.: U.S. Government Printing Office.

Lord, W. B. (1994). Flood hazard delineation: The one percent standard. *Water Resources Update, 95* (Spring), 36–39.

Lowrance, W. W. (1976). *Of acceptable risk: Science and the determination of safety.* Los Altos, Calif.: William Kaufmann.

Luhr, J. F., & Simkin, T. (Eds.). (1993). *Paricutin: The volcano born in a Mexican cornfield.* Phoenix, Ariz.: Geosciences Press.

MacDonald, D. N., Murdoch, J. L., & White, H. L. (1987). Uncertain hazards, insurance, and consumer choice: Evidence from housing markets. *Land Economics, 63*(4), 361–371.

Madakasira, S., & O'Brien, K. F. (1987). Acute posttraumatic stress disorder in victims of a natural disaster. *The Journal of Nervous and Mental Disease, 175*(5), 286–290.

Magistro, J. V. (1993). Ecology and production in the Middle Senegal Valley wetlands. Ph.D. diss., Binghamton University.

Magnusson, D., & Torestad, B. (1992). The individual as an interactive agent in the environment. In W. B. Walsh, K. H. Craik, & R. H. Price (Eds.), *Person–environment psychology: Models and perspectives* (pp. 89–126). Hillsdale, N.J.: Erlbaum.

Manni, C., Magalini, S., & Proiett, R. (1988). Volcanoes. In P. Baskett & R. Weller (Eds.), *Medicine for disasters* (pp. 308-317). London: Wright.

Marx, W. (1977). *Acts of God, acts of men.* New York: Coward, McGann, and Geohegan.

May, C. D. (1987). The role of the media in identifying and publicizing drought. In D. A. Wilhite & W. E. Easterling, with D. A. Wood (Eds.), *Planning for drought: Toward a reduction of societal vulnerability* (pp. 515-518). Boulder, Colo.: Westview.

May, P. J. (1989). Social science perspectives: Risk and disaster preparedness. *International Journal of Mass Emergencies and Disasters, 7*(3), 281-303.

May, P. J., & Williams, W. (1986). *Disaster policy implementation: Managing programs under shared governance.* New York: Plenum.

McFarlane, A. C. (1988). The phenomenology of posttraumatic stress disorders following a natural disaster. *Journal of Nervous and Mental Disease, 176*(1), 22-29.

McHarg, I. (1969). *Design with nature.* New York: Natural History Press.

Meistrell, F. G. (1957). Protection for communities against flood hazards. *American City, 72* (April), 191-195.

Mileti, D. S., Drabek, T. E., & Haas, J. E. (1975). *Human systems in extreme environments: A sociological perspective* (Monograph 21). Boulder, Colo.: Institute of Behavioral Science, University of Colorado.

Mileti, D. S., & Fitzpatrick, C. (1993). *The great earthquake experiment: Risk communication and public action.* Boulder, Colo.: Westview.

Mileti, D. S., & Krane, S. (1973). Countdown: Response to the unlikely—Warning and response to impending system stress. Paper presented at annual meeting of the American Sociological Association, New York.

Milliman, J. W., & Roberts, R. B. (1985). Economic issues in formulation policy for earthquake hazard mitigation. *Policy Studies Review, 4*(4), 645-654.

Mitchell, J. K. (1984). Hazard perception studies: Convergent concerns and divergent approaches during the past decade. In T. F. Saarinen, D. Seamon, & J. L. Sell (Eds.), *Environmental perception and behavior: An inventory and prospect* (Department of Geography Research Paper 209, pp. 33-59). Chicago: University of Chicago Press.

Mitchell, J. K. (1989a). Hazards research. In G. L. Gaile & C. J. Willmott (Eds.), *Geography in America* (pp. 410-424). Columbus, Ohio: Merrill.

Mitchell, J. K. (1989b). *Where might the International Decade for Natural Disaster Reduction concentrate its activities?* (Decade for Natural Disaster Research DNDR 2). Boulder, Colo.: Institute of Behavioral Science, University of Colorado.

Mitchell, J. K. (1990). Human dimensions of environmental hazards, complexity, disparity, and the search for guidance. In A. Kirby (Ed.), *Nothing to fear* (pp. 131-175). Tucson: University of Arizona Press.

Mitchell, J. K. (1993). Natural hazards predictions and responses in very large cities. In J. Nemec et al. (Eds.), *Prediction and perception of natural hazards* (pp. 29-37). Dordrecht, Netherlands: Kluwer.

Mitchell, J. K., Devine, N., & Jagger, K. (1989). A contextual model of natural hazard. *The Geographical Review, 79*(4), 391-409.

Mitchell, J. K., & Ericksen, N. J. (1992). Effects of climate change on weather-related disasters. In I. M. Mintzer (Ed.), *Confronting climate change: Risks, implications, and responses* (pp. 141-151). Cambridge: Cambridge University Press.

Montz, B. E. (1983). The effects and effectiveness of flood insurance requirements: Agent perspectives. *The Environmental Professional, 5,* 116-123.

Montz, B. E. (1987). Floodplain delineation and housing submarkets. *Professional Geographer, 39,* 59-61.

Montz, B. E. (1992). The effects of flooding on residential property values in three New Zealand communities. *Disasters: The Journal of Disaster Studies and Management, 16*(4), 283-298.

Montz, B. E. (1993). Hazard area disclosure in New Zealand: The impacts on residential property values in two communities. *Applied Geography, 13*(3), 225-242.

Montz, B. E. (1994). *Methodologies for analysis of multiple hazard probabilities: An application in Rotorua, New Zealand* (Unpublished report on Fulbright Research Scholar Project). Centre for Environmental and Resource Studies, University of Waikato, Hamilton, New Zealand.

Montz, B. E., & Gruntfest, E. (1986). Changes in American urban floodplain occupancy since 1958: The experiences of nine cities. *Applied Geography, 6,* 325-338.

Montz, B. E., & Tobin, G. A. (1988). The spatial and temporal variability of residential real estate values in response to flooding. *Disasters, 12,* 345-355.

Montz, B. E., & Tobin, G. A. (1990). The impacts of Tropical Storm Agnes on residential property values in Wilkes-Barre, Pennsylvania. *Pennsylvania Geographer, 28*(2), 55-67.

Montz, B. E., & Tobin, G. A. (1992). *Market gatekeepers: Their impact on property values following flooding in Liberty County, Texas* (Quick Response Program, Final Field Report, 52). Boulder, Colo.: Institute of Behavioral Science, University of Colorado.

Moore, H. E., Bates, F. L., Layman, M. V., & Parenton, V. J. (1963). *Before the wind: A study of response to Hurricane Carla* (National Research Council Disaster Study 19). Washington, D.C.: National Academy of Sciences.

Morgan, J. (1979). The tsunami hazard in Tohoku and the Hawaiian Islands. *Science Reports of Tohoku University, 7th Series (Geography), 29,* 149-159.

Morgan, M. G. (1993). Risk analysis and management. *Scientific American,* July, 32-41.

Muckleston, K. W. (1983). The impact of floodplain regulations on residential land values in Oregon. *Water Resources Bulletin, 19,* 1-7.

Muckleston, K., Turner, M., & Brainerd, R. (1981). *Floodplain regulations and residential land values in Oregon* (Water Resources Research Institute Report WRRI-73). Corvallis: Oregon State University.

Mukerjee, T. (1971). *Economic analysis of natural hazards: A preliminary study of adjustments to earthquakes and their costs* (Natural Hazard Research Working Paper 17). Boulder, Colo.: Institute of Behavioral Science, University of Colorado.

Murphy, F. C. (1958). *Regulating floodplain development* (Department of Geography Research Paper 56). Chicago: University of Chicago Press.

Murton, B. J., & Shimabukuro, S. (1974). Human adjustment to volcanic hazard in Puna District, Hawaii. In G. F. White (Ed.), *Natural hazards: Local, national, global* (pp. 151-159). New York: Oxford University Press.

Murty, T. S. (1988). List of major natural disasters, 1960-1987. *Natural Hazards, 1,* 303-304.

Mushkin, S. J. (1962). Health as an investment. *Journal of Political Economy, 20,* 129-157.

Myers, M. F., & White, G. F. (1993). The challenge of the Mississippi flood. *Environment, 35*(10), 6-9, 25-35.

Nakano, T. (1974). Natural hazards: Report from Japan. In G. F. White (Ed.), *Natural hazards: Local, national, global* (pp. 231-243). New York: Oxford University Press.

National Oceanic Atmospheric Administration. (1995). *Deadliest, costliest, and most intense U.S. hurricanes of this century.* Miami: National Hurricane Center. [Received from http://www.nhc.noaa.gov.].

Nakano, T., & Matsuda, I. (1984). Earthquake damage, damage prediction, and countermeasures in Tokyo, Japan. *Ekistics, 51,* 415-420.

National Research Council. (1989). *Reducing disasters' toll: The United States Decade for Natural Disaster Reduction.* Washington, D.C.: National Academy Press.

National Science Foundation. (1980). *A report on flood hazard mitigation.* Washington, D.C.: U.S. Government Printing Office.

National Weather Service. (1993). Update on the Midwestern floods: Heat and drought in the East. *Special Climate Summary, 93*(2). Administrative message, 9-12 August, National Weather Service Headquarters.

Natsios, A. S. (1991). Economic incentives and disaster mitigation. In A. Kreimer & M. Munasinghe (Eds.), *Managing natural disasters and the environment* (pp. 111-114). Washington, D.C.: World Bank.

Natural Hazards Research and Applications Information Center. (1976). Voluntary relocation. *Natural Hazards Observer, 1,* December 1.

Natural Hazards Research and Applications Information Center. (1985). Alaska earthquake mitigation study released. *Natural Hazards Observer, 9*(4), 4.

Natural Hazards Research and Applications Information Center. (1996). FYI—Summary of Kobe earthquake one year later. *Disaster Research, 187,* 5.

Nelsen, E. W. (1994). The International Decade for Natural Disaster Reduction (IDNDR): A global concept yet to reach its potential. Unpublished independent research paper, Environmental Studies Program, Binghamton University, Binghamton, New York.

Nelson, K. (1980). Are we ready for the big one? *San Francisco,* March, 34-35.

Neumann, C. J., Jarvinen, B. R., & Pike, A. C. (1987). *Tropical cyclones of the North Atlantic Ocean, 1871-1986.* Coral Gables, Fla.: National Climate Data Center and National Hurricane Center.

Newhall, C. G. and Self, S. (1982). The volcanic explosivity index (VEI): An estimate of explosive magnitude for historical volcanism. *Journal of Geophysical Research, 87*(C2), 1231-1238.

Newmark, N. M. (1970). Current trends in the seismic analysis and design of high-rise structures. In R. L. Wiegel (Ed.), *Earthquake engineering* (pp. 403–414). Englewood Cliffs, N.J.: Prentice-Hall.

Nicholls, N. (1987). Prospects for drought prediction in Australia and Indonesia. In D. A. Wilhite & W. E. Easterling, with D. A. Wood (Eds.), *Planning for drought: Toward a reduction of societal vulnerability* (pp. 61–72). Boulder, Colo.: Westview.

Nichols, T. C. (1974). Global summary of human response to natural hazards: Earthquakes. In G. F. White (Ed.), *Natural hazards: Local, national, global* (pp. 274–284). New York: Oxford University Press.

Nilsen, T. H., Wright, R. H., Vlasic, T. C., Spangle, W. E., & Spangle, W. (1979). *Relative slope stability and land-use planning* (U.S. Geological Survey Report 944). Washington, D.C.: U.S. Government Printing Office.

Noji, E. K. (1991). Natural disasters. *Disaster Management, 7*(2), 271–292.

Noji, E. K., & Sivertson, K. T. (1987). Injury prevention in natural disasters: A theoretical framework. *Disasters, 11*(4), 290–296.

Norris, F. H., & Murrell, S. A. (1988). Prior experience as a moderator of disaster impact on anxiety symptoms in older adults. *American Journal of Community Psychology, 16*(5), 664–683.

Novoa, J. I., & Halff, A. H. (1977). Management of flooding in a fully developed, low-cost housing neighborhood. *Water Resources Bulletin, 13*, 1237–1252.

Nuhfer, E. B., Proctor, R. J., & Moser, P. H. (1993). *The citizens' guide to geologic hazards*. Arvada, Colo.: American Institute of Professional Geologists.

Office of Emergency Preparedness. (1972). *Disaster preparedness* (Vol. 1). Washington, D.C.: U.S. Government Printing Office.

Office of U.S. Foreign Disaster Assistance. (1988). *Disaster history: Significant data on major disasters worldwide, 1900–present*. Washington, D.C.: Office of Foreign Disaster Assistance, Agency for International Development.

Office of U.S. Foreign Disaster Assistance. (1990). *Disaster history: Significant data on major disasters worldwide, 1900–Present*. Washington, D.C.: Office of Foreign Disaster Assistance, Agency for International Development.

Office of U.S. Foreign Disaster Assistance. (1993). *Disaster history: Significant data on major disasters worldwide, 1900–Present*. Washington, D.C.: Office of Foreign Disaster Assistance, Agency for International Development.

Ohls, J. C., Weisberg, J. C., & White, M. J. (1974). The effect of zoning on land values. *Journal of Urban Economics, 1*, 428–444.

Ollenburger, J. C. (1981). Criminal victimization and fear of crime. *Research on Aging, 3*(1), 101–118.

Oliver, J. (1978). *Natural hazard response and planning in tropical Queensland* (Natural Hazard Research Working Paper 33). Boulder, Colo.: Institute of Behavioral Science, University of Colorado.

O'Riordan, T. (1971). *The New Zealand Earthquake and War Damage Commission : A study of a national natural hazard insurance scheme* (Natural Hazard Research Working Paper 20). Toronto: University of Toronto.

O'Riordan, T., & More, R. J. (1969). Choice in water use. In R. J. Chorley (Ed.), *Introduction to geographical hydrology* (pp. 175–201). London: Methuen.

Oya, M. (1970). Land use control and settlement plans in the flooded area of the city of Nagoya and its vicinity, Japan. *Geoforum, 4,* 27–35.

Palm, R. I. (1981). *Real estate agents and Special Studies Zones disclosure: The response of California home buyers to earthquake hazards information* (Program on Environment and Behavior Monograph 32). Boulder, Colo.: Institute of Behavioral Science, University of Colorado.

Palm, R. (1990). *Natural hazards: An integrative framework for research and planning.* Baltimore: Johns Hopkins University Press.

Palm, R. (1995). *Earthquake insurance: A longitudinal study of California homeowners.* Boulder, Colo.: Westview.

Palm, R. I., Marston, S., Kellner, P., Smith, D., & Budetti, M. (1983). *Home mortgage lenders, real property appraisers and earthquake hazards* (Program on Environment and Behavior Monograph 38). Boulder, Colo.: Institute of Behavioral Science, University of Colorado.

Pararas-Carayannis, G. (1986). The effects of tsunami on society. In R. H. Maybury (Ed.), *Violent forces of nature* (pp. 157–168). Mt. Airy, Md.: Lomond.

Parrett, C., Melcher, N. B., & James, R. W., Jr. (1993). *Flood discharges in the Upper Mississippi River basin, 1993* (U.S. Geological Survey Circular 1120-A). Washington, D.C.: Government Printing Office.

Penick, J., Jr. (1976). *The New Madrid earthquake of 1811 and 1812.* Columbia: University of Missouri Press.

Penning-Rowsell, E. C. (1986). Themes, speculations, and agendas for landscape research. In E. C. Penning-Rowsell and D. Lowenthal (Eds.), *Landscape meanings and values* (pp. 114–128). London: Allen and Unwin.

Penning-Rowsell, E. C., & Chatterton, J. B. (1977). *The benefits of flood alleviation: A manual of assessment techniques.* Westmead, England: Saxon House.

Penning-Rowsell, E. C., & Parker, D. J. (1974). Improving floodplain development control. *The Planner, 60,* 540–543.

Pereau, J. (1990). *First World/Third World and disasters in context: A study of the Saragosa and Wichita Falls, Texas, tornadoes.* College Station, Tex.: Hazard Reduction and Recovery Center, Texas A & M University.

Perry, A. H. (1981). *Environmental hazards in the British Isles.* London: Allen and Unwin.

Perry, R. W., & Mushkatel, A. H. (1984). *Disaster management: Warning response and community relocation.* Westport, Conn.: Quorum.

Petak, W. J., & Atkisson, A. A. (1982). *Natural hazard risk assessment and public policy: Anticipating the unexpected.* New York: Springer-Verlag.

Petrovski, J. T. (1978). Seismic zoning and related problems. In UNESCO, *The assessment and mitigation of earthquake risk* (pp. 48–65). Paris: UNESCO.

Petts, G., & Foster, I. (1985). *Rivers and landscape.* London: Edward Arnold.

Phifer, J. F. (1990). Psychological distress and somatic symptoms after natural disaster: Differential vulnerability among older adults. *Journal of Psychology and Aging, 5*(3), 412–420.

Phifer, J. F., & Norris, F. H. (1989). Psychological symptoms in older adults following natural disaster: Nature, timing, duration and course. *Journal of Gerontology, 44*(6), S207–S217.

Pierce, E. T. (1977). Lightning, location and warning systems. In R. H. Golde (Ed.), *Lightning* (Vol. 2, pp. 497-519). London: Academic Press.

Pielke, R. A. (1990). *The hurricane.* New York: Routledge.

Platt, R. H. (1976). *Land use control: Interface of law and geography* (Resource Paper 75-1). Washington, D.C.: Association of American Geographers.

Platt, R. H. (1992). Life after Lucas. *Natural Hazards Observer, 17*(1), 8.

Platt, R. H. (1994). Parsing Dolan. *Environment, 36*(8), 4-5, 43.

Platt, R. H., Beatley, T., & Miller, H. C. (1991). The folly at Folly's Beach and other failings of U.S. coastal policy. *Environment, 33*(9), 6-9.

Powell, M. D. (1994). Meteorology. In Committee on Natural Disasters (Ed.), *Hurricane Hugo: Puerto Rico, the U.S. Virgin Islands, and South Carolina, September 17-22, 1989* (Natural Disaster Studies, Vol. 6, pp. 172-201). Washington, D.C.: National Research Council.

Press, F. (1975). Earthquake prediction. *Scientific American, 232*(5), 14-23.

Prest A. R., & Turvey, R. (1965). Cost-benefit analysis: A survey. *Economic Journal, 75*, 683-735.

Price, L. W. (1972). *The periglacial environment, permafrost and man* (Resource Paper 14). Washington, D.C.: Association of American Geographers

Prowse, T. D., Owens, I. F., & McGregor, G. R. (1981). Adjustment to avalanche hazard in New Zealand. *New Zealand Geographer, 37*(1), 25-31.

Quarantelli, E. L. (1994). *Disaster stress.* Paper presented at the After Everyone Leaves: Preparing for, Managing and Monitoring Mid- and Long-Term Effects of Large-Scale Disasters conference, Minneapolis, Minnesota.

Quarantelli, E. L., &. Dynes, R. R. (1976). Community conflict: Its absence and its presence in natural disasters. *Mass Emergencies, 1*(2), 139-152.

Ramachandran, R., & Thakur, S. C. (1974). India and the Ganga floodplains. In G. F. White (Ed.), *Natural hazards: Local, national, global* (pp. 36-42). New York: Oxford University Press.

Ramsli, G. (1974). Avalanche problems in Norway. In G. F. White (Ed.), *Natural hazards: Local, national, global* (pp. 175-180). New York: Oxford University Press.

Raphael, B. (1986). *When disaster strikes: How communities and individuals cope with disasters.* New York: Basic Books.

Rasmussen, E. M. (1987). Global prospects for the prediction of drought: A meteorological perspective. In D. A. Wilhite & W. E. Easterling, with D. A. Wood (Eds.), *Planning for drought: Toward a reduction of societal vulnerability* (pp. 31-43). Boulder, Colo.: Westview.

Rees, J. D. (1979). Effects of the eruption of Paricutín Volcano on landforms, vegetation, and human occupancy. In P. D. Sheets & D. K. Grayson (Eds.), *Volcanic activity and human ecology* (pp. 249-292). New York: Academic Press.

Reiter, L. (1990). *Earthquake hazard analysis.* New York: Columbia University Press.

Relph, E. C. (1968). *Methods of flood loss reduction–With reference to the Devon River Authority.* M.Phil. thesis, University of London.

Rettger, M. J., & Boisvert, R. N. (1979). Flood insurance or disaster loans: An economic evaluation. *American Journal of Agricultural Economics,* August, 496-505.

Riebau, M. (1990). Is the NFIP insuring but not regulating? *Natural Hazards Observer*, *14*(4), 4-5.

Riebsame, W. E., Diaz, H. F., Moses, T., & Price, M. (1986). The social burden of weather and climate hazards. *American Meteorological Society Bulletin*, *67*(11), 1378-1388.

Ritchie, D. (1988). *Superquake: Why earthquakes occur and when the big one will hit.* New York: Crown.

Roder, W. (1961). Attitudes and knowledge in the Topeka Floodplain. In G. F. White (Ed.), *Papers on flood problems* (Department of Geography Research Paper 70, pp. 62-83). Chicago: University of Chicago Press.

Rommel, C. (1960). Principles and forms of international elemental hazard insurance. In H. F. Glass (Ed.), *International insurance* (pp. 44-72). New York: International Insurance Monitor.

Rossi, P. H., Wright, J. D., & Weber-Burdin, E. (1982). *Natural hazards and public choice: The state and local politics of hazard mitigation.* New York: Academic Press.

Rowntree, R. A. (1974). Coastal erosion: The meaning of a natural hazard in the cultural and ecological context. In G. F. White (Ed.), *Natural hazards: Local, national, global* (pp. 70-79). New York: Oxford University Press.

Rubin, C., & Palm, R. (1987). National origin and earthquake response: Lessons from the Whittier Narrows earthquake of 1987. *Journal of Mass Emergencies and Disasters*, *5*(3), 347-355.

Rubin, C. B., Saperstein, M. D., & Barbee, D. G. (1985). *Community recovery from a major natural disaster* (Program on Environment and Behavior Monograph 41). Boulder, Colo.: Institute of Behavioral Science, University of Colorado.

Ruffner, J. A., & Bair, F. E. (1984). *The weather almanac* (4th ed.). Detroit: Gale.

Russell, D. W., & Cutrona, C. E. (1991). Social support, stress and depressive symptoms among the elderly: Test of a process model. *Journal of Psychology and Aging*, *6*(2), 190-201.

Saarinen, T. F. (1966). *Perception of the drought hazard on the Great Plains.* Chicago: University of Chicago Press.

Sachanski, S. (1978). Buildings: Codes, materials, design. In UNESCO, *The assessment and mitigation of earthquake risk* (pp. 157-170). Paris: UNESCO.

Sagoff, M. (1992). Technological risk: A budget of distinctions. In D. E. Cooper & J. A. Palmer (Eds.), *The environment in question: Ethics and global issues* (pp. 194-211). New York: Routledge.

Salanave, L. E. (1980). *Lightning and its spectrum: An atlas of photographs.* Tucson: University of Arizona Press.

Sanderson, L. M. (1989). Tornadoes. In M. B. Gregg (Ed.), *The public health consequences of disasters, 1989* (pp. 39-49). Atlanta: Centers for Disease Control, U.S. Department of Health and Human Services.

Sapir, D. G., & Misson, C. (1992). The development of a database on disasters. *Disasters*, *16*(1), 75-80.

Scawthorn, C., Iemura, H., & Yamada, Y. (1982). The influence of natural hazards on urban housing location. *Journal of Urban Economics*, *11*, 242-251.

Schmitz, C. A. (1987). *Disaster! The United Nations and international relief management* (Studies Program Background Papers). New York: Council on Foreign Relations.

Seaman, J. (1990). Disaster epidemiology: Or why most international disaster relief is ineffective. *Injury: The British Journal of Accident Surgery, 21,* 5–8.

Segoe, L. (1937). Flood control and the cities. *American City, 52* (March), 55–56.

Semple, E. C. (1911). *The influences of geographic environment: On the basis of Ratzel's System of Anthropo-Geography.* New York: Henry Holt.

Sewell, W. R. D., Davis, J., Scott, A. D., & Ross, D. W. (1962). A guide to benefit-cost analysis. In I. Burton & R. W. Kates (Eds.), *Readings in resource management and conservation.* Chicago: University of Chicago Press.

Shah, B. V. (1983). Is the environment becoming more hazardous? A global survey 1947 to 1980. *Disasters, 7*(3), 202–209.

Sheaffer, J. R. (1969). The interaction of urban redevelopment and floodplain management in Waterloo, Iowa. In M. D. Dougal (Ed.), *Floodplain management: Iowa's experience* (pp. 123–136). Ames: Iowa State University Press.

Sheaffer and Roland, Inc. (1981). *Evaluation of the economic, social, and environmental effects of floodplain regulation.* Washington, D.C.: U.S. Department of Housing and Urban Development.

Sheehan, L., & Hewitt, K. (1969). *A pilot survey of global national disasters of the past twenty years* (Natural Hazard Research Working Paper 11). Boulder, Colo.: Institute of Behavioral Science, University of Colorado.

Sheets, P. D., & Grayson, D. K (1979). Introduction. In P. D. Sheets & D. K. Grayson (Eds.), *Volcanic activity and human ecology* (pp. 1–8). New York: Academic Press.

Sheets, R. C. (1995). Stormy weather. *Forum: For Applied Research and Public Policy, 10*(1), 5–15.

Shilling, J. D., Benjamin, J. D., & Sirmans, C. F. (1985). Adjusting comparable sales for floodplain location. *Appraisal Journal,* July, 429–436.

Showalter, P. S. (1991). *Field observations in Memphis during the New Madrid earthquake "projection" of 1990: How pseudoscience affected a region* (Natural Hazard Research Working Paper 71). Boulder, Colo.: Institute of Behavioral Science, University of Colorado.

Showalter, P. S. (1994). A guide for responding to unconventional earthquake predictions. In *Proceedings, Fifth U.S. national conference on earthquake engineering* (Vol. 3, pp. 1035–1044). Oakland, Calif.: Earthquake Engineering Research Institute.

Showalter, P. S., & Myers, M. F. (1994). Natural disasters in the United States as release agents of oils, chemicals, or radiological materials between 1980–1989: Analysis and recommendations. *Risk Analysis, 14*(2), 169–182.

Sigurdsson, H. (1988). Gas bursts from Cameroon Crater Lakes: A new natural hazard. *Disasters, 12*(2), 130–146.

Simpson, R. L. (1965). Sociology of the community: Current status and prospects. *Rural Sociology, 30*(2), 127–149.

Simpson-Housley, P., & Bradshaw, P. (1978). Personality and the perception of earthquake hazard. *Australian Geographical Studies, 16,* 65–72.

Singer, S. J. (1990). Flooding the Fifth Amendment: The National Flood Insurance Program and the "takings" clause. *Environmental Affairs, 17,* 323–370.

Slovic, P. (1987). Perception of risk. *Science, 236,* 280–286.

Slovic, P., Fischhoff, B., & Lichtenstein, S. (1979). Rating the risks. *Environment, 21*, 14-20, 36-39.

Slovic, P., Fischhoff, B., & Lichtenstein, S. (1980). Facts and fears: Understanding perceived risk. In R. Schwing & W. Albers (Eds.), *Societal risk assessment: How safe is safe enough?* (pp. 181-216). New York: Plenum.

Slovic, P., Fischhoff, B., & Lichtenstein, S. (1985). Characterizing perceived risk. In R. W. Kates, C. Hohenemser, & J. X. Kasperson (Eds.), *Perilous progress: Technology as hazard* (pp. 91-125). Boulder, Colo.: Westview.

Slovic, P., Kunreuther, H., & White, G. F. (1974). Decision processes, rationality, and adjustment to natural hazards. In G. F. White (Ed.), *Natural hazards: Local, national, global* (pp. 187-205). New York: Oxford University Press.

Smith, D. I. (1981). Actual and potential flood damage: A case study for urban Lismore, NSW, Australia. *Applied Geography, 1*, 31-39.

Smith, D. I. (1988). The assessment of direct and indirect flood damage, actual and potential. In N. J. Ericksen, J. Handmer, & D. I. Smith (Eds.), *Anuflood: Evaluation of a computerised urban flood-loss assessment policy for New Zealand* (Water and Soil Conservation Authority Miscellaneous Publication 115, pp. 46-60). Wellington, New Zealand: National Water and Soil Conservation Authority.

Smith, D. I. (1989). Urban flood damages, flood insurance, and relief in Australia: A survey of the data. In D. I. Smith & J. W. Handmer (Eds.), *Flood insurance and relief in Australia* (pp. 43-59). Canberra: Centre for Resource and Environmental Studies, Australian National University.

Smith, D. I., & Handmer, J. W. (Eds.). (1989). *Flood insurance and relief in Australia.* Canberra: Centre for Resource and Environmental Studies, Australian National University.

Smith, K. (1992). *Environmental hazards: Assessing risk and reducing disaster.* London: Routledge.

Smith, K., & Tobin, G. A. (1979). *Human adjustment to the flood hazard.* London: Longman Group.

Smith, N. (1971). *A history of dams.* London: Peter Davies.

Smith, P. J. (1990). Redefining decision: Implications for managing risk and uncertainty. *Disasters, 14*(3), 230-240.

Solomon, S. D., Regier, D. A., & Burke, J. D. (1989). Role of perceived control in coping with disaster. *Journal of Clinical Psychology, 8*(4), 376-392.

Solomon, S. D., Smith, E. M., Robins, L. N., & Fischbach, R. L. (1987). Social involvement as a mediator of disaster-induced stress. *Journal of Applied Social Psychology, 17*(12), 1092-1112.

Spangle, W. E., Mader, G. G., & Blair, M. L. (1980). *Land use planning after earthquakes.* Portola Valley, Calif.: William Spangle Associates.

Spilka, B. (1985). *The psychology of religion.* Englewood Cliffs, N.J.: Prentice-Hall.

Starr, C. (1969). Social benefit versus technological risk: What is our society willing to pay for safety? *Science, 165*, 1232-1236.

Starr, C., Rudman, R., &. Whipple, C. (1976). Philosophical basis for risk analysis. *Annual Review of Energy, 1*, 692-762.

Steinglass, P., & Gerrity, E. (1990). Natural disasters and post-traumatic stress disorder: Short-term versus long-term recovery in two disaster affected communities. *Journal of Applied Psychology, 20*(21), 1746–1765.

Stevens, A. E. (1988). Earthquake hazard and risk in Canada. In M. I. El-Sabh & T. S. Murty (Eds.), *Natural and man-made hazards* (pp. 43–61). Boston: Reidel.

Subrahmanyan, V. P. (1967). *Incidence and spread of continental drought* (World Meteorological Organization/International Hydrological Decade Report 2). Geneva: World Meteorological Organization.

Susman, P., O'Keefe, P., & Wisner, B. (1983). Global disasters, a radical interpretation. In K. Hewitt (Ed.), *Interpretations of calamity from the viewpoint of human ecology* (pp. 263–283). Boston: Allen and Unwin.

Talcott, F. W. (1992). How certain is that environmental risk estimate? *Resources,* Spring, 10–15.

Taylor, A. J. W. (1989). *Disasters and disaster stress* [Stress in Modern Society Series, No. 10]. New York: AMS Press.

Taylor, A. J. (1990). A pattern of disasters and victims. *Disasters, 14*(4), 291–300.

Taylor, J., Greenaway, M. A., & Smith, D. I. (1983). *Anuflood, programmer's guide, and user's manual.* Canberra: Centre for Resource and Environmental Studies, Australian National University.

Taylor, J. G., Stewart, T. R., & Downton, M. (1988). Perceptions and drought in the Ogallala aquifer region. *Environment and Behavior, 20*(2), 150–175.

Tazieff, H. (1986). Seismic and volcanic hazards. In R. H. Maybury (Ed.), *Violent forces of nature* (pp. 103–109). Mt. Airy, Md.: Lomond.

ten Hoopen, H. G. H., & Bakker, W. T. (1974). Erosion problems of the Dutch island of Goree. In American Society of Civil Engineers (Ed.), *Fourteenth Coastal Engineering Research Conference* (pp. 1213–1231). New York: American Society of Civil Engineers.

Tennessee Valley Authority. (1967). *TVA–1969.* Washington, D.C.: Government Printing Office.

Theiler, D. F. (1969). Effects of flood protection on land-use in Coon Creek, Wisconsin watershed. *Water Resources Research, 5,* 1216–1222.

Thomas, G., & Witts, M. M. (1971). *The San Francisco earthquake.* New York: Stein and Day.

Thompson, J. D., & Hawkes, R. W. (1962). Disaster, community organization, and administrative processes. In G. W. Baker & D. W. Chapman (Eds.), *Man and society in disaster* (pp. 268–300). New York: Basic Books.

Thompson, S. A. (1982). *Trends and developments in global natural disasters, 1947–1981* (Natural Hazard Research Working Paper 45). Boulder, Colo.: Institute of Behavioral Sciences, University of Colorado.

Tierney, K. J., & Baisden, B. (1979). *Crisis intervention programs for disaster victims: A source book and manual for smaller communities.* Rockville, Md.: U.S. Department of Health, Education and Welfare, National Institute of Mental Health.

Thorarinsson, S. (1979). On damage caused by volcanic eruptions with special reference to tephra and gases. In P. D. Sheets & D. K. Grayson (Eds.), *Volcanic activity and human ecology* (pp. 125–159). New York: Academic Press.

Tobin, G. A. (1978). Some aspects of flood hazard assessment and response with particular reference to Cumbria. Ph.D. diss., Strathclyde University, Scotland.

Tobin, G. A. (1992). Community response to floodplain relocation in Soldiers

Grove, Wisconsin. *Transactions of the Wisconsin Academy of Sciences, Arts, and Letters, 80,* 87–99.

Tobin, G. A. (1995). The levee love affair: A stormy relationship? *Water Resources Bulletin, 31*(3), 359–367.

Tobin, G. A., & Montz, B. E. (1988). Catastrophic flooding and the response of the residential real estate market. *Social Science Journal, 25*(2), 167–177.

Tobin, G. A., & Montz, B. E. (1990). Response of the real estate market to frequent flooding: The case of Des Plaines, Illinois. *Bulletin of the Illinois Geographical Society, 33*(2), 11–21.

Tobin, G. A., & Montz, B. E. (1994a). The flood hazard and dynamics of the urban residential land market. *Water Resources Bulletin, 30*(4), 673–685.

Tobin, G. A., & Montz, B. E. (1994b). *The great midwestern floods of 1993.* Fort Worth, Tex.: Saunders Press.

Tobin, G. A., & Ollenburger, J. C. (1994). An examination of stress in a flood-prone environment. *Papers and Proceedings of Applied Geography Conferences, 17,* 74–81.

Tobin, G. A., & Ollenburger, J. C. (1996). Predicting levels of postdisaster stress in adults following the 1993 floods in the Upper Midwest. *Environment and Behavior, 28*(3), 340–357.

Tobin, G. A., & Peacock, T. (1982). Problems and issues in comprehensive planning for a small community: The case of Soldiers Grove, Wisconsin. *The Environmental Professional, 4*(1), 43–50.

Toole, M. J., & Foster, S. (1989). Famines. In M. B. Gregg (Ed.), *The public health consequences of disasters, 1989* (pp. 79–89). Atlanta: Centers for Disease Control, U.S. Department of Health and Human Services.

Torry, W. I. (1980). Urban earthquake hazard in developing countries: Squatter settlements and the outlook for Turkey. *Urban Ecology, 4,* 317–327.

Tuan, Y.-F. (1974). *Topophilia: A study of environmental perception, attitudes, and values.* Englewood Cliffs, N.J.: Prentice-Hall.

Turner, J. F. C. (1969). Uncontrolled urban settlement: Problems and policies. In G. Breese (Ed.), *The city in newly developing nations* (pp. 507–534). Engelwood Cliffs, N.J.: Prentice-Hall.

Tversky, A., & Kahneman, D. (1982). Judgment under uncertainty: Heuristics and biases. In D. Kahneman, P. Slovic, & A. Tversky (Eds.), *Judgment under uncertainty: Heuristics and biases* (pp. 3–20). New York: Cambridge University Press.

UNDRO (United Nations Disaster Relief Organization). (1977). *Disaster prevention and mitigation. Vol. 5, Land use aspects.* New York: United Nations.

UNDRO (United Nations Disaster Relief Organization). (1978). *Disaster prevention and mitigation. Vol. 3. Seismological aspects.* New York: United Nations.

UNDRO (United Nations Disaster Relief Organization). (1979). *Disaster prevention and mitigation. Vol. 7. Economic aspects.* New York: United Nations.

UNDRO (United Nations Disaster Relief Organization). (1980). *Disaster prevention and mitigation. Vol. 9. Legal aspects.* New York: United Nations.

UNECE (United Nations Economic Commission for Europe). (1980). *United Nations Economic Commission for Europe symposium on human settlements planning and development in the Arctic.* New York: Pergamon.

United Nations. (1979). *Everyone's United Nations*. New York: Department of Public Information, United Nations.

United Nations. (1993). *Yearbook of the United Nations, 1992* (Vol. 46). New York: Department of Public Information, United Nations.

U.S. Army Corps of Engineers (1976). *A perspective on flood plain regulations for flood plain management*. Washington, D.C.: Department of the Army, Office of the Chief of Engineers.

U.S. Department of Commerce. (1973). *A study of earthquake losses in the Los Angeles, California area* (Reports prepared for the Federal Disaster Assistance Administration). Washington, D.C.: U.S. Government Printing Office.

U.S. Department of Housing and Urban Development. (1979). *FIRM (Flood Insurance Rate Map), City of West Des Moines, Iowa* (Community Panel Number 190231 0005 B). Washington, D.C.: U.S. Government Printing Office.

U.S. Water Resources Council. (1971). *Regulation of flood hazard areas to reduce flood losses* (Vol. 1, Pts. 1–4). Washington, D.C.: U.S. Government Printing Office.

U.S. Water Resources Council. (1972). *Regulation of flood hazard areas to reduce flood losses* (Vol. 2, Pts. 5–7). Washington, D.C.: U.S. Government Printing Office.

U.S. Water Resources Council. (1979). *A unified national program for flood plain management*. Washington, D.C.: U.S. Government Printing Office.

U.S. Weather Bureau. (1957). *Rainfall intensity-frequency regime: Part I, the Ohio Valley* (Technical Paper 29). Washington, D.C.: U.S. Government Printing Office.

Utterback, J. A. (1977). Geologic hazards and land-use decisions in Routt County, Colorado. In D. C. Shelton (Ed.), *Proceedings of the governor's third conference on environmental geology* (pp. 95–96). Denver, Colo.: Colorado Geological Survey.

Van Dissen, R., & McVerry, G. (1994). Earthquake hazard and risk in New Zealand. In A. G. Hull & R. Coory (Eds.), *Proceedings of the Natural Hazards Management Workshop, Wellington, 8–9 November, 1994* (Institute of Geological and Nuclear Sciences Information Series 31, pp. 67–71). Lower Hutt, New Zealand: Institute of Geological and Nuclear Sciences.

Vaughn, C. K. (1971). *Notes on insurance against loss from natural hazards* (Natural Hazard Research Working Paper 21). Boulder, Colo.: Institute of Behavioral Science, University of Colorado.

Verstappen, H. T. (1992). Volcanic hazards in Colombia and Indonesia: Lahars and related phenomena. In G. J. H. McCall, D. J. C. Laming, & S. C. Scott (Eds.), *Geohazards: Natural and man-made* (pp. 33–42). London: Chapman Hall.

Visvader, H., & Burton, I. (1974). Natural hazards and hazard policy in Canada and the United States. In G. F. White (Ed.), *Natural hazards: Local, national, global* (pp. 219–231). New York: Oxford University Press.

Waananen, A. O., & Spangle, W. E. (1977). *Flood prone areas and land use planning* (U.S. Geological Survey Professional Paper 492). Washington, D.C.: Government Printing Office.

Waddell, E. (1977). The hazards of scientism: A review article. *Human Ecology, 5*, 69–77.

Wahl, K. L., Vining, K. C., & Wiche, G. J. (1993). *Precipitation in the Upper Mississippi*

River basin, January 1 through July 31, 1993 (U.S. Geological Survey Circular 1120-B). Washington, D.C.: U.S. Government Printing Office.

Walker, A. B., Redmayne, D. W., & Browitt, C. W. A. (1992). Seismic monitoring of Lake Nyos, Cameroon, following the gas release disaster of August 1986. In G. J. H. McCall, D. J. C. Laming, & S. C. Scott (Eds.), *Geohazards: Natural and man-made* (pp. 65–79). London: Chapman Hall.

Walters, R. C. S. (1971). *Dam geology*. London: Butterworths.

Ward, P. L., & Page, R. A. (1990). *The Loma Prieta earthquake of October 17, 1989: What happened, what is expected, what can be done?* Washington, D.C.: U.S. Geological Survey.

Ward, R. (1978). *Floods: A geographical perspective*. London: Macmillan.

Ward, R. A., LaGory, M., & Sherman, S. R. (1985). Neighborhood and network age concentration: Does age homogeneity matter for older people? *Social Psychology Quarterly, 48*(2), 138–149.

Warrick, R. A. (1975). *Drought hazard in the United States: A research assessment* (Program on Technology, Environment and Man Monograph NSF-RA-E-75-004). Boulder, Colo.: Institute of Behavioral Science, University of Colorado.

Weathers, J. W. (1965). Comprehensive flood damage prevention. *Journal of the Hydraulics Division (ASCE), 91,* 17–27.

Weems, J. E. (1977). *The tornado*. Garden City, N.Y.: Doubleday.

Wells, H. G. (1924). *The outline of history: Being a plain history of life and mankind*. New York: Macmillan.

Wenger, D. E. (1978). Community response to disaster: Functional and structural alterations. In E. L. Quarantelli. (Ed.), *Disasters, theory and research* (pp. 17–47). Beverly Hills, Calif.: Sage.

White, G. F. (1945). *Human adjustments to floods* (Department of Geography Research Paper 29). Chicago: University of Chicago Press.

White, G. F. (Ed.). (1961). *Papers on flood problems* (Department of Geography Research Paper 70). Chicago: University of Chicago Press.

White, G. F. (1964). *Choice of adjustments to floods* (Department of Geography Research Paper 93). Chicago: University of Chicago Press.

White, G. F. (1966). Optimal flood damage management: Retrospect and prospect. In A. V. Kneese & S. C. Smith (Eds.), *Water research* (pp. 251–269). Baltimore: Johns Hopkins University Press.

White, G. F. (Ed.). (1974). *Natural hazards: Local, national, global*. New York: Oxford University Press.

White, G. F. (1975). *Flood hazard in the United States: A research assessment* (Program on Technology, Environment, and Man Monograph NSF-RA-E-75-006). Boulder, Colo.: Institute of Behavioral Science, University of Colorado.

White, G. F., Calef, W. C., Hudson, J. W., Mayer, H. M. Scheaffer, J. R., & Volk, D. J. (1958). *Changes in urban occupance of flood plains in the United States* (Department of Geography Research Paper 57). Chicago: University of Chicago Press.

White, G. F., & Haas, J. E. (1975). *Assessment of research on natural hazards*. Cambridge, Mass.: MIT Press.

Whittow, J. (1979). *Disasters: The anatomy of environmental hazards*. Athens, Ga.: University of Georgia Press.

Whyte, A. V. (1982). Probabilities, consequences, and values in the perception of risk. In *Risk assessment and perception symposium*. Toronto: Royal Society of Canada.

Wijkman, A., & Timberlake, L. (1984). *Natural disasters: Acts of God or acts of man?* London: International Institute for Environment and Development.

Wilhite, D. A., & Glantz, M. H. (1987). Understanding the drought phenomenon: The role of definitions. In D. A. Wilhite, & W. E. Easterling, with D. A. Wood (Eds.), *Planning for drought: Toward a reduction of societal vulnerability* (pp. 11–27). Boulder, Colo.: Westview.

Williams, M. (1986). Emergency food aid to Africa: High risk of permanent dependence. *UNDRO News*, March/April, 17–20.

Wilson, R., & Crouch, E. A. C. (1987). Risk assessment and comparisons: An introduction. *Science, 236*, 267–270.

Wirth, L. (1938). Urbanism as a way of life. *American Journal of Sociology, 44*(1), 1–24.

Wisner, B., & Mbithi, P. M. (1974). Drought in Eastern Kenya: Nutritional status and farmer activity. In G. F. White (Ed.), *Natural hazards: Local, national, global* (pp. 87–97). New York: Oxford University Press.

Wood, D. P., & Cowan, M. L. (1991). Crisis intervention following disasters: Are we doing enough? (A second look). *American Journal of Emergency Medicine, 9*(6), 598–602.

Wood, H. O., & Neumann, F. (1931). Modified Mercalli intensity scale of 1931. *Bulletin of the Seismological Society of America, 21*, 277–283.

Working Group on California Earthquake Probabilities. (1988). *Probabilities of large earthquakes occurring in California on the San Andreas Fault* (U.S. Geological Survey Open-File Report 88-398). Washington, D.C.: U.S. Government Printing Office.

Working Group on California Earthquake Probabilities. (1995). Seismic hazards in southern California: Probable earthquakes, 1994–2024. *Bulletin of the Seismological Society of America, 85*(2), 379–439.

World Meteorological Organization. (1975). *Drought and agriculture: Report of the working group on assessment of drought* (Technical Note 138, World Meteorological Organization Publication 392). Geneva: Secretariat of the World Meteorological Organization.

World Resources Institute. (1994). *World resources, 1994–1995: A guide to the global environment.* New York: Oxford University Press.

Wright, J. D., Rossi, P. H., Wright, S. R., & Weber-Burdin, E. (1979). *After the clean-up: Long-range effects of natural disasters.* Beverly Hills, Calif.: Sage.

Yezer, A. M., & Rubin, C. B. (1987). *The local economic effects of natural disasters* (Natural Hazard Research Working Paper 61). Boulder, Colo.: Institute of Behavioral Science, University of Colorado.

Zupka, D. (1988). Economic impact of disasters. *UNDRO News*, January/February, 19–22.

Index

Printed in the United States
123812LV00006B/13-30/A